Carbon Fiber Composites

Carbon Fiber Composites

Editors

Jiadeng Zhu
Guoqing Li
Lixing Kang

Basel • Beijing • Wuhan • Barcelona • Belgrade • Novi Sad • Cluj • Manchester

Editors

Jiadeng Zhu
Brewer Science Inc.
Springfield, IL
USA

Guoqing Li
North Carolina
State University
Raleigh, NC
USA

Lixing Kang
Chinese Academy
of Sciences
Suzhou
China

Editorial Office
MDPI
St. Alban-Anlage 66
4052 Basel, Switzerland

This is a reprint of articles from the Special Issue published online in the open access journal *Journal of Composites Science* (ISSN 2504-477X) (available at: https://www.mdpi.com/journal/jcs/special_issues/Carbon_Fiber_Composites).

For citation purposes, cite each article independently as indicated on the article page online and as indicated below:

Lastname, A.A.; Lastname, B.B. Article Title. *Journal Name* **Year**, *Volume Number*, Page Range.

ISBN 978-3-7258-1063-5 (Hbk)
ISBN 978-3-7258-1064-2 (PDF)
doi.org/10.3390/books978-3-7258-1064-2

© 2024 by the authors. Articles in this book are Open Access and distributed under the Creative Commons Attribution (CC BY) license. The book as a whole is distributed by MDPI under the terms and conditions of the Creative Commons Attribution-NonCommercial-NoDerivs (CC BY-NC-ND) license.

Contents

About the Editors ix

Jiadeng Zhu, Guoqing Li and Lixing Kang
Editorial for the Special Issue on Carbon Fiber Composites
Reprinted from: *J. Compos. Sci.* **2024**, *8*, 113, doi:10.3390/jcs8030113 1

Chandreyee Manas Das, Lixing Kang, Guang Yang, Dan Tian and Ken-Tye Yong
Multifaceted Hybrid Carbon Fibers: Applications in Renewables, Sensing and Tissue Engineering
Reprinted from: *J. Compos. Sci.* **2020**, *4*, 117, doi:10.3390/jcs4030117 4

Hassan Almousa, Qing Peng and Abduljabar Q. Alsayoud
A Molecular Dynamics Study of the Stability and Mechanical Properties of a Nano-Engineered Fuzzy Carbon Fiber Composite
Reprinted from: *J. Compos. Sci.* **2022**, *6*, 54, doi:10.3390/jcs6020054 27

Olusanmi Adeniran, Weilong Cong, Eric Bediako and Victor Aladesanmi
Additive Manufacturing of Carbon Fiber Reinforced Plastic Composites: The Effect of Fiber Content on Compressive Properties
Reprinted from: *J. Compos. Sci.* **2021**, *5*, 325, doi:10.3390/jcs5120325 37

Mengyang Li, Jiayi Rong, Ning Guo, Susu Chen, Meiqi Gao, Feng Cao and Guoqing Li
Controllable Synthesis of Graphene-Encapsulated NiFe Nanofiber for Oxygen Evolution Reaction Application
Reprinted from: *J. Compos. Sci.* **2021**, *5*, 314, doi:10.3390/jcs5120314 51

Yongbum Choi, Xuan Meng and Zhefeng Xu
Manufacturing and Performance of Carbon Short Fiber Reinforced Composite Using Various Aluminum Matrix
Reprinted from: *J. Compos. Sci.* **2021**, *5*, 307, doi:10.3390/jcs5120307 61

Raffael Bogenfeld, Christopher Gorsky
An Experimental Study of the Cyclic Compression after Impact Behavior of CFRP Composites
Reprinted from: *J. Compos. Sci.* **2021**, *5*, 296, doi:10.3390/jcs5110296 71

Imad Hanhan and Michael D. Sangid
Design of Low Cost Carbon Fiber Composites via Examining the Micromechanical Stress Distributions in A42 Bean-Shaped versus T650 Circular Fibers
Reprinted from: *J. Compos. Sci.* **2021**, *5*, 294, doi:10.3390/jcs5110294 92

Reza Moazed, Mohammad Amir Khozeimeh and Reza Fotouhi
Simplified Approach for Parameter Selection and Analysis of Carbon and Glass Fiber Reinforced Composite Beams
Reprinted from: *J. Compos. Sci.* **2021**, *5*, 220, doi:10.3390/jcs5080220 101

Frank Manis, Georg Stegschuster, Jakob Wölling and Stefan Schlichter
Influences on Textile and Mechanical Properties of Recycled Carbon Fiber Nonwovens Produced by Carding
Reprinted from: *J. Compos. Sci.* **2021**, *5*, 209, doi:10.3390/jcs5080209 122

Yucheng Peng, Ruslan Burtovyy, Rajendra Bordia and Igor Luzinov
Fabrication of Porous Carbon Films and Their Impact on Carbon/Polypropylene Interfacial Bonding
Reprinted from: *J. Compos. Sci.* **2021**, *5*, 108, doi:10.3390/jcs5040108 138

Robin James, Roshan Prakash Joseph and Victor Giurgiutiu
Impact Damage Ascertainment in Composite Plates Using In-Situ Acoustic Emission Signal Signature Identification
Reprinted from: *J. Compos. Sci.* **2021**, *5*, 79, doi:10.3390/jcs5030079 154

Niraj Kumbhare, Reza Moheimani and Hamid Dalir
Analysis of Composite Structures in Curing Process for Shape Deformations and Shear Stress: Basis for Advanced Optimization
Reprinted from: *J. Compos. Sci.* **2021**, *5*, 63, doi:10.3390/jcs5020063 169

Antoine Lemartinel, Mickaël Castro, Olivier Fouché, Julio-César De Luca and Jean-François Feller
Strain Mapping and Damage Tracking in Carbon Fiber Reinforced Epoxy Composites during Dynamic Bending Until Fracture with Quantum Resistive Sensors in Array
Reprinted from: *J. Compos. Sci.* **2021**, *5*, 60, doi:10.3390/jcs5020060 188

Farzin Azimpour-Shishevan, Hamit Akbulut and M.A. Mohtadi-Bonab
Thermal Shock Behavior of Twill Woven Carbon Fiber Reinforced Polymer Composite
Reprinted from: *J. Compos. Sci.* **2021**, *5*, 33, doi:10.3390/jcs5010033 200

Patricio Martinez, Bo Cheng Jin and Steven Nutt
Droplet Spreading on Unidirectional Fiber Beds
Reprinted from: *J. Compos. Sci.* **2021**, *5*, 13, doi:10.3390/jcs5010013 215

Arivazhagan Selvam, Suresh Mayilswamy, Ruban Whenish, Rajkumar Velu and Bharath Subramanian
Preparation and Evaluation of the Tensile Characteristics of Carbon Fiber Rod Reinforced 3D Printed Thermoplastic Composites
Reprinted from: *J. Compos. Sci.* **2021**, *5*, 8, doi:10.3390/jcs5010008 231

Haichen Zhang, Lili Tong and Michael Anim Addo
Mechanical Analysis of Flexible Riser with Carbon Fiber Composite Tension Armor
Reprinted from: *J. Compos. Sci.* **2021**, *5*, 3, doi:10.3390/jcs5010003 246

Sunil C. Joshi
Boosting Inter-ply Fracture Toughness Data on Carbon Nanotube-Engineered Carbon Composites for Prognostics
Reprinted from: *J. Compos. Sci.* **2020**, *4*, 170, doi:10.3390/jcs4040170 261

Mohamed Ali Charfi, Ronan Mathieu, Jean-François Chatelain, Claudiane Ouellet-Plamondon and Gilbert Lebrun
Effect of Graphene Additive on Flexural and Interlaminar Shear Strength Properties of Carbon Fiber-Reinforced Polymer Composite
Reprinted from: *J. Compos. Sci.* **2020**, *4*, 162, doi:10.3390/jcs4040162 275

Ahmed S. AlOmari, Khaled S. Al-Athel, Abul Fazal M. Arif and Faleh. A. Al-Sulaiman
Experimental and Computational Analysis of Low-Velocity Impact on Carbon-, Glass- and Mixed-Fiber Composite Plates
Reprinted from: *J. Compos. Sci.* **2020**, *4*, 148, doi:10.3390/jcs4040148 289

Alexander Kyriazis, Kais Asali, Michael Sinapius, Korbinian Rager and Andreas Dietzel
Adhesion of Multifunctional Substrates for Integrated Cure Monitoring Film Sensors to Carbon Fiber Reinforced Polymers
Reprinted from: *J. Compos. Sci.* **2020**, *4*, 138, doi:10.3390/jcs4030138 308

M.H. Khan, Bing Li and K.T. Tan
Impact Performance and Bending Behavior of Carbon-Fiber Foam-Core Sandwich Composite Structures in Cold Arctic Temperature
Reprinted from: *J. Compos. Sci.* **2020**, *4*, 133, doi:10.3390/jcs4030133 . . . . **327**

Frederik Wilhelm, Sebastian Strauß, Raffael Weigant and Klaus Drechsler
Effect of Power Ultrasonic on the Expansion of Fiber Strands
Reprinted from: *J. Compos. Sci.* **2020**, *4*, 50, doi:10.3390/jcs4020050 . . . . **342**

About the Editors

Jiadeng Zhu

Jiadeng Zhu received his B.S. degree in Chemical Engineering and Materials Science from Soochow University and his Ph.D. in Fiber and Polymer Science from North Carolina State University. He is now a Principal Investigator/Team Lead in the Discovery and Proof-of-Concept group for emerging polymer/carbon-based sensing materials and technology at Brewer Science Inc. He has published 10 book chapters and over 100 peer-reviewed journal papers with an h-index of 44. He is an enthusiastic, confident, and creative scholar who has a fruitful and consistent history of research on applications of advanced polymers and carbon materials in energy and environmental areas, including, but not limited to, energy storage/conversion, lightweight structural materials, printed/wearable electronics, smart textiles, filtration, sensor fabrication and testing, and sensor integration.

Guoqing Li

Guoqing Li obtained a Ph.D. in Fiber and Polymer Science in 2017 and started his post-doctoral Materials Science training at North Carolina State University. Li's research focuses on atomistic designing and manufacturing of two-dimensional materials and exploring this control for developing novel electronic, photonic, and energy-harvesting devices. He developed a method to manufacture wafer-scale, high-quality, two-dimensional materials, attracting great academic and industry interest. He has published 40+ papers with 3000+ citations, including in *JACS*, *Science Advance*, *ACS Nano*, etc. He has also held seven patents/provisional patents.

Lixing Kang

Lixing Kang received his Ph.D. degree (2017) in Physical Chemistry from the Suzhou Institute of Nanotech and Nanobionics, Chinese Academy of Sciences (CAS). After his post-doctoral studies from 2017 to 2021, he joined the Suzhou Institute of Nanotech and Nanobionics, CAS, as a full professor. He mainly focuses on CVD techniques and controls the growth of low-dimensional materials (e.g., carbon nanotubes, graphene, MoS_2, and boron nitride). He has published 24 papers in *Nature* (1), *Nature Commun.* (3), *Science Adv.* (1), *J. Am. Chem. Soc.* (4), *Nano Lett.* (3) as a co-author and corresponding author and has participated in 60 publications with >3000 citations. He has applied for 11 patents (one international and ten Chinese patents); four patents have been granted, and a monograph on the growth and application of carbon nanotubes has been authored in English. He has also served on the Carbon Energy (IF > 20) and SmartMat (IF > 20) Young Editorial Boards.

Editorial

Editorial for the Special Issue on Carbon Fiber Composites

Jiadeng Zhu [1,*], Guoqing Li [2,*] and Lixing Kang [3,*]

1. Smart Devices, Brewer Science Inc., Springfield, MO 65810, USA
2. Department of Materials Science and Engineering, North Carolina State University, Raleigh, NC 27606, USA
3. Division of Advanced Materials, Suzhou Institute of Nano-Tech and Nano-Bionics, Chinese Academy of Sciences, Suzhou 215123, China
* Correspondence: zhujiadeng@gmail.com (J.Z.); guoqingli36@gmail.com (G.L.); lxkang2013@sinano.ac.cn (L.K.)

Citation: Zhu, J.; Li, G.; Kang, L. Editorial for the Special Issue on Carbon Fiber Composites. *J. Compos. Sci.* **2024**, *8*, 113. https://doi.org/10.3390/jcs8030113

Received: 11 March 2024
Accepted: 19 March 2024
Published: 21 March 2024

Copyright: © 2024 by the authors. Licensee MDPI, Basel, Switzerland. This article is an open access article distributed under the terms and conditions of the Creative Commons Attribution (CC BY) license (https://creativecommons.org/licenses/by/4.0/).

Carbon fibers (CFs) have received tremendous attention since their discovery in the 1860s due to their unique properties, including outstanding mechanical properties, low density, excellent chemical resistance, good thermal conductivity, etc. [1–3]. CFs are widely applied in energy storage/conversion, sports, wind energy, electronics, etc. [4,5]. Additionally, with continuous efforts and ever-growing demand, CFs are widely utilized to reinforce composite materials because of the abovementioned characteristics, which have received remarkable interest [6].

The collection of papers in this Special Issue may provide new insights regarding the development of CFs and their composites from both experimental and simulation perspectives, advancing technology and facilitating the practical application of these devices. Various candidates, including metal, polymer, inorganic materials, etc., have been explored and implemented in composites [7–9]. Meanwhile, factors such as starting materials, structural designs, compositions, etc., which may affect the overall properties of the resultant composites, have been thoroughly investigated [10–13]. For example, Adeniran et al. studied the influence of fiber content on the compressive properties of the prepared composites [14]. A method proposed by Martinez et al. was used to analyze the spread-flow kinetic effect of fluid drops on the unidirectional fiber beds [15]. Additionally, computational studies have been performed to assist in better understanding the impact of different parameters [16,17].

Preparing these composites (i.e., curing, cyclic compression, cyclic temperature, etc.) is still crucial since processing parameters also play critical roles in determining their final properties [18–22]. The curing reaction progress, which can help to monitor the quality of prepared parts, was measured by Kyriazis et al. [23]. In addition to studying these parameters, different strategies have been discovered and used to prepare the composites [24–26]. For instance, Moazed et al. developed and plotted structural indices and efficiency metrics in design charts in order to better select parameters [27]. In contrast to the traditional structural function, Li et al. used the carbon fiber composite for catalysis [28]. Moreover, a critical review of the broad applications of carbon fibers and their composites in renewables, sensing, and tissue engineering is also presented [29].

This Special Issue covers experimental designs and computational studies, moving from discussions of principles, parameter optimization, and manufacturing to end uses. It may provide new methods and advanced technologies that could help us to better understand these approaches to the unique characterization and modeling of carbon fiber composites, facilitating their practical application.

Conflicts of Interest: The author Jiadeng Zhu was employed by the Brewer Science Inc. company. All authors declare that the research was conducted in the absence of any commercial or financial relationships that could be construed as a potential conflict of interest.

References

1. Zhu, J.; Yan, C.; Li, G.; Cheng, H.; Li, Y.; Liu, T.; Mao, Q.; Cho, H.; Gao, Q.; Gao, C.; et al. Recent Developments of Electrospun Nanofibers for Electrochemical Energy Storage and Conversion. *Energy Storage Mater.* **2023**, *65*, 103111. [CrossRef]
2. Zhu, J.; Gao, Z.; Kowalik, M.; Joshi, K.; Ashraf, C.M.; Arefev, M.I.; Schwab, Y.; Bumgardner, C.; Brown, K.; Burden, D.E.; et al. Unveiling carbon ring structure formation mechanisms in polyacrylonitrile-derived carbon fibers. *ACS Appl. Mater. Interfaces* **2019**, *11*, 42288–42297. [CrossRef]
3. Mao, Q.; Rajabpour, S.; Talkhoncheh, M.K.; Zhu, J.; Kowalik, M.; van Duin, A.C.T. Cost-Effective Carbon Fiber Precursor Selections of Polyacrylonitrile-Based Blend Polymers: Carbonization Chemistry and Structural Characterizations. *Nanoscale* **2022**, *14*, 6357–6372. [CrossRef]
4. Zhu, J.; Park, S.W.; Joh, H.-I.; Kim, H.C.; Lee, S. Study on the stabilization of isotropic pitch-based fibers. *Macromol. Res.* **2015**, *23*, 79–85. [CrossRef]
5. Zhu, J.; Park, S.W.; Joh, H.-I.; Kim, H.C.; Lee, S. Preparation and characterization of isotropic pitch-based carbon fiber. *Carbon Lett.* **2013**, *14*, 94–98. [CrossRef]
6. Demchuk, Z.; Zhu, J.; Li, B.; Zhao, X.; Islam, N.M.; Yang, G.; Zhou, H.; Jiang, Y.; Choi, W.; Advincula, R.; et al. Unravel the Influence of Surface Modification on the Ultimate Performance of Carbon Fiber Epoxy Composites. *ACS Appl. Mater. Interfaces* **2022**, *14*, 45775–45787. [CrossRef]
7. Choi, Y.; Meng, X.; Xu, Z. Manufacturing and Performance of Carbon Short Fiber Reinforced Composite Using Various Aluminum Matrix. *J. Compos. Sci.* **2021**, *5*, 307. [CrossRef]
8. Ali Charfi, M.; Mathieu, R.; Chatelain, J.F.; Ouellet-Plamondon, C.; Lebrun, G. Effect of graphene additive on flexural and interlaminar shear strength properties of carbon fiber-reinforced polymer composite. *J. Compos. Sci.* **2020**, *4*, 162. [CrossRef]
9. Joshi, S.C. Boosting Inter-ply Fracture Toughness Data on Carbon Nanotube-Engineered Carbon Composites for Prognostics. *J. Compos. Sci.* **2020**, *4*, 170. [CrossRef]
10. Khan, M.H.; Li, B.; Tan, K.T. Impact performance and bending behavior of carbon-fiber foam-core sandwich composite structures in cold arctic temperature. *J. Compos. Sci.* **2020**, *4*, 133. [CrossRef]
11. Wilhelm, F.; Strauß, S.; Weigant, R.; Drechsler, K. Effect of power ultrasonic on the expansion of fiber strands. *J. Compos. Sci.* **2020**, *4*, 50. [CrossRef]
12. Manis, F.; Stegschuster, G.; Wölling, J.; Schlichter, S. Influences on textile and mechanical properties of recycled carbon fiber nonwovens produced by carding. *J. Compos. Sci.* **2021**, *5*, 209. [CrossRef]
13. Zhang, H.; Tong, L.; Addo, M.A. Mechanical analysis of flexible riser with carbon fiber composite tension armor. *J. Compos. Sci.* **2020**, *5*, 3. [CrossRef]
14. Adeniran, O.; Cong, W.; Bediako, E.; Aladesanmi, V. Additive manufacturing of carbon fiber reinforced plastic composites: The effect of fiber content on compressive properties. *J. Compos. Sci.* **2021**, *5*, 325. [CrossRef]
15. Martinez, P.; Jin, B.C.; Nutt, S. Droplet Spreading on Unidirectional Fiber Beds. *J. Compos. Sci.* **2021**, *5*, 13. [CrossRef]
16. Almousa, H.; Peng, Q.; Alsayoud, A.Q. A Molecular Dynamics Study of the Stability and Mechanical Properties of a Nano-Engineered Fuzzy Carbon Fiber Composite. *J. Compos. Sci.* **2022**, *6*, 54. [CrossRef]
17. AlOmari, A.S.; Al-Athel, K.S.; Arif, A.F.; Al-Sulaiman, F.A. Experimental and computational analysis of low-velocity impact on carbon-, glass- and mixed-fiber composite plates. *J. Compos. Sci.* **2020**, *4*, 148. [CrossRef]
18. Azimpour-Shishevan, F.; Akbulut, H.; Mohtadi-Bonab, M.A. Thermal shock behavior of twill woven carbon fiber reinforced polymer composites. *J. Compos. Sci.* **2021**, *5*, 33. [CrossRef]
19. Bogenfeld, R.; Gorsky, C. An experimental study of the cyclic compression after impact behavior of CFRP composites. *J. Compos. Sci.* **2021**, *5*, 296. [CrossRef]
20. James, R.; Joseph, R.P.; Giurgiutiu, V. Impact damage ascertainment in composite plates using in-situ acoustic emission signal signature identification. *J. Compos. Sci.* **2021**, *5*, 79. [CrossRef]
21. Lemartinel, A.; Castro, M.; Fouché, O.; De Luca, J.C.; Feller, J.F. Strain mapping and damage tracking in carbon fiber reinforced epoxy composites during dynamic bending until fracture with quantum resistive sensors in array. *J. Compos. Sci.* **2021**, *5*, 60. [CrossRef]
22. Kumbhare, N.; Moheimani, R.; Dalir, H. Analysis of composite structures in curing process for shape deformations and shear stress: Basis for advanced optimization. *J. Compos. Sci.* **2021**, *5*, 63. [CrossRef]
23. Kyriazis, A.; Asali, K.; Sinapius, M.; Rager, K.; Dietzel, A. Adhesion of multifunctional substrates for integrated cure monitoring film sensors to carbon fiber reinforced polymers. *J. Compos. Sci.* **2020**, *4*, 138. [CrossRef]
24. Hanhan, I.; Sangid, M.D. Design of Low Cost Carbon Fiber Composites via Examining the Micromechanical Stress Distributions in A42 Bean-Shaped versus T650 Circular Fibers. *J. Compos. Sci.* **2021**, *5*, 294. [CrossRef]
25. Peng, Y.; Burtovyy, R.; Bordia, R.; Luzinov, I. Fabrication of Porous carbon films and their impact on carbon/polypropylene interfacial bonding. *J. Compos. Sci.* **2021**, *5*, 108. [CrossRef]
26. Selvam, A.; Mayilswamy, S.; Whenish, R.; Velu, R.; Subramanian, B. Preparation and evaluation of the tensile characteristics of carbon fiber rod reinforced 3D printed thermoplastic composites. *J. Compos. Sci.* **2020**, *5*, 8. [CrossRef]
27. Moazed, R.; Khozeimeh, M.A.; Fotouhi, R. Simplified approach for parameter selection and analysis of carbon and glass fiber reinforced composite beams. *J. Compos. Sci.* **2021**, *5*, 220. [CrossRef]

28. Li, M.; Rong, J.; Guo, N.; Chen, S.; Gao, M.; Cao, F.; Li, G. Controllable Synthesis of Graphene-Encapsulated NiFe Nanofiber for Oxygen Evolution Reaction Application. *J. Compos. Sci.* **2021**, *5*, 314. [CrossRef]
29. Das, C.M.; Kang, L.; Yang, G.; Tian, D.; Yong, K.T. Multifaceted hybrid carbon fibers: Applications in renewables, sensing and tissue engineering. *J. Compos. Sci.* **2020**, *4*, 117. [CrossRef]

Disclaimer/Publisher's Note: The statements, opinions and data contained in all publications are solely those of the individual author(s) and contributor(s) and not of MDPI and/or the editor(s). MDPI and/or the editor(s) disclaim responsibility for any injury to people or property resulting from any ideas, methods, instructions or products referred to in the content.

Review

Multifaceted Hybrid Carbon Fibers: Applications in Renewables, Sensing and Tissue Engineering

Chandreyee Manas Das [1,†], Lixing Kang [1,†], Guang Yang [1], Dan Tian [2,*] and Ken-Tye Yong [1,*]

1. School of Electrical and Electronic Engineering, Nanyang Technological University, 50 Nanyang Avenue, Singapore 639798, Singapore; CHANDREY001@e.ntu.edu.sg (C.M.D.); lxkang@ntu.edu.sg (L.K.); YANG0389@e.ntu.edu.sg (G.Y.)
2. College of Materials Science and Engineering, Nanjing Forestry University, Nanjing 210037, China
* Correspondence: tiandan@njfu.edu.cn (D.T.); ktyong@ntu.edu.sg (K.-T.Y.)
† The authors contributed equally to this work.

Received: 27 July 2020; Accepted: 14 August 2020; Published: 16 August 2020

Abstract: The field of material science is continually evolving with first-class discoveries of new nanomaterials. The element carbon is ubiquitous in nature. Due to its valency, it can exist in various forms, also known as allotropes, like diamond, graphite, one-dimensional (1D) carbon nanotube (CNT), carbon fiber (CF) and two-dimensional (2D) graphene. Carbon nano fiber (CNF) is another such material that falls within the category of CF. With much smaller diameters (around hundreds of nanometers) and lengths in microns, CNFs have higher aspect (length to diameter) ratios than CNTs. Because of their unique properties like high electrical and thermal conductivity, CNFs can be applied to many matrices like elastomers, thermoplastics, ceramics and metals. Owing to their outstanding mechanical properties, they can be used as reinforcements that can enhance the tensile and compressive strain limits of the base material. Thus, in this short review, we take a look into the dexterous characteristics of CF and CNF, where they have been hybridized with different materials, and delve deeply into some of the recent applications and advancements of these hybrid fiber systems in the fields of sensing, tissue engineering and modification of renewable devices since favorable mechanical and electrical properties of the CFs and CNFs like high tensile strength and electrical conductivity lead to enhanced device performance.

Keywords: carbon nano fiber; sensing; tissue engineering; renewables

1. Introduction

The science behind materials plays an important role in almost every aspect of engineering, medicine and industry. To be more precise, the dimension of these materials can completely alter the way they interact with other physical aspects like electromagnetic radiation. For instance, when it comes to bulk materials, these do not display any spectacular phenomenon upon interaction with light. However, when we go to the nano level in terms of size, the material property changes entirely and can give rise to many novel physical principles like surface effect, quantum size effect, quantum tunneling effect, dielectric confinement effect and many more. The multifarious carbon nano fibers (CNF) have excellent thermal, mechanical and electrical properties and they are being utilized in a variety of fields like aerospace, transportation, civil engineering and green technology. However, they also possess some limitations like diminished specific surface area, lipophobic surface and low chemical activity. Thus, to take the maximum advantage as well as reduce the physical restrictions, many researchers make use of hybrid materials.

CNFs are mainly prepared by two methods namely the catalytic thermal chemical vapor deposition (CVD) and the electrospinning process followed by heat treatment. There are two types of CNFs that can be prepared by the CVD method: cup-stacked CNF and platelet CNF. In the CVD method many types of metals that can dissolve carbon to form metal carbide are used like Iron, Nickel, Cobalt, Chromium, Vanadium. To obtain carbon sources in the range of 700 K to 1200 K, Molybdenum, Methane, Carbon Monoxide, Synthesis gas (H_2/CO), Ethyne or Ethene are used. In the electrospinning process, polymer nanofibers are required as precursors. The properties of the obtained CNF depend on the polymer solution used and the processing parameters. Polyacrylonitrile (PAN), Poly (vinyl alcohol) (PVA), Polyimides (PIs), Polybenzimidazole (PBI), Poly (vinylidene fluoride) (PVDF), Phenolic resin and lignin are the common polymers that are used. After successfully preparing the polymer nanofibers, heat treatment is used to carbonize the polymer nanofibers to form CNFs. The shape, porosity, diameter and other structural characteristics are governed by the physical conditions of the heat treatment process like temperature and pressure [1].

The main physical properties that make CNF utility unique in sensors and other devices is due to their electrical, thermal and mechanical behavior. The addition of CNF in polymers enhances its mechanical properties. In general, by just a minute addition of CNF, the resistance to fracture is greatly enhanced. The addition of 0.5 wt% and 1.0 wt% CNF in epoxy enhances its fracture resistance by 66% and 78% respectively. In another application, by 4.0 wt% addition of CNF in thermal-plastic polyurethane (TPU), the tensile strength increased by 49% as compared to neat TPU. However, increased addition of CNF can be counterintuitive as it can result in void formation and other defects due to development of bundles of CNF that lead to stress concentration and easy fatigue resulting in premature failure. Coming to the thermal properties, the addition of CNF in the matrix material can enhance its thermal conductivity leading to better heat dissipation and reduced chances of thermal failure. Many models have been developed for prediction of thermal conductivity of CNF composite, like the Maxwell model and the series/parallel model, as a function of individual thermal conductivities of the matrix and filler materials. The enhanced electrical conductivity of CNF composites is due to the tunneling effect, where conductive pathways are developed in the matrix material as a result of the addition of CNF. Many theories and models have been developed to ascertain the reason behind the enhanced electrical conductivity of these composites [1].

CNT and Graphene have been used in many applications of bio-sensing, therapeutics and flexible electronics because of their favorable electrical and mechanical characteristics. However, CNF has its own benefits and is unique in its own way. They are discontinuous and highly graphitic in nature and can be easily blended with polymer processing techniques because of which the development of CNF composites is considerably less complicated. They can be hybridized easily with a wide range of matrix materials including thermoplastics, thermosets, elastomers and others. With just minute additions of CNF, the mechanical, thermal and electrical properties of the matrix material can be significantly improved. Hence, it is quite cost effective to manufacture CNF composites and they can be commercialized easily.

The structure of the review is divided into three parts. In the coming sections we discuss the applications of CF and CNF in renewable energy sector in Section 2, especially concentrating on batteries, supercapacitors, solar cells and fuel cells. Section 3 will deal with sensing applications of CF and CNF and Section 4 will be about usage of CF and CNF in tissue engineering and finally we end with a concluding note in Section 5.

Below we take a brief look into some efforts made by researchers in developing CF composites and reinforced polymers with enhanced mechanical, thermal and electrical properties.

Carbon fiber (CF) composites display poor interfacial adhesion due to which stresses are not transferred well from the matrix to the reinforcing fibers. Thus, these composites are prone to interfacial failures. Semitekolos et al. used poly methacrylic acid (PMAA) to modify carbon fiber (CF) fabric in order to attain better fiber–matrix interfacial strength [2]. With strong hydrogen bonding between the respective carboxyl and hydroxyl groups of PMAA and epoxy resin, an interlayer was created that showed maximum enhancement in Interlaminar Shear Strength (ILSS). Structural defects pose significant hindrance in the path of enhancing mechanical properties of CFs. In order to improve the electrical conductivity and mechanical strength of these fibers, Sui et al. used Polyacrylonitrile (PAN) nanofibers embedded with 0–20 wt% Multi-walled Carbon Nanotubes (MWCNT) to generate hybrid nano-scale CFs with the help of electrospinning [3]. They observed significant improvement in electrical conductivity of about 26 Scm^{-1} even with 3 wt% addition of MWCNT. In another interesting application of hybrid CNF, Lui et al. used CNF as a lubricating filler [4]. They modified high strength glass fabric (HSGF)/phenolic laminates with 1 to 3% CNF and enhanced the ILSS. In addition to high modulus and strength, because of the self-lubricating properties of CNF, the composite material displayed superior resistivity towards corrosion in water-based environments. Ulus et al. prepared hybrid nanocomposites made of CF, Boron Nitride Nano Particles (BNNP) and CNT [5]. The addition of BNNP and CNT enhanced the tensile, flexural and shear strengths of epoxy resin and CF. Scanning Electron Microscopy (SEM) images showed minimum damage and maximum improvement in mechanical properties for BNNP–CNT hybrid nanocomposites. The enhancements in bending stiffness and shear strengths for BNNP–CNT-Epoxy/CF were 38.5% and 90%, respectively, in comparison to plain CF/Epoxy composites. Sui et al. prepared a PAN–nano CF composite by multi-step hot stretching that displayed higher mechanical strength, was lightweight and had less structural deformations [6]. For polyetherimide composite membranes containing 1 wt% nano CF, there were 21% and 60% improvements in tensile strength and Young's modulus, respectively.

Gabr et al. added nano-clay as filler material into CF polymer composites in order to improve the strength [7]. Using dynamic mechanical analysis (DMA) and SEM they studied CF/compatibilized polypropylene (PPc)/organoclay composites and found that at 3% of the filler material, mode I initiation and propagation interlaminar fracture toughness improved by 64% and 67%, respectively. The SEM images showed that the fibers pulled themselves at the tip during initiation delamination. Additionally, Zhang et al. prepared a ternary biocomposite comprising of nano-hydroxyapatite/polyamide66 (HA/PA) and CF [8]. They observed that CF bonded well with the HA/PA matrix. Higher wt% of CF lead to better enhancements in compressive, bending and tensile strengths that ranged in between 116–212 MPa, 89–138 MPa and 109–181 MPa, respectively. The HA/PA/CF composite also showed high cytocompatibility towards MG-63 cells and thus they demonstrated that the composite can also have potential applications as bone repair materials.

CFs show high thermal and electrical conductivity. However, there are certain applications that simultaneously require good thermal conductivity and electrical insulating properties. To obtain this, Zhang et al. prepared MgO nanoparticle-decorated CFs and blended them into Nylon 66 [9]. The addition of CF lead to enhancements in thermal conductivity. However, MgO nanoparticles helped in quenching the electrical conductivity. Osouli-Bostanabad et al. fabricated nano-magnetite coated CFs that not only had enhanced strength, but also acted as an electromagnetic (EM) shield [10]. Briefly, they manufactured a double-layered composite material where the first layer provided good mechanical strength and the second layer aided in absorbing stray EM radiations. Thermal protection systems (TPS) are extensively incorporated in spacecraft where ablative polymer composites are majorly used to manufacture the system components. Naderi et al. prepared a nano Zirconia modified phenolic resin/CF composite that showed superior ablative properties in addition to greater mechanical strength [11]. Due to its high carbon content, outstanding thermal stability and reduced thermal expansion, CFs are usually employed as ablative materials. With the addition of nano Zirconia as a filler material, the thermal, insulative and ablative properties of the composite material were significantly enhanced.

CF reinforced polymers are used in various industries. However, due to certain restrictions of CF like smooth graphitic surface and diminished surface energy, they cannot form a good bond with the matrix. Thus, to enhance the interfacial adhesion between the fiber and the matrix, Jager et al. used halloysite nanotubes (HNT) and they observed significant improvement in interfacial shear strength (IFSS) with just 5% HNT [12]. Blugan et al. modified Alumina matrix with different concentrations of CNF to enhance the electrical conductivity of the matrix [13]. MWCNTs have long been used to modify and enhance the performance characteristics of the matrix material. However, they possess some technical challenges like non-uniform dispersion throughout the matrix and weak bonding with the matrix. The use of CNF not only aids in eliminating these issues but also provides a more economical solution since they are cheaper than MWCNT. Li et al. performed tensile creep studies of CF/epoxy resin and MWCNT/CF/epoxy resin composites and found that the latter fared well in all the mechanical tests performed on them [14]. They proved that addition of MWCNT into CF composites can significantly boost their performance. Pervin et al. prepared nano composites using CNF and SC-15 epoxy [15]. Increased percentage of CNF resulted in higher values of modulus and mechanical strength parameters. With 4% addition of CNF, the modulus and strength of the composite material improved by 27% and 17%, respectively. Additionally, the composite also displayed better thermal response characteristics like enhanced glass transition temperature (T_g). Charles et al. modified CF–epoxy resin composites using a triblock copolymer of poly (styrene)-b-poly (butadiene)-b-poly(methylmethacrylate) [16]. The results showed that the initiation fracture toughness improved by 88% and there was a 6-degree Celsius rise in T_g. Furthermore, there was a 121% increment in shear strain with just an 8% decrement in shear strength. The superior performance of the triblock copolymer toughened composite was due to nano-structuring that took place inside the resin system and led to matrix cavitation, which resulted in easy dissipation of strain energy.

Figure 1 below shows the SEM image of electrochemically treated CF, apparatus used for carbonization of the nano sheets, schematic representation of the process used for fabrication of CF–MgO hybrid nanocomposites and stress–strain curves for SC-15 epoxy nanocomposite with different percentages of CNF.

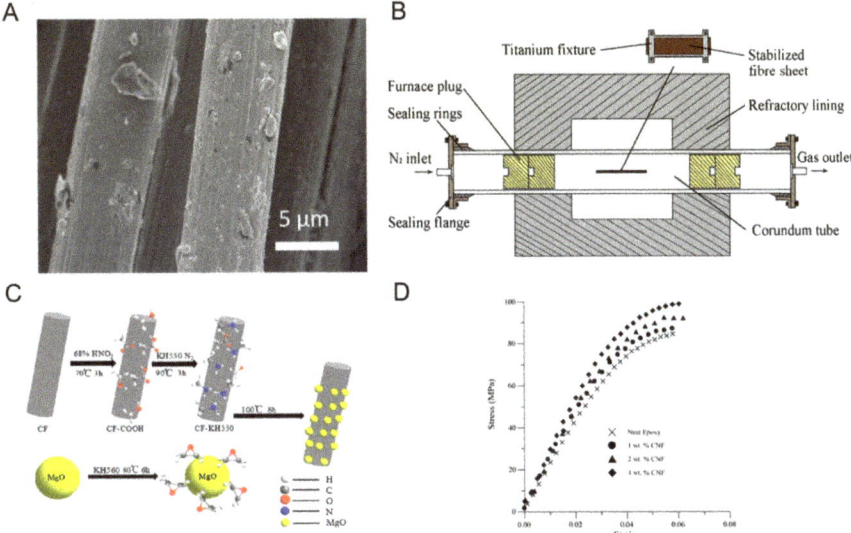

Figure 1. (**A**) SEM image of electrochemically treated Carbon Fiber. Reproduced with permission from [2]. (**B**) Schematic of the apparatus used for carbonization of nanosheets. Reproduced with permission from [6]. (**C**) Illustration of the process for generation of carbon fiber (CF)-MgO hybrid nanocomposite. Reproduced with permission from [9]. (**D**) Stress–strain curves for SC–15 epoxy nanocomposite for different weight percentages of carbon nano fiber (CNF). Reproduced with permission from [15].

2. Carbon Fibers in Renewables

The renewable energy sector has evolved to a great extent with alternative energy sources like secondary rechargeable batteries, supercapacitors, solar cells and fuel cells. In this section we will discuss hybrid CNF-based renewable devices that have enhanced the performance characteristics of these devices.

2.1. Rechargeable Batteries

Secondary rechargeable batteries can provide an excellent replacement for conventional sources of energy. Additionally, they can serve as a back-up to other sources in a distributed generation system. Here, we briefly take a look at the various hybrid CNF-based vanadium redox flow [17,18], sodium-ion [19], lithium oxygen [20], lithium-ion [21–25] and zinc–carbon26 batteries that have been designed by researchers.

Vanadium flow batteries (VFB) provide many benefits like high efficiency, durability, being environment-friendly, and they also avoid the common problem of ion crossover. However, the commercialization process of these batteries is a challenging aspect because of technical barriers associated with the electrode, separator and electrolyte. A bipolar plate is an important aspect of these batteries since it gives a path for conduction of electrons and it also separates cells. Nam et al. fabricated a CF/fluoroelastomer composite bipolar plate [18]. Compared to earlier versions, the present bipolar plate manufactured by them gave high electrical conductivity and chemical stability against oxidizing and acidic conditions. The fluoroelastomer enabled the composite to have a high volume that led to enhanced electrical conductivity. Mechanical tests performed on the composite plate revealed good strength with a Young's modulus of 48.5 GPa and a Poisson ratio of 0.34. Moreover, the composite showed good electrical performance characteristics where the energy efficiency was 80.4% at a current density of 1000 A/m^2. Sodium-ion batteries (SIB) have recently gained high popularity and they can perform similar to lithium-ion batteries (LIB) and sometimes even outperform them. Electrochemically, both Na and Li function similarly. However, due to high availability of Na, the cost of SIBs is much cheaper than LIBs. However, due to larger radius and smaller diffusion of Na ions as compared to Li ions, it becomes difficult for Na ions to enter graphite, which is the common electrode used for LIB. Due to the unique electrical and chemical properties of Transition Metal Dichalcogenides (TMDCs) like MoS$_2$, they have been intensively incorporated as electrode materials. However, due to large volume change during sodiation/desodiation process, MoS$_2$ suffers from weak intrinsic conductivity and fast capacity reduction. To combat these difficulties, Chen et al. used MoS$_2$/electrospun CNF as a new composite material for SIBs [19]. The use of CNF restricts the volume change of MoS$_2$ and also enhances its conductivity. The fabricated electrode had good interfacial contact between the two materials used in the composite which led to improved cycling stability and better rate performances. The modified SIB had a high charge capacity of 380 mA h g^{-1} after 50 charge–discharge cycles. Additionally, after 500 cycles, it could still provide a charge capacity of 198 mA h g^{-1} at a high current density of 1 A g^{-1}.

Figure 2 below shows the fabrication procedure and SEM image of the CS/GF electrode used in VFB, and generation method as well as sodium-storage mechanism of the MoS$_2$@CNF electrode incorporated in SIB.

Non-aqueous lithium–oxygen batteries (LOB) have many advantages over LIB like the usage of oxygen from the atmosphere instead of having heavy reactants that are normally used in LIB. However, due to the slow oxygen reduction reaction (ORR) and oxygen evolution reaction (OER), LOB have low cycling performance, large over-potential, poor round-trip efficiency and restricted rate capability. Since both ORR and OER occur at the cathode, it is essential to have high conductive electrodes. Cao et al. fabricated Co loaded CNF as a free-standing cathode for LiO$_2$ batteries [20]. The addition of Co in proper amounts can result in a continuous and porous CNF that enhances its surface area and facilitates electron transfer and oxygen diffusion. The fabricated Co/CNF films have more active sites that encourage reactant diffusion and can also store vast amounts of Li$_2$O. Due to good catalytic performance of Co, the Co/CNF electrode enhances the capacity, rate capability and cyclic stability of LOB. 7.4 wt% of Co can give maximum benefits. However, overloading the CNF with 11.1 wt% of Co can result in disorientation of the film structure and can make it fragile. The LOB based on Co/CNF electrode had a charge capacity of 4583 mA h g^{-1} at a current density of 100 mA g^{-1}. Additionally, after 40 charge–discharge cycles, it still displayed a charge capacity of 500 mA h g^{-1} at 100 mA g^{-1}.

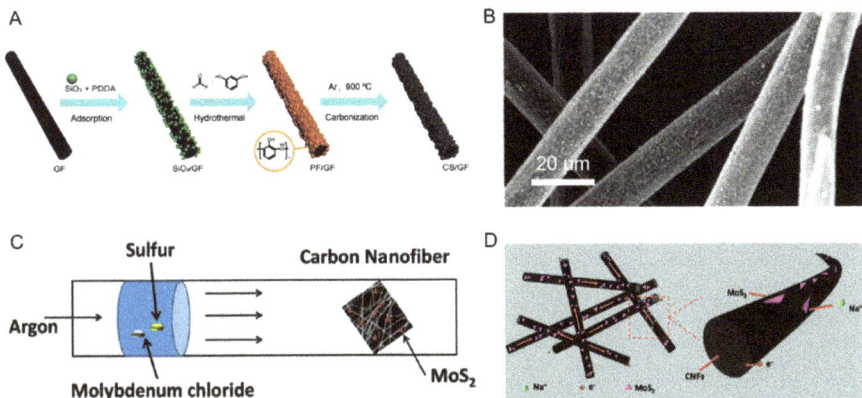

Figure 2. (**A**) Fabrication method of Carbon Sphere(CS)/glass fabric (GF) electrode. (**B**) SEM image of CS/GF electrode. Reproduced with permission from [17]. (**C**) Schematic depiction of the method of Figure 2. CNF composite. (**D**) Pictorial representation of sodium storage mechanism in MoS2@CNFs. Reproduced with permission from [19].

Recently, the demand for flexible battery systems has increased dramatically because of the introduction of smart devices and wearable electronics in the market. Hence, it is essential to develop flexible electrodes. Most LIBs make use of graphite as the electrode material. However, because of its low specific capacity of 372 mA h g^{-1}, its use is limited. Due to the advantage of high theoretical specific capacity of Si, Si-based alloys are being researched as the new electrode material for LIBs. However, during the charge–discharge cycles, the entry of Lithium ions into the Si electrode causes it to expand by almost 400%. This results in the loss of ohmic contact between the Carbon conductor and Si that further causes degraded electrical performance. To meet this drawback, Dirican et al. fabricated a free-standing and flexible Silicon/Silica/Carbon (Si/SiO$_2$/C) nano fiber composite as an anode material for LIBs [21]. The addition of SiO$_2$ helped in controlling the expansion of Si during multiple charge–discharge cycles. The generated nano fiber composite was further coated with nanoscale carbon by chemical vapor deposition (CVD). This further helped in maintaining the Si nanoparticles within the composite. The CVD carbon-coated fiber composite showed high capacity retention and coulombic efficiency of 86.7% and 96.7%, respectively, at the 50th charge cycle.

In another fiber-based battery system for flexible electronic applications, Yu et al. designed CF electrodes for Zinc–Carbon (Zn–C) batteries [26]. Zn–C batteries have grown enormously since their inception into the market because of their low cost due to frugal raw materials, low internal resistance and high energy density. These batteries utilize Zn as the cathode and MnO$_2$-graphite powder as the anode. The researchers modified the composition of the electrode materials and coated the conventionally used Zn and MnO$_2$-graphite powder on CF. The modified battery had an open circuit voltage of 1.5 V. Additionally, at the discharge density of 70 mA g^{-1}, the battery had a discharge capacity of 158 mA h g^{-1}. Moreover, the fiber added flexibility to the battery. Upon changing the bending radius from 3 cm to 0.7 cm, the battery did not show any degradation in its performance. Furthermore, when the fiber length expanded to 8 cm from 2 cm, the discharge capacity remained intact.

2.2. Supercapacitors

Similar to batteries, supercapacitors provide higher power but have lower specific energy. Owing to the conductive nature of CF, they have been widely employed in making modern-day supercapacitors that have enhanced output performance characteristics [27–36].

MnO_2 electrode materials provide many benefits due to their low cost, environmental compatibility and abundance. Chi et al. fabricated Boron-doped MnO_2/CF composites intended to be used as electrodes in supercapacitors [27]. The addition of Boron enhanced the growth rate of MnO_2 crystals. Doping improves the specific capacitance and cyclic stability of supercapacitors. Thus, the MnO_2/CF composite electrode supported the electrochemical reactions and enhanced the surface charge storage and rate capabilities of the supercapacitor. Even after 1000 charge–discharge cycles, the supercapacitor retained 80% of its initial capacitance. The composite fiber had a worm-like structure that led to an increased specific capacitance of 364.8 F g^{-1} and a surface charge density of 19.5 C g^{-1}.

Polymeric capacitors that are deposited on carbon materials show enhanced performance. Davoglio et al. prepared thin films of Polypyrrole (PPy) and poly-2,5-dimercapto-1,3,4-thiadiazole (poly (DMcT)) coated on CF cloth [28]. The bilayer composite was prepared by coating PPy on poly (DMcT)-functionalized CF. The addition of PPy helped in preserving poly (DMcT) and hence bettered its charge-storage capabilities after 1000 charge–discharge cycles. The composite was lightweight and had a high surface area. The CF/poly (DMcT)/PPy composite had a high specific capacity value of 320 mA h g^{-1}. The CF/PPy and CF/poly (DMcT)/PPy composites had specific capacitance values of 460 ± 50 and 1130 ± 100 Fg^{-1}, respectively.

Xie et al. generated coaxial micro fibers (CMF) that were comprised of Ni and CNT coated on CF [29]. The purpose of functionalizing CF was to increase its specific surface area. To increase the energy storage capacity, they coated Ni on CF and then added CNT. The fabricated CMF had higher surface area, electrical conductivity and capacitance. Additionally, the CMF had good tensile strength to be used as electrode material for flexible electronic applications. The CF–Ni–CNT composite had 1400 and 100 times higher specific surface area (SSA) and capacitance, respectively, as compared to bare CF.

Bare Carbon fiber papers (CFP) have been widely used in batteries, supercapacitors and fuel cells. However, because of weak ionic conductivity and hydrophobic surface, they cannot store a high amount of charge. Suktha et al. designed functionalized CFP in order to overcome these shortcomings [30]. They functionalized it mainly with carboxyl (-COOH) and hydroxyl (-OH) groups. The functionalized CFP (f-CFP) exhibited good performance characteristics. It displayed areal, volumetric and specific energies of 49 µW h cm^{-2}, 1960 mW h L^{-1} and 5.2 W h kg^{-1} and powers of 3 mW cm^{-2}, 120 W L^{-1} and 326.2 W kg^{-1}, respectively.

Figure 3 below shows the X-ray diffraction (XRD) patterns for CF, undoped electrode and Boron-doped electrode, the SEM images of CF/poly(DMcT)/PPy composite and CNTs grown on CFs and the Fourier Transform Infrared spectroscopy (FTIR) spectra of bare and functionalized CFP.

Yin et al. fabricated a flexible and conductive thin film made of Polydimethylsiloxane (PDMS), Ag nanowires (AgNWs) and CFs [31]. Along with improving the electrical conductivity, CF helped in strengthening the composite by helping it to resist any mechanical deformations. Moreover, the addition of AgNWs assisted in increasing the surface area of CF and in reducing the contact resistance between the adjacent CFs. The composite film had a low sheet resistance of 0.99 Ω m^{-2}. Additionally, even after 275 consecutive cycles of bending and releasing processes, the sheet resistance decreased by 3%. Interestingly, the increased addition of CF enhanced the load bearing capability of the composite. Increasing the amount of CF from 50 mg to 200 mg resulted in better stress–strain curves. The resistance of the composite with a higher content of CF was also lower. The resistance decreased from 50.1 Ω for 50 mg CF to 19.1 Ω for 100 mg CF and 9.6 Ω for 200 mg CF. In another application of developing flexible electrodes for supercapacitor application, Ma et al. fabricated a unique composite by depositing Nickel Hexacyanoferrate nanocubes (NiHCF-NCs) on flexible CFs [32]. The generated electrode had a high capacitance of 476 F g^{-1} at 0.2 A g^{-1}. Additionally, the electrode was able to retain 92.5% of its capacitance even after 8000 charge/discharge cycles. Nanostructured NiHCF helps in overcoming the shortcomings of volume expansion and extreme agglomeration that are usually faced by bulk-sized NiHCF.

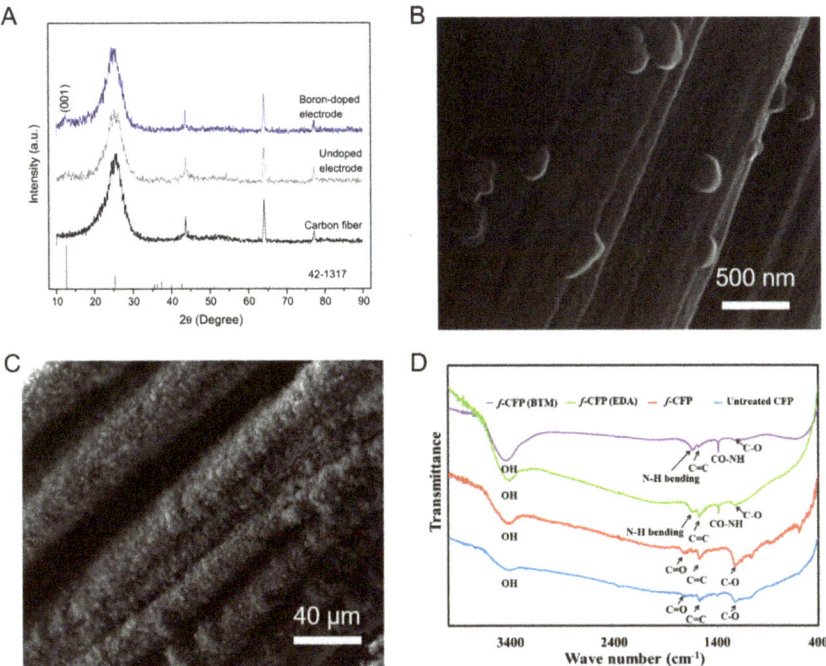

Figure 3. (**A**) XRD patterns for CF, undoped electrode and Boron-doped electrode. Reproduced with permission from [27]. (**B**) SEM image of CF/poly-2,5-dimercapto-1,3,4-thiadiazole (DMcT)/Polypyrrole (PPy) composite. Reproduced with permission from [28]. (**C**) SEM image of carbon nanotubes CNTs grown on CFs. Reproduced with permission from [29]. (**D**) FTIR spectra of bare and functionalized carbon fiber paper CFP. Reproduced with permission from [30].

The poor electrical conductivity of MnO_2 restricts its usage in supercapacitors as the rate capabilities are significantly lowered because of fall of specific capacitance. Thus, Zhao et al. fabricated ZnO@Au@MnO_2 nanosheets on CF paper [34]. Due to this hierarchical structure, the electrical conductivity and specific capacitance is enhanced as there is large electric contact that increases ion and electron transport rate and also shortens the ion diffusion path. The specific capacitance of the composite material was 654 F g^{-1}, calculated using cyclic voltammetry (CV), and 478 F g^{-1} at a current density of 2.6 A g^{-1}. Furthermore, it could retain 80% of its capacitance after 2500 charge–discharge cycles. The presence of CF paper makes the composite light weight and thus makes it a promising candidate for applications in smart electronics. Dong et al. prepared mesoporous graphitic carbon fibers that had large surface areas and high pore volumes of 870–1790 m^2g^{-1} and 0.729–1.308 $cm^{-3}g^{-1}$ [35]. The prepared fibers had a high specific capacitance of 303 Fg^{-1} at 0.7 Ag^{-1} that could be retained well after several charge/discharge cycles. The enhanced performance was due to the morphology of the fiber, better porous structure and degree of graphitization.

2.3. Solar Cells

Harnessing solar energy has been an enormous blessing as it has reduced the burden on conventional generators and has also helped in lessening the negative impacts of non-renewable sources of energy. Here, we look into some structural configurations of dye-sensitized solar cells (DSSCs) that utilize hybrid CF materials as electrodes [37–46].

Fibers made of Carbon nanomaterials have high-performance characteristics. Fang et al. fabricated a unique core-sheath carbon nanostructure fiber [37]. The structure comprised of a CNT core that provided high tensile strength and electrical conductivity and gold nanoribbons (GNRs) that gave high electrocatalytic activity. Similar to the bare CNT fiber, the core-sheath type composite fiber had electrical conductivities and tensile strengths of 10^2–10^3 S cm^{-1} and 10^2–10^3 MPa, respectively. However, as an added advantage, the sheath helped in achieving high catalytic activity as the atomic edges were exposed on the surface. As compared to other structures (CNT/Graphene Oxide (GO) fiber, CNT fiber, GNR fiber), CNT/GNR displayed the best output characteristics: Open Circuit voltage, V_{oc} of 0.7 V, current density, J_{sc} of 12.07 mA cm^{-2}, fill factor (FF) of 60.95% and an efficiency, η of 5.16%.

Veerappan et al. replaced conventional Pt electrodes with CNFs as the counter electrode (CE) for DSSCs [38]. Because of the nano structured morphology, the CNFs-CE had a faster I_3^- reduction rate and low charge transfer resistance R_{CT} of 0.5 Ω cm^{-2} as compared to Pt. Because of specific characteristics like terminating graphitic layers on the fiber surface, large defects on edge planes and big pore diameters as well as rough and large surface areas, CNFs have faster electron transfer mechanics that lead to enhanced photovoltaic output characteristics.

The improved performance of CEs in DSSCs requires low charge-transfer resistance and high electrocatalytic activity of the active material used as the CE. Due to the high cost of Pt, other transition metal structures are being researched. Yousef et al. fabricated NiCu nanoparticles (NP) that were coated with CNF and used them as the CE for DSSCs [39]. CNF provided good shielding against corrosion and also helped in enhancing electrical conductivity and adsorption capacity. The DSSC based on the Cu/Ni CNF composite CE had good photovoltaic output characteristics: V_{oc} of 0.7 V, J_{sc} of 7.67 mA cm^{-2}, FF of 65% and η of 3.5%. In another Pt-free application, Yousef et al. prepared Co-TiC NPs embedded on CNF [40]. The composite was used as a CE in DSSCs and fuel cells (FCs). The photovoltaic output characteristics of the DSSC were V_{oc} of 0.758 V, J_{sc} of 9.98 mA cm^{-2}, FF of 50.7% and η of 3.87%. The researchers attributed the enhanced electrocatalytic activity of the CE to the synergetic effects of its individual components.

Figure 4 below displays the SEM image of the core-sheath nanostructured fiber, the photovoltaic output characteristics for antler Carbon nanofiber (CNF–LSA) and Pt CE-based DSSCs and the Voltage-Current (V-I) performance for NiCu–CNF composite CE-based DSSC and Co-TiC CNF CE-based DSSC.

Chen et al. used a composite of PtNPs and vapor grown carbon fibers (VGCFs) coated on Fluorine-doped Tin Oxide (FTO) glass as a CE for DSSC [43]. Compared to CNT, VGCFs have poor mechanical characteristics. However, VGCFs have more structural defects due to which they have more active sites for electrocatalytic reactions. In order to have higher efficiency at a reduced cost, the researchers focused on preparing PtNPs/VGCF composites. The PtNPs increased the thermal stability of VGCFs and they were uniformly distributed over VGCFs, which aided in increasing the surface area that facilitated the redox reactions taking place inside the films. The hybrid PtNP/VGCF CE with a weight ratio of 3:7 for PtNPs to VGCFs, had a higher photovoltaic conversion efficiency of 7.77% as compared to 7.31% and 3.97% for the conventional Pt CE and bare VGCF CE.

To produce a low-cost alternative to conventional Pt-based electrodes, Chen et al. used transition metal compounds that had good catalytic activity and electronic structures that resembled that of Pt [44]. They fabricated $CoNi_2S_4$ nanoribbons on CFs as a CE for fiber-shaped DSSCs (FDSSCs). The CE made up of the $CoNi_2S_4$/CF composite material had an efficiency of 7.03%. The researchers also fabricated a different versions of the composite where they used $CoNi_2S_4$ nanorods on CFs. However, it displayed a low photovoltaic conversion efficiency of 4.10%. Additionally, while comparing the I-V curves of bare CF, $CoNi_2S_4$ nanorod-CF, Pt wire and $CoNi_2S_4$ nanoribbon-CF, the nanoribbon morphology gave the highest current density. The photovoltaic output characteristics of the DSSC made up of $CoNi_2S_4$ nanorod-CF were V_{oc} of 0.68 V, J_{sc} of 15.3 mA cm^{-2} and FF of 67.7%.

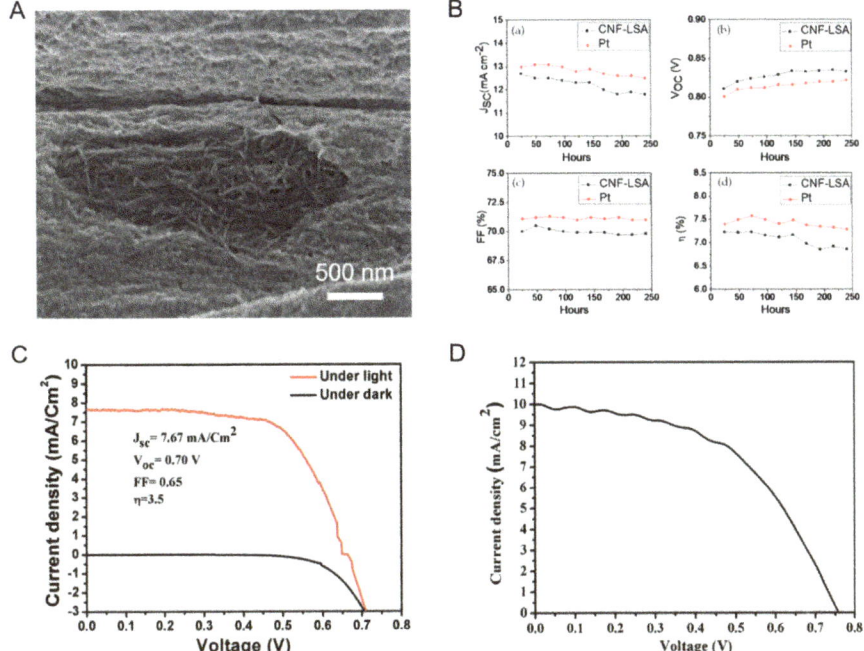

Figure 4. (**A**) SEM image of the core-sheath nanostructured fiber. Reproduced with permission from [37]. (**B**) Photovoltaic output characteristics ((a) short-circuit current density, J_{SC}, (b) open-circuit voltage, V_{OC}, (c) fill-factor, FF, (d) energy conversion efficiency, µ) for CNF-LSA and Pt counter electrode (CE)-based dye-sensitized solar cells (DSSC)s. Reproduced with permission from [38]. (**C**) V-I performance and cyclic voltammetery curves for NiCu-CNF composite CE-based DSSC. Reproduced with permission from [39]. (**D**) V-I performance curve of the Co-TiC CNF CE-based DSSC. Reproduced with permission from [40].

Guo et al. fabricated TiO_2 nanorod (NR) arrays grown on CF as photoanode for DSSC [45]. The NR-based solar cell had V_{oc} of 0.63 V, J_{sc} of 2.57 mA cm^{-2}, FF of 47% and η of 0.76%. The CF-based DSSC was tube-shaped and it could capture light from all directions. Additionally, it showed high electrical conductivity, anti-corrosive property towards I_2 and high reactivity for triiodide reduction. Combining all these with the economical price of carbon materials, the CF-based solar cell could be a good alternative to Pt electrodes. In another hybrid application, Guo et al. fabricated Pt/CF composites for a CE that could be used for redox reactions of Co^{3+}/Co^{2+}, T_2/T^- and I_3^-/I^- [46]. With a low 1 wt% loading of Pt, the best output characteristics were obtained. With just bare CF, the output characteristics were V_{oc} of 0.846 V, J_{sc} of 13.49 mA cm^{-2}, FF of 56% and η of 6.39%. With 1 wt% of Pt, the output characteristics of the composite were J_{sc} of 15.52 mA cm^{-2}, FF of 68% and η of 8.97%. The enhanced features of the DSSC could be attributed to the high catalytic activity of the Pt/CF composite towards the redox couples.

Table 1 below summarizes the output characteristics of the CF-based DSSCs discussed above.

Table 1. Summary of output characteristics of CF-based DSSCs.

S.No.	Active Materials	Output Performance	Reference
1.	CNT/GNR fiber	V_{oc} = 0.846 V J_{sc} = 13.49 mA cm^{-2} FF = 56% η = 6.39%	[37]
2.	CNF-LSA	V_{oc} = 0.779 V J_{sc} = 12.6 mA cm^{-2} FF = 55.2% η = 5.4%	[38]
3.	CuNi NPs-CNF	V_{oc} = 0.7 V J_{sc} = 7.67 mA cm^{-2} FF = 65% η = 3.5%	[39]
4.	Co-TiC NPs-CNF	V_{oc} = 0.758 V J_{sc} = 9.98 mA cm^{-2} FF = 50.7% η = 3.87%	[40]
5.	Hollow activated CNF	V_{oc} = 0.73 V J_{sc} = 14.5 mA cm^{-2} FF = 62% η = 6.58%	[41]
6.	NiCo$_2$S$_4$-CF	V_{oc} = 0.63 V J_{sc} = 17.78 mA cm^{-2} FF = 56% η = 6.31%	[42]
7.	PtNPs/VGCF	V_{oc} = 0.55 V J_{sc} = 15.47 mA cm^{-2} FF = 45% η = 3.79%	[43]
8.	CoNi$_2$S$_4$ nanoribbon-CF	V_{oc} = 0.68 V J_{sc} = 15.3 mA cm^{-2} FF = 67.7% η = 7.03%	[44]
9.	TiO$_2$ NR-CF	V_{oc} = 0.63 V J_{sc} = 2.57 mA cm^{-2} FF = 47% η = 0.76%	[45]
10.	Pt/CF	J_{sc} = 15.52 mA cm^{-2} FF = 68% η = 8.97%	[46]

2.4. Fuel Cells

In addition to the above-mentioned energy storage devices, fuel cells are also equally gaining popularity because of their compact size, easy decentralization and low maintenance requirements. Here, we summarize some recently developed hybrid CF-based fuel cells that have been designed to give enhanced performance.

In a unique application, Li et al. developed a soil microbial fuel cell (MFC) by mixing CF with petroleum hydrocarbon contaminated soil [47]. The addition of CF helped the anode in collecting more electrons and thus the maximum current and power density and accumulated charge output was enhanced 10, 22 and 16 times as compared to fuel cell made without the hybrid material. Moreover, the internal resistance of the cell reduced by 58% which lead to improvements in efficiency.

The researchers hence found that the use of conductive CF was beneficial for bioelectricity recovery from soil.

Gas diffusion layers (GDL) in proton exchange membrane fuel cells (PEMFCs) are important for maintaining H_2/air system performance in regions of high current density. It performs crucial tasks of distributing reactants to active sites, managing water supply and enhancing electrical contact between the electrode and bipolar plates. Kannan et al. prepared a micro-porous GDL with the aid of CNF and carbon nano-chain Pureblack [48]. The researchers found that addition of CNF enhanced the mechanical properties of the GDL. The fuel cell made with CNF-based GDL gave a high power density of 0.55 Wm^{-2}.

Okada et al. used CNF interlayer in a direct methanol fuel cell at the interface of carbon paper and a PtRu NP catalyst layer [49]. The dense and crackless CNF layer reduced catalyst loss and led to increased active reaction sites on the anode. The CNF layer helped in enhancing the electrical conductivity of the fuel cell and also enhanced the power density.

Lim et al. developed a CF/Poly Ether–ether Ketone (PEEK) composite bipolar plate for a high temperature PEMFC [50]. The composite enhanced the electrical conductivity and mechanical strength of the fuel cell. Furthermore, environmental durability tests confirmed the sustainability of the device.

Shu et al. prepared GDL made up of CF felt and Polytetrafluoroethylene (PTFE) for Mg-air fuel cells [51]. The prepared GDL showed enhanced mechanical properties and improved electrical conductivity along with water-repellent properties and high gas permeability as compared to Mg-air fuel cells based on a conventional carbon powder-based cathode.

CNF and activated CNF (ACNF) have been used as the cathode catalyst in MFCs [52]. The ACNFs enhance the catalytic activity sites because of their large surface area. Additionally, they are much more economical as compared to standard Pt cathodes. With enhanced physical properties, the CNFs provide great help in increasing the device performance.

Abdelkareem et al. used Ni–Cd CNFs as a catalyst for urea fuel cells [53]. The low commercialization of these cells is due to low catalytic activity of the anode. With the composite catalyst developed by the researchers, they were able to increase the number of active sites for urea oxidation and hence improve the device's electrical and mechanical output characteristics.

Hence, CF and CNF-based composites have been applied as electrode materials in renewable energy devices like batteries, supercapacitors, solar cells and fuel cells. With just minute additions of CF/CNF, the electrical, thermal and mechanical properties of the base material can be greatly enhanced. Due to features like high surface area, the number of catalytically active sites and the electron flow rate is improved to a high extent, leading to enhanced efficiency and power output of these devices.

3. Sensing Using Carbon Fibers

Several hybrid sensing schemes have recently been developed by researchers that incorporate CF. The inclusion of CF enhances the sensing properties of the composite material. In this section, we describe in depth some of the glucose [54], cortisol [55], neurotransmitter [56], Bisphenol-A [57], Ethanol [58], NO and CO [59], Hydrogen [60], Uracil and 5-Fluorouracil [61], Hydrogen Peroxide [62], DNA [63] and strain [64–67] and chemoresistive [68] gas sensors that are based on hybrid CF materials.

Weina et al. fabricated a glucose sensor by coating β-MnO_2 micro/nanorod arrays on a CF fabric [54]. β-MnO_2 was chosen as the active material for the bio-sensor because of its favorable physical and electrochemical properties. The coating of β-MnO_2 on CF further enhanced its desirable properties. CF led to higher porous channels for electrolyte diffusion and also facilitated ion transport. The sensitivity of the glucose sensor was 1650.6 $\mu A\ mM^{-1}\ cm^{-2}$. The detection limit was 1.9 μM and the linear range was 0.01–4.5 mM. The composite sensor showed high sensitivity, selectivity and stability. The improved performance of the CF-based sensor was due to an increased number of electrochemically active sites, appropriate utilization of β-MnO_2, faster mass transport and direct contact with the current collector.

Sekar et al. used conductive carbon yarn (CCY) to manufacture a wearable sensor for measurement of sweat cortisol [55]. Due to the crystalline nature of CCY, the fiber showed high strength-to-volume ratio. Briefly, they functionalized CCY with α-Fe$_2$O$_3$ in order to immobilize antibodies specific to cortisol. The cortisol sensor had a detection limit of 0.005 fg mL^{-1} and had a linear range from 1 fg to 1 µg with a high correlation factor of R^2 = 0.998. Additionally, the sensor results were repeatable, reliable and reproducible. Moreover, it showed good selectivity towards cortisol when compared to other analogous compounds like cortisone, progesterone and cholesterol. Moreover, it had a short response time of 120 s.

Fiber-reinforced polymer composites often undergo complex and multi-phase failures. The initiation of the damage process begins with the appearance of microcracks. Incorporation of carbon materials in the composites can help in self-diagnosis of cracks. In this regard, Gallo et al. modelled and characterized self-sensing of microcracks of CNT and CF-based composites [64] There was a direct correlation between electrical conductivity of the composite and applied mechanical stress. It was observed that increased stress resulted in lower electric conductivity. The self-sensing capability of the CF-based polymer composites can be a huge benefit for timely detection of cracks, which if undetected can pose a significant threat to civil and mechanical structures and can adversely affect and put several lives in danger.

Figure 5 below shows a schematic depiction of the fabrication process of a β-MnO$_2$ nanorod array on CF fabric and the XRD pattern of β-MnO$_2$ nanorod array for the glucose sensor, an illustration of the hydrothermal process of generating Fe$_2$O$_3$/CCY hybrid electrode and the Field Emission SEM (FESEM) image of the Fe$_2$O$_3$/CCY electrode for the cortisol sensor and a schematic of the transverse crack insertion between the electrode gage length and electrical conductivity summary for different carbon-based composite polymers for the self-sensing CF-based micro-crack detection sensor.

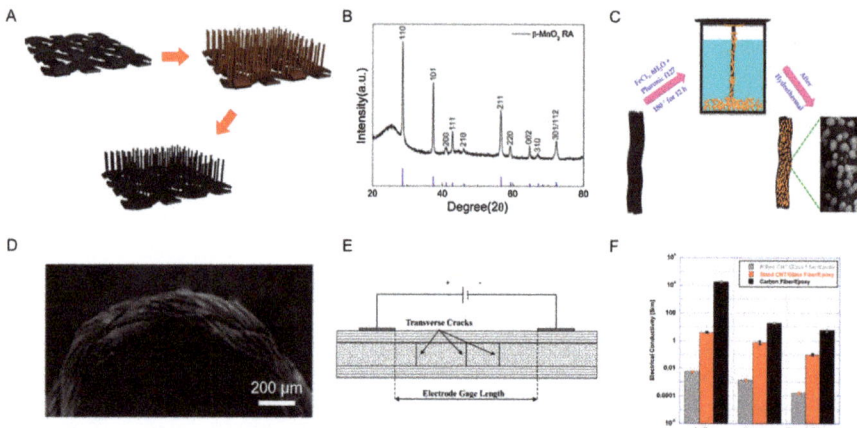

Figure 5. (**A**) Schematic depiction of fabrication of β-MnO$_2$ nanorod array on CF fabric. (**B**) XRD pattern of β-MnO$_2$ nanorod array. Reproduced with permission from [54]. (**C**) Illustration of the hydrothermal process of generating Fe$_2$O$_3$/conductive carbon yarn (CCY) hybrid electrode. (**D**) the Field Emission SEM (FESEM) image of the Fe$_2$O$_3$/CCY electrode. Reproduced with permission from [55]. (**E**) Schematic of transverse crack insertion between the electrode gage length. (**F**) Electrical conductivity summary for different carbon-based composite polymers. Reproduced with permission from [64].

Using electrospinning and heat treatment, Im et al. fabricated a CF composite structure for sensing NO and CO [59]. The generated CF was functionalized to improve the number of gas adsorption sites. Additionally, Carbon Black (CB) derivates were added to enhance the electrical conductivity. Due to the functionalization and addition of CB, the sensitivity was improved five times. The CB additives formed an electrically conductive network inside the fiber structure and the porous structure on the outside of the fiber which resulted in higher gas adsorption. Thus, all these features led to higher sensitivity and selectivity of the sensor towards NO and CO gases.

Ou et al. prepared Pd–Ni nanofilms and NPs of various sizes and electrodeposited them on CF for hydrogen gas sensing [60]. In the concentration ranges of 0–2.8% and 3.6–6% H_2 gas, the response of the sensor increases with increasing concentration. However, for concentrations between 2.8–3.6% H_2 gas, the response decreases for the nanofilm sensor because of rearrangements in the molecular structure. Alternatively, the NP-based sensor displayed consistent results for the entire 0–6% H_2 gas concentration range. Increasing the concentration of H_2 gas results in higher conductivity. This was attributed to the diffusion of hydrogen gas inside the sensor that causes the NPs to expand, resulting in higher contact points between the particles.

Prasad et al. fabricated a silica-molecularly imprinted polymer (MIP) composite fiber using carboxylated MWCNTs for sensing ultra-trace levels of Uracil (Ura) and 5-Fluorouracil (5-FU) [61]. The carboxylated MWCNT enhanced the electrochemical reaction because of its edged plane-like surface area. The use of composite carbon-based electrodes can offer many benefits like greater binding sites and electrochemical activity as compared to traditional electrodes.

The generated sensor was cost-effective, disposable and reliable and had detection limits of 1.3 ng mL^{-1} and 0.56 ng mL^{-1} for Uracil and 5-Fluorouracil, respectively.

Figure 6 below shows the resistive response and FESEM image of the CF-based electrode for NO gas sensing, the SEM image and response curve of the CF-composite Pd–Ni alloy NP and nanofilm hydrogen sensor and the schematic and SEM image of the Silica–MWCNT–MIP composite fiber for Ura and 5-FU sensing.

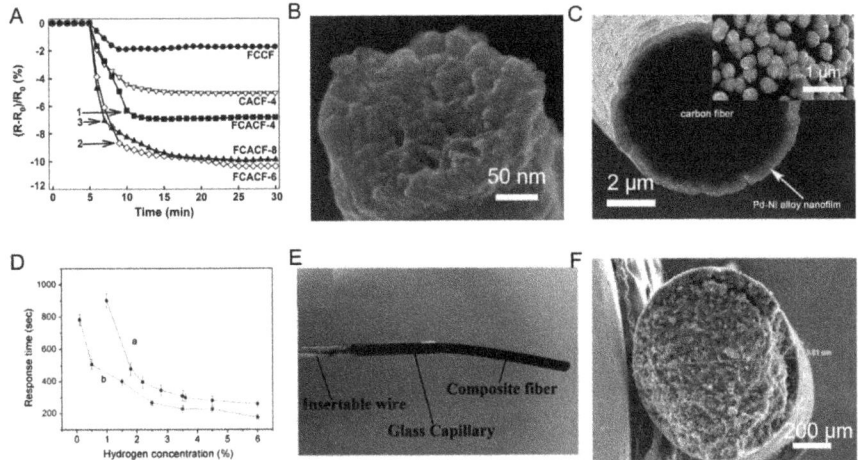

Figure 6. (**A**) Resistive response for NO gas sensing. (**B**) FESEM image of the CF-based electrode. Reproduced with permission from [59]. (**C**) SEM image of Pd–Ni alloy nanoparticle (NP) and nanofilm (insert). (**D**) Response curve of (a) nanofilm type hydrogen sensor and (b) CF-composite Pd–Ni NP. Reproduced with permission from [60]. (**E**) Schematic of Silica– Multi-walled Carbon Nanotube (MWCNT)– molecularly imprinted (MIP) composite fiber. (**F**) SEM image of MIP–5-Fluorouracil (5-FU) adduct. Reproduced with permission from [61].

Wu et al. fabricated a Pt NP doped carbon fiber ultramicroelectrode (Pt/CFUME) for an amperometric biosensor for detecting H_2O_2 [62] The response was measured in terms of output current. The response current was linearly related to H_2O_2 concentration in the range of 0.64 µM to 3.6 mM with a correlation factor of 0.9953 and the detection limit was 0.35 µM. The addition of Pt NPs helped in faster electron transport that led to higher sensitivity. Additionally, the CF ultramicroelectrode increased the effective surface area leading to enhanced mass transport.

In another gas-sensing application, Calestani et al. prepared CFs that were functionalized by ZnO nanowires [68]. They generated a smart material that could be used as a piezoelectric strain as well as a chemoresistive gas sensor. Due to the crossing nature of the CF, the sensor can be modified into a non-invasive integrated array of structures having dimensions in microns and each array can be utilized for sensing a different signal. Hence, it is possible to achieve a wide number of sensing applications with just one single CF composite material.

As a biosensing application, Dogru et al. prepared a DNA biosensor using CF as a microelectrode and Methylene Blue (MB) as an indicator of hybridization [63]. The response was measured in milliamperes. The sensitivity of the biosensor was 0.19 mA µM^{-1} and the linear range was from 0.08–1.6 µM with a correlation factor of 0.995. Additionally, the detection limit was 0.12 µM.

Figure 7 below shows the SEM image of Pt/CFUME, the response curve for the H_2O_2 sensor, and a schematic and an SEM image of the two crossing CFs functionalized with ZnO NWs.

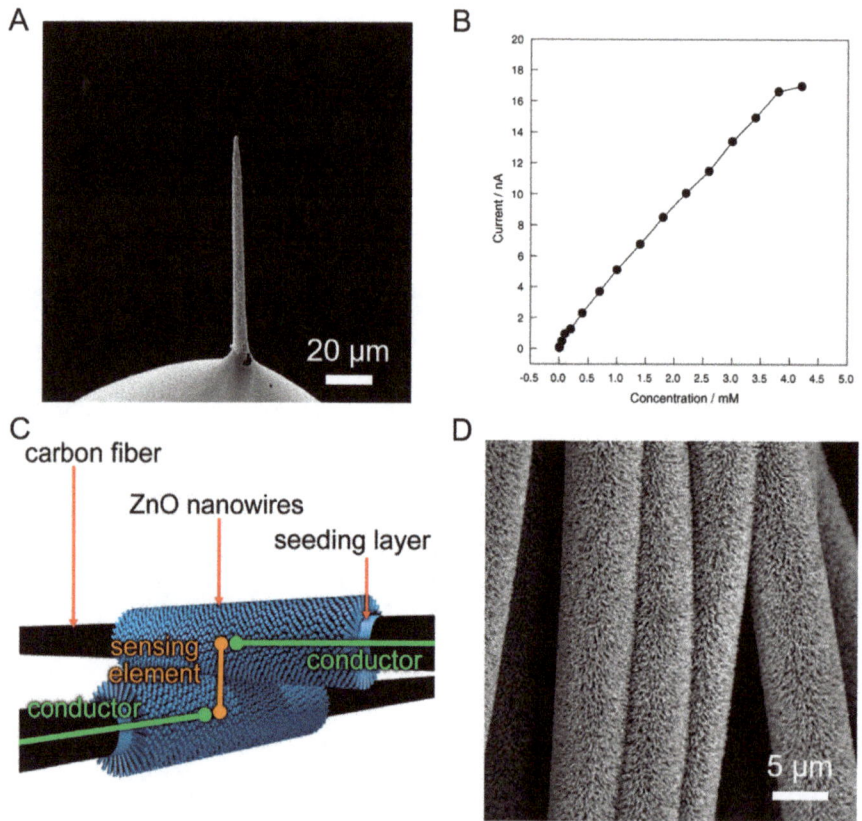

Figure 7. (**A**) SEM image of Pt NP doped carbon fiber ultramicroelectrode (Pt/CFUME). (**B**) Response curve for the H_2O_2 sensor. Reproduced with permission from [62]. (**C**) Schematic of the two crossing CFs functionalized with ZnO nanowires (NWs). (**D**) SEM image of CF functionalized with ZnO NWs. Reproduced with permission from [68].

Again, favorable thermal, mechanical and electrical features of CF and CNF can be highly beneficial for sensing applications. With more surface area, the electrical conductivity is enhanced and even heat dissipation is better due to which the thermal limit and operating range of the sensor is improved. Moreover, with an increased number of sensing sites, the detection limit of the sensors can also be reduced.

4. Tissue Engineering Using Carbon Fibers

Tissue Engineering is a field of regenerative medicine and nanomaterials play an essential role as they impart sturdiness to various prosthetics and implants that are designed for the ailment of affected limbs and other human body parts. Carbon Fiber offers multiple advantages for repair of damaged cells and tissues. In this section, we briefly take a review of several composite CF-based materials that have been fabricated for the purpose of healing injured tissues.

Dentistry makes use of several glass fiber-based implants. However, CF can give several advantages over glass because of higher stiffness and strength, low abrasion and thermal expansion and chemical inertness towards most materials. Menini et al. examined several CF-based dental implants and performed destructive and non-destructive tests on them and found that they provided similar results to conventional metal-based implants [69]. Additionally, biological compatibility was tested using 3-(4,5-dimethylthiazol-2-yl)-2,5-diphenyltetrazolium bromide (MTT) assay and they found that the cell vitality for the CF-based implant was 91.4%. Peterson fabricated a bisphenyl-polymer/carbon fiber reinforced composite and found that it could replace Titanium alloy bone implants quite effectively [70]. The composite could offer electrical conductivity properties that were similar to bones and were much higher than metallic implants. Radiation therapy for orthopedic patients having bone implants can have certain limitations like back scattering and beam attenuation due to the implants that can compromise and reduce the therapeutic effect of radiation. Laux et al. fabricated Carbon Fiber Polyether Ether Ketone (CF:PEEK) composite implants and found that they led to much lower attenuation for the radiation as compared to standard Titanium implants [71]. Rajzer et al. prepared CNFs by using PAN/hydroxyapatite (HAp) precursors and used them for bone tissue engineering [72]. The researchers found that upon insertion of the composite in simulated body fluid (SBF), the surface of the composite was covered with bone-like apatite. Huang et al. fabricated a CF-based platform for wound healing [73]. They performed in vivo analysis and found that fibroblast cells that had the three-dimensional (3D) CF scaffold could accelerate treatment of the wound. Furthermore, the fibroblast cells migrated to the wound location and increased the production of fibronectin and type I collagen that resulted in faster healing than other conventional treatment methods.

Figure 8 below shows the dental implant made of CF composite, a pictorial image of the Bisphenyl-polymer/CF implant and tissue and also the bone prepared with the CF/PEEK implant, and a schematic of the electrospinning apparatus for generating CNF.

Naskar et al. fabricated a biocompatible composite using CNF and nonmulberry silk fibroin and used them for regeneration of bones [74]. The compressive modulus of the composite is 46.54 MPa, which is quite a bit higher than the minimum required human trabecular bone modulus of 10 MPa. The composite showed minimum toxicity both in vitro and in vivo as there was minimum release of pro-inflammatory cytokines. Carbon fiber-reinforced polyetheretherketone (CFRPEEK) composites are ideal for implants. However, due to biological inertness and poor osteogenic properties they have limited applications. Xu et al. modified CFRPEEK composites by adding hydroxyapatite (HA) and produced nano-sized PEEK/CF/n-HA ternary composites [75]. The new composite demonstrated outstanding biocompatibility and also bonded well with the bones. In another application, Srivastava et al. developed tricalcium phosphate–polyvinyl alcohol doped CF reinforced polyester resin composites and tested them by implanting them in the artificially created bone defects in the bone marrow of rabbits [76]. The composite showed high compressive, tensile and bending strengths. They found the formation of cancellous bone around the implant region after 12–32 weeks of implantation. In an altogether different study, PROKIJ et al. used Magnetic Resonance Imaging (MRI) to evaluate its efficacy in monitoring the biocompatibility of surface functionalized CF implants [77]. They analyzed the interaction of CF implants with subcutaneous and muscular tissues of rabbits and found that the MRI technique was quite beneficial in studying the biocompatibility of CF-based implants. Garcia-Ruiz et al. developed 3D structures using CFs and functionalized it with human mesenchymal stem cells (h-MSC) and found that the scaffold supported cell adhesion, proliferation and viability [78]. The 3D structure has the potential for not only repairing tendons and ligaments but also for regenerating cartilage and endochondral bone.

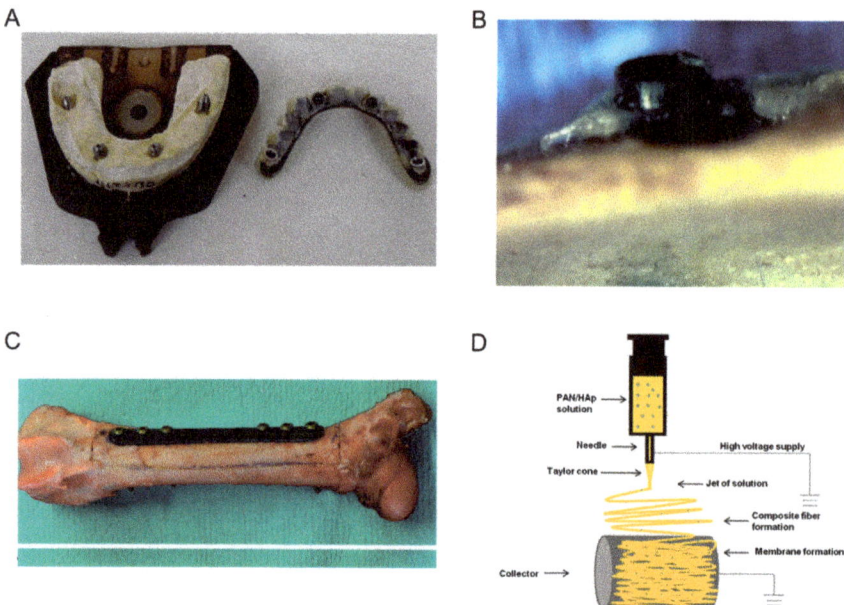

Figure 8. (**A**) Dental implant made of CF composite. Reproduced with permission from [69]. (**B**) Pictorial image of the Bisphenyl-polymer/CF implant and tissue. Reproduced with permission from [70]. (**C**) Pictorial representation of the bone prepared with the Carbon Fiber Polyether Ether Ketone (CF/PEEK) implant. Reproduced with permission from [71]. (**D**) Schematic of the electrospinning apparatus for generating CNF. Reproduced with permission from [72].

Figure 9 below shows the differential scanning calorimetry heat flow thermographs of the CF-based nanocomposite, the Alkaline Phosphatase activities of MG-63 cells cultured on different implant materials, the histopathology of the newly formed bone in a rabbit and the computer aided designs of the 3D composite implant.

In another application of CF in dentistry, Pesce et al. evaluated the mechanical characteristics of multi and uni-directional CF structures. They found that the dynamic elastic modulus was higher for the uni-directional fibers while the static elastic modulus was higher for the multi-directional fibers [79]. In a unique application, Wan et al. fabricated 3D CNF scaffolds with bacterial cellulose as the starting source [80]. The 3D composite is made up of CNF and HAp. The produced CNFs had diameters in the range of 10–20 nm and they showed high biocompatibility during in vitro tests. Deng et al. prepared biocompatible and mechanically strong PEEK/n-HA/CF composites for dental implants [81]. The ternary composite shows excellent in vivo bioactivity and promotes fast and effective osseointegration with canine tooth defects. Chen et al. prepared electrically conductive scaffolds from CF for tissue engineering of electroactive tissues [82]. Upon in vitro tests on nerve cells, they found that electrical stimulation promoted cell proliferation and differentiation. Araoye et al. examined the role of CF intramedullary nails in hindfoot fusion [83]. They found that the implant was quite biocompatible and could be safely used for foot and ankle surgery without any considerable risk of failure or further complications.

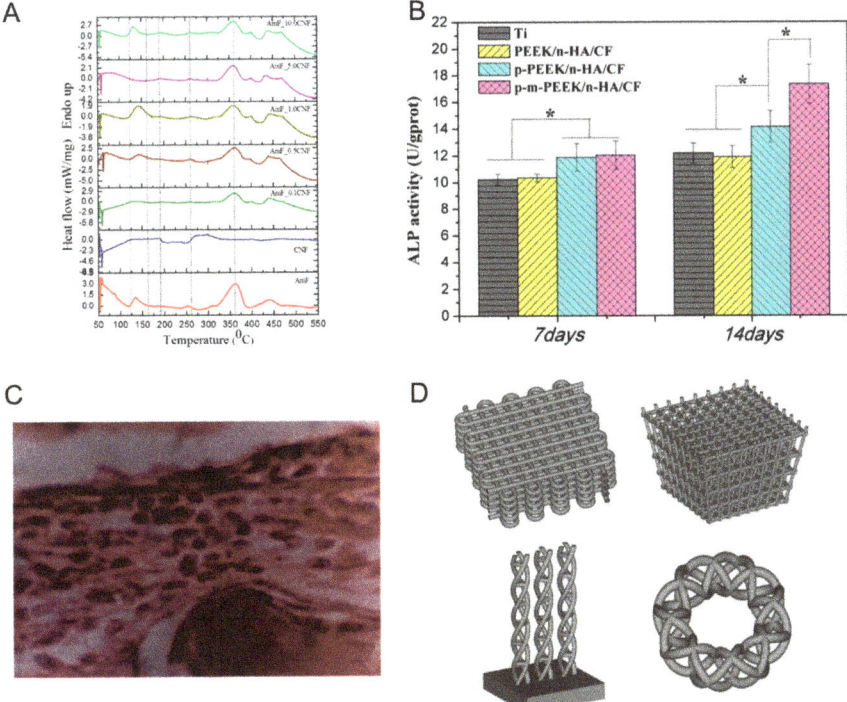

Figure 9. (**A**) Differential scanning calorimetry heat flow thermographs of the CF-based nanocomposite. Reproduced with permission from [74]. (**B**) Alkaline Phosphatase activities of MG-63 cells cultured on different implant materials. Reproduced with permission from [75]. (**C**) Histopathology of the newly formed bone in a rabbit. Reproduced with permission from [76]. (**D**) Computer aided designs of the 3D composite implant. Reproduced with permission from [78]. * Represents $p < 0.05$.

5. Conclusions

The aviation industry, automotive sector and even sporting goods, all need durable yet light-weight materials so that it can have enhanced performance characteristics. Carbon fiber has contributed enormously to these areas as well as other frontline domains of engineering and medicine. Nano CF reinforced composites have been applied in diverse areas since they display high thermal and electrical conductivity, and superior tensile and compressive strengths. In this short review, we briefly discuss some recent developments in the fabrication of new-age nano CF hybrid materials that have been applied in the fields of sensing, renewables (energy storage systems like batteries and supercapacitors, and solar cells), and tissue engineering. We focus on the working principle of the sensors/renewable devices and how hybridized nano CF alters and boosts the device output characteristics because of superior electrical (increased conductivity) and mechanical (enhanced tensile and compressive strengths) properties.

Author Contributions: C.M.D. and L.K. prepared the main draft of the manuscript. G.Y., D.T. and K.-T.Y. helped in revising the manuscript. All authors have read and agreed to the published version of the manuscript.

Funding: This work was supported by the Singapore National Research Foundation (NRF) and French National Research Agency (ANR), grant number (NRF2017–ANR002 2DPS). D.T. is grateful for support from the Nanjing Forestry University Startup Fund (163030164).

Conflicts of Interest: The authors declare no conflict of interest.

References

1. Feng, L.; Xie, N.; Zhong, J. Carbon nanofibers and their composites: A Review of synthesizing, properties and applications. *Materials* **2014**, *7*, 3919–3945. [CrossRef] [PubMed]
2. Semitekolos, D.; Kainourgios, P.; Jones, C.; Rana, A.; Koumoulos, E.P.; Charitidis, C.A. advanced carbon fibre composites via poly methacrylic acid surface treatment; surface analysis and mechanical properties investigation. *Compos. Part B Eng.* **2018**, *155*, 237–243. [CrossRef]
3. Sui, G.; Xue, S.S.; Bi, H.T.; Yang, Q.; Yang, X.P. Desirable electrical and mechanical properties of continuous hybrid nano-scale carbon fibers containing highly aligned multi-walled carbon nanotubes. *Carbon* **2013**, *64*, 72–83. [CrossRef]
4. Liu, N.; Wang, J.; Yang, J.; Han, G.; Yan, F. Effects of nano-sized and micro-sized carbon fibers on the interlaminar shear strength and tribological properties of high strength glass fabric/phenolic laminate in water environment. *Compos. Part B Eng.* **2015**, *68*, 92–99. [CrossRef]
5. Ulus, H.; Sahin, O.S.; Avci, A. Enhancement of flexural and shear properties of carbon fiber/epoxy hybrid nanocomposites by boron nitride nano particles and carbon nano tube modification. *Fibers Polym.* **2015**, *16*, 2627–2635. [CrossRef]
6. Sui, G.; Sun, F.; Yang, X.; Ji, J.; Zhong, W. Highly aligned polyacrylonitrile-based nano-scale carbon fibres with homogeneous structure and desirable properties. *Compos. Sci. Technol.* **2013**, *87*, 77–85. [CrossRef]
7. Gabr, M.H.; Okumura, W.; Ueda, H.; Kuriyama, W.; Uzawa, K.; Kimpara, I. Mechanical and thermal properties of carbon fiber/polypropylene composite filled with nano-clay. *Compos. Part B Eng.* **2015**, *69*, 94–100. [CrossRef]
8. Zhang, X.; Zhang, Y.; Zhang, X.; Wang, Y.; Wang, J.; Lu, M.; Li, H. Mechanical properties and cytocompatibility of carbon fibre reinforced nano-hydroxyapatite/polyamide66 ternary biocomposite. *J. Mech. Behav. Biomed. Mater.* **2015**, *42*, 267–273. [CrossRef]
9. Zhang, J.; Du, Z.; Zou, W.; Li, H.; Zhang, C. Mgo nanoparticles-decorated carbon fibers hybrid for improving thermal conductive and electrical insulating properties of nylon 6 composite. *Compos. Sci. Technol.* **2017**, *148*, 1–8. [CrossRef]
10. Bostanabad, K.O.; Hosseinzade, E.; Kianvash, A.; Entezami, A. Modified nano-magnetite coated carbon fibers magnetic and microwave properties. *Appl. Surf. Sci.* **2015**, *356*, 1086–1095. [CrossRef]
11. Naderi, A.; Mazinani, S.; Javad Ahmadi, S.; Sohrabian, M.; Arasteh, R. Modified thermo-physical properties of phenolic resin/carbon fiber composite with nano zirconium dioxide. *J. Therm. Anal. Calorim.* **2014**, *117*, 393–401. [CrossRef]
12. Jäger, M.; Zabihi, O.; Ahmadi, M.; Li, Q.; Depalmeanar, A.; Naebe, M. Nano-enhanced interface in carbon fibre polymer composite using halloysite nanotubes. *Compos. Part A Appl. Sci. Manuf.* **2018**, *109*, 115–123. [CrossRef]
13. Blugan, G.; Michalkova, M.; Hnatko, M.; Šajgalík, P.; Minghetti, T.; Schelle, C.; Graule, T.; Kuebler, J. Processing and properties of alumina–carbon nano fibre ceramic composites using standard ceramic technology. *Ceram. Int.* **2011**, *37*, 3371–3379. [CrossRef]
14. Li, Y.-L.; Shen, M.-Y.; Chen, W.-J.; Chiang, C.-L.; Yip, M.-C. tensile creep study and mechanical properties of carbon fiber nano-composites. *J. Polym. Res.* **2012**, *19*, 19. [CrossRef]
15. Pervin, F.; Zhou, Y.; Rangari, V.K.; Jeelani, S. Testing and evaluation on the thermal and mechanical properties of carbon nano fiber reinforced Sc-15 epoxy. *Mater. Sci. Eng. A* **2005**, *405*, 246–253. [CrossRef]
16. Charles, A.D.M.; Rider, A.N. Triblock copolymer toughening of a carbon fibre-reinforced epoxy composite for bonded repair. *Polymers* **2018**, *10*, 888. [CrossRef] [PubMed]
17. Zhao, Y.; Yu, L.; Qu, X.; Xi, J. Carbon layer-confined sphere/fiber hierarchical electrodes for efficient durable vanadium flow batteries. *J. Power Sources* **2018**, *402*, 453–459. [CrossRef]
18. Nam, S.; Lee, D.; Lee, D.G.; Kim, J. Nano carbon/fluoroelastomer composite bipolar plate for a vanadium redox flow battery (VRFB). *Compos. Struct.* **2017**, *159*, 220–227. [CrossRef]
19. Chen, C.; Li, G.; Lu, Y.; Zhu, J.; Jiang, M.; Hu, Y.; Cao, L.; Zhang, X. Chemical vapor deposited MoS_2/electrospun carbon nanofiber composite as anode material for high-performance sodium-ion batteries. *Electrochim. Acta* **2016**, *222*, 1751–1760. [CrossRef]

20. Cao, Y.; Lu, H.; Hong, Q.; Bai, J.; Wang, J.; Li, X. Co decorated n-doped porous carbon nanofibers as a free-standing cathode for Li-O2 battery: Emphasis on seamlessly continuously hierarchical 3d nano-architecture networks. *J. Power Sources* **2017**, *368*, 78–87. [CrossRef]
21. Dirican, M.; Yildiz, O.; Lu, Y.; Fang, X.; Jiang, H.; Kizil, H.; Zhang, X. Flexible binder-free silicon/silica/carbon nanofiber composites as anode for lithium-ion batteries. *Electrochim. Acta* **2015**, *169*, 52–60. [CrossRef]
22. Wu, H.; Hou, C.; Shen, G.; Liu, T.; Shao, S.; Xiao, R.; Wang, H. MoS$_2$/C/C nanofiber with double-layer carbon coating for high cycling stability and rate capability in lithium-ion batteries. *Nano Res.* **2018**, *11*, 5866–5878. [CrossRef]
23. Kim, S.Y.; Kim, B.-H.; Yang, K.S. Preparation and electrochemical characteristics of a polyvinylpyrrolidone-stabilized Si/carbon composite nanofiber anode for a lithium ion battery. *J. Electroanal. Chem.* **2015**, *705*, 52–56. [CrossRef]
24. Chen, Y.; Hu, Y.; Shen, Z.; Chen, R.; He, X.; Zhang, X.; Zhang, Y.; Wu, K. Sandwich structure of graphene-protected silicon/carbon nanofibers for lithium-ion battery anodes. *Electrochim. Acta* **2016**, *210*, 53–60. [CrossRef]
25. Kim, Y.S.; Shoorideh, G.; Zhmayev, Y.; Lee, J.; Li, Z.; Patel, B.; Chakrapani, S.; Park, J.H.; Lee, S.; Joo, Y.L. The critical contribution of unzipped graphene nanoribbons to scalable silicon–carbon fiber anodes in rechargeable li-ion batteries. *Nano Energy* **2015**, *16*, 446–457. [CrossRef]
26. Yu, X.; Fu, Y.; Cai, X.; Kafafy, H.; Wu, H.; Peng, M.; Hou, S.; Lv, Z.; Ye, S.; Zou, D. Flexible fiber-type zinc–carbon battery based on carbon fiber electrodes. *Nano Energy* **2013**, *2*, 1242–1248. [CrossRef]
27. Chi, H.Z.; Zhu, H.; Gao, L. Boron-doped MnO$_2$/carbon fiber composite electrode for supercapacitor. *J. Alloy. Compd.* **2015**, *645*, 199–205. [CrossRef]
28. Davoglio, R.A.; Biaggio, S.R.; Bochhi, N.; Filho, R.C.R. Flexible and high surface area composites of carbon fiber, polypyrrole, and poly(DMcT) for supercapacitor electrodes. *Electrochim. Acta* **2013**, *93*, 93–100. [CrossRef]
29. Xie, Y.; Lu, L.; Tang, Y.; Zhang, F.; Shen, C.; Zang, X.; Ding, X.; Cai, W.; Lin, L. Hierarchically nanostructured carbon fiber-nickel-carbon nanotubes for high-performance supercapacitor electrodes. *Mater. Lett.* **2017**, *186*, 70–73. [CrossRef]
30. Suktha, P.; Chiochan, P.; Iamprasertkun, P.; Wutthiprom, J.; Phattharasupakun, N.; Suksomboon, M.; Kaewsongpol, T.; Sirisinudomkit, P.; Pettong, T.; Sawangphruk, M. High-performance supercapacitor of functionalized carbon fiber paper with high surface ionic and bulk electronic conductivity: Effect of organic functional groups. *Electrochim. Acta* **2015**, *176*, 504–513. [CrossRef]
31. Yin, J.; Kim, J.; Lee, H.U.; Park, J.Y. Highly conductive and flexible thin film electrodes based on silver nanowires wrapped carbon fiber networks for supercapacitor applications. *Thin Solid Film.* **2018**, *660*, 564–571. [CrossRef]
32. Ma, X.; Du, X.; Li, X.; Hao, X.; Jagadale, A.D.; Abudula, A.; Guan, G. In situ unipolar pulse electrodeposition of nickel hexacyanoferrate nanocubes on flexible carbon fibers for supercapacitor working in neutral electrolyte. *J. Alloy. Compd.* **2017**, *695*, 294–301. [CrossRef]
33. Li, X.; Zang, X.; Li, Z.; Li, X.; Li, P.; Sun, P.; Lee, X.; Zhang, R.; Huang, Z.; Wang, K.; et al. Large-area flexible core-shell graphene/porous carbon woven fabric films for fiber supercapacitor electrodes. *Adv. Funct. Mater.* **2013**, *23*, 4862–4869. [CrossRef]
34. Zhao, Y.; Jiang, P. MnO$_2$ nanosheets grown on the ZnO-nanorod-modified carbon fibers for supercapacitor electrode materials. *Colloids Surf. A Physicochem. Eng. Asp.* **2014**, *444*, 232–239. [CrossRef]
35. Dong, Y.; Lin, H.; Zhou, D.; Niu, H.; Jin, Q.; Qu, F. Synthesis of mesoporous graphitic carbon fibers with high performance for supercapacitor. *Electrochim. Acta* **2015**, *159*, 116–123. [CrossRef]
36. Ren, J.; Li, L.; Chen, C.; Chen, X.; Cai, Z.; Qiu, L.; Wang, Y.; Zhu, X.; Peng, H. Twisting carbon nanotube fibers for both wire-shaped micro-supercapacitor and micro-battery. *Adv. Mater.* **2013**, *25*, 1155–1159, 1224. [CrossRef]
37. Fang, X.; Yang, Z.; Qiu, L.; Sun, H.; Pan, S.; Deng, J.; Luo, Y.; Peng, H. Core-sheath carbon nanostructured fibers for efficient wire-shaped dye-sensitized solar cells. *Adv. Mater.* **2014**, *26*, 1694–1698. [CrossRef]
38. Veerappan, G.; Kwon, W.; Rhee, S.-W. Carbon-nanofiber counter electrodes for quasi-solid state dye-sensitized solar cells. *J. Power Sources* **2011**, *196*, 10798–10805. [CrossRef]

39. Yousef, A.; Akhtar, M.S.; Barakat, N.A.M.; Motlak, M.; Yang, O.B.; Kim, H.Y. Effective nicu nps-doped carbon nanofibers as counter electrodes for dye-sensitized solar cells. *Electrochim. Acta* **2013**, *102*, 142–148. [CrossRef]
40. Yousef, A.; Brooks, R.M.; Newehy, M.H.E.; Deyab, S.S.A.; Kim, H.Y. Electrospun co-tic nanoparticles embedded on carbon nanofibers: Active and Chemically stable counter electrode for methanol fuel cells and dye-sensitized solar cells. *Int. J. Hydrog. Energy* **2017**, *42*, 10407–10415. [CrossRef]
41. Park, S.-H.; Jung, H.-R.; Lee, W.-J. Hollow activated carbon nanofibers prepared by electrospinning as counter electrodes for dye-sensitized solar cells. *Electrochim. Acta* **2013**, *102*, 423–428. [CrossRef]
42. Chi, Z.; Shen, J.; Zhang, H.; Chen, L. $NiCo_2S_4$ nanosheets in situ grown on carbon fibers as an efficient counter electrode for fiber-shaped dye-sensitized solar cells. *J. Mater. Sci. Mater. Electron.* **2017**, *28*, 10640–10644. [CrossRef]
43. Chen, L.-C.; Lee, K.-J.; Chen, J.-H.; Pan, T.-C.; Huang, C.-M. Novel platinum nanoparticle/vapor grown carbon fibers composite counter electrodes for high performance dye sensitized solar cells. *Electrochim. Acta* **2013**, *112*, 698–705. [CrossRef]
44. Chen, L.; Zhou, Y.; Dai, H.; Yu, T.; Liu, J.; Zou, Z. One-step growth of $coni_2s_4$ nanoribbons on carbon fibers as platinum-free counter electrodes for fiber-shaped dye-sensitized solar cells with high performance: Polymorph-dependent conversion efficiency. *Nano Energy* **2015**, *11*, 697–703. [CrossRef]
45. Guo, W.; Xu, C.; Wang, X.; Wang, S.; Pan, C.; Lin, C.; Wang, Z.L. Rectangular bunched rutile Tio_2 nanorod arrays grown on carbon fiber for dye-sensitized solar cells. *J. Am. Chem. Soc.* **2012**, *134*, 4437–4441. [CrossRef]
46. Guo, H.; Zhu, Y.; Li, W.; Zheng, H.; Wu, K.; Ding, K.; Ruan, B.; Hagfeldt, A.; Ma, T.; Wu, M. Synthesis of highly effective Pt/Carbon fiber composite counter electrode catalyst for dye-sensitized solar cells. *Electrochim. Acta* **2015**, *176*, 997–1000. [CrossRef]
47. Li, X.; Wang, X.; Zhao, Q.; Wan, L.; Li, Y.; Zhou, Q. Carbon fiber enhanced bioelectricity generation in soil microbial fuel cells. *Biosens. Bioelectron.* **2016**, *85*, 135–141. [CrossRef]
48. Kannan, A.M.; Munukutla, L. Carbon nano-chain and carbon nano-fibers based gas diffusion layers for proton exchange membrane fuel cells. *J. Power Sources* **2007**, *167*, 330–335. [CrossRef]
49. Okada, M.; Konta, Y.; Nakagawa, N. Carbon nano-fiber interlayer that provides high catalyst utilization in direct methanol fuel cell. *J. Power Sources* **2008**, *185*, 711–716. [CrossRef]
50. Lim, J.W.; Kim, M.; Yu, Y.H.; Lee, D.G. Development of carbon/PEEK composite bipolar plates with nano-conductive particles for High-Temperature PEM fuel cells (HT-PEMFCs). *Compos. Struct.* **2014**, *118*, 519–527. [CrossRef]
51. Shu, C.; Wang, E.; Jiang, L.; Sun, G. High performance cathode based on carbon fiber felt for magnesium-air fuel cells. *Int. J. Hydrog. Energy* **2013**, *38*, 5885–5893. [CrossRef]
52. Ghasemi, M.; Daud, W.R.W.; Hassan, S.H.A.; Oh, S.-E.; Ismail, M.; Rahimnejad, M.; Jahim, J.M. Nano-structured carbon as electrode material in microbial fuel cells: A comprehensive review. *J. Alloy. Compd.* **2013**, *580*, 245–255. [CrossRef]
53. Abdelkareem, M.A.; Haj, Y.A.; Alajami, M.; Alawadhi, H.; Barakat, N.A.M. Ni-Cd carbon nanofibers as an effective catalyst for urea fuel cell. *J. Environ. Chem. Eng.* **2018**, *6*, 332–337. [CrossRef]
54. Weina, X.; Guanlin, L.; Chuanshen, W.; Hu, C.; Wang, X. A novel β-MnO_2 micro/nanorod arrays directly grown on flexible carbon fiber fabric for high-performance enzymeless glucose sensing. *Electrochim. Acta* **2017**, *225*, 121–128. [CrossRef]
55. Sekar, M.; Pandiaraj, M.; Bhansali, S.; Ponpandian, N.; Viswanathan, C. Carbon fiber based electrochemical sensor for sweat cortisol measurement. *Sci Rep.* **2019**, *9*, 403. [CrossRef]
56. Yang, C.; Trikantzopoulos, E.; Jacobs, C.B.; Venton, B.J. Evaluation of carbon nanotube fiber microelectrodes for neurotransmitter detection: Correlation of electrochemical performance and surface properties. *Anal. Chim. Acta* **2017**, *965*, 1–8. [CrossRef]
57. Shim, K.; Wang, Z.L.; Mou, T.H.; Bando, Y.; Alshehri, A.A.; Kim, J.; Hossain, M.S.A.; Yamauchi, Y.; Kim, J.H. Facile synthesis of palladium-nanoparticle-embedded n-doped carbon fibers for electrochemical sensing. *Chempluschem* **2018**, *83*, 401–406. [CrossRef]
58. Wei, N.; Cui, H.; Wang, X.; Xie, X.; Wang, M.; Zhang, L.; Tian, J. Hierarchical assembly of In_2O_3 nanoparticles on ZnO hollow nanotubes using carbon fibers as templates: Enhanced photocatalytic and gas-sensing properties. *J. Colloid Interface Sci.* **2017**, *498*, 263–270. [CrossRef]

59. Im, J.S.; Kang, S.C.; Lee, S.-H.; Lee, Y.-S. Improved gas sensing of electrospun carbon fibers based on pore structure, conductivity and surface modification. *Carbon* **2010**, *48*, 2573–2581. [CrossRef]
60. Ou, Y.J.; Si, W.W.; Yu, G.; Tang, L.L.; Zhang, J.; Dong, Q.Z. Nanostructures of Pd–Ni alloy deposited on carbon fibers for sensing hydrogen. *J. Alloy. Compd.* **2013**, *569*, 130–135. [CrossRef]
61. Prasad, B.B.; Kumar, D.; Madhuri, R.; Tiwari, M.P. Nonhydrolytic Sol–Gel derived imprinted polymer–multiwalled carbon nanotubes composite fiber sensors for electrochemical sensing of uracil and 5-fluorouracil. *Electrochim. Acta* **2012**, *71*, 106–115. [CrossRef]
62. Wu, Z.; Chen, L.; Shen, G.; Yu, R. Platinum nanoparticle-modified carbon fiber ultramicroelectrodes for mediator-free biosensing. *Sens. Actuators B Chem.* **2006**, *119*, 295–301. [CrossRef]
63. Doğru, E.; Erhan, E.; Arikan, O.A. The using capacity of carbon fiber microelectrodes in DNA biosensors. *Electroanalysis* **2017**, *29*, 287–293. [CrossRef]
64. Gallo, G.J.; Thostenson, E.T. Electrical characterization and modeling of carbon nanotube and carbon fiber self-sensing composites for enhanced sensing of microcracks. *Mater. Today Commun.* **2015**, *3*, 17–26. [CrossRef]
65. Ahmed, S.; Thostenson, E.T.; Schumacher, T.; Doshi, S.M.; McConnell, J.R. Integration of carbon nanotube sensing skins and carbon fiber composites for monitoring and structural repair of fatigue cracked metal structures. *Compos. Struct.* **2018**, *203*, 182–192. [CrossRef]
66. Kalashnyk, N.; Faulques, E.; Thomsen, J.S.; Jensen, L.R.; Rauhe, J.C.M.; Pyrz, R. Strain sensing in single carbon fiber epoxy composites by simultaneous in-situ raman and piezoresistance measurements. *Carbon* **2016**, *109*, 124–130. [CrossRef]
67. Vaidya, S.; Allouche, E.N. Strain sensing of carbon fiber reinforced geopolymer concrete. *Mater. Struct.* **2011**, *44*, 1467–1475. [CrossRef]
68. Calestani, D.; Villani, M.; Culiolo, M.; Delmonte, D.; Coppedè, N.; Zappettini, A. Smart composites materials: A new idea to add gas-sensing properties to commercial carbon-fibers by functionalization with Zno nanowires. *Sens. Actuators B Chem.* **2017**, *245*, 166–170. [CrossRef]
69. Menini, M.; Pesce, P.; Pera, F.; Barberis, F.; Lagazzo, A.; Bertola, L.; Pera, P. Biological and mechanical characterization of carbon fiber frameworks for dental implant applications. *Mater. Sci. Eng. C* **2017**, *70*, 646–655. [CrossRef]
70. Peterson, R.C. Bisphenyl-polymer/carbon-fiber-reinforced composite compared to titanium alloy bone implant. *Int. J. Polym. Sci.* **2011**, 168924. [CrossRef]
71. Laux, C.J.; Villefort, C.; Ehrbar, S.; Wilke, L.; Guckenberger, M.; Muller, D.A. Carbon fiber/polyether ether ketone (CF/PEEK) implants allow for more effective radiation in long bones. *Materials* **2020**, *13*, 1754. [CrossRef] [PubMed]
72. Rajzer, I.; Kwiatkowski, R.; Piekarczyk, W.; Binias, W.; Janicki, J. Carbon nanofibers produced from modified electrospun PAN/hydroxyapatite precursors as scaffolds for bone tissue engineering. *Mater. Sci. Eng. C* **2012**, *32*, 2562–2569. [CrossRef]
73. Huang, W.-Y.; Yeh, C.-L.; Lin, J.-H.; Yang, J.-S.; Ko, T.-H.; Lin, Y.-H. Development of fibroblast culture in three-dimensional activated carbon fiber-based scaffold for wound healing. *J. Mater. Sci. Mater. Med.* **2012**, *23*, 1465–1478. [CrossRef] [PubMed]
74. Naskar, D.; Ghosh, A.K.; Mandal, M.; Das, P.; Nandi, S.K.; Kundu, S.C. Dual growth factor loaded nonmulberry silk fibroin/carbon nanofiber composite 3D scaffolds for in vitro and in vivo bone regenration. *Biomaterials* **2017**, *136*, 67–85. [CrossRef]
75. Xu, A.; Liu, X.; Gao, X.; Deng, F.; Deng, Y.; Wei, S. Enhancement of osteogenesis on micro/nano-topographical carbon fiber-reinforced polyetheretherketone-nanohydroxyapatite biocomposite. *Mater. Sci. Eng. C* **2015**, *48*, 592–598. [CrossRef]
76. Srivastava, V.K.; Rastogi, A.; Goel, S.C.; Chukowry, S.K. Implantation of tricalcium phosphate-polyvinyl alcohol filled carbon fibre reinforced polyester resin composites into bone marrow of rabbits. *Mater. Sci. Eng. A* **2007**, *448*, 335–339. [CrossRef]
77. Prokić, B.B.; Bačić, G.; Prokić, B.; Kalijadis, A.; Todorović, V.; Puškaš, N.; Vidojević, D.; Laušević, M.; Laušević, Z. In vivo MRI biocompatibility of functionalized carbon fibers in reaction with soft tissues. *Acta Vet.* **2012**, *62*, 683–696. [CrossRef]

78. Ruiz, J.P.G.; Lantada, A.D. 3D printed structures filled with carbon fibers and functionalized with mesenchymal stem cell conditioned media as in vitro cell niches for promoting chondrogenesis. *Materials* **2018**, *11*, 23. [CrossRef]
79. Pesce, P.; Lagazzo, A.; Barbeis, F.; Repetto, L.; Pera, F.; Baldi, D.; Menini, M. Mechanical characterisation of multi vs. uni-directional carbon fiber frameworks for dental implant applications. *Mater. Sci. Eng. C* **2019**, *102*, 186–191. [CrossRef]
80. Wan, Y.; Zuo, G.; Yu, F.; Huang, Y.; Ren, K.; Luo, H. Preparation and mineralization of three-dimensional carbon nanofibers from bacterial cellulose as potential scaffolds for bone tissue engineering. *Surf. Coat. Technol.* **2011**, *205*, 2938–2946. [CrossRef]
81. Deng, Y.; Zhou, P.; Liu, X.; Wang, L.; Xiong, X.; Tang, Z.; Wei, J.; Wei, S. Preparation, characterization, cellular response and in vivo osseointegration of polyetheretherketone/nano-hydroxyapatite/carbon fiber ternary biocomposite. *Colloids Surf. B Biointerfaces* **2015**, *136*, 64–73. [CrossRef] [PubMed]
82. Chen, X.; Wu, Y.; Ranjan, V.D.; Zhang, Y. Three-dimensional electrical conductive scaffold from biomaterial-based carbon microfiber sponge with bioinspired coating for cell proliferation and differentiation. *Carbon* **2018**, *134*, 174–182. [CrossRef]
83. Araoye, I.B.; Chodaba, Y.E.; Smith, K.S.; Hadden, R.W.; Shah, A.B. Use of intramedullary carbon fiber nail in hindfoot fusion: A small cohort study. *Foot Ankle Surg.* **2019**, *25*, 2–7. [CrossRef] [PubMed]

© 2020 by the authors. Licensee MDPI, Basel, Switzerland. This article is an open access article distributed under the terms and conditions of the Creative Commons Attribution (CC BY) license (http://creativecommons.org/licenses/by/4.0/).

Article

A Molecular Dynamics Study of the Stability and Mechanical Properties of a Nano-Engineered Fuzzy Carbon Fiber Composite

Hassan Almousa [1,2], Qing Peng [1,3,4] and Abduljabar Q. Alsayoud [4,5,*]

1 Physics Department, King Fahd University of Petroleum & Minerals, Dhahran 31261, Saudi Arabia; hassan.almousa@aramco.com (H.A.); qing.peng@kfupm.edu.sa (Q.P.)
2 Research & Development Center, Saudi Aramco, Dhahran 31311, Saudi Arabia
3 KACARE Energy Research & Innovation Center at Dhahran, Dhahran 31261, Saudi Arabia
4 Hydrogen Energy and Storage Center, King Fahd University of Petroleum & Minerals, Dhahran 31261, Saudi Arabia
5 Material Science and Engineering Department, King Fahd University of Petroleum & Minerals, Dhahran 31261, Saudi Arabia
* Correspondence: sayoudaq@kfupm.edu.sa

Citation: Almousa, H.; Peng, Q.; Alsayoud, A.Q. A Molecular Dynamics Study of the Stability and Mechanical Properties of a Nano-Engineered Fuzzy Carbon Fiber Composite. *J. Compos. Sci.* **2022**, *6*, 54. https://doi.org/10.3390/jcs6020054

Academic Editors: Jiadeng Zhu, Gouqing Li and Lixing Kang

Received: 10 January 2022
Accepted: 2 February 2022
Published: 10 February 2022

Publisher's Note: MDPI stays neutral with regard to jurisdictional claims in published maps and institutional affiliations.

Copyright: © 2022 by the authors. Licensee MDPI, Basel, Switzerland. This article is an open access article distributed under the terms and conditions of the Creative Commons Attribution (CC BY) license (https://creativecommons.org/licenses/by/4.0/).

Abstract: Carbon fiber-reinforced polymer composites are used in various applications, and the interface of fibers and polymer is critical to the composites' structural properties. We have investigated the impact of introducing different carbon nanotube loadings to the surfaces of carbon fibers and characterized the interfacial properties by molecular dynamics simulations. The carbon fiber (CF) surface structure was explicitly modeled to replicate the graphite crystallites' interior consisting of turbostratic interconnected graphene multilayers. Then, single-walled carbon nanotubes and polypropylene chains were packed with the modeled CFs to construct a nano-engineered "fuzzy" CF composite. The mechanical properties of the CF models were calculated through uniaxial tensile simulations. Finally, the strength to peel the polypropylene from the nano-engineered CFs and interfacial energy were calculated. The interfacial strength and energy results indicate that a higher concentration of single-walled carbon nanotubes improves the interfacial properties.

Keywords: fuzzy carbon fiber; turbostratic interconnected graphene; molecular dynamics; interfacial property; uniaxial deformation

1. Introduction

Greenhouse gases trap heat in the atmosphere, increasing Earth's surface temperature from 1–4 °C by 2100 [1]. One of the significant contributors to emitting these gases is the transportation sector [2]. Energy consumption in transport is dependent on a vehicle's mass. Therefore, reducing the vehicle's weight will have an immediate positive impact on reducing the emitted gases. A study showed that reducing the weight of a passenger car by 27% can reduce its emissions by 40% [3]. New materials with low density and high strength need to be utilized to replace conventional metals for weight reduction, and carbon fiber-reinforced polymer (CFRP) composites are the most fitting candidate for this job. Carbon fibers (CFs) have high strength, high modulus, and low density [4]. These outstanding properties have made them essential as reinforcement in structural materials for many areas, such as aerospace, automotive, high-grade sports, and many other applications [5]. However, under compression, fiber-reinforced composites suffer a range of failures typically associated with fiber micro-buckling or kinking linked to the interfacial issues. Additionally, the anisotropic nature of composite materials and their internal interface tend to drive the accumulation of internal damage and result in a possible catastrophic failure [6].

A conventional solution is the dispersion of additives such as carbon nanotubes (CNTs) in a polymer matrix. Computational studies indicated improvement in the elastic-plastic response, fracture toughness, and fiber/matrix stiffness by altering the load path under debonding. However, local agglomeration of CNTs causes microscopic stress concentrations that induce premature damage [7,8]. Therefore, new constituents and processes became available to create nano-engineered interphases of CNTs on the CF surface. This concept has been experimentally demonstrated using a chemical vapor deposition process and electro-deposition methods for linking CNTs to CF surfaces via van der Waals interaction [9,10]. The so-called "fuzzy" CFs contain a microfiber core made of carbon coated with CNTs prior to matrix impregnation. The fuzzy fiber is hierarchical and shows enhanced fracture toughness, interlaminar shear strength, and thermal and electrical properties [11]. To further understand this type of hierarchical fiber and its potential without the challenging experimental parametrization, an atomistic model framework was constructed for investigating the interface interaction between the fuzzy CF and the polymer matrix.

Molecular dynamics (MD) simulation is a well-established and reliable atomistic model for various applications [12–14]. A series of MD simulations were conducted first to construct the following components individually: CFs, single-walled carbon nanotubes (SWCNTs), and polypropylene (PP) as the matrix. This model was then subjected to uniaxial stress deformation through MD simulation. In MD simulations, materials are a combination of atoms that are regarded as classical mass particles. This technique is based on tracking the trajectories of atoms in a modeled system that can provide data with the time evolution of a nanostructured material subjected to external conditions. Ultimately, the computation involves position, velocity, and acceleration as a solution to Newton's second law [15]. Additionally, force fields are used to describe the interatomic interactions. The selection of an appropriate force field is critical for the validity of MD simulations. Therefore, several force fields have been developed for different material systems [16–19]. Subsequently, the motivation of this study is to demonstrate efficiently, via MD computational methods, the significance of nano-engineering CF surfaces with CNTs in thermoplastic-based composites as a valuable lightweight structure for transportation applications.

2. Methods

2.1. Model

To model the complex nano-engineered fuzzy CFs embedded in the polymer matrix, subsequent simulation steps were required to construct the system and then perform mechanical and interfacial analysis. The CF microstructure has interior graphite crystallites consisting of turbostratic interconnected graphene multilayers. Several atomistic approaches and computational tools were used to generate 2D and 3D microstructures of CF. The simulations range from surface studies with simplified graphene layers [20–22] and graphite blocks annealing [23] to complex precursor and reactive simulations [24,25] for the analysis of CF chemistry and mechanical properties. Therefore, an MD simulation (See Section 2.2.1) was used to model the range of CF microstructures with varied degrees of graphitic region. The thermoplastic PP chain of 40 monomers with an atactic configuration was built through the Polymer Modeler tool [26]. In addition, an (8,6) SWCNT with a length of 2.5 nm and a diameter of 0.95 nm was built by the Visual Molecular Dynamics (VMD) extension of the nanobuilder tool [27]. The experimental diameter and length of nanotubes grown on a CF surface are typically about 10 nm and 125 nm, respectively [28]. Nonetheless, the interaction energy decreases as the nanotube diameter increases [29]. For the following study, the chosen dimensions are sufficient to investigate the interfacial properties. After building the above constitutes, the CFs were modified by introducing both CNTs on the surface and including 20 PP chains through the PACKMOL software package [30]. A tolerance level of 2.0 Å was selected to perform the PACKMOL packing through several iterations to find the optimum configuration. Additionally, the x- and y-axis dimensions were set based on the built CF models as the center of mass. Moreover, the SWCNT was varied in loading percentages as 0, 1, and 3 wt% based on the assumption that the model was a 60 wt% and 40 wt% of matrix and fibers, respectively. Finally, the nano-engineered CF composites were visualized by the Open Visualization Tool (OVITO) software [31], as shown in Figure 1.

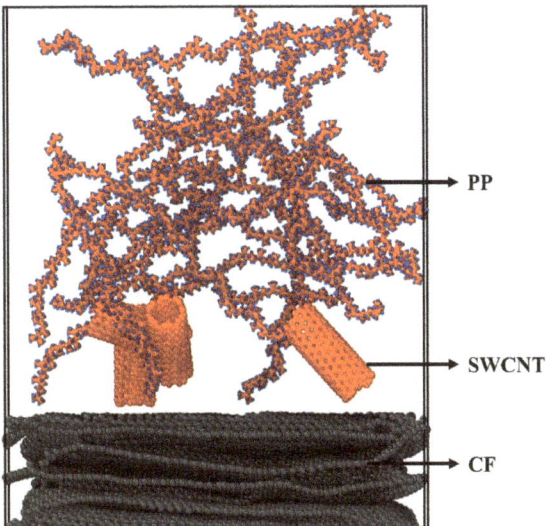

Figure 1. Nano-engineered carbon fiber composite model built with PACKMOL consisting of carbon fibers and turbostratic interconnected graphene multilayered structure with a density of 1.69 g/cm^3 coated with four (8, 6) 2.5 nm in length and 0.95 nm in diameter single-walled carbon nanotubes entangled with twenty atactic 40-monomer polypropylene chains.

2.2. Molecular Dynamic Simulations

The MD simulations were conducted using a large-scale atomic/molecular massively parallel simulator (LAMMPS) [32]. Through time and ensemble average of motion trajectory, the dynamic properties of the system were obtained. It can model extensive systems using a variety of interatomic potentials and boundary conditions. For CF model simulation, Tersoff empirical interatomic potential described the total binding energy. This potential incorporates a dependency on the number of bonds between a pair of atoms. The general form of the Tersoff formula is given in Equation (1), where ij indicates the interacting atoms, B_{ij} is the bond order between i and j, and $\varnothing_R$ and $\varnothing_A$ are the repulsive and attractive potentials, respectively [33]. The remaining simulations have been conducted with the modified potential AIREBO-m. It replaces the singular Lennard–Jones potential with a Morse potential, compared to the original AIREBO potential. The Morse potential is presented in Equation (2), where α is a curvature modification parameter, and ϵ and r^{eq} correspond to the minimum energy depth and location, respectively [34]. Researchers have modified the cutoff distances based on specified system studies to adopt this potential for large deformation. This modification was mainly applied to count for the tilt grain boundaries effect on the strength of graphene [35–37]. In this paper, an r_{cc}^{min} parameter of the value of 1.92 Å was adopted based on previous work done on deformation of graphene structures [13,14].

$$\varnothing_{ij}(r_{ij}) = [\varnothing_R(r_{ij}) - B_{ij}\varnothing_A(r_{ij})] \tag{1}$$

$$U_{ij}(r) = -\epsilon_{ij}\left[1 - \left(1 - e^{-\alpha_{ij}(r - r_{ij}^{eq})}\right)^2\right] \tag{2}$$

2.2.1. Carbon Fiber Simulation

The inspiration for CF models was taken from Zhu et al.'s work [38]. The concept is to stack several graphene layers and anneal them at high temperatures to disturb the crystalline structure and create turbostratic interconnected graphene multilayers that define a CF microstructure. A (56.3 × 109.6 × 28.0) Å^3 simulation box consisting of eight graphene layers with 15,739 or 16,120 carbon atoms to vary the final densities was built. The system was first equilibrated at a temperature of 300 K in 10 ps using a Nose–Hoover thermostat [39] with periodic boundary conditions in a canonical ensemble (NVT). Next, the temperature was ramped to 3500–5500 K at a rate of 110 K/ps, followed by relaxation for 60 ps at the same temperature. Afterward, extremely rapid quenching of 2750 K/ps to a temperature of 300 K was performed to preserve the turbostratic arrangements and graphene regions. Finally, the system was fully relaxed in an isothermal-isobaric ensemble (NPT) for 10 ps at a temperature of 300 K to release any internal stresses. The simulation process is summarized in Figure 2 with illustrations of the structure evolution throughout the process in a polyhedral template matching to show the graphene regions and regular visualization via OVITO.

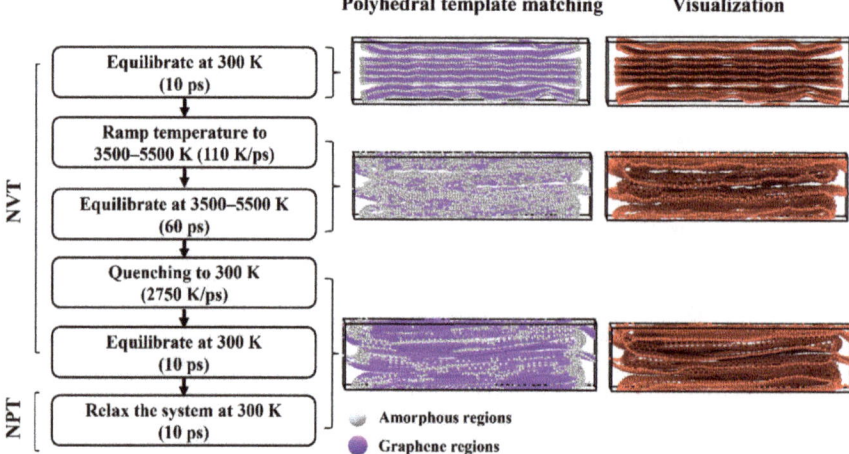

Figure 2. Molecular dynamic simulation steps for building a carbon fiber structure via annealing methodology using the NVT and NPT ensembles with different graphene regions percentages and turbostratic arrangements. Polyhedral template matching represents the amorphous regions in white and the graphene regions in purple.

2.2.2. Uniaxial Stress Simulation

Atomistic simulation of uniaxial tensile loading can be conducted with a parallel molecular dynamics code. After equilibrating any of the developed structures, the simulation box was deformed in the targeted direction at a strain rate of 0.001/ps. First, MD deformation in the long direction with periodic boundary conditions in all directions was conducted to analyze the strength of the developed CF models. The simulation was conducted at a temperature of 300 K using the Nose–Hoover thermostat and the AIREBO-m potential. Second, the interfacial analysis for the nano-engineered CF composite system was conducted by deformation in the z-direction with a fixed boundary and periodic boundary condition in the x- and y-directions, respectively. Moreover, the simulation was conducted at 50 K after an equilibration time of 20 ps using the AIREBO-m potential. Then, uniaxial stress of 15% strain was applied in the z-axis to peel the PP from the nano-engineered CF surface for theoretical strength and interfacial energy calculations.

3. Results and Discussions

3.1. Carbon Fiber Model Properties

After completing the CF microstructure simulation, four different models (Table 1) were developed by varying the number of carbon atoms and the maximum temperature. This resulted in densities close to the density of real CFs and graphene region percentages. All the developed models showed clear stress-strain curves (Figure 3), with a range of strength and Young's modulus caused by the differences in densities and graphene region percentages. The simulated strength values ranged from 9.9 to 26.7 GPa and Young's modulus of 129.6 to 277 GPa. In comparison, the experimental values of carbon fiber strength and Young's modulus are 3.0–7.0 GPa and 200–450 GPa, respectively [4,40]. The computational tensile strength values were higher than the experimental values because of the absence of defects inside the components, the extremely high loading rate, and length-scale discrepancies. The simulated microstructures were around three orders of magnitude smaller than the experimental samples, and thus the maximum simulated pore size is small compared to the experimental. The larger pores in experimental samples act as stress concentration sites and thus lead to brittle fracture and lower strength. Moreover, the published work of computational large-scale carbon fiber microstructure stress-strain curves by Zhu et al. [25] for three different structures have resulted in a comparable range of strength values between 6.73–6.90 GPa and Young's modulus of 95.9–284.5 GPa. Last, the simulated CF models showed crystallite arrangement around the longitudinal axis with turbostratic stacks of graphite structure, as shown in Figures S6–S9. This structure was reported for physical CF via X-ray diffraction and TEM images [41,42].

Table 1. The resulting physical and mechanical properties of the used carbon fiber models.

Model	Annealing Temperature (K)	Number of Carbon Atoms	Density (g/cm^3)	Graphene Region (%)	Strength (GPa)	Young's Modulus (GPa)
B1	5500	15,739	1.64	70	9.9	138.1
B2	4500	15,739	1.63	71	10.5	129.6
B3	3500	16,120	1.68	84	26.7	277.0
B4	4500	16,120	1.69	85	24.6	261.0

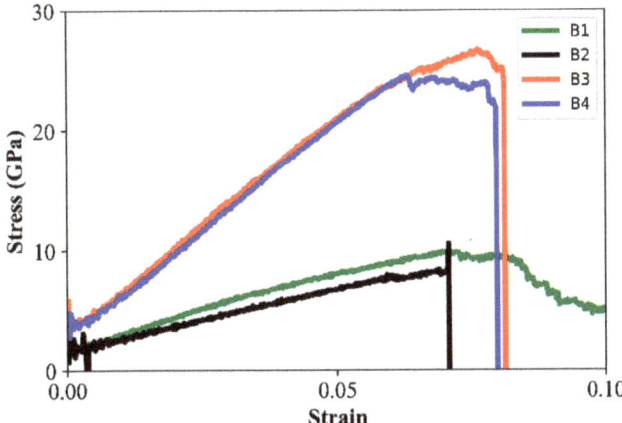

Figure 3. Stress-strain curves predicted in axial deformation of simulated carbon fiber structures with a maximum value of Young's modulus equal to 277.0 GPa and tensile strength of 26.7 GPa for the B3 model.

3.2. Nano-Engineered CF Composites Interfacial Properties

The four developed CF models were transformed into nano-engineered composites by introducing SWCNTs and PP via PACKMOL. The result was the construction of 12 different structures, with each CF model loaded with 0, 1, and 3 wt% SWCNTs. The calculations were conducted for the complex structures with the theoretical assumptions of low-temperature deformation after equilibration. However, the SWCNTs' positions can represent a minimized arrangement as local minimum energy and not the global minimum energy. Nonetheless, the non-bonding van der Waals (vdW) interactions were the main interactions that governed the three different constitutes, and among the PP chains, SWCNTs, and graphene layers. The calculated interfacial strength of each CF model was the stress required to peel PP chains away from the nano-engineered CFs, plotted against the averaged loading percentages of SWCNTs, as shown in Figure 4. The required strength increased linearly with the addition of SWCNTs in comparison to the pristine CFs, and independently from the CF model. The enhancement of interfacial strength resulted in an improved load transfer to the CFs [43]. This observation can be attributed to the increase in non-bonded vdW interactions, as the PP chains can entangle and interact with a larger carbon surface area. Moreover, mechanical interlocking can also contribute to the increase in strength. Experimentally, S. Yumitori et al. [44] showed that the interfacial shear strength improved by 35% in a CF/PP composite after grafting CFs with CNTs by performing a fiber pull-out test.

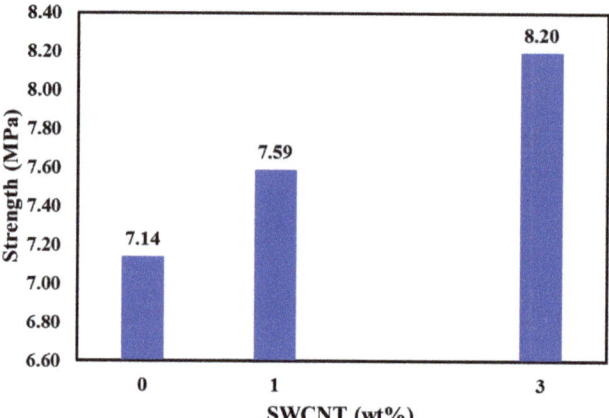

Figure 4. Average value of peeling strength to separate polypropylene from nano-engineered carbon fibers for the four developed CF structures with different loading percentages of 0, 1, and 3 wt% for single-walled carbon nanotubes after a 15% strain in the z-direction.

Potential energy is a function of several components, such as torsion, Lennard–Jones (LJ), pair, and REBO energies. Therefore, the calculated interfacial energy is the delta between the total potential energy and other components. Figure 5 shows the interfacial energy and other components as a function of displacement of the nano-engineered CFs from PP chains in the z-direction. This indicates that the dominant interfacial energy response to the complex model was based strictly on the non-bonded vdW interactions between the PP chains and the nano-engineered CFs. As a result, the interfacial energy increased linearly and reached a plateaus trend. The same behavior was reported previously on SWCNT interaction with polyethylene (PE) chains [45].

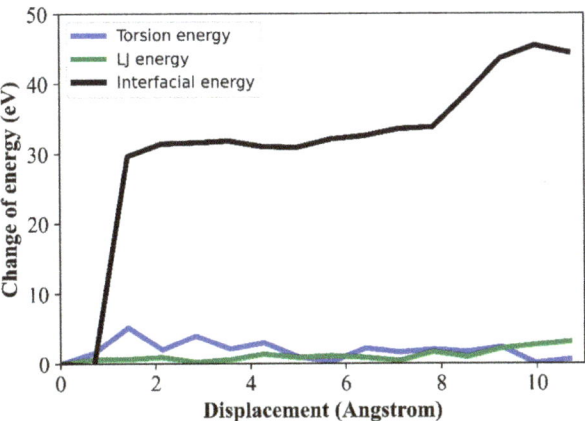

Figure 5. Change of the potential energy as a function of the displacement of nano-engineered carbon fiber-B4 with 3 wt% (8,6) 2.5 nm in length and 0.95 nm in diameter single-walled carbon nanotubes from twenty atactic 40-monomer polypropylene chains in the z-direction.

Accordingly, the interfacial energy of different SWCNT loadings was averaged and plotted in Figure 6. The models with 3 wt% loading showed a significant increase in the interfacial energy because of the increase in the non-bonded vdW interactions. Below the two angstroms mark, the interfacial energy peaked, and energy values increased with the SWCNTs' higher loadings. This observation indicates an improvement in the interfacial properties of nano-engineered CFs compared to pristine CFs.

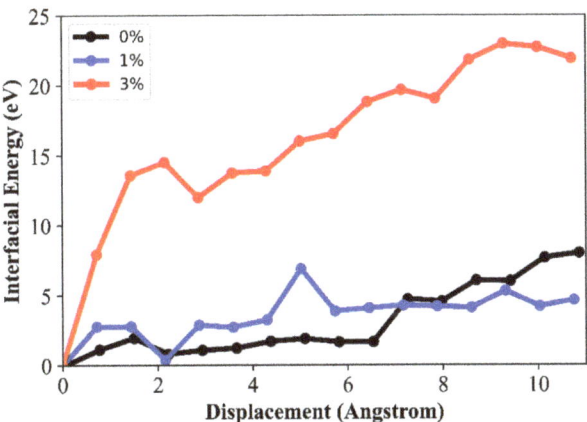

Figure 6. Averaged interfacial energy of four models of nano-engineered carbon fibers for each single-walled carbon nanotube loading percentage as a function of displacement in the z-direction from the twenty atactic 40-monomer polypropylene chains.

4. Conclusions

We have assessed the interfacial properties of nano-engineered CF composites using MD simulations. Four different CF models were developed with graphene sheets and an annealing process governed by Tersoff potential to produce models with a density range of 1.63–1.69 g/cm^3, and graphene region percentage ranging from 70–85%. The built CF structures showed the turbostratic interconnected graphene multilayer arrangement with strengths ranging from 9.9–26.7 GPa and Young's modulus of 129.6–277.0 GPa. The tensile strength values of the CFs were higher than the experimental because of the high loading rate and length-scale discrepancy.

Moreover, a nano-engineered CFRP was constructed with PP and SWCNTs using PACKMOL. The interfacial energy just below two angstroms increased by 180% and 700% for the 1 and 3 wt% SWCNT loading, respectively. The stress required to peel PP from the nano-engineered CFs also showed an increasing trend due to the enhancement in non-bonded vdW interactions. Finally, the nano-engineered CFRP is considered to be a multi-functional composite that can serve beyond structural applications [11]. This advancement in structural components for the automotive industry can contribute to emissions reduction. Therefore, the explored theoretical atomistic modeling in this paper may lay a building block for more advanced computational research to design and optimize nano-engineered CFs.

Supplementary Materials: The following supporting information can be downloaded at: https://www.mdpi.com/article/10.3390/jcs6020054/s1.

Author Contributions: H.A., A.Q.A. and Q.P. contributed conception and design of the study. H.A. performed the calculations. H.A., A.Q.A. and Q.P. performed the statistical analysis. H.A. wrote the first draft of the manuscript. All authors have read and agreed to the published version of the manuscript.

Funding: Q. P. would like to acknowledge the support provided by the Deanship of Scientific Research (DSR) at King Fahd University of Petroleum & Minerals (KFUPM) through the K.A.CARE project (KACARE211-RFP-02). For computer time, this research used the resources of the Supercomputing Laboratory at King Abdullah University of Science Technology (KAUST) in Thuwal, Saudi Arabia, through project #1532.

Data Availability Statement: The datasets generated and/or analyzed during the current study are available from the corresponding author on reasonable request.

Acknowledgments: We would like to thank Hassan Shoab and Jafar Albinmousa for the useful discussions.

Conflicts of Interest: There are no conflicts of interest to declare.

References

1. Aizebeokhai, A.P. Global Warming and Climate Change: Realities, Uncertainties and Measures. *Int. J. Phys. Sci* **2009**, *4*, 868–879. [CrossRef]
2. Albuquerque, F.D.B.; Maraqa, M.A.; Chowdhury, R.; Mauga, T.; Alzard, M. Greenhouse Gas Emissions Associated with Road Transport Projects: Current Status, Benchmarking, and Assessment Tools. In *Transportation Research Procedia*; Elsevier B.V.: Amsterdam, The Netherlands, 2020; Volume 48, pp. 2018–2030. [CrossRef]
3. Windisch, E.; Benezech, V.; Chen, G.; Kauppila, J. *Lightening Up: How Less Heavy Vehicles Can Help Cut CO_2 Emissions*; OECD: Paris, France, 2017.
4. Newcomb, B.A. Processing, Structure, and Properties of Carbon Fibers. *Compos. Part A* **2016**, *91*, 262–282. [CrossRef]
5. Das, T.K.; Ghosh, P.; Das, N.C. Preparation, Development, Outcomes, and Application Versatility of Carbon Fiber-Based Polymer Composites: A Review. *Adv. Compos. Hybrid Mater.* **2019**, *2*, 214–233. [CrossRef]
6. Kyriakides, S.; Arseculeratne, R.; Perry, E.J.; Liechti, K.M. On the compressive failure of fiber reinforced composites. *Int. J. Solids Structures* **1995**, *32*, 689–738. [CrossRef]
7. Imani Yengejeh, S.; Kazemi, S.A.; Öchsner, A. Carbon Nanotubes as Reinforcement in Composites: A Review of the Analytical, Numerical and Experimental Approaches. *Comput. Mater. Sci.* **2017**, *136*, 85–101. [CrossRef]
8. Clancy, A.J.; Anthony, D.B.; de Luca, F. Metal Mimics: Lightweight, Strong, and Tough Nanocomposites and Nanomaterial Assemblies. *ACS Appl. Mater. Interfaces* **2020**, *12*, 15955–15975. [CrossRef]
9. Zhao, J.; Liu, L.; Guo, Q.; Shi, J.; Zhai, G.; Song, J.; Liu, Z. Growth of Carbon Nanotubes on the Surface of Carbon Fibers. *Carbon* **2008**, *46*, 380–383. [CrossRef]
10. Bekyarova, E.; Thostenson, E.T.; Yu, A.; Kim, H.; Gao, J.; Tang, J.; Hahn, H.T.; Chou, T.W.; Itkis, M.E.; Haddon, R.C. Multiscale Carbon Nanotube-Carbon Fiber Reinforcement for Advanced Epoxy Composites. *Langmuir* **2007**, *23*, 3970–3974. [CrossRef]
11. Qian, H.; Greenhalgh, E.S.; Shaffer, M.S.P.; Bismarck, A. Carbon Nanotube-Based Hierarchical Composites: A Review. *J. Mater. Chem.* **2010**, *20*, 4751–4762. [CrossRef]
12. Peng, Q.; Meng, F.; Yang, Y.; Lu, C.; Deng, H.; Wang, L.; De, S.; Gao, F. Shockwave Generates <100> Dislocation Loops in Bcc Iron. *Nat. Commun.* **2018**, *9*, 4880. [CrossRef]

13. Deng, B.; Hou, J.; Zhu, H.; Liu, S.; Liu, E.; Shi, Y.; Peng, Q. The Normal-Auxeticity Mechanical Phase Transition in Graphene. *2D Materials* **2017**, *4*, 021020. [CrossRef]
14. Hou, J.; Deng, B.; Zhu, H.; Lan, Y.; Shi, Y.; De, S.; Liu, L.; Chakraborty, P.; Gao, F.; Peng, Q. Magic Auxeticity Angle of Graphene. *Carbon* **2019**, *149*, 350–354. [CrossRef]
15. Rapaport, D.C. *The Art of Molecular Dynamics Simulation*, 2nd ed.; Cambridge University Press: Cambridge, UK, 2004.
16. Ritschl, F.; Fait, M.; Fiedler, K.; Khler, J.E.H.; Kubias, B.; Meisel, M. An Extension of the Consistent Valence Force Field (CVFF) with the Aim to Simulate the Structures of Vanadium Phosphorus Oxides and the Adsorption of n-Butane and of 1-Butene on Their Crystal Planes. *Z. Anorg. Allg. Chem.* **2002**, *628*, 1385–1396. [CrossRef]
17. Sun, H. Ab Initio Calculations and Force Field Development for Computer Simulation of Polysilanes. *Macromolecules* **1995**, *28*, 701–712. [CrossRef]
18. Sun, H. Compass: An Ab Initio Force-Field Optimized for Condensed-Phase Applications—Overview with Details on Alkane and Benzene Compounds. *J. Phys. Chem. B* **1998**, *102*, 7338–7364. [CrossRef]
19. Van Duin, A.C.T.; Dasgupta, S.; Lorant, F.; Goddard, W.A. ReaxFF: A Reactive Force Field for Hydrocarbons. *J. Phys. Chem. A* **2001**, *105*, 9396–9409. [CrossRef]
20. Allington, R.D.; Attwood, D.; Hamerton, I.; Hay, J.N.; Howlin, B.J. A Model of the Surface of Oxidatively Treated Carbon Fibre Based on Calculations of Adsorption Interactions with Small Molecules. *Compos. Part A* **1998**, *29*, 1283–1290. [CrossRef]
21. Allington, R.D.; Attwood, D.; Hamerton, I.; Hay, J.N.; Howlin, B.J. Developing Improved Models of Oxidatively Treated Carbon Fibre Surfaces, Using Molecular Simulation. *Compos. Part A* **2004**, *35*, 1161–1173. [CrossRef]
22. Awan, I.S.; Xiaoqun, W.; Pengcheng, H.; Shanyi, D. Developing an Approach to Calculate Carbon Fiber Surface Energy Using Molecular Simulation and Its Application to Real Carbon Fibers. *J. Compos. Mater.* **2012**, *46*, 707–715. [CrossRef]
23. Penev, E.S.; Artyukhov, V.I.; Yakobson, B.I. Basic Structural Units in Carbon Fibers: Atomistic Models and Tensile Behavior. *Carbon* **2015**, *85*, 72–78. [CrossRef]
24. Rajabpour, S.; Mao, Q.; Gao, Z.; Khajeh Talkhoncheh, M.; Zhu, J.; Schwab, Y.; Kowalik, M.; Li, X.; van Duin, A.C.T. Low-Temperature Carbonization of Polyacrylonitrile/Graphene Carbon Fibers: A Combined ReaxFF Molecular Dynamics and Experimental Study. *Carbon* **2021**, *174*, 345–356. [CrossRef]
25. Zhu, J.; Gao, Z.; Kowalik, M.; Joshi, K.; Ashraf, C.M.; Arefev, M.I.; Schwab, Y.; Bumgardner, C.; Brown, K.; Burden, D.E.; et al. Unveiling Carbon Ring Structure Formation Mechanisms in Polyacrylonitrile-Derived Carbon Fibers. *ACS Appl. Mater. Interfaces* **2019**, *11*, 42288–42297. [CrossRef] [PubMed]
26. Haley, B.; Wilson, N.; Li, C.; Arguelles, A.; Jaramillo, E.; Strachan, A. Polymer Modeler. 2010. Available online: https://nanohub.org/resources/polymod (accessed on 14 June 2021).
27. Humphrey, W.; Dalke, A.; Schulten, K. VMD—Visual Molecular Dynamics. *J. Molec. Graphics* **1996**, *14*, 33–38. [CrossRef]
28. Anthony, D.B.; Sui, X.M.; Kellersztein, I.; de Luca, H.G.; White, E.R.; Wagner, H.D.; Greenhalgh, E.S.; Bismarck, A.; Shaffer, M.S.P. Continuous Carbon Nanotube Synthesis on Charged Carbon Fibers. *Compos. Part A* **2018**, *112*, 525–538. [CrossRef]
29. Rouhi, S.; Alizadeh, Y.; Ansari, R. Molecular Dynamics Simulations of the Interfacial Characteristics of Polypropylene/Single-Walled Carbon Nanotubes. *Proc. IMechE Part L J. Mater. Des. Appl.* **2016**, *230*, 190–205. [CrossRef]
30. Martinez, L.; Andrade, R.; Birgin, E.G.; Martínez, J.M. PACKMOL: A Package for Building Initial Configurations for Molecular Dynamics Simulations. *J. Comput. Chem.* **2009**, *30*, 2157–2164. [CrossRef]
31. Stukowski, A. Visualization and Analysis of Atomistic Simulation Data with OVITO-the Open Visualization Tool. *Modelling Simul. Mater. Sci. Eng.* **2010**, *18*, 015012. [CrossRef]
32. Plimpton, S. Fast Parallel Algorithms for Short-Range Molecular Dynamics. *J. Comput. Phys.* **1995**, *117*, 1–19. [CrossRef]
33. Tersoff, J. Empirical Interatomic Potential for Carbon, with Applications to Amorphous Carbon. *Phys. Rev. Lett.* **1988**, *61*, 2879–2882. [CrossRef]
34. O'Connor, T.C.; Andzelm, J.; Robbins, M.O. AIREBO-M: A Reactive Model for Hydrocarbons at Extreme Pressures. *J. Chem. Phys.* **2015**, *142*, 024903. [CrossRef]
35. Grantab, R.; Shenoy, V.B.; Ruoff, R.S. Anomalous Strength Characteristics of Tilt Grain Boundaries in Graphene. *Science* **2010**, *330*, 946–948. [CrossRef]
36. Wei, Y.; Wu, J.; Yin, H.; Shi, X.; Yang, R.; Dresselhaus, M. The Nature of Strength Enhancement and Weakening by Pentagong-Heptagon Defects in Graphene. *Nat. Mater.* **2012**, *11*, 759–763. [CrossRef]
37. Huhtala, M.; Krasheninnikov, A.V.; Aittoniemi, J.; Stuart, S.J.; Nordlund, K.; Kaski, K. Improved Mechanical Load Transfer between Shells of Multiwalled Carbon Nanotubes. *Phys. Rev. B* **2004**, *70*, 045404. [CrossRef]
38. Zhu, C.; Liu, X.; Yu, X.; Zhao, N.; Liu, J.; Xu, J. A Small-Angle X-Ray Scattering Study and Molecular Dynamics Simulation of Microvoid Evolution during the Tensile Deformation of Carbon Fibers. *Carbon* **2012**, *50*, 235–243. [CrossRef]
39. Hoover, W.G. Canonical Dynamics: Equilibrium Phase-Space Distributions. *Phys. Rev. A* **1985**, *31*, 1695–1697. [CrossRef]
40. Carbon Fiber. Toray Composite Materials America, Inc. Available online: https://www.toraycma.com/products/carbon-fiber/ (accessed on 30 July 2021).
41. Tse-Hao, K.; Tzy-Chin, D.; Jeng-An, P.; Ming-Fong, L. The Characterization of PAN-Based Carbon Fibers Developed by Two-Stage Continuous Carbonization. *Carbon* **1993**, *31*, 765–771. [CrossRef]
42. Zhu, C.Z.; Yu, X.L.; Liu, X.F.; Mao, Y.Z.; Liu, R.G.; Zhao, N.; Zhang, X.L.; Xu, J. 2D SAXS/WAXD Analysis of Pan Carbon Fiber Microstructure in Organic/Inorganic Transformation. *Chin. J. Polym. Sci.* **2013**, *31*, 823–832. [CrossRef]

43. Thostenson, E.T.; Li, W.Z.; Wang, D.Z.; Ren, Z.F.; Chou, T.W. Carbon Nanotube/Carbon Fiber Hybrid Multiscale Composites. *J. Appl. Phys.* **2002**, *91*, 6034–6037. [CrossRef]
44. Yumitori, S.; Arao, Y.; Tanaka, T.; Naito, K.; Tanaka, K.; Katayama, T. Increasing the Interfacial Strength in Carbon Fiber/Polypropylene Composites by Growing CNTs on the Fibers. *Comput. Methods Exp. Meas. XVI* **2013**, *55*, 275–284. [CrossRef]
45. Zhang, Z.Q.; Ward, D.K.; Xue, Y.; Zhang, H.W.; Horstemeyer, M.F. Interfacial Characteristics of Carbon Nanotube-Polyethylene Composites Using Molecular Dynamics Simulations. *ISRN Mater. Sci.* **2011**, *2011*, 145042. [CrossRef]

Article

Additive Manufacturing of Carbon Fiber Reinforced Plastic Composites: The Effect of Fiber Content on Compressive Properties

Olusanmi Adeniran [1,*], Weilong Cong [1], Eric Bediako [1] and Victor Aladesanmi [2]

[1] Department of Industrial, Manufacturing and Systems Engineering, Texas Tech University, Lubbock, TX 79409, USA; weilong.cong@ttu.edu (W.C.); eric.bediako@ttu.edu (E.B.)

[2] Department of Physics, Eberhard Karl University of Tubingen, 72076 Tübingen, Germany; victor-ifetayo.aladesanmi@student.uni-tuebingen.de

* Correspondence: Olusanmi.adeniran@ttu.edu; Tel.: +1-(682)-561-4015

Abstract: The additive manufacturing (AM) of carbon fiber reinforced plastic (CFRP) composites continue to grow due to the attractive strength-to-weight and modulus-to-weight ratios afforded by the composites combined with the ease of processibility achievable through the AM technique. Short fiber design factors such as fiber content effects have been shown to play determinant roles in the mechanical performance of AM fabricated CFRP composites. However, this has only been investigated for tensile and flexural properties, with no investigations to date on compressive properties effects of fiber content. This study examined the axial and transverse compressive properties of AM fabricated CFRP composites by testing CF-ABS with fiber contents from 0%, 10%, 20%, and 30% for samples printed in the axial and transverse build orientations, and for axial tensile in comparison to the axial compression properties. The results were that increasing carbon fiber content for the short-fiber thermoplastic CFRP composites slightly reduced compressive strength and modulus. However, it increased ductility and toughness. The 20% carbon fiber content provided the overall content with the most decent compressive properties for the 0–30% content studied. The AM fabricated composite demonstrates a generally higher compressive property than tensile property because of the higher plastic deformation ability which characterizes compression loaded parts, which were observed from the different failure modes.

Keywords: additive manufacturing; compressive properties; carbon fiber reinforced plastics; composites; thermoplastics; polymers; mechanical performance; materials science

Citation: Adeniran, O.; Cong, W.; Bediako, E.; Aladesanmi, V. Additive Manufacturing of Carbon Fiber Reinforced Plastic Composites: The Effect of Fiber Content on Compressive Properties. *J. Compos. Sci.* **2021**, *5*, 325. https://doi.org/10.3390/jcs5120325

Academic Editor: Jiadeng Zhu

Received: 1 December 2021
Accepted: 14 December 2021
Published: 16 December 2021

Publisher's Note: MDPI stays neutral with regard to jurisdictional claims in published maps and institutional affiliations.

Copyright: © 2021 by the authors. Licensee MDPI, Basel, Switzerland. This article is an open access article distributed under the terms and conditions of the Creative Commons Attribution (CC BY) license (https://creativecommons.org/licenses/by/4.0/).

1. Introduction

1.1. Additive Manufacturing of Carbon Fiber Composites

Developments in additive manufacturing (AM) of carbon fiber reinforced plastic (CFRP) composites continue to grow because of their increasing acceptance as alternatives to metallic materials in several applications, mainly due to their attractive strength-to-weight and modulus-to-weight ratios [1]. They also benefit from easier processability and manufacturing flexibilities because of their lower melting temperatures, which are usually high enough to meet operational temperature requirements and low enough for flexible manufacturing outside of dedicated facilities [2]. Recent developments in the AM of CFRP composites are leading to increasing applications, especially where material or fabrication methods were not previously available or are more complicated [3,4].

Synthetic fibers provide the bulk of the short fibers used as reinforcement materials in the AM fabrication of FRP composites due to their superior mechanical properties and thermal stability over natural fibers [1]. Short fibers are preferred to continuous fibers in certain AM applications because of the ease of fabrication into intricate shapes while still providing improved mechanical properties. Carbon fibers make up most of these short-fiber applications followed by glass fiber (GF) and Kevlar fiber. The carbon fibers are preferred because of their higher strength, stiffness, chemical resistance, and thermal stability, etc., and are by itself is the strongest and stiffest synthetic fiber, with a strength-to-weight ratio nearly twice that of 6061 Aluminum [5].

Various investigations on the AM fabrication of FRP composites concluded on the higher mechanical and thermal preference of carbon fibers over the other synthetic fibers. Goh et al. [6] in the characterization of the mechanical properties of AM fabricated CFRP and GFRP reported higher tensile strength and modulus for CFRP composite at 600 MPa and 13.0 GPa, respectively, whereas the glass fiber was 450 MPa and 7.2 GPa, respectively. They also obtained higher flexural properties where the carbon-fiber-reinforced-plastic (CFRP) composite showed almost triple that of the glass-fiber-reinforced-plastic (GFRP) composite. Zhang et al. [7] in their evaluation of the bond strength of printed acrylonitrile-butadiene-styrene (ABS), carbon nanotube-reinforced ABS (CNT-ABS), and short carbon-fiber-reinforced ABS (CF-ABS) samples found the printed CF-ABS to offer better tensile strength and modulus than the CNT-ABS and pure ABS for most of the print orientations. Ning et al. [8] compared graphite (GR) flakes as alternative reinforcement instead of chopped fibers and the result were that the AM fabricated CF-ABS exhibited better tensile properties than GR-ABS composite parts, even though the GRFP may offer less porosity. The reason they claimed was that the carbon fiber has a much higher aspect ratio which allows for better interfacial bonding compared to the graphite. Mohammadizadeh and Fidan [9] reported on the higher thermomechanical properties of AM fabricated carbon-fiber-reinforced Nylon compared to glass-fiber and Kevlar reinforced, where their Dynamic Mechanical Analysis results showed the highest properties improvement under 150 °C for the carbon-fiber-reinforced Nylon over the other fiber-reinforced composites.

A highlight of the use of thermoplastic CFRP composites is their ease of processability, repairability, and maintenance offering compared to their thermoset versions which are more difficult to repair. Their physical state reversibility on the application of processing temperatures often allows for easier service repairability and recyclability which makes them preferred whenever they meet process requirements [10]. Tekinalp et al. [11] described the thermoplastic matrixes' relatively high melt viscosities as translating to rapid manufacturing, repairability, recyclability, and significant cost reductions. Hence, the justification for more understanding of AM fabricated CFRP composites from conducting this investigation to determine the effects of fiber content on the compressive properties.

1.2. Compressive Properties of AM Fabricated CFRP Composites

Observing short fiber design factors has been shown to improve the mechanical properties of CFRP composites. Material factors such as fiber content [12–14], fiber orientation [12], fiber length [13], aspect ratio, fiber-diameter, fiber-matrix adhesion [7], etc. have been reported as major contributors to mechanical performance. However, in studying these, only the tensile properties improvements have been the hallmark of AM fabricated CFRP investigations, with the comprehension of compressive properties still lagging. Table 1 shows the investigations conducted to date on the compressive properties of short fiber AM fabricated CFRP composites, but still no investigations on fiber content effects on compressive properties.

Understanding the effect of carbon fiber content on compressive properties is important to the sustainable application of short fiber AM fabricated CFRP composites in applications such as wind energy, where they have been used to fabricate wind turbine molds and are currently being explored to fabricate turbine components [15,16]. Such components operate at different times under static and dynamic loading, where they are subjected to a combination of compressive and tensile stresses. Other applications such as aerospace trim tools, hand layup tools, and vacuum-assisted resin transfer (VARTM) tools made from short fiber AM fabricated CFRP composites which are subjected to some forms of compressive stresses have also been proposed [17,18]. Table 1 shows the limited investigations conducted so far on the compressive properties of short fiber thermoplastic AM fabricated CFRP composites, with missing reports on fiber content effects.

Table 1. Summary of Reported Investigations on the Compressive Properties of AM Fabricated CFRP Composites.

Short Fiber CFRP Sample	Investigations	Carbon Fiber Content (%)	Maximum Tensile Strength (Mpa)	Author
CF-ABS	Microstructural Evaluation through X-ray CT Technique	15	67.0	[19,20]
CF-ABS	Chopped CF-ABS	20%	N/A	[21]
CF-PA	Chopped CF-ABS	N/A	101	[22]

To date, the only four reported investigations on the compressive properties of short fiber AM fabricated CFRP composites did not investigate the fiber content effects [19–22]. Quan et al. tested at different strain levels the damage evolution of short fiber AM fabricated under compression loading using the X-ray CT technique. They also investigated the compressive behavior of 3D orthogonal short fiber AM fabricated CFRP composites and that of its silicone infused composite to demonstrate the feasibility of AM fabricated 3D orthogonal preform. Meraz et al. analyzed the compressive and rheology properties of AM fabricated CFRP composites in which they used optical imaging to create a relationship effect of scanning strategies and printing temperatures on the compressive behaviors of 3D Printed polyamide (PA) based composites. Wang et al. [22] considered the effects of printing temperature and orientations on the compressive properties of AM fabricated CF-PA composites but did not test the effect of fiber content.

Other investigations also either addressed pure thermoplastics or continuous fiber-reinforced composites. Lee et al. [23] reported the anisotropic compressive strength of AM fabricated thermoplastics but did not consider the effects of short fiber content composition. M. Araya-Calvo et al. [24] investigated the effect of continuous fiber content on CF-PA6 but did not address short carbon fibers. They explained the effects of reinforcement pattern, reinforcement distribution, print orientation, and percentage of fiber on the compressive properties and claimed optimum compressive stress and modulus at 24% fiber reinforcement. However, this was reported for continuous fibers. J. Chacon et al. [24] investigated 59.2% continuous fiber CF-PA6, which also did not address short carbon fibers.

Previous investigations which addressed short fiber content only considered tensile and flexural properties but not compressive properties. Tekinalp et al. [12] evaluated highly oriented CF-ABS composites up to 40% carbon fiber wt.% and reported tensile strength and modulus to increase by up to 115% and 700%, respectively. Ning et al. [13] investigated the effect of carbon fiber wt.% up to 15% in ABS, with the conclusion that adding carbon fiber would increase the tensile strength and modulus but decrease toughness and ductility. Duty et al. [14] investigated carbon fiber content effects on tensile properties in which they reported improvements in the CF-ABS tensile strength and modulus up to ~60 MPa and ~12 GPa, respectively, but with some degree of anisotropy. Love et al. [25] investigated 13% short fiber AM fabricated CFRP composite in which they claimed the addition of carbon fiber to a thermoplastic matrix significantly increases the strength, stiffness, and thermal conductivity, and greatly reduced the distortion of the parts. However, these only addressed tensile and flexural properties.

1.3. Research Motivations

The current state of knowledge has only addressed short carbon fiber content effects in the tensile and flexural modes, but no reported investigations in the compressive mode. It is important to conduct this study to understand how compressive properties fill in the research gap of mechanical performance improvements with carbon fiber additions to thermoplastic matrixes for AM composite applications. The study will help to rule out any possible bias from the general hypothesis that increasing carbon fiber content in short fiber AM fabricated CFRP composites should improve material compressive properties. It will also help to make more technically sound judgments on materials selection decisions that may arise for applications where parts are subjected to a combination of tensile and compression loading.

2. Materials and Methods

2.1. AM Workpiece Fabrication Procedure

2.1.1. Material Compounding and Filamenting

Chopped carbon fiber averaging 7 µm average diameter and 150 µm length (Pantex 30, Zoltek, St. Louis, MO, USA) and MG94 natural ABS pellets (Filastruder, Snellville, GA, USA) was used to fabricate the samples. The 10%, 20%, and 30% fiber contents were compounded with the ABS pellets by melt mixing in a twin-screw Prep-Center Mixer (CWB Brabender Instruments Inc., Duisburg, Germany), with the working temperature and rotor speed set to 220 °C and 60 rpm, respectively, for one hour to ensure stable temperature and speed before introducing the carbon fiber. The thermoplastic fraction of the composite was first introduced into the mixer and allowed to maintain the glass transition phase for 5 min before the carbon fiber fraction was added, and subsequently melt-mixed for another 7 min for effective compounding. The compounded mixture was then removed from the hopper and allowed to cool for 30 min after which they were crushed into smaller bits using a Pelletizer (Fritsch, Idar-Oberstein, Germany). The same process was repeated for each of the 10%, 20%, and 30% carbon fiber content compositions. The pelletized composite contents were made individually made into filaments using a filamenting extruder machine (Felfil Evo, Turin, Italy), with a single screw extruder of 1.75 mm Die-head with the temperature set to 220 °C and speed at 7 rpm. Figure 1 illustrates the sample fabrication process.

2.1.2. Fabricated Workpiece

A Fused Deposition Modelling (FDM) 3D printer (Prusa Mk3 i3, Prague, Czech Republic), with a modified enclosure system (Creality 3D, Shenzhen, China) to control the temperature to 45 ± 5 °C and relative humidity to less than 25% RH was used to fabricate the CFRP composite compression and tensile workpieces. The modified printing setup was influenced by studies that reported that controlling temperature within a printing chamber reduces porosity and improves material properties [26]. Table 2 lists the printing parameters used to fabricate the test samples, some of which were determined from reported studies [22,27]. The compression samples illustrated in Figure 2a,b consist of the axial and transverse orientations printed as 25.4 mm × 12.7 mm (height × diameter) dimensions, respectively, while Figure 2c illustrates the axial tensile orientation printed as 165 mm × 13 mm × 3.2 mm (length × width × thickness). The axial axis is printed parallel to the gage section of the sample while the transverse axis is printed perpendicular to the gage.

Table 2. Printing Processing Parameters.

Parameter	Value	Unit
Infill density	100	%
Nozzle Temperature	270	°C
Bed Temperature	100	°C
Printing Enclosure	50 ± 5	°C
Layer Thickness	0.25	mm
Printing Speed	30	mm/s
Raster angle	0 and 90	°

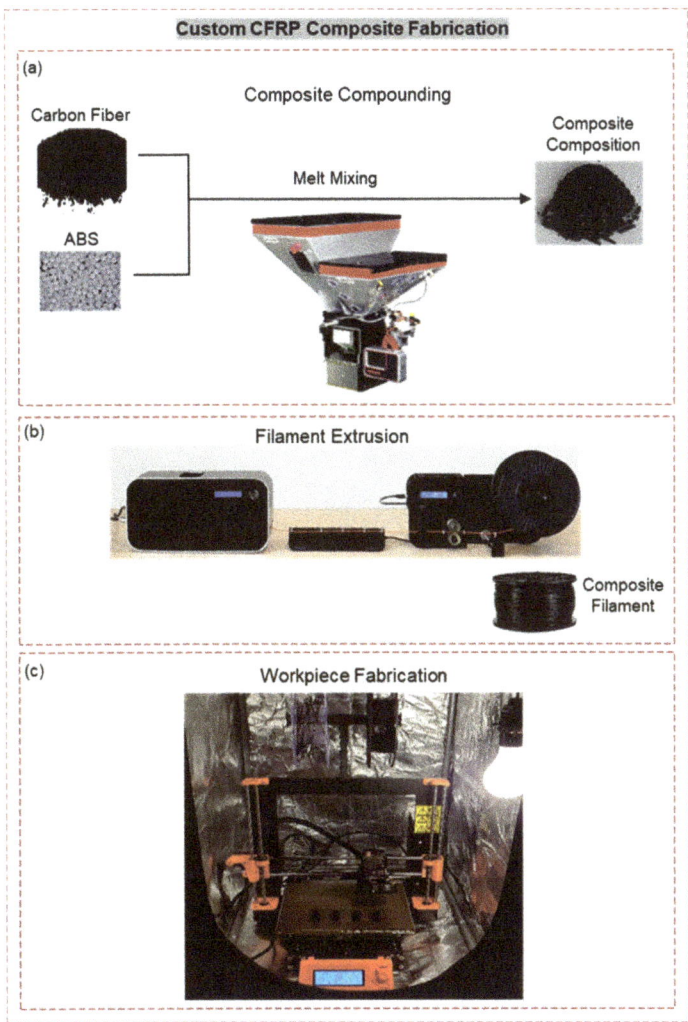

Figure 1. Custom CFRP Composite Fabrication Process (**a**) Melt Mix Compounding (**b**) Filament Extrusion (**c**) Sample Fabrication.

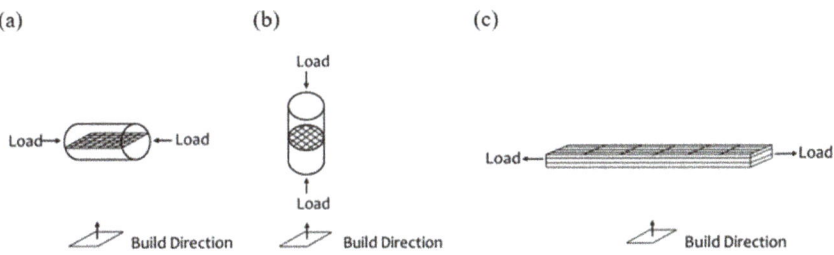

Figure 2. Test Samples Printing Layers Orientation (**a**) Axial Compression (**b**) Transverse Compression (**c**) Axial Tensile.

2.2. Measurement Procedure

Axial and transverse compression (ASTM D695) [28] and axial tensile (ASTM D638) [29] samples consisting of five workpieces each were tested at the normal ambient condition to determine the effects of carbon fiber content on the compressive properties in comparison to the already established tensile properties for AM fabricated CFRP composites. The compression samples tested 0%, 10%, 20%, and 30% carbon fiber contents, while the tensile was only able to test 0%, 10%, and 20% carbon fiber contents because of the difficulty of processing tensile samples beyond 20% content. Mechanical testing was conducted using a universal test machine (AGS-J, Shimadzu Co., Kyoto, Japan) with a 100 kN load cell fixture for the compression at a test speed of 1.3 mm/s and a 10 kN for the tensile at a test speed of 5.0 mm/s. Figure 3 shows the mechanical test setup and the quality of the fabricated workpieces.

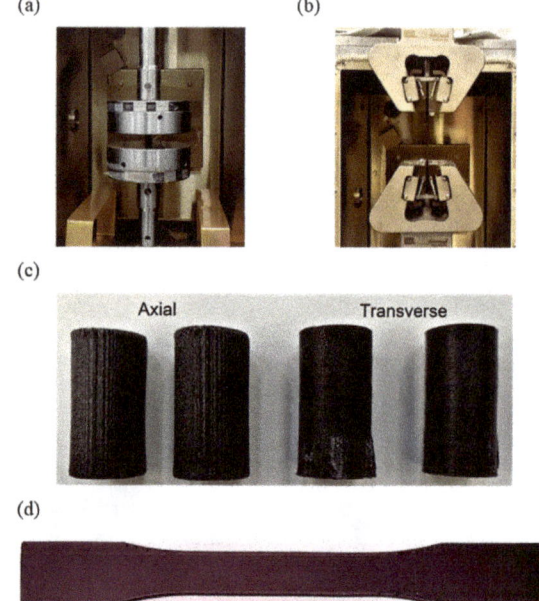

Figure 3. Test Setup and Workpiece Quality (**a**) ASTM D695 Compression (**b**) ASTM D638 Tensile (**c**) Axial and Transverse Compressive Workpieces (**d**) Axial Tensile Workpiece.

The strength, ductility, modulus, and toughness values were determined from the plot of the stress-strain curves using the Origin pro graphing and analysis software. The compressive strength was determined from an average of maximum stress values from the plots of four test specimens. The compressive modulus was calculated from the ratio of initial stress to the corresponding strain. The modulus values were specifically determined from the slope of the tangent to the elastic region of the curve from the (0,0) origin. The ductility values were determined as the strain values at the maximum strength, while the toughness values were determined from the area under the stress-strain curve using the integration function under the 'peak and baseline' tab of the software. Microscopic evaluation of the ABS and CFRP composite compressive and tensile specimens were examined at 1000× magnifications using a Scanning Electron Microscope (Thermo Fisher Phenon XL, Waltham, MA, USA) to determine the influence of carbon fiber on the failure mode of the composites.

3. Results

The result evaluated the compressive and tensile properties of AM fabricated CFRP composites comparing 0% (ABS), 10%, 20%, and 30% carbon fiber contents. This was used to establish the effect of fiber content on the compressive strength, modulus, ductility, and toughness and to compare compressive to tensile properties.

3.1. Compressive Properties
3.1.1. Compressive Strength

The material compressive strength measures the maximum stress that can be applied before the materials buckling failure. Figure 4a shows the comparison of the compressive strength across the axial and transverse axes of the composites, while Figure 4b shows the comparison of the axial compressive and axial tensile strength.

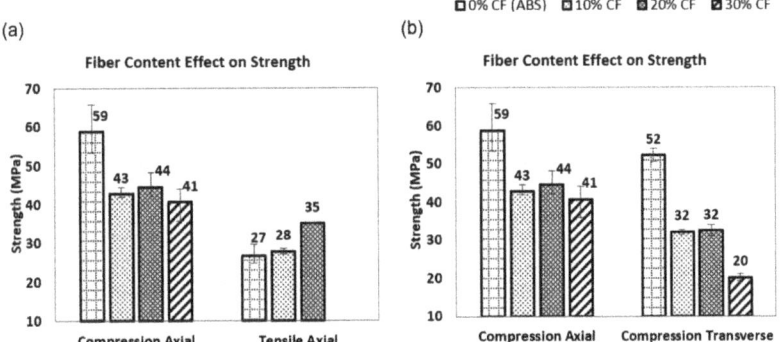

Figure 4. Fiber Content Effect on the Compressive Strength of AM Fabricated CFRP Composites (a) Axial versus Transverse Compression (b) Axial Compression versus Axial Tensile.

As seen in Figure 4a, the AM fabricated composite exhibited higher compressive than tensile strength from the plastic deformation characteristics in the compressive mode against elastic deformation which is typical of the tensile mode. The plastic deformation more easily redistributes stress at existing flaws, thus, limiting crack propagation which often fails at lower strengths. However, unlike the tensile strength which reported improved strength with increasing fiber content, the compression reported reduced strength. A 27% strength reduction across the axial axis of the composite was observed for the 10–20% fiber content, while a reduction by up to 31% was observed for the 30% fiber content. Larger strength reduction was observed across the transverse axis where a reduction up to 38% was observed for the 10–20% fiber content and up to a 62% reduction for the 30% content.

A comparison of compressive strength across the axial and transverse axes for up to 30% carbon fiber content as seen in Figure 4b showed the composite exhibiting higher strength in the axial than in the transverse axis like reported investigations for the tensile mode [14,30]. This is because of the better bonding across the continuous print bead which characterizes the axial axis versus the interlayer stacking which characterizes the transverse axis of AM prints. It is interesting to note the increasing anisotropy (differences in the compressive strength) across the axial and transverse axis with increasing carbon fiber content, which is important to understand the optimal mechanical performance of the material.

This reduction in compressive strength corresponds to previously reported studies [31,32] for compression molded parts which evaluated the roles of small imperfections, such as porosity and fiber misalignment in the formation of kink bands and material strength. C. Pascual-González et al. [33] highlighted the volumetric contraction resulting from the solidification process at fabrication that could be affected by fiber content to result in microstructure and interlaminar effects which influence mechanical properties.

Pukanszky [34] determined interfacial interactions effects in multicomponent materials and recommended the understanding of the effects as key to material optimization. The theory of fiber-matrix interfacial effects on compressively loaded composites proposed by Grezczuk [35] can be applied to explain the reducing compressive strength with increasing fiber content in AM fabricated CFRP composites. This is because compression loading more easily affects the fiber-matrix interface because of the shearing nature of failure propagation typical of the loading mode. Grezczuk presented both experimental and theoretical studies on the failure modes compressively loaded composites to support his theory that the composites exhibit much lower strength than predicted by the micro buckling theory. This he claims to be the result of the lower fiber-matrix interface properties which is much lower than that of the average properties of the entire composite. Dharan and Lin's [36] three-phase model which proposed the effects of the fiber, matrix, and fiber-matrix interface properties on the compressive strength can also be applied to explain the results. The studies using boron, carbon, and glass composites showed clearly that the strength/modulus of the fiber-matrix interface can have a significant effect on the compressive strength of the composites. Thin fiber-matrix interface regions between the fiber and matrix characterize much lower strength than the fiber and matrix which could be used to explain the lower compressive strength with increasing fiber content since increasing fiber content means increasing fiber-matrix interface regions. According to the model, and overall fiber-matrix interfacial strength less than 1/50th of the overall matrix will result in reduced compressive strength. Similarly, fiber-interface width above 1/1000th of the fiber width is proposed to reduce compressive strength. These are considerable features of AM fabricated short-fiber CFRP composites, given the typical 7 μm diameter and the fiber-matrix interface width which are expected to be higher than 1/1000th of 7 μm (7 nm). Hence, they account for the reductions in compressive strength with increasing fiber content.

3.1.2. Compressive Modulus

The compressive modulus results which measure the stiffness of the material under compression is presented in Figure 5. Figure 5a shows the comparison of the modulus across the axial and transverse axes of the composites, while Figure 4b shows the comparison of the axial compressive and axial tensile properties.

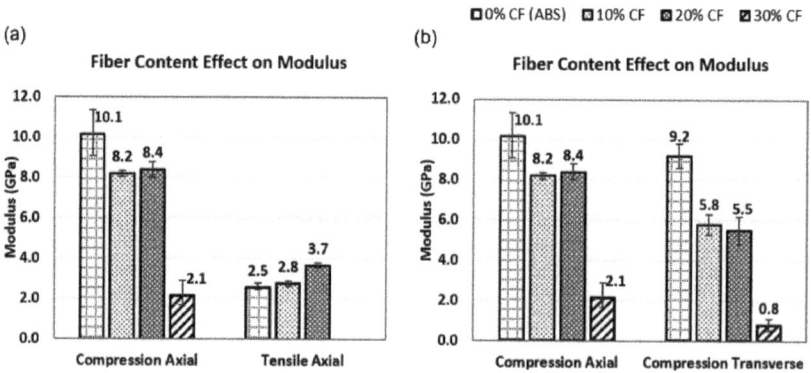

Figure 5. Fiber Content Effect on the Compressive Modulus of AM Fabricated CFRP Composites (a) Axial versus Transverse Compression (b) Axial Compression versus Axial Tensile.

In comparing the compressive to the tensile modulus as seen in Figure 5a, the AM composite exhibited a generally higher modulus in compression than in tension. This is similar to the material strength where the difference in the compressive and tensile values was related to the plasticity of the material under compression as against elastic characteristics exhibited in tension. The modulus was also observed to reduce with increasing fiber content, but unlike the tensile modulus which exhibits improving properties.

Across the axial axes, an 18% modulus reduction was observed at 10–20% fiber content and a reduction up to 79% for the 30% fiber content. Larger modulus reduction was observed in the transverse direction at 38% reduction for the 10–20% fiber content and up to 91% reduction for the 30% fiber content. The modulus reduction with increasing fiber content can also be related to the larger amounts of fiber-matrix interface regions which are characterized by lower properties and are the cause of the reducing modulus in line with the theory of the compressive loading failure of short-fiber thermoplastic matrix composites [34–36]. The higher modulus in the axial axes as seen in Figure 5b can be related to the higher bond strength through the longitudinal length of the print bead which characterizes the axial axis compared to the weaker interlayer bond strength across the print layers of the transverse axis in a similar trend as obtained for the compressive strength.

3.1.3. Compressive Ductility

The described ductility result shown in Figure 6 measures the ability of the AM composite to deform plastically before fracture. The lower strain to fracture indicates brittleness while higher strains indicate ductile characteristics.

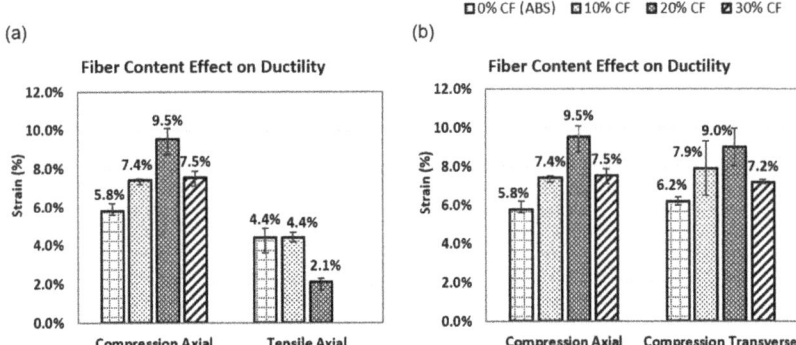

Figure 6. Fiber Content Effect on the Compressive Ductility of AM Fabricated CFRP Composites (**a**) Axial versus Transverse Compression (**b**) Axial Compression versus Axial Tensile.

In comparing the compressive to the tensile ductility as seen in Figure 6a, the composite exhibited in compression more than the 5% strain which is used to set the benchmark for brittle to ductile characteristics, while the tensile exhibited below the 5% benchmark value. Thus, the material features a ductile characteristic in compression and brittle characteristics in tension. The material is seen to exhibit increasing compressive ductility with increasing fiber content, with values peaking at 20% fiber content, against the tensile ductility which exhibited reduced values with increasing carbon fiber content. As discussed by Junaid et al. [37] in their ductile to brittle transitions theory of CFRP composites, a balance of ductility, strength, and modulus properties are important characteristics for the effective application of the material.

In comparing the ductility across the axial and transverse axes as seen in Figure 6b, the axial compression showed a 28% increase in ductility at 10% carbon fiber content, a 64% increase for the 20% carbon fiber content, and a 29% increase for the 30% carbon fiber content. A similar ductility trend was observed in the transverse compression but at a lower rate of increase, at 27%, for the 10% fiber content, 45% for the 20% fiber content, and a 10% increase for the 30% fiber content. The compressive ductility increase with fiber content can be related to the ability of the increased fiber content to support more plastic deformation before buckling and eventual shear failure. The ductile thermoplastic matrix dominates to support the compressive strain to failure, while the fiber fills any porosities upon compression before buckling failure.

3.1.4. Compressive Toughness

The toughness property which measures the ability of the material to absorb energy and plastically deforms without fracturing is presented in Figure 7. Though determined from the area under the stress-strain curve, it is generally provided by a balance of the strength and ductility properties.

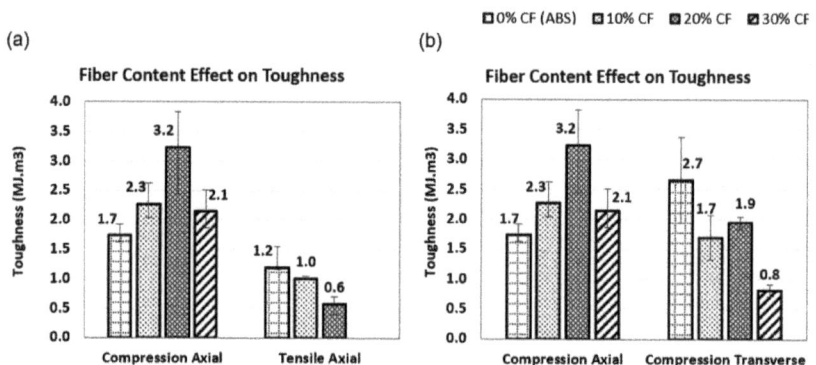

Figure 7. Fiber Content Effect on the Compressive Toughness of AM Fabricated CFRP Composites (a) Axial versus Transverse Compression (b) Axial Compression versus Axial Tensile.

The large plastic deformation behaviors exhibited under compression relate to their energy absorption performance, which influences their toughness. Figure 7a shows the material exhibiting higher toughness in compression than in tension in line with the plastic deformation exhibited under compression compared to the elastic deformation under tension.

The compressive toughness increased up to 20% fiber content after which it stabilized. However, the tensile reduction with increasing fiber content is in line with the ductile to brittle transition theory which relates to the elastic deformation failure in tension. This is in line with the investigations by Ning et al. [13], where increasing fiber content was found to reduce the toughness, ductility, yield strength in the tensile mode. The more brittle fiber reinforcement is more active in tensile loading while the ductile thermoplastic matrix with its toughness features dominates the compressive loading. The mix of the increasing strength from the increased fiber content and ductility from the dominating thermoplastic matrix offers the material the increasing energy absorption before micro-buckling, thus, increasing toughness under compression.

When comparing compressive toughness across the axes of the composite as seen in Figure 7b, the axial compression showed a 35% increase in toughness at 10% carbon content, a 47% increase for the 20% carbon fiber content, and a 24% increase for the 30% carbon fiber content. However, a different trend was observed in the transverse axis, where toughness was found to reduce with increasing fiber content. An approximate 37% reduction was observed for the 10% fiber content, a 30% reduction for the 20% fiber content, while a reduction up to 70% was observed for the 30% fiber content.

The toughness increase in the axial axis can be related to the higher ductility across the longitudinal print layers with increasing fiber contents which are better able to withstand micro-buckling. This is different in the transverse axis where increasing fiber (reducing thermoplastic matrix) contents results in lesser bond strength of the interlayers. Hence, the result of reducing toughness with increasing fiber content is experienced across the axis. The toughness reduction trend can also be related to the more significant strength reduction across the axes which is a result of the weaker interlayer bonding across adjacent print layers that characterizes the transverse axis.

3.2. Evaluation of Failure Modes

The compressive failure mode observed during testing started as small wavelength buckling of the print layers with the eventual barreling of the samples, which can be related to reported investigations in the compressive failure of thermoplastic composites [38] Adjacent fiber-matrix and matrix-matrix interfaces buckled in similar manners such that the deformation of adjacent layers within the composite was primarily of shear deformation. The failure mode in tension was different from that observed in compression due to the difference in the effective fiber and matrix functions under the two different loading modes. While compression deformed the materials plastically by barreling before eventual shear deformation, tension deformed the materials elastically, which is due to the more active brittle carbon fiber function under tension. Figure 8 evaluates the fracture surfaces of the samples at final failure to compare material features and evaluate failure modes under compression and tension.

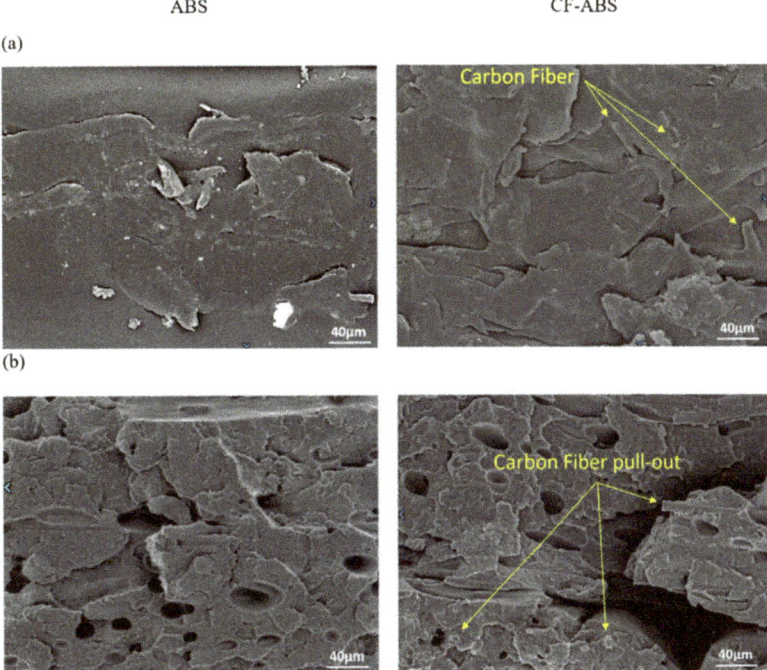

Figure 8. Microscopic Evaluation of Mechanical Tested ABS versus CF-ABS [1000×] (a) Tensile Loading (b) Compressive Loading.

Figure 8a shows the internal failure images typical for AM fabricated ABS and CF-ABS under compression. The compressive loading is seen to close the intra-layers and inter-layers porosities generally exhibited within the print beads and adjacent layers of the print. The CF-ABS was observed to show coarser shearing due to the effective carbon fiber after the porosity closure which was resisting plastic deformation, hence, generating the higher ductility and toughness observed. In Figure 8b the surface failure mode of the CF-ABS composite loaded in tension showed fiber pull-out and matrix cracking dominating the failure mode, with the carbon fiber pull-out of the matrix indicating the additional impediment the breaking forces must overcome in CFRP composites before yielding. Hence, the increasing carbon fiber content can be related to increasing the reinforcing impediment, thus, increasing tensile strength and modulus as observed in testing.

4. Conclusions

In this study, the effects of fiber content on the compressive properties of AM fabricated CFRP composites using CF-ABS samples were investigated. Compressive properties were compared along the axial and transverse axis of the composite, and the axial compression was also compared to axial tensile properties. The results were reinforced with the fracture surface failure modes and the following conclusions were drawn:

1. The increasing carbon fiber content in AM fabricated short-fiber thermoplastic CFRP composites slightly reduces compressive strength and modulus. However, they increase ductility and toughness.
2. The 20% carbon fiber content provided an optimized fiber volume for AM fabricated short-fiber thermoplastic CFRP composite applications subjected to compressive loading.
3. AM fabricated CFRP composites generally demonstrate a higher compressive than tensile properties because the higher plastic deformation characterizes compressive loading and more easily redistributes stress at existing flaws.
4. The failure mode in compression starts with the fiber-matrix debonding, followed by a unidirectional buckling of the interlayers and then barreling of the composite before an eventual shear deformation of the composite.
5. The material properties trends and failure mode in compression differ from tension due to the differences in the active functions of the fiber and matrix under the different loading conditions.

Author Contributions: Conceptualization, O.A. and W.C.; methodology, O.A., W.C. and E.B.; validation, O.A., W.C. and V.A.; formal analysis, O.A. and V.A.; investigation, O.A., E.B. and V.A.; resources, O.A. and W.C.; data curation, O.A., W.C. and E.B.; writing—original draft preparation, O.A.; writing—review and editing, O.A., W.C. and E.B.; supervision, W.C. All authors have read and agreed to the published version of the manuscript.

Funding: This research received no external funding.

Data Availability Statement: The data presented in this study are available on request from the corresponding author.

Conflicts of Interest: The authors declare no conflict of interest.

References

1. Bin Hamzah, H.H.; Keattch, O.; Covill, D.; Patel, B.A. The Effects of Printing Orientation on the Electrochemical Behaviour of 3D Printed Acrylonitrile Butadiene Styrene (ABS)/Carbon Black Electrodes. *Sci. Rep.* **2018**, *8*, 9135. [CrossRef]
2. Tymrak, B.M.; Kreiger, M.; Pearce, J.M. Mechanical Properties of Components Fabricated with Open-Source 3-D Printers under Realistic Environmental Conditions. *Mater. Des.* **2014**, *58*, 242–246. [CrossRef]
3. Dimas, L.S.; Bratzel, G.H.; Eylon, I.; Buehler, M.J. Tough Composites Inspired by Mineralized Natural Materials: Computation, 3D Printing, and Testing. *Adv. Funct. Mater.* **2013**, *23*, 4629–4638. [CrossRef]
4. Martin, J.J.; Fiore, B.E.; Erb, R.M. Designing Bioinspired Composite Reinforcement Architectures via 3D Magnetic Printing. *Nat. Commun.* **2015**, *6*, 8641. [CrossRef]
5. Unterweger, C.; Brüggemann, O.; Fürst, C. Synthetic Fibers, and Thermoplastic Short-Fiber-Reinforced Polymers: Properties and Characterization. *Polym. Compos.* **2014**, *35*, 227–236. [CrossRef]
6. Goh, G.D.; Dikshit, V.; Nagalingam, A.P.; Goh, G.L.; Agarwala, S.; Sing, S.L.; Wei, J.; Yeong, W.Y. Characterization of Mechanical Properties and Fracture Mode of Additively Manufactured Carbon Fiber and Glass Fiber Reinforced Thermoplastics. *Mater. Des.* **2018**, *137*, 79–89. [CrossRef]
7. Zhang, W.; Cotton, C.; Sun, J.; Heider, D.; Gu, B.; Sun, B.; Chou, T.W. Interfacial Bonding Strength of Short Carbon Fiber/Acrylonitrile-Butadiene-Styrene Composites Fabricated by Fused Deposition Modeling. *Compos. Part B Eng.* **2018**, *137*, 51–59. [CrossRef]
8. Ning, F.; Cong, W.; Hu, Z.; Huang, K. Additive Manufacturing of Thermoplastic Matrix Composites Using Fused Deposition Modeling: A Comparison of Two Reinforcements. *J. Compos. Mater.* **2017**, *51*, 3733–3742. [CrossRef]

9. Mohammadizadeh, M.; Fidan, I. Thermal Analysis of 3D Printed Continuous Fiber Reinforced Thermoplastic Polymers for Automotive Applications. In Proceedings of the 30th Annual International Solid Freeform Fabrication Symposium—An Additive Manufacturing Conference, SFF 2019, Austin, TX, USA, 12–14 August 2019; pp. 899–906.
10. Mohamed, O.A.; Masood, S.H.; Bhowmik, J.L. Optimization of Fused Deposition Modeling Process Parameters: A Review of Current Research and Future Prospects. *Adv. Manuf.* **2015**, *3*, 42–53. [CrossRef]
11. van de Werken, N.; Tekinalp, H.; Khanbolouki, P.; Ozcan, S.; Williams, A.; Tehrani, M. Additively Manufactured Carbon Fiber-Reinforced Composites: State of the Art and Perspective. *Addit. Manuf.* **2020**, *31*, 100962. [CrossRef]
12. Tekinalp, H.L.; Kunc, V.; Velez-Garcia, G.M.; Duty, C.E.; Love, L.J.; Naskar, A.K.; Blue, C.A.; Ozcan, S. Highly Oriented Carbon Fiber-Polymer Composites via Additive Manufacturing. *Compos. Sci. Technol.* **2014**, *105*, 144–150. [CrossRef]
13. Ning, F.; Cong, W.; Qiu, J.; Wei, J.; Wang, S. Additive Manufacturing of Carbon Fiber Reinforced Thermoplastic Composites Using Fused Deposition Modeling. *Compos. Part B Eng.* **2015**, *80*, 369–378. [CrossRef]
14. Duty, C.E.; Kunc, V.; Compton, B.; Post, B.; Erdman, D.; Smith, R.; Lind, R.; Lloyd, P.; Love, L. Structure and Mechanical Behavior of Big Area Additive Manufacturing (BAAM) Materials. *Rapid Prototyp. J.* **2017**, *23*, 181–189. [CrossRef]
15. Post, B.K. Additive Manufacturing in Wind Turbine Components and Tooling Project ID #: T13. 2016. Available online: https://www.energy.gov/sites/prod/files/2019/05/f62/ORNL-T13-Post%204_2_19.pdf (accessed on 8 November 2021).
16. Post, B.K.; Richardson, B.; Lind, R.; Love, L.J.; Lloyd, P.; Kunc, V.; Rhyne, B.J.; Roschli, A.; Hannan, J.; Nolet, S.; et al. Big Area Additive Manufacturing Application in Wind Turbine Molds. In Proceedings of the 28th Annual International Solid Freeform Fabrication Symposium—An Additive Manufacturing Conference, SFF 2017, Austin, TX, USA, 7–9 August 2017; pp. 2430–2446.
17. Kunc, V.; Hassen, A.A.; Lindahl, J.; Kim, S.; Post, B. Large Scale Additively Manufactured Tooling for Composites. In Proceedings of the 15th Japan International Sampe Symposium and Exhibition, Tokyo, Japan, 27–29 November 2017; pp. 1–6.
18. Brenken, B.; Barocio, E.; Favaloro, A.; Kunc, V.; Pipes, R.B. Fused Filament Fabrication of Fiber-Reinforced Polymers: A Review. *Addit. Manuf.* **2018**, *21*, 1–16. [CrossRef]
19. Meraz Trejo, E.; Jimenez, X.; Billah, K.M.M.; Seppala, J.; Wicker, R.; Espalin, D. Compressive Deformation Analysis of Large Area Pellet-Fed Material Extrusion 3D Printed Parts in Relation to in Situ Thermal Imaging. *Addit. Manuf.* **2020**, *33*, 101099. [CrossRef]
20. Quan, Z.; Larimore, Z.; Qin, X.; Yu, J.; Mirotznik, M.; Byun, J.H.; Oh, Y.; Chou, T.W. Microstructural Characterization of Additively Manufactured Multi-Directional Preforms and Composites via X-Ray Micro-Computed Tomography. *Compos. Sci. Technol.* **2016**, *131*, 48–60. [CrossRef]
21. Quan, Z.; Larimore, Z.; Wu, A.; Yu, J.; Qin, X.; Mirotznik, M.; Suhr, J.; Byun, J.H.; Oh, Y.; Chou, T.W. Microstructural Design and Additive Manufacturing and Characterization of 3D Orthogonal Short Carbon Fiber/Acrylonitrile-Butadiene-Styrene Preform and Composite. *Compos. Sci. Technol.* **2016**, *126*, 139–148. [CrossRef]
22. Wang, J.; Xiang, J.; Lin, H.; Wang, K.; Yao, S.; Peng, Y.; Rao, Y. Effects of Scanning Strategy and Printing Temperature on the Compressive Behaviors of 3D Printed Polyamide-Based Composites. *Polymers* **2020**, *12*, 1783. [CrossRef] [PubMed]
23. Lee, C.S.; Kim, S.G.; Kim, H.J.; Ahn, S.H. Measurement of Anisotropic Compressive Strength of Rapid Prototyping Parts. *J. Mater. Process. Technol.* **2007**, *187–188*, 627–630. [CrossRef]
24. Araya-Calvo, M.; López-Gómez, I.; Chamberlain-Simon, N.; León-Salazar, J.L.; Guillén-Girón, T.; Corrales-Cordero, J.S.; Sánchez-Brenes, O. Evaluation of Compressive and Flexural Properties of Continuous Fiber Fabrication Additive Manufacturing Technology. *Addit. Manuf.* **2018**, *22*, 157–164. [CrossRef]
25. Love, L.J.; Kunc, V.; Rios, O.; Duty, C.E.; Elliott, A.M.; Post, B.K.; Smith, R.J.; Blue, C.A. The Importance of Carbon Fiber to Polymer Additive Manufacturing. *J. Mater. Res.* **2014**, *29*, 1893–1898. [CrossRef]
26. Chacón, J.M.; Caminero, M.A.; Núñez, P.J.; García-Plaza, E.; García-Moreno, I.; Reverte, J.M. Additive Manufacturing of Continuous Fibre Reinforced Thermoplastic Composites Using Fused Deposition Modelling: Effect of Process Parameters on Mechanical Properties. *Compos. Sci. Technol.* **2019**, *181*, 107688. [CrossRef]
27. Costa, A.E.; Ferreira da Silva, A.; Sousa Carneiro, O. A Study on Extruded Filament Bonding in Fused Filament Fabrication. *Rapid Prototyp. J.* **2019**, *25*, 555–565. [CrossRef]
28. ASTM D695-15. In *Annual Book of ASTM Standards*; ASTM International: West Conshohocken, PA, USA, 2015; pp. 1–8. [CrossRef]
29. ASTM D638-14. In *Annual Book of ASTM Standards*; ASTM International: West Conshohocken, PA, USA, 2014; pp. 1–17. [CrossRef]
30. Duty, C.E. *Material Development for Tooling Applications Using Big Area Additive Manufacturing (BAAM)*; Oak Ridge National Laboratory: Oak Ridge, TN, USA, 2015. [CrossRef]
31. Fleck, N.; Deng, L.; Budiansky, B. Prediction of Kink Width in Compressed Fiber Composites. *J. Appl. Mech.* **1995**, *62*, 329–337. [CrossRef]
32. Lo, K.H.; Chim, E.S.-M. Compressive Strength of Unidirectional Composites. *J. Reinf. Plast. Compos.* **1992**, *11*, 838–896. [CrossRef]
33. Pascual-González, C.; San Martín, P.; Lizarralde, I.; Fernández, A.; León, A.; Lopes, C.S.; Fernández-Blázquez, J.P. Post-Processing Effects on Microstructure, Interlaminar and Thermal Properties of 3D Printed Continuous Carbon Fibre Composites. *Compos. Part B Eng.* **2021**, *210*, 108652. [CrossRef]
34. Pukánszky, B. Interfaces and Interphases in Multicomponent Materials: Past, Present, Future. *Eur. Polym. J.* **2005**, *41*, 645–662. [CrossRef]
35. Greszczuk, L.B. On Failure Modes of Unidirectional Composites Under Compressive Loading. In *Fracture of Composite Materials*; Springer: Singapore, 1982; pp. 231–244. ISBN 9024726999.

36. Dharan, C.K.H.; Lin, C.L. Longitudinal Compressive Strength of Continuous Fiber Composites. *J. Compos. Mater.* **2007**, *41*, 1389–1405. [CrossRef]
37. Junaedi, H.; Albahkali, E.; Baig, M.; Dawood, A.; Almajid, A. Ductile to Brittle Transition of Short Carbon Fiber-Reinforced Polypropylene Composites. *Adv. Polym. Technol.* **2020**, *2020*, 6714097. [CrossRef]
38. Thompson, R. Compressive Strength of Continuous Fiber Unidirectional Composites. 2012. Available online: https://tigerprints.clemson.edu/all_dissertations/953 (accessed on 8 November 2021).

Article

Controllable Synthesis of Graphene-Encapsulated NiFe Nanofiber for Oxygen Evolution Reaction Application

Mengyang Li [1,†], Jiayi Rong [1,†], Ning Guo [1], Susu Chen [1], Meiqi Gao [1], Feng Cao [1,*] and Guoqing Li [2,*]

[1] Key Lab for Anisotropy and Texture of Materials (MoE), School of Materials Science and Engineering, Northeastern University, Shenyang 110819, China; 2100534@stu.neu.edu.cn (M.L.); 20182924@stu.neu.edu.cn (J.R.); guoning_0226@163.com (N.G.); 17854173815@163.com (S.C.); gaomq@smm.neu.edu.cn (M.G.)
[2] Department of Materials Science and Engineering, North Carolina State University, Raleigh, NC 27695, USA
* Correspondence: caof@atm.neu.edu.cn (F.C.); gli4@ncsu.edu (G.L.)
† These authors contributed equally to this work.

Abstract: Carbon-Encapsulated NiFe Nanofiber Ni_xFe_y@C-CNFs have been demonstrated to be promising candidates to replace conventional nobel metals-based catalysts for oxygen evolution reaction. Here, we developed a facile method of electrospinning and high temperature carbonization to synthesize Ni_xFe_y@C-CNFs catalysts. It is proved that Ni_3Fe_7@C-CNFs exhibited low overpotential (245 mV) and excellent stability in alkaline electrolyte for OER. This work provides a good platform for the synthesis and design of graphene-encapsulated alloy catalysts.

Keywords: graphene-encapsulated; electrospinning; OER; nanoparticles

1. Introduction

Oxygen evolution reaction (OER) is one of the most basic and important electrochemical reactions for the production of renewable energy, which can be used in many green energy systems such as water splitting, fuel cells and metal-air batteries [1,2]. The multi-electron transfer reaction mechanism makes it challenging to develop such highly efficient catalysts [3,4]. Noble metals oxides (RuO_2/IrO_2) have been used as OER catalysts in practical applications [5,6]. However, poor stability, low selectivity and high cost hinder their large-scale development [7]. Therefore, there is an urgent need to develop efficient OER non-noble metal electrocatalysts. Researchers demonstrated that transition metal (TM) with empty d-electron orbital participates in the reaction as catalyst [8,9]. In particular, Ni-based transition metals are highly efficient OER catalysts [10,11]. The researchers improved the electrocatalytic properties of Ni by alloying with other metals, hydroxylation, and forming compounds with nonmetal based on non-nobel metals [12,13]. In addition, graphene layers encapsulated metal or metal alloy nanoparticles is an effective method to improve the activity and stability of OER catalyst [14–17]. Thereby, it is a promising strategy to encapsulate Ni-based transition metals alloys in a graphite layer [18,19]. By doing so, not only the synergistic effects between the graphite shell and the NiFe nanoparticles core enhance the OER activity, but also the graphene shell can protect the alloy from the corrosion of alkaline electrolyte [15,17,20].

Although graphene layers encapsulated metal materials have been widely studied, there is still a lack of an effective method to synthesize them. First of all, the composition and proportion of the core metal is difficult to accurately control [21,22]. Secondly, the size of graphene layers encapsulated metal nanoparticles and the thickness of carbon layer are difficult to meet the expected requirements [23]. Finally, the activity and stability of the catalysts cannot be guaranteed [24]. Therefore, it will be of great interest to develop a more facile method to synthesize graphene layers encapsulated metal materials.

Herein, we report a simple method to synthesize graphene layers encapsulating Ni_3Fe_7 alloy materials supported by carbon nanofibers (Ni_3Fe_7@C-CNFs) by electrospinning and subsequent high temperature carbonization. In this way, we can precisely control the proportion of the core metal. The nanoparticles were about 12 nm in size and distributed uniformly along the carbon nanofibers. In the alkaline OER reaction, Ni_3Fe_7@C-CNFs shows excellent catalytic activity, the overpotential is only 245 mV to reach the current density of 10 mA·cm^{-2}, even better than that of the commercial catalyst RuO_2. Moreover, the Tafel slope of Ni_3Fe_7@C-CNFs is only 67 mV·dec^{-1} and it has a relatively fast chemical kinetics. In addition, the catalyst has long-term stability. This work provides a possible way for the design and synthesis of OER catalysts.

2. Materials and Methods

2.1. Materials

Nickel nitrate hexahydrate (Ni $(NO_3)_2 \cdot 6H_2O$), ferric nitrate nonahydrate ($Fe(NO_3)_3 \cdot 9H_2O$), N,N-dimethylformamide (DMF) and polyvinylpyrrolidone (K30) were purchased from Sinopharm Chemical Reagent Co., Ltd., Shanghai, China. Anhydrous ethanol and potassium hydroxide were purchased from Yongda Chemical Reagent Co., Ltd., Tianjin, China. The Nafion solution was purchased from Minnesota Minerals and Manufacturing Company. All chemical reagents are of analytical grade and no further purification is required.

2.2. Preparation of Ni_xFe_y@C-CNFs

The brown precursor solution with a certain viscosity was obtained by adding 2 mmol of nitrate and 5.6 g PVP to 10 mL DMF and stirring at a constant speed for 12 h. The prepared precursor solution was slowly inhaled into a 20 mL Syringe. Then switch on the injection pump at the rate of 0.5 mL/h. When the temperature reaches 30 °C, turn on the high voltage power supply, adjusting the voltage to 17 kV, and observe that there were fine fibers sprayed on the aluminum foil paper. After 12 h of spinning, the nanofiber precursor was obtained. The precursor was pre-oxidized in muffle furnace. It was heated to 250 °C at the rate of 1 °C/min in the air, and then furnace cooled to room temperature. The pre-oxidized nanofiber precursor was carbonized at high temperature in a tube furnace with a heating rate of 5 °C/min to 800 °C for 2 h. After grinding and sieving, it was installed in the sample tube and sealed for spare.

2.3. Preparation of Ni_xFe_y@C-CNFs Electrode

During the electrochemical performance test, the catalyst samples need to be uniformly loaded on the carbon paper electrode. Cut the carbon paper into a small rectangle of 1 × 1.5 cm^2, bond the wires and the carbon paper together with conductive silver glue, letting it dry completely in the air for an hour. Take the same amount of A and B glue and mix them evenly. Cover the conductive silver glue on both the front and back and apply it around to form a groove. Weigh 5 mg of the catalyst to be tested, add 600 μL of ultrapure water, 200 μL of absolute ethanol and 10 μL of Nafion solution, sonicating for 30 min to make the sample uniformly dispersed. Finally, use a pipette to suck 30 μL of the catalyst solution into the groove part of the carbon paper, when it is completely dried out, repeatting the dripping five times. And then perform the electrochemical performance test after it is completely dried.

2.4. Material Characterization

X-ray diffraction (XRD) is an effective means to determine the crystal structure of materials. The phase constitutions of the alloys were examined by Smartlab X-ray diffraction from Japanese company Rigaku. The test working current is 200 mA and the working voltage is 40 kV, with Cu Kα Ray (λ = 1.54 Å) as the incident source, filtered by graphite monochromator, using θ/2θ, the linkage mode is used for continuous scanning in the range of 10–90°, and the scanning speed is 5°/min.

Scanning electron microscopy (SEM) can interact with materials through incident electrons, and then obtain the morphology information of materials. The field emission scanning electron microscope JSM-7001F produced by Japan Electronics Corporation (JEOL) was used in this experiment to obtain surface morphology.

Transmission electron microscopy (TEM) is a method to observe more detailed crystal structure and phase composition. In this experiment, JEM-2100F transmission electron microscope produced by Japan Electronics Corporation (JEOL) was used to obtain the information of material morphology and crystal structure.

2.5. Electrochemical Measurements

The electrochemical measurements using IM6e electrochemical workstation produced by Zahner company of Germany were carried out by a standard three-electrode cell in 1 M KOH solution (without oxygen). Among them, the glassy carbon electrode is the working electrode, the mercury oxide (Hg/HgO) electrode is the reference electrode, and the carbon rod is the counter electrode.

During the Linear voltammetric scanning (LSV) test, the scanning rate is 5 mV/s, the voltage range of the oxygen evolution reaction (OER) is 0.1–1.1 V. Tafel slope is calculated from LSV polarization curve according to the equation as follows:

$$\eta = a + b \log j \qquad (1)$$

where j was the current density, b was the Tafel slope, η was the overpotential.

Electrochemical surface areas (ECSA) is proportional to the electric double layer capacitance (Cdl) of the electrode. Cdl is measured by cyclic voltammetry curve, and the appropriate potential range is selected in the non-Faraday zone, with 10, 20, 30, 40, 50 mV·s^{-1} speed, scan 20 circles, read the current density difference at a specific potential, plot with the scan speed, and the slope of the curve obtained by fitting is the electric double layer capacitance.

Electrochemical impedance spectroscopy (EIS) can be used to measure the impedance of the catalyst under applied voltage, which can reflect the speed of electron transfer. The disturbance potential is set at 5 mV, and the frequency range is set from 100 MHz to 100 kHz.

Electrochemical stability test: In this experiment, the electrochemical stability test is to maintain the sample at Constant potential for 12 h in the i-t working mode of the electrochemical workstation, and observe the current density change of the sample.

The electrochemical experiments were all conducted at room temperature.

The reference electrode was calibrated to reversible hydrogen potential (RHE) using platinum electrode for both working and counter electrodes and converted into RHE according Nernst equation RHE = 0.0591 pH + 0.197.

3. Result and Discussion

3.1. Characterizations of Samples

Ni$_x$Fe$_y$@C-CNFs is synthesized by a simple and versatile method using electrospinning and subsequent heat treatment, as shown in Figure 1. Firstly, the metal salt was dissolved in DMF, and then PVP was added to form the spinning solution. The precursor was electrospun from the prepared spinning solution mixture. Then Ni$_x$Fe$_y$@C-CNFs was obtained by pre-oxidation and high temperature carbonization of the precursor.

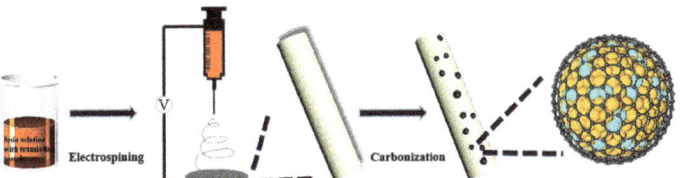

Figure 1. Schematic illustration of the synthesis process of the Ni$_x$Fe$_y$@C-CNFs.

The controllable synthesis and compositions are confirmed by XRD measurements as Figure 2. Compared with the diffraction peaks of Ni and Fe, the diffraction peaks of NiFe alloy shift to some extent, which proves the formation of the alloy structure. The Ni_3Fe_7@C-CNFs samples showed characteristic peaks at 43.7°, 51.0°, and 74.8°, corresponding to the (111), (200) and (220) faces of the NiFe alloy, respectively. Except for the characteristic diffraction peaks corresponding to face centered cubic phase alloy, there are no other diffraction peaks, which proves that there are only single-phase alloys. The magnified portion of XRD pattern in the 2 θ range of 42–48° is provided on the right. The strongest diffraction peak shifts slightly at a small angle in a regular trend with the change of alloy composition, which proves that the lattice constant changes due to the formation of ideal alloy structure. All the composites show a major peak at 2 θ = 26°, corresponding to the typical (002) plane of graphitic carbon.

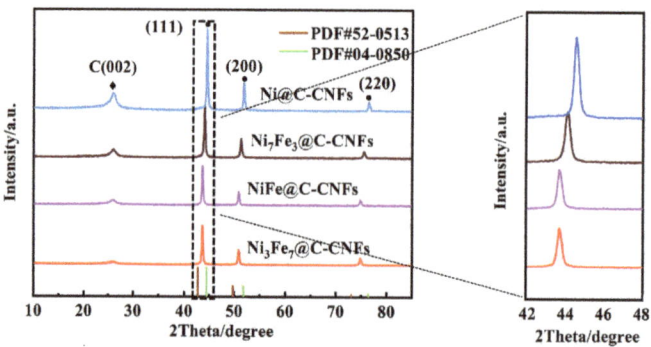

Figure 2. XRD patterns of Ni_xFe_y@C-CNFs and magnified portion in the 2 θ range of 42°–48°.

After mixing the nitrate solution and the PVP uniformly, the precursor of $NixFey$@C-CNFs can be formed after electrospinning for 12 h. As shown in Figure 3a, the SEM morphology of the prepared precursor was characterized, from which the produced nanofibers have a diameter of about 100 nm. The surface of carbon nanofibers is smooth and its morphology is uniform. Figure 3b shows that the precursor remains fibrous after high temperature carbonization, at which point the fiber becomes a highly conductive carbon material. Carbon nanofibers have the advantages of small size, large specific surface area and large aspect ratio. Therefore, electrons and protons can transfer in a controlled direction. However, compared with the precursors before high temperature treatment, the surface of the nanofibers is no longer smooth. It can be clearly observed that both the inner and surface of the carbon nanofibers are formed into round alloy particles. The nano-alloy particles adhere uniformly, and there is no obvious agglomeration of the alloy particles. At the same time, to determine the distribution of each element in the prepared sample, part of the region was selected for mapping detection in the experiment. The Ni and Fe elements in the sample are evenly distributed, and there is no obvious element agglomeration, indicating that the formation of single-phase alloys as shown in the illustration.

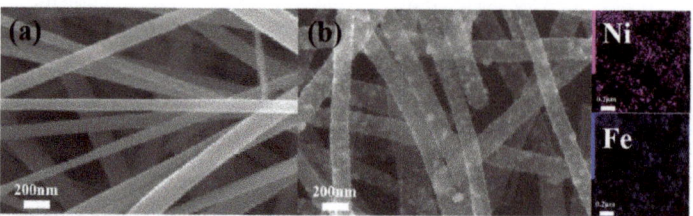

Figure 3. (a) SEM images of Precursor; (b) SEM images of Ni_3Fe_7@C-CNFs and the corresponding elemental mapping of Ni_3Fe_7@C-CNFs.

To clarify the Ni_xFe_y@C-CNFs crystal structure and the nano particle size distribution, TEM images were analyzed further. Computer software Nano Measurer was used to count and calculate the nano particle size distribution. First, the scale is specified by selecting the length in the imported figure and setting the actual length and unit (Figure 4a). Next, select the particles in the figure one by one and their lengths will be automatically recorded and stored. Finally, the statistical table of nano particle sizes and the distribution figure are exhibited. The results show that metal alloy particles with a size of about 12 nm are distributed along the inside and outside of the carbon nanofibers uniformly (inset in Figure 4a). The size of alloy particles is small and the specific surface area is large, which provides rich active sites for catalytic reaction. A nano particle growing on carbon nanotubes was selected for observation, and the core-shell structure could be clearly observed again. High-resolution transmission electron microscopy (HRTEM) characterization of the Ni_3Fe_7@C-CNFs indicates that the alloy particles are completely encapsulated by the graphene layer (the red curve area) with a thickness of 3.2 nm (Figure 4b). Not only can the core—shell structure separate the alloy from the external electrolyte to avoid the corrosion of the alloy in a strong alkaline environment, but also it will prevent the agglomeration of alloy particles in the reaction process to improve the stability of the catalyst.

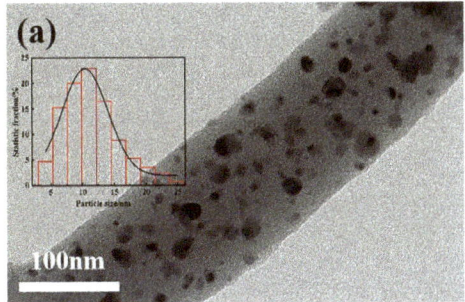 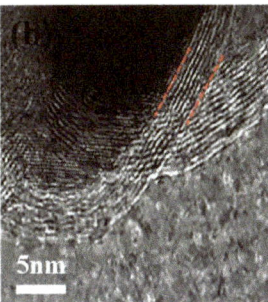

Figure 4. (a) TEM image (the inset shows Particle size distribution); (b) HRTEM image of Ni_3Fe_7@C-CNFs.

3.2. Electrochemical Analysis

The electrocatalytic OER activity of the catalysts is investigated in Nitrogen saturated alkaline solution (1 M KOH) in a standard three-electrode system. The polarization curves of LSV are shown in Figure 5a. At a current density of 10 mA cm^{-2}, the overpotential of Ni@C-CNFs is the highest 348 mV. Then the metal element Fe is introduced to form Ni-Fe alloy. Compared with pure nickel, the over-potential of bimetallic alloys has a decreasing tendency, which indicates introducing Fe into Ni can improve the OER catalytic performance. In addition to the necessity of qualitative analysis of the existence of Fe, we also designed experiments to quantitatively analyze the influence of Fe on the catalytic activity. We designed NiFe alloys with different proportions of Ni_7Fe_3@C-CNFs, NiFe@C-CNFs and Ni_3Fe_7@C-CNFs. As shown in Figure 5b, it can be observed visually the overpotential of Ni_3Fe_7@C-CNFs (245 mV), is better than NiFe@C-CNFs (308 mV) and Ni_7Fe_3@C-CNFs (300 mV). In addition, we also tested commercial catalyst RuO_2, the results show that the catalytic activity of Ni_3Fe_7@C-CNFs is even better than RuO_2 (290 mV). Previous studies and relevant calculations show that adjusting the proportion of the core metals is an effective means to adjust the catalytic performance [25–29]. The main effect is that the difference of Ni/Fe ratio leads to the transfer of electrons to the active sites on the catalyst surface, which increases the free energy of O *, making it easier to be adsorbed and desorbed, which reduce the over-potential of OER reaction. This is consistent with the results of our OER electrochemical measurements. When the ratio is 3/7, the catalyst has the lowest Tafel slope (67.0 mV dec^{-1}) and the charge-transfer resistance (Figure 5c), suggesting that introduction of different ratio of NiFe affects the kinetics of the electrode reactions. The catalyst of Ni_3Fe_7@C-CNFs has the largest electrochemical double-layer capacitance (73.02 mF cm^{-2}) indicating that the maximum number of active sites are on its surface among the five catalysts. To verify the synergistic effect of graphene and metal core can significantly affect the catalytic performance, Ni_3Fe_7 alloy without graphene and the carbon fiber, which did not support any metals or alloys were prepared respectively for comparison. The electrochemical test results indicated the bare carbon fiber sample did not show any catalytic activity. Thus, the positive effect of carbon fiber on performance can be excluded. The OER activity of the prepared samples with graphene layer removed decreases greatly, which indicates that the synergistic effect of graphene and metal core is the main reason for the improvement of its activity. According to previous relevant studies, the transfer of electrons from the core metal to graphene and the increase of the electron density of states near the Fermi level is one of the main reasons for the materials [30]. Graphene encapsulated transition non noble metal catalyst can effectively reduce the reaction energy barrier of each unit of OER reaction by adjusting the electronic structure of outer graphene through the inner metal to reduce OER overpotential and enhance the activity of the catalyst [19]. Thus, it can be concluded that the excellent catalytic performance results from the transfer of electrons from the internal metal to graphene shell. The study has proved that changing the relative proportion of metals in the alloy is a common method, which can provide a greater opportunity to change the catalytic activity. The electronic structure of the carbon shell can be adjusted by changing the proportion of the core metal, so as to optimize the catalytic activity [26]. In order to have a more comprehensive understanding of the electrocatalytic properties of the catalysts, we fit the Tafel slope from the LSV curve (Figure 5c). This value can be used to reflect the reaction kinetics of materials in catalytic reaction. The Tafel slope of Ni_3Fe_7@C-CNFs is 67.0 mV dec^{-1}, which is smaller than that of Ni@C-CNFs (82.9 mV dec^{-1}), NiFe@C-CNFs (79.1 mV dec^{-1}), and Ni_7Fe_3@C-CNFs (73.8 mV dec^{-1}). Thus, the catalyst Ni_3Fe_7@C-CNFs has the fastest reaction kinetics of our ferronickel alloy catalysts for the OER application. In order to further determine how graphene can enhance their performance on its OER activity, the electrochemical reaction kinetics of the sample was analyzed by Tafel slope. Figure 5c shows the Tafel curve. It can be seen that removing the graphene can increase its Tafel slope and bring a negative effect. It is inferred that after the graphene capsulated alloy structure is formed, the electronic structure is optimized due to the transfer of electrons from the internal metal to graphene, to accelerate the OER kinetic rate. It is inferred that after the graphene encapsulated alloy structure is formed, the electronic structure is optimized due to the transfer of electrons from the internal metal to graphene, so as to accelerate OER kinetic rate. Optimizing the Ni/Fe ratio can further reduce the Tafel slope, which shows that the ability of electron transfer to graphene can be adjusted by changing the internal alloy composition [14]. The electronic coupling between the internal alloy and graphene accelerates its kinetic rate, which is also one of the main reasons for the improvement of its OER activity. In addition, when compared with the recently reported OER electrocatalysts (Figure 5d), the Ni_3Fe_7@C-CNFs exhibit the remarkable activity with low overpotential at 10 mA cm^{-2} of 245 mV. All the compared electrocatalysts were summarized in Table S1.

The source of the differences between NiFe alloy catalysts with different Ni/Fe ratios in OER activity can be further explored through the change of electrochemical active area. Since the electrochemical surface areas (ECSA) are proportional to the electric double layer capacitance (C_{dl}) of the electrode, the ECSA can be expressed by C_{dl}. The CV curves of Ni_3Fe_7@C-CNFs at different scanning rates are shown in Figure 6a. The others are shown in Figure S1. All samples were subjected to CV cycle under the same conditions, and the electrochemical active area could be obtained by data processing and fitting (Figure 6b). Compared with the others catalysts, the electric double layer capacitance of Ni_3Fe_7@C-CNFs is the largest, which is 73.02 mF cm^{-2}. Large electrochemical active area can provide abundant active sites for catalytic reaction. How graphene can enhance their performance can be further explored through the change of electrochemical active area. Ni_3Fe_7@C has the larger electrochemical active area than Ni_3Fe_7 alloy, indicating that electrons transferred from the internal metal to the outer layer graphene, activating more active sites, to improve its OER activity. Subsequently, the electrochemical impedance spectroscopy (EIS) of the samples was measured to further explain the difference of charge transfer ability between different catalysts. The surface kinetics and conductivity of the catalyst can be analyzed by impedance spectroscopy. All samples show a small charge transfer resistance Rct, and its change trend is consistent with the polarization curves of LSV, Tafel slope and the electrochemical surface areas of the corresponding electrode. Ni_3Fe_7@C-CNFs has the smallest charge transfer resistance in comparison with others, which means that it has faster interfacial charge transfer efficiency and reaction kinetic rate. From the Figure 6c, Ni_3Fe_7 alloy without graphene sample exhibited a larger charge transfer resistance, which means it has lower interfacial charge transfer efficiency and reaction kinetic rate and has a negative effect on OER catalytic activity. The above comparison results between Ni_3Fe_7 alloy and Ni_3Fe_7@C-CNFs show that the electronic coupling between the internal alloy and graphene accelerates its kinetic rate, which is also one of the main reasons for the improvement of its OER activity.

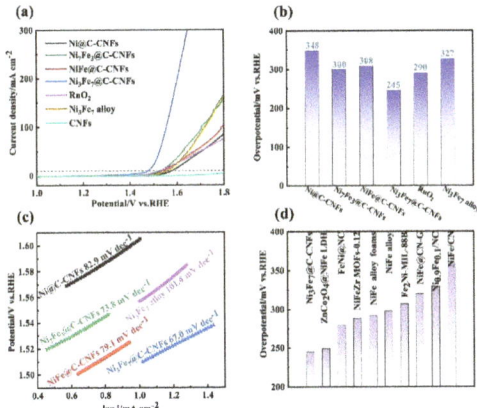

Figure 5. (a) LSV polarization curves after iR correction; (b) The overpotentials obtained from polarization curves at the current density 10 mA cm^{-2}; (c) The corresponding Tafel plots; (d) Comparison of overpotentials at 10 mA cm^{-2} of Ni_3Fe_7@C-CNFs with reported literatures.

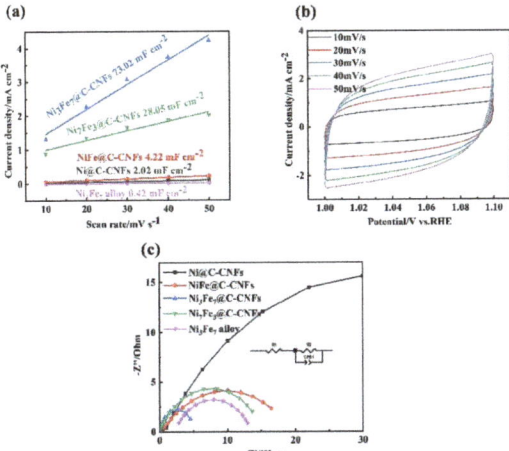

Figure 6. (a) Cyclic voltammetry curves of samples; (b) ECSA measurements under different treatment conditions; (c) Nyquist plots.

Researchers have tended to focus on decreasing the overpotential when designing catalysts, while focusing less on the stability. In addition to the catalytic activity, the stability is also a necessary condition to judge the performance of the catalyst. We tested the i-t curve of the sample Ni_3Fe_7@C-CNFs by potentiostatic method, as shown in Figure 7a. It can be seen from the figure that the current density of the sample hardly decreased after 12 h test at high current density of 100 mA cm^{-2}. The LSV curve before and after stability is almost consistent, which indicates that Ni_3Fe_7@C-CNFs has excellent stability (Figure 7b). The stability of the sample after scanning electron microscope characterization, as shown in Figure 7c. The morphology of the carbon fiber-loaded alloy was not destroyed after the stability test, which showed that Ni_3Fe_7@C-CNFs had excellent properties and morphological stability. The good stability results from the core-shell structure. The carbon shell can not only prevent the agglomeration of the nanoparticles, but also separate the inner alloy from the strong alkaline environment of the electrolyte. Therefore, its excellent activity and good stability make it possible to replace the practical application of noble metal based electrocatalysts in OER.

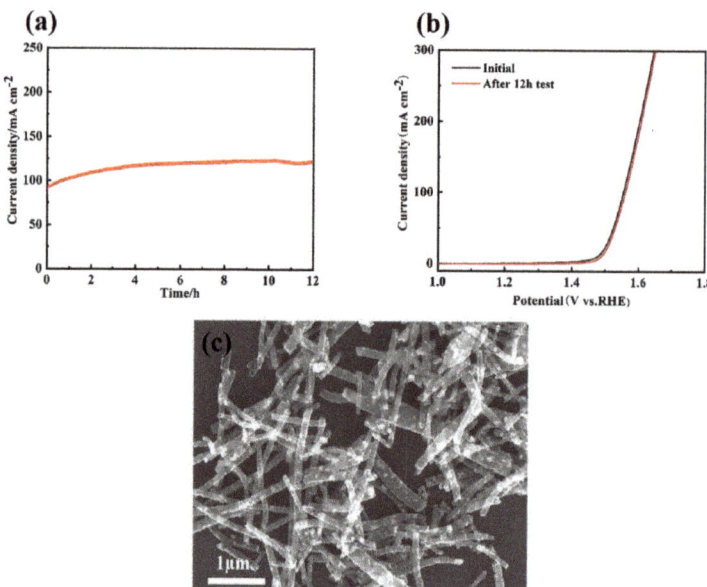

Figure 7. (a) LSV image after stability test; (b) i-t image; (c) SEM image after stability test.

4. Conclusions

In this paper, Ni_xFe_y@C-CNFs catalysts for OER were prepared by a facile method of electrospinning combined with pre-oxidation and high temperature carbonization. This method can precisely regulate the compositions of different samples. Due to the synergistic effects of NiFe nanoparticles metal core and the graphene layer, superior OER activity was achieved successfully. By adjusting the proportion of Ni and Fe, it is proved that Ni_3Fe_7@C-CNFs exhibited excellent catalytic performance for OER in 1.0 M KOH with an overpotential of 245 mV to reach a current density of 10 mA cm^{-2}. After 12 h reaction, the LSV curve can still coincide with the one before reaction, and there is no obvious decline, which proves that it shows remarkable stability. This work provides a new idea for OER application of nano particles prepared by electrospinning combined with pre-oxidation and high temperature carbonization and broadens the horizon of design graphene-encapsulated alloy catalysts.

Supplementary Materials: The following are available online at https://www.mdpi.com/article/10.3390/jcs5120314/s1, Figure S1: Cyclic voltammetry curves of other Ni_xFe_y@C-CNFs. (a) Ni@C-CNFs. (b) NiFe@C-CNFs. (c) Ni_7Fe_3@C-CNFs; Table S1: Electrocatalytic performance comparison of reported electrocatalysts.

Author Contributions: Methodology, experiment, validation, writing—original draft preparation, writing—review and editing, M.L. and J.R.; investigation, resources, and data curation, N.G., S.C. and M.G.; conceptualization, visualization, supervision, project administration, and funding acquisition, F.C. and G.L. All authors have read and agreed to the published version of the manuscript.

Funding: This research was funded by National Training Program of Innovation and Entrepreneurship for Undergraduates, grant 201910145104, and "the Fundamental Research Funds for the Central Universities", grant N182410001.

Acknowledgments: (1) Project 201910145104 supported by National Training Program of Innovation and Entrepreneurship for Undergraduates. (2) Supported by "the Fundamental Research Funds for the Central Universities" (N182410001).

Conflicts of Interest: The authors declare no conflict of interest.

References

1. Song, J.J.; Wei, C.; Huang, Z.F.; Liu, C.; Zeng, L.; Wang, X.; Xu, Z.J. A review on fundamentals for designing oxygen evolution electrocatalysts. *Chem. Soc. Rev.* **2020**, *49*, 2196–2214. [CrossRef] [PubMed]
2. Zhu, B.J.; Xia, D.G.; Zou, R.Q. Metal-organic frameworks and their derivatives as bifunctional electrocatalysts. *Coord. Chem. Rev.* **2018**, *376*, 430–448. [CrossRef]
3. Zou, X.X.; Zhang, Y. Noble metal-free hydrogen evolution catalysts for water splitting. *Chem. Soc. Rev.* **2015**, *44*, 5148–5180. [CrossRef]
4. Zhao, J.; Zhang, J.J.; Li, Z.Y.; Bu, X.H. Recent Progress on NiFe-Based Electrocatalysts for the Oxygen Evolution Reaction. *Small* **2020**, *16*, 2003916. [CrossRef] [PubMed]
5. Antolini, E. Iridium as catalyst and cocatalyst for oxygen evolution/reduction in acidic polymer electrolyte membrane electrolyzers and fuel cells. *ACS Catal.* **2014**, *4*, 1426–1440. [CrossRef]
6. Pi, Y.; Zhang, N.; Guo, S.; Guo, J.; Huang, X. Ultrathin laminar Ir superstructure as highly efficient oxygen evolution electrocatalyst in broad pH range. *Nano Lett.* **2016**, *16*, 4424–4430. [CrossRef]
7. Raman Abhinav, S.; Patel, R.; Vojvodic, A. Surface stability of perovskite oxides under OER operating conditions: A first principles approach. *Faraday Discuss.* **2021**, *229*, 75–88. [CrossRef] [PubMed]
8. Hu, C.; Hu, Y.; Fan, C.; Yang, L.; Zhang, Y.; Li, H.; Xie, W. Surface-Enhanced Raman Spectroscopic Evidence of Key Intermediate Species and Role of NiFe Dual-Catalytic Center in Water Oxidation. *Angew. Chem. Int. Ed.* **2021**, *133*, 19774–19778. [CrossRef]
9. Lee, Y.; Suntivich, J.; May, K.J.; Perry, E.E.; Shao-Horn, Y. Synthesis and activities of rutile IrO_2 and RuO_2 nanoparticles for oxygen evolution in acid and alkaline solutions. *Phys. Chem. Lett.* **2012**, *3*, 399–404. [CrossRef]
10. Gao, M.; Sheng, W.; Zhuang, Z.; Fang, Q.; Gu, S.; Jiang, J.; Yan, Y. Efficient water oxidation using nanostructured α-nickel-hydroxide as an electrocatalyst. *Am. Chem. Soc.* **2014**, *136*, 7077–7084. [CrossRef]
11. Huang, G.; Zhang, C.; Liu, Z.; Yuan, S.; Yang, G.; Li, N. Ultra-small NiFe-layered double hydroxide nanoparticles confined in ordered mesoporous carbon as efficient electrocatalyst for oxygen evolution reaction. *Appl. Surf. Sci.* **2021**, *565*, 150533. [CrossRef]
12. Yu, Z.-Y.; Duan, Y.; Liu, J.-D.; Chen, Y.; Liu, X.-K.; Liu, W.; Ma, T.; Li, Y.; Zheng, X.-S.; Yao, T.; et al. Unconventional CN vacancies suppress iron-leaching in Prussian blue analogue pre-catalyst for boosted oxygen evolution catalysis. *Nat. Commun.* **2019**, *10*, 358–368. [CrossRef]
13. Feng, X.; Jiao, Q.; Cui, H.; Yin, M.; Li, Q.; Zhao, Y.; Li, H.; Zhou, W.; Feng, C. One-Pot synthesis of $NiCo_2S_4$ hollow spheres via sequential ion-exchange as an enhanced oxygen bifunctional electrocatalyst in alkaline solution. *ACS Appl. Mater. Inter.* **2018**, *10*, 29521–29531. [CrossRef]
14. Deng, J.; Ren, P.; Deng, D.; Yu, L.; Yang, F.; Bao, X. Highly active and durable non-precious-metal catalysts encapsulated in carbon nanotubes for hydrogen evolution reaction. *Energy Environ. Sci.* **2014**, *7*, 1919–1923. [CrossRef]
15. Hu, K.; Ohto, T.; Chen, L.; Han, J.; Wakisaka, M.; Nagata, Y.; Fujita, J.-i.; Ito, Y. Graphene Layer Encapsulation of NonNoble Metal Nanoparticles as Acid-Stable Hydrogen Evolution Catalysts. *ACS Energy Lett.* **2018**, *3*, 1539–1544. [CrossRef]
16. Mahmood, J.; Anjum, M.A.R.; Shin, S.; Ahmad, I.; Noh, H.; Kim, S.; Jeong, H.Y.; Lee, J.S.; Baek, J. Baek, Encapsulating Iridium Nanoparticles Inside a 3D Cage-Like Organic Network as an Efficient and Durable Catalyst for the Hydrogen Evolution Reaction. *Adv. Mater.* **2018**, *30*, 1805606. [CrossRef]
17. Chen, Z.; Wu, R.; Liu, Y.; Ha, Y.; Guo, Y.; Sun, D.; Liu, M.; Fang, F. Ultrafine Co Nanoparticles Encapsulated in Carbon-Nanotubes-Grafted Graphene Sheets as Advanced Electrocatalysts for the Hydrogen Evolution Reaction. *Adv. Mater.* **2018**, *30*, e1802011. [CrossRef]
18. Gao, X.; Zhang, H.; Li, Q.; Yu, X.; Hong, Z.; Zhang, X.; Liang, C.; Lin, Z. Hierarchical $NiCo_2O_4$ Hollow microcuboids as bifunctional eectrocatalysts for overall water-splitting. *Angew. Chem. Int. Ed.* **2016**, *128*, 6290–6294. [CrossRef]

19. Cui, X.; Ren, P.; Deng, D.; Deng, J.; Bao, X. Single layer graphene encapsulating non-precious metals as high-performance electrocatalysts for water oxidation. *Energy Environ. Sci.* **2016**, *9*, 123–129. [CrossRef]
20. Zhou, Y.; Chen, W.; Cui, P.; Zeng, J.; Lin, Z.; Kaxiras, E.; Zhang, Z. Enhancing the Hydrogen Activation Reactivity of Nonprecious Metal Substrates via Confined Catalysis Underneath Graphene. *Nano Lett.* **2016**, *16*, 6058–6063. [CrossRef] [PubMed]
21. Liu, Z.; Tan, H.; Liu, D.; Liu, X.; Xin, J.; Xie, J.; Zhao, M.; Song, L.; Dai, L.; Liu, H. Promotion of overall water splitting activity over a wide pH range by interfacial electrical effects of metallic NiCo-nitrides nanoparticle/$NiCo_2O_4$ nanoflake/graphite fibers. *Adv. Sci.* **2019**, *6*, 1801829. [CrossRef]
22. Tong, Y.; Chen, P.; Zhou, T.; Xu, K.; Chu, W.; Wu, C.; Xie, Y. A bifunctional hybrid electrocatalyst for oxygen reduction and evolution: Cobalt oxide nanoparticles strongly coupled to B, N-decorated graphene. *Angew. Chem. Int. Ed.* **2017**, *129*, 7227–7231. [CrossRef]
23. Wang, C.; Yang, H.; Zhang, Y.; Wang, Q. NiFe alloy nanoparticles with hcp crystal structure stimulate superior oxygen evolution reaction electrocatalytic activity. *Angew. Chem. Int. Ed.* **2019**, *131*, 6099–6103. [CrossRef]
24. Hu, L.; Zhang, R.; Wei, L.; Zhang, F.; Chen, Q. Synthesis of FeCo nanocrystals encapsulated in nitrogen-doped graphene layers for use as highly efficient catalysts for reduction reactions. *Nanoscale* **2015**, *7*, 450–454. [CrossRef]
25. Song, Q.; Li, J.; Wang, S.; Liu, J.; Liu, X.; Pang, L.; Li, H.; Liu, H. Enhanced Electrocatalytic Performance through Body Enrichment of Co-Based Bimetallic Nanoparticles In Situ Embedded Porous N-Doped Carbon Spheres. *Small* **2019**, *15*, 1903395. [CrossRef]
26. Sun, H.; Min, Y.; Yang, W.; Lian, Y.; Lin, L.; Feng, K.; Deng, Z.; Chen, M.; Zhong, J.; Xu, L.; et al. Morphological and Electronic Tuning of Ni_2P through Iron Doping toward Highly Efficient Water Splitting. *ACS Catal.* **2019**, *9*, 8882–8892. [CrossRef]
27. Tian, X.; Zhao, X.; Su, Y.-Q.; Wang, L.; Wang, H.; Dang, D.; Chi, B.; Liu, H.; Hensen, E.J.; Lou, X.W.; et al. Engineering bunched Pt-Ni alloy nanocages for efficient oxygen reduction in practical fuel cells. *Science* **2019**, *366*, 850–856. [CrossRef]
28. Kitchin, J.R.; Nørskov, J.K.; Barteau, M.A.; Chen, J.G. Role of strain and ligand effects in the modification of the electronic and chemical Properties of bimetallic surfaces. *Phys. Rev. Lett.* **2004**, *93*, 4–7. [CrossRef]
29. Yang, Y.; Lun, Z.; Xia, G.; Zheng, F.; He, M.; Chen, Q. Non-precious alloy encapsulated in nitrogen-doped graphene layers derived from MOFs as an active and durable hydrogen evolution reaction catalyst. *Energy Environ. Sci.* **2015**, *8*, 3563–3571. [CrossRef]
30. Deng, J.; Deng, D.; Bao, X. Robust catalysis on 2D materials encapsulating metals: Concept, application, and perspective. *Adv. Mater.* **2017**, *29*, 1606967. [CrossRef]

Article

Manufacturing and Performance of Carbon Short Fiber Reinforced Composite Using Various Aluminum Matrix

Yongbum Choi [1,*], Xuan Meng [2] and Zhefeng Xu [3]

1. Mechanical Engineering Program, Graduate School of Advanced Science and Engineering, Hiroshima University, Hiroshima 739-8527, Japan
2. Department of Mechanical Science and Engineering, Graduate School of Engineering, Hiroshima University, Hiroshima 739-8527, Japan; d173809@hiroshima-u.ac.jp
3. State Key Laboratory of Metastable Materials Science and Technology, Yanshan University, Hebei Street 438#, Qinhuangdao 066004, China; zfxu@ysu.edu.cn
* Correspondence: ybchoi@hiroshima-u.ac.jp; Tel.: +81-82-424-5752

Abstract: A new fabrication process without preform manufacturing has been developed for carbon short fiber (CSF) reinforced various aluminum matrix composites. And their mechanical and thermal properties were evaluated. Electroless Ni plating was conducted on the CSF for improving wettability between the carbon fiber (CF) and aluminum. It was confirmed that pores in Ni plated CSF/Al and Al alloy matrix composites prepared by applied pressure, 0.8 MPa, had some imperfect infiltration regions between the CF/CF and CF/matrix in all composites. However, pores size in the region between the CF/CF and CF/matrix to use the A336 matrix was about 1 µm. This size is smaller than that of other aluminum-based composites. Vickers hardness of Ni plated CSF/A1070, A356 alloy, and A336 alloy composites were higher as compared to matrix. However, the A1070 pure aluminum matrix composite had the highest hardness improvement. The Ultimate tensile strength of the A1070 and A356 aluminum matrix composite was increased due to carbon fiber compared to only aluminum, but the Ultimate tensile strength of the A336 aluminum matrix composite was rather lowered due to the highest content of Si precipitate and large size of Al_3Ni compounds. The Thermal Conductivity of Ni plated CSF/A1070 composite has the highest value (167.1 $W \cdot m^{-1} \cdot K^{-1}$) as compared to composites.

Keywords: metal matrix composite; carbon short fiber; preform-less; low pressure infiltration; thermal and mechanical property

1. Introduction

In the fields of thermal management and engineering applications, composite materials with high thermal conductivity (TC) and high mechanical properties that are lightweight are expected to be utilized as an alternative to existing materials such as heat sink material [1,2] or internal engine parts [3,4]. Among the various reinforcements, carbon fibers (CFs) possess high thermal and electrical conductivity, extremely low coefficients of thermal expansion (CTE), and excellent mechanical properties, which have attracted much attention and have been particularly considered as an ideal candidate for reinforcement in multifunctional composites and engineering applications [5]. Aluminum is a dominant matrix for fabricating composite, CF-reinforced aluminum composite can combine the superior characteristics of CF and Al matrix; therefore, CF/Al matrix composites can be the promising materials for the heat sink components or structural material with high TC, are lightweight, and have good workability. Recently, much work has been devoted to mainly investigating the effect of CF amount on the modified TC and the enhancement of the mechanical properties [6]. However, the analysis of the effect of the matrix with various alloying elements on the properties of CF/Al matrix composites has been rarely studied. The alloying elements of the matrix (such as silicon and nickel) make an important impact thermal and mechanical properties of composites. A suitable manufacturing process for AMCs is the point for achieving high-performance composites with a good combination of reinforcement and matrix. The low-pressure infiltration (LPI) method is the useful fabrication process for composites with cost-saving and the possibility of large or complex structures [7–9].

In addition, the problems associated with the fabrication of CF-reinforced Al matrix composites are the poor wettability and chemical reactions between the carbon and Al matrix. An effective way to solve problems is the surface coating on CFs [10]. The coating materials used electroless nickel (Ni) plating, which is improving the wettability. A new process without manufacturing preform process was developed for CF/Al and Al alloy matrix composites. In this study, a new process without preform manufacturing was developed for carbon short fiber (CSF) reinforced various aluminum matrix composites, and the effect of the matrix with various alloying elements on the thermal and mechanical properties of composites has been investigated.

2. Materials and Manufacturing Methods

A1070 (purity, 99.7%), A356 alloy (6.5–7.5% Si—0.25-0.45% Mg) and A336 alloy (11–13% Si—0.8–1.5% Ni—0.5–1.5% Cu) are used as matrix for composites. The three types of Al-based matrix contain different kinds and contents of alloying elements. In order to infiltrate the molten Al into porous CF easily by the low-pressure infiltration (LPI) process, a good wettability between the CFs and molten Al is required. Electroless Ni plating was conducted on the CSFs for improving wettability [11]. As received CSFs and electroless nickel plated CSFs by 60 to 180 s plating time (pH of 6.5 and temperature of 313 K). The CSFs deposited by the electroless Ni plated were observed by EDS analysis (Genesis XM2).

Ni plated CSFs were put into a graphite mold of diameter 10 mm with adjustable height. To achieve a volume fraction of 10 vol.%, the height was adjusted to 10 mm. Al and Al alloy ingot were placed on the fibers, then the total mold was put into LPI equipment and heated up to the temperature of 623 K holding for 2 h. (in order to recrystallization of nickel [12]), and the temperature of the matrix was 1073 K. The infiltration pressure was 0.8 MPa [13]. Figure 1a shows the schematic illustration of the new fabrication process without preform manufacturing by low-pressure infiltration method. The microstructures of the Ni plated CSF and composites were observed using an Electron Probe Micro-Analyzer (EPMA, JXA-8900RL). The elemental analysis was carried out by energy-dispersive X-ray spectroscopy (EDS, S-5200 Energy Dispersive X-ray Analyzer, Hitachi, Japan). X-ray diffraction (XRD) was carried out with a Rigaku Smart Lab, Japan. Using Cu Kα radiation at 40 kV and 0.1 A, a scan speed of 2°/min was used in the range of 20–80°. Element analysis of surrounding carbon fibers was observed by Electron Probe Micro-Analyzer (EPMA, JXA-8900RL).

The hardness of the CSF/Al matrix composites was measured in a Vickers hardness tester. The values reported for Vickers hardness represented the average of ten separate measurements taken at randomly selected points using a load of 3 kg for 10 s. Tensile specimens with a thickness of 2 mm and a gauge length of 18 mm (Figure 1b) were cut from the composites in accordance with ASTM test method E8M-11. Tensile tests were carried out using a precision universal testing machine (AG-50KNX STD, SHIMADZU Inc., Kyoto, Japan) with a strain rate of 0.5 mm/min, the sampling time of 1 s at room temperature. The strain gauge (YEFLA-2, t = Tokyo Measuring Instruments Lab) was used for strain analysis. Thermal conductivity of CSF/Al matrix composites was measured by laser flash method thermal constants measuring system (TC-9000H, ULVAC-RIKO Inc., Yokohama, Japan) at the room temperature in air. The specimen size of CF/Al composites was ø 10 × 1.5 h. Thermal conductivity was evaluated according to ASTM E 1461-01.

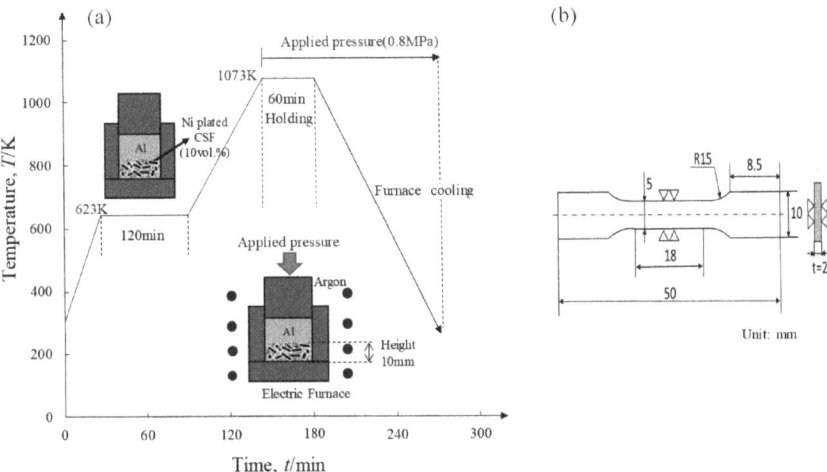

Figure 1. (a) Schematic illustration of the new fabrication process without preform manufacturing by low-pressure infiltration method and (b) shape and dimension of tensile specimens (unit: mm).

3. Results

3.1. Thickness of Ni Plated CSFs by Plating Time

Figure 2 shows SEM images of as-received CSFs and electroless nickel-plated CSFs by 60 to 180 s plating time (pH of 6.5 and temperature of 313 K). As-received CSF presented a clean and smooth surface in Figure 2a. In Figure 2b,c,e, CSFs were not uniformly plated with Ni layer. In Figure 2d, a perfect uniformly Ni-plated layer was obtained over the CSFs. In the case of CSFs plating for 120 s that the Ni particles on fiber surface grew gradually, met each other particles, and formed a uniform metallic layer by the process of "layer formation", which provided uniform wettability, and protection over CSFs. Figure 2f shows that there was no oxygen or other atoms atomic peak which meant the surface of CSF (plating time, 120 s) was wholly even covered by nickel layer. Therefore, Ni plated CSFs (plating time, 120 s) were selected for fabrication composites. The diameters of Ni plated CSFs by different plating times; diameters were calculated by Image-Pro Plus 5.0, which were marked in the images. The volume fraction of electroless plated Ni to CSF, the density of Ni plated CSF and the Ni plated thickness are summarized in Table 1.

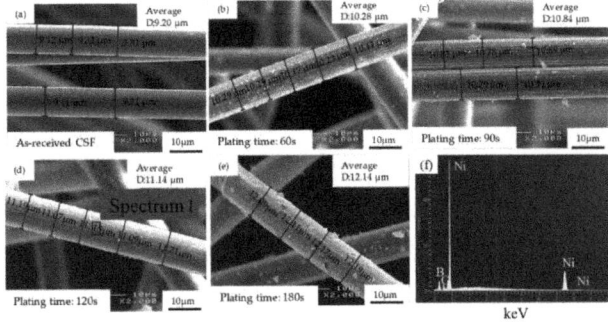

Figure 2. SEM images of (**a**) as-received CSF, (**b**) Ni plated CSF (60 s), (**c**) Ni plated CSF (90 s), (**d**) Ni plated CSF (120 s), (**e**) Ni plated CSF (180 s) and (**f**) EDS data of Ni plated CSF (120 s).

Table 1. Results of Electroless Ni plated CFs with different plating times.

Plating Time (s)	Ni/CSF (Vol.%)	Ni/CSF (Wt.%)	Density of Ni Plated CSF (g/cm^3)	Ni Plated Thickness (µm)
0	—	—	—	—
60	2.8	11.5	2.37	$0.54^{+0.05}_{-0.02}$
90	4.3	17.5	2.47	$0.82^{+0.03}_{-0.13}$
120	5.3	22.5	2.53	$0.97^{+0.11}_{-0.02}$
180	7.9	32.0	2.70	$1.45^{+0.12}_{-0.01}$

3.2. Microstructures of Ni plated CSF Reinforced Aluminum Matrix Composites

Figure 3 shows the microstructures on the surface of the Ni plated CSF/Al matrix composites fabricated under 0.8 MPa. The surface and cross-section of the composites have almost the same structure, so this paper shows only the surface structure of the composites. The volume fraction of carbon fiber is 10 vol.%. In Figure 3a–c, it is observed that the CSFs were randomly distributed in the composites, white phases were formed around CSFs and some were dispersed in the matrix. The white phases were supposed to be Al-Ni compounds that were generated by not only the Ni plated layers dissolving and reacting with the Al matrix, but also the Ni element in the matrix promoted the formation of compounds. Therefore, the content of the Al-Ni compound in Ni plated CSF/A336 alloy composites were much higher than the other two composites due to a higher amount of Ni element of A336 alloy. Si phases were observed in Figure 3b,c due to the existence of Si elements in A356 and A336 alloys. Table 2 shows the image analysis of the volume fraction of CSFs, Al-Ni compounds and Si phases of the Ni plated CSF/Al and Al alloy matrix composites fabricated under 0.8 MPa. Ni plated CSF/1070 and A356 alloy composites exhibited 3.6 vol.% of Al-Ni compounds, Ni plated CSF/A336 alloy composite exhibited 8.4 vol.% of Al-Ni compounds. The amount of Si phases in A356 and A336 alloys composites was about 8.0 vol.% and 15.0 vol.%, respectively, which was caused by 7 mass% and 12 mass% Si content in A356 and A336 alloys.

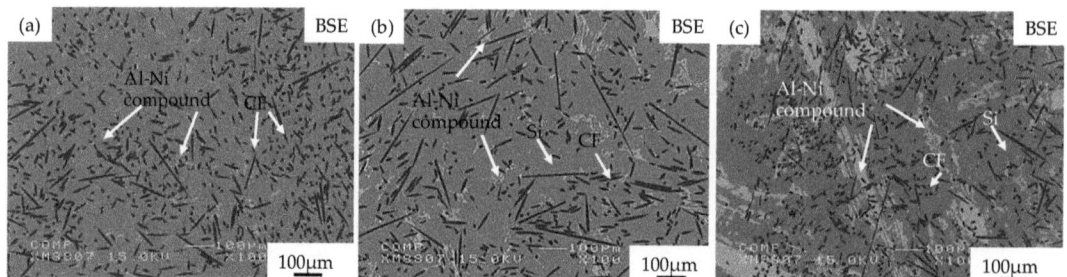

Figure 3. Microstructures of composites: (a) Ni plated CSF/A1070 composite, (b) Ni plated CSF/A356 alloy composite, and (c) Ni plated CSF/A336 alloy composite.

Table 2. Volume fraction of CSFs, Al-Ni compounds and Si phases in Ni plated CSF/Al matrix composites fabricated under 0.8 MPa.

CSF/Al1070 Composite		CSF/A356 Composite			CSF/A336 Composite		
CSFs Vol.%	Al-Ni Compounds Vol.%	CSFs vol.%	Al-Ni Compounds Vol.%	Si Vol.%	CSFs vol.%	Al-Ni Compounds Vol.%	Si Vol.%
13.7	3.6	13.0	3.6	8.0	13.5	8.4	15.2

Figure 4 shows the SEM images and pore size of the Ni plated CSF/Al and Al alloy matrix composites fabricated 0.8 MPa. It can be found that there existed some imperfect infiltration regions (defects) such as pores between the CF/CF and CF/matrix in all composites. It is supposed that the molten Al preferentially infiltrated a largely spaced region of the adjacent CSFs. At the same time, the narrow space between the CSFs was difficult to infiltrate because of the existence of capillary resistance. Meanwhile, the fluidity of metals also had a great effect on infiltration. At the condition of 0.8 MPa, some large-size pores were observed in Figure 4a in the region of CF/CF (2.3 µm) and CF/matrix (1.9 µm); because there is no alloying element in 1070 Al lead to a poor metal fluidity, the CSFs cannot be fully infiltrated result in large pores between the CF/CF and CF/matrix. Figure 4b,c showed an apparent decrease in the size of pores than 1070 composite due to the existence of Si element in Al alloys, which could increase the metal fluidity. Especially, Figure 4c possessed the smallest pore size compared to Figure 4a,b due to the highest amount of Si element (13 mass%) of A336 alloy. Compared with Figure 4a,b, Figure 4c showed an obvious decrease of pore size in the region between the CF/CF and CF/matrix of each composite by applied infiltration pressure of 0.8 MPa.

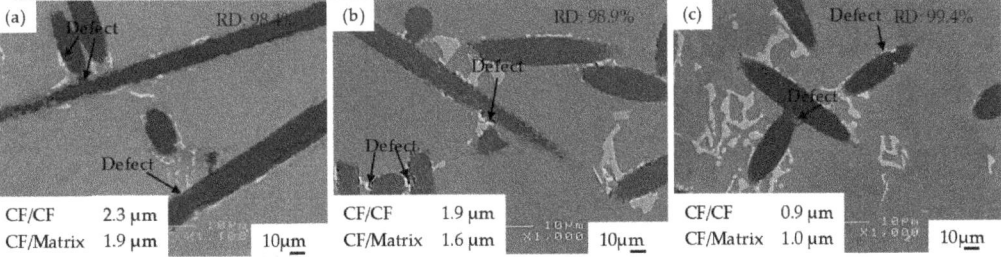

Figure 4. SEM images of pores of composites (0.8 MPa): (**a**) Ni plated CSF/A1070 composite, (**b**) Ni plated CSF/A356 alloy composite, and (**c**) Ni plated CSF/A336 alloy composite.

Element area mapping by EMPA and EDS analysis were conducted to Ni plated CSF/A1070 composite as shown in Figure 5 and Table 3. In Figure 5, Intermetallic compounds (IMC) were identified around CSF by the reaction of Ni plating and aluminum. The EDX analysis data of IMC showed the ratio of Al to Ni was 3:1, which confirmed the Al-Ni compound was Al$_3$Ni phase (Table.3). The Al3Ni phase was generated as the following process: the infiltration temperature was 1073 K, from the Ni plated layer on the surface of CSFs was in contact with molten aluminum and reacted with aluminum directly. It was observed that the Al$_3$Ni phases were separated from the surface of CSFs in Figure 5, in light of the densities of the Ni and Al$_3$Ni, i.e., 8.9 and 4.0 kg/m^3, the formation of Al$_3$Ni was thought to be an expansion reaction and finally, the fine Al$_3$Ni phases were dispersed into the matrix, which was separated from the CSFs surface.

Table 3. Point analysis of Ni plated CSF/A1070 composite by EDS in Figure 5c.

Position	Composition (at%)		Al:Ni	Al-Ni Phase
	Al	Ni		
1	72.45	23.36	03:01	Al$_3$Ni
2	74.86	22.58		

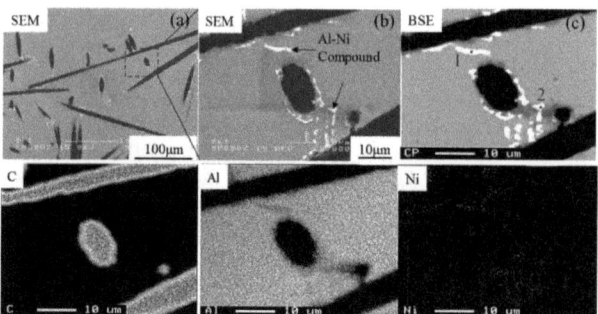

Figure 5. Element mapping analysis of surrounding carbon fiber and Al matrix in Ni plated CSF/A1070 composite.

Figure 6a–c depicts the X-ray diffraction results of electroless Ni plated CSF/Al matrix composites. The detected crystals are only Al, C, Al_3Ni, Mg_2Si, and Si. It is worth noting that no peaks for Al_4C_3 were detected in all electroless Ni plated CSF/Al matrix composites which indicated that the Ni plated layer acted as a barrier layer to effectively prevent the matrix from chemically reacting with the CSFs to cause damage to the strength of the fibers. The peak of Al_3Ni, Mg_2Si, and Si phase in Figure 6c is higher than that of other composites, which meant a higher content of IMCs and precipitate in electroless Ni plated CSF/A336 alloy composite.

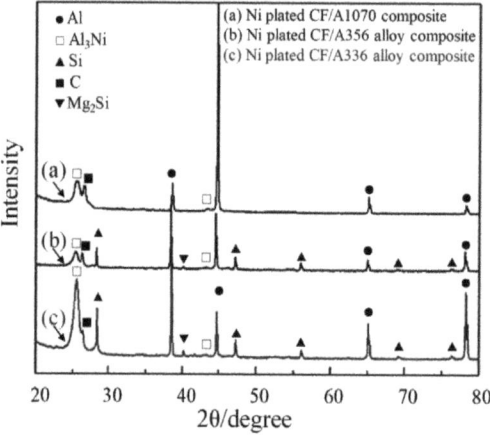

Figure 6. Results of X-ray diffraction spectra of composites: (a) Ni plated CSF/A1070 composite, (b) Ni plated CSF/A356 alloy composite, and (c) Ni plated CSF/A336 alloy composite.

3.3. Hardness and Tensile Strength of Each Matrix and CSF Reinforced Al and Al Alloy Matrix Composites

Figure 7 shows the Vickers hardness of each matrix and Ni plated CSF/Al and Al alloy matrix composites fabricated without preform manufacturing. In addition, the properties of the CSF/Al and Al alloy matrix composites produced by fabricating a preform (SiO_2 binder) used in the conventional composite material were compared. Vickers hardness of Ni plated CSF/A1070, A356 alloy, and A336 alloy composites were 41.3 Hv, 58.8 Hv, and 88.4 Hv, respectively, which presented an improvement of 116.2%, 23.1%, and 31.4% as compared to matrix. The introduction of CSF, generation of Al_3Ni phase, and the solid solution strengthening caused by Si phase precipitation in Al alloys composites contributed to the increase of hardness. A336 alloy possessed the highest content of Si (12 mass%)

and Ni (1.5 mass%) elements, therefore, the highest content of IMCs and precipitate in Ni plated CSF/A336 alloy composite enables it to possess the highest hardness. In addition, the hardness of the Ni plated CSF/Al and Al alloy matrix composite was slightly increased compared to the CSF/Al and Al alloy matrix composite fabricated by preform (SiO$_2$ binder). The measurement of hardness is the global hardness of the composites. In measuring the hardness of the three types of composite, the hardness indentations photograph shows that they were measured including carbon fiber and base material.

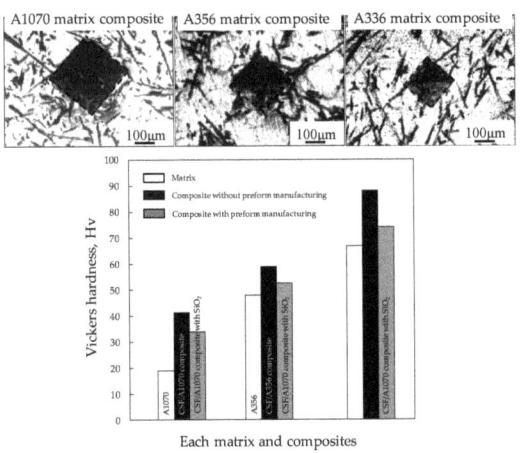

Figure 7. Vickers hardness of Ni plated CSF/Al and Al alloy matrix composites compared to composite of preform manufacturing with SiO$_2$ binder.

Figure 8 shows the tensile properties of each matrix and Ni plated CSF/Al and Al alloy matrix composites fabricated without preform manufacturing. In addition, the properties of the CSF/Al and Al alloy matrix composites produced by fabricating a preform (SiO$_2$ binder) used in the conventional composite material were compared. The 0.2% yield strength and ultimate tensile strength (UTS) increased by changing the matrix from A1070 to A336 alloy, and the CSF/A336 alloy composite possessed the highest UTS than the other two kinds of composites due to the highest content of Si precipitate and Al$_3$Ni compounds. As for the comparison of Ni plated CSF/Al alloys composites and CSF/Al and Al alloy matrix composite fabricated by preform (SiO$_2$ binder), the 0.2% yield strength and UTS of Ni plated CSF/A356 alloy composite was slightly increased compared to the CSF/Al and Al alloy matrix composite fabricated by preform (SiO$_2$ binder). However, there was no trend in elastic modulus or elongation. The manufactured composites without a preform show better tensile strength characteristics than a composite made by the conventional manufacturing method with preform. However, it can be seen that the elongation of the composites is much weaker than that of the general aluminum. This is due to the weak interfacial strength between the CF and aluminum. Figure 9 shows tensile fracture surfaces of each matrix and Ni plated CSF/Al matrix composites: (a), (d) A1070 and composite, (b), (e) A356 alloy and composite, (e), (f) A336 alloy and composite. The fracture surface of A1070 shows the presence of many dimples. However, for the A356 and A336 alloys, the main fracture mechanism was a ductile fracture, while the brittle fracture occurred at the place of Si and IMC, a lot of cleavage stages were observed. The fracture mechanism of Ni plated CSF/A1070 composite was changed to brittle fracture, and lots of large size pores were also observed as shown in Figure 9d. The A1070 matrix possesses no alloying elements with relatively limited tensile properties. The increased UTS of the A1070 composite than A1070 matrix was attributed to the presence of CSFs and the formation of Al$_3$Ni in composites, which could increase the resistance to crack propagation, as a result of enhancement of the tensile property of composite. Figure 9d shows fibers pull out and debonding generated by the high number of pores between the CSF/CSF and CSF/matrix. Therefore, the enhancement effect on the tensile property was weakened accordingly. The fracture mechanism of Ni plated CSF/Al alloys composites were brittle fracture, while cleavage stages were seldom observed by the addition of CSFs as shown in Figure 9e,f. Pores with reduced size were observed compared to A1070 composites. However, lots of fibers debonding and fibers pull-out were observed at the fracture surfaces of CSF/Al alloys composites. Al alloys matrix possess a high content of alloying elements which fulfill their high tensile strength. The pores in composites increased the possibility of minute cracks generation, also the interface of CSFs and Al matrix shows poor bonding characteristics, cracks propagated along with the interface that rarely propagated through the fibers, which inevitably led to the failure of the material with low strength than Al alloys matrix.

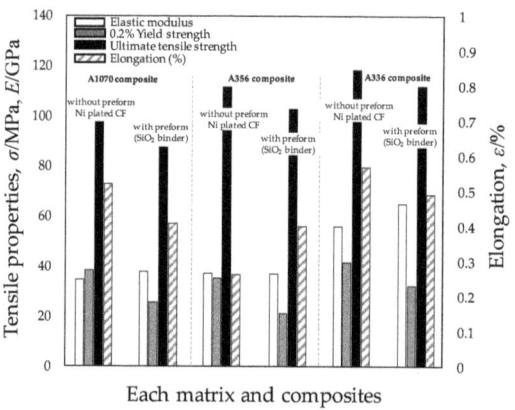

Figure 8. Tensile properties of Ni plated CSF/Al and Al alloy matrix composites compared to composite of preform manufacturing with SiO_2 binder.

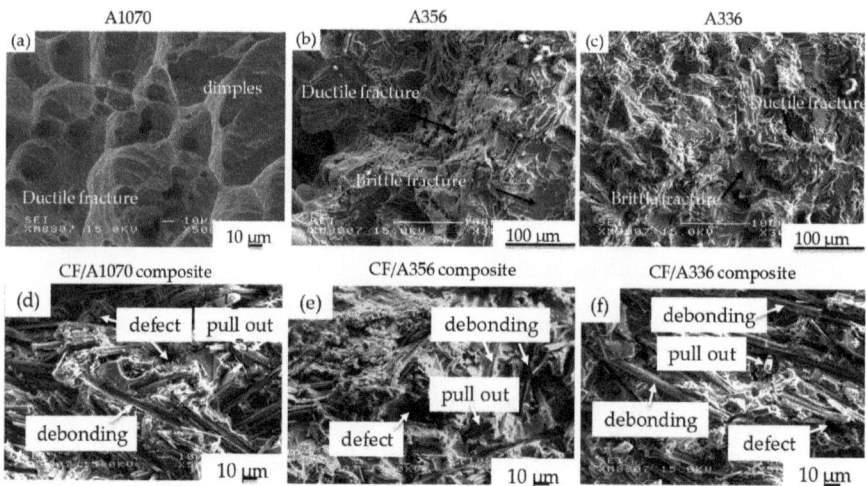

Figure 9. Tensile fracture surfaces of each matrix and Ni plated CSF/Al matrix composites: (**a,d**) A1070 and composite, (**b,e**) A356 alloy and composite, (**c,f**) A336 alloy and composite.

3.4. Thermal Conductivity of CSF Reinforced Al and Al Alloy Matrix Composites

Figure 10 shows the thermal conductivity of Ni plated CSF/Al and Al alloy composites. TC of Ni plated CSF/A1070 composite was 167.1 W·m^{-1}·K^{-1}. However, the TC of Ni plated CSF/A356 alloy and A336 alloy composites were 147.6 and 121. 5 W·m^{-1}·K^{-1} receptivity. Especially, Ni plated CSF/A336 alloy composite presented the lowest TC compared with other composites. Ni plated CSF/A336 alloy composite possessed the highest content of Al$_3$Ni compounds than other composites. The highest amount of Al$_3$Ni compounds was by not only the Ni plated layer but also the high content of Ni (1.5 mass%) in the A336 alloy matrix. The Al$_3$Ni compound with low TC to be approximately 35 W·m^{-1}·K^{-1} acted a role of the thermal resistance to the heat flow and leads to degradation of the thermal properties of composites [14]. In addition, the precipitate was also a factor that weakens the TC of composites. A336 alloy possessed the highest content of Si than A356 alloy, resulting in a higher degree of Si phase precipitation in Ni plated CSF/A336 alloy composite. The higher degree of supersaturation of the Si phase results in severe distortion of the crystal lattice and provides additional interfacial resistance, which would further obstruct effective thermal conduction through the matrix. As a result, the Ni plated CSF/A336 alloy composite exhibited the lowest TC.

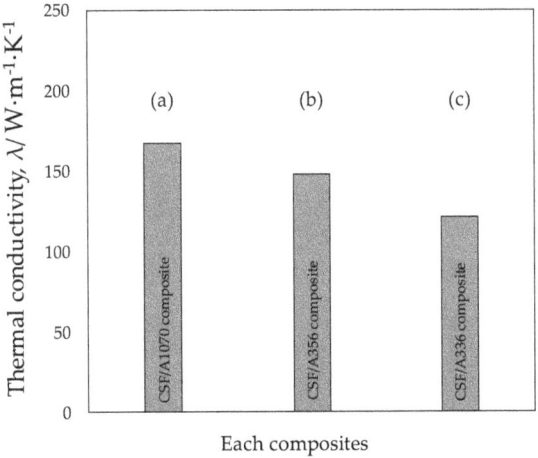

Figure 10. Thermal conductivity of Ni plated CSF/Al matrix composites: (**a**) Ni plated CSF/A1070 composite, (**b**) Ni plated CSF/A356 alloy composite and (**c**) Ni plated CSF/A336 alloy composite.

4. Conclusions

A new fabrication process without preform manufacturing has been developed for CSF/Al and Al alloy matrix composites. And their mechanical and thermal properties were evaluated. Pores of the Ni plated CSF/Al and Al alloy matrix composites fabricated 0.8 MPa. It can be found that there existed some imperfect infiltration regions such as infiltration pores between the CF/CF and CF/matrix in all composites. However, pore size (1 μm) in the region between the CF/CF and CF/matrix to used A336 matrix. Vickers hardness of Ni plated CSF/A1070, A356 alloy, and A336 alloy composites were higher as compared to matrix. Al$_3$Ni by electroless Ni plating and the solid solution strengthening caused by Si phase precipitation in Al alloys contributed to the increase of hardness. A336 alloy possessed the highest content of Si (12 mass%) and Ni (1.5 mass%) elements; therefore, the highest content of IMCs and precipitate in Ni plated CSF/A336 alloy composite enables it to possess the highest hardness. Ultimate tensile strength (UTS) increased by changing the matrix from A1070 to A336 alloy, and the CSF/A336 alloy composite possessed the highest UTS than the other two kinds of composites due to the highest content of Si precipitate and Al$_3$Ni compounds. However, the tensile strength of Ni plated CSF/A336 alloy composite was decreased than the A336 alloy matrix because of the large size of IMCs. Furthermore, the 0.2% yield strength and UTS of Ni plated CSF/A356 alloy composite was slightly increased compared to the CSF/Al and Al alloy matrix composite fabricated by preform (SiO$_2$ binder). The thermal conductivity of Ni plated CSF/A1070 composite has higher value (167.1 W·m^{-1}·K^{-1}) because of small contains the IMCs.

Author Contributions: Y.C. and Z.X. conceived the idea and designed experiments. X.M. carried out the experiments and data analysis. They also contributed to the manuscript writing and Y.C. prepare the response to the reviewer's comments. Z.X. checked the data and discussed it. All authors have read and agreed to the published version of the manuscript.

Funding: This study was supported by JSPS KAKENHI Grant Number 18K03839.

Conflicts of Interest: The authors declare no competing interests.

References

1. Molina, J.; Narciso, J.; Weber, L.; Mortensen, A.; Louis, E. Thermal conductivity of Al-SiC composites with monomodal and bimodal particle size distribution. *Mater. Sci. Eng. A* **2008**, *480*, 483–488. [CrossRef]
2. Euh, K.J.; Kang, S.B. Effect of rolling on the thermon-physical properteis of SiCp/Al composites fabricated by plasma spraying. *Mater. Sci. Eng. A* **2005**, *395*, 47–52. [CrossRef]
3. Ames, W.; Alpas, A.T. Wear mechanisms in hybrid composites of graphite-20 Pct SiC in A 356 alumium alloy. *Metall. Mater. Trans. A* **1995**, *26*, 85–98. [CrossRef]
4. Tjong, S.C.; Lau, K.C.; Wu, S.Q. Wear of al-badsed hybird composites contating BN and SiC particulates. *Metall. Mater. Trans. A* **1999**, *30*, 2551–2555. [CrossRef]
5. Laffont, L.; Monthioux, M.; Serin, V. Plasmon as a tool for in situ evaluation of physical properties for carbon materials. *Carbon* **2002**, *40*, 767–780. [CrossRef]
6. Chang, K.C.; Matsugi, K.; Sasaki, G.; Yanagisawa, O. Infiluence of fiber surface structure on interface reaction between carbon and alumminum. *J. Jpn. Institure Met.* **2005**, *48*, 205–209.
7. Choi, Y.B.; Sasaki, G.; Matsugi, K.; Yanagisawa, O. Mater. Effect of ultrasonic vibration on infiltration of nickel porous preform with molten aluminum alloy. *Mater. Trans.* **2005**, *46*, 2156–2158. [CrossRef]
8. Choi, Y.B.; Sasaki, G.; Matsugi, K.; Sorida, N.; Kondoh, S.; Fujii, T.; Yanagisawa, O. Simulation of infiltration of molten alloy to porous prefrom using low pressure. *Jpn. Soc. Mechanical Eng. A* **2006**, *49*, 20–24.
9. Choi, Y.B.; Matsugi, K.; Sasaki, G.; Arita, K.; Yanagisawa, O. Analysis of manufacturing process for metal fiver reinforced aluminum alloy composite fabricated by low pressure casting. *Mater. Trans.* **2006**, *47*, 1227–1231. [CrossRef]
10. Bakshi, S.R.; Keshri, A.K.; Singh, V.; Seal, S.; Agarwal, A. Inerface in carbon nanotube reinforced aluminum silicon composites: Thermodynamic analysis and experimental verification. *J. Alloys Compouds* **2009**, *481*, 207–213. [CrossRef]
11. Cui, G.F.; Li, N.; Li, D.; Zheng, J.; Wu, Q.L. The physical and electrochemical properties of eletroless deposited nickel-phosphorus black coating. *Surf. Coat. Technol.* **2006**, *200*, 6808–6814. [CrossRef]
12. Ashassi-Sorkhabi, H.; Rafizadeh, S.H. Effect of coating time and heat treatment on structures and corrosion characteristics of electroless Ni-P alloy deposits. *Surf. Coat. Technol.* **2004**, *176*, 318–326. [CrossRef]
13. Meng, X.; Choi, Y.B.; Matsugi, K.; Liu, W. Development of carbon short fiber reinforced al based composite without preform manufacturing. *Mater. Trans.* **2020**, *61*, 1041–1044. [CrossRef]
14. Terada, Y.; Ohkubo, K.; Mohri, T.; Suzuki, T. Thermal conductivity. *Metall. Mater. Trans. A* **2003**, *34*, 3167–3176. [CrossRef]

Article

An Experimental Study of the Cyclic Compression after Impact Behavior of CFRP Composites

Raffael Bogenfeld * and Christopher Gorsky

German Aerospace Center, Institute for Composite Materials and Adaptive Systems, Lilienthalplatz 7, 38108 Braunschweig, Germany; christopher.gorsky@dlr.de
* Correspondence: raffael.bogenfeld@dlr.de

Abstract: The behavior of impact damaged composite laminates under cyclic load is crucial to achieve a damage tolerant design of composite structures. A sufficient residual strength has to be ensured throughout the entire structural service life. In this study, a set of 27 impacted coupon specimens is subjected to quasi-static and cyclic compression load. After long intervals without detectable damage growth, the specimens fail through the sudden lateral propagation of delamination and fiber kink bands within few load cycles. Ultrasonic inspections were used to reveal the damage size after certain cycle intervals. Through continuous dent depth measurements during the cyclic tests, the evolution of the dent visibility was monitored. These measurements revealed a relaxation of the indentation of up to 90% before ultimate failure occurs. Due to the distinct relaxation and the short growth interval before ultimate failure, this study confirms the no-growth design approach as the preferred method to account for the damage tolerance of stiffened, compression-loaded composite laminates.

Keywords: damage tolerance; delamination; damage growth; impact; fatigue; fatigue after impact

Citation: Bogenfeld, R.; Gorsky, C. An Experimental Study of the Cyclic Compression after Impact Behavior of CFRP Composites. *J. Compos. Sci.* **2021**, *5*, 296. https://doi.org/10.3390/jcs5110296

Academic Editors: Jiadeng Zhu, Gouqing Li and Lixing Kang

Received: 14 October 2021
Accepted: 8 November 2021
Published: 10 November 2021

Publisher's Note: MDPI stays neutral with regard to jurisdictional claims in published maps and institutional affiliations.

Copyright: © 2021 by the authors. Licensee MDPI, Basel, Switzerland. This article is an open access article distributed under the terms and conditions of the Creative Commons Attribution (CC BY) license (https://creativecommons.org/licenses/by/4.0/).

1. Introduction

To achieve a damage tolerant composite structure, it is vital to consider the influence of damage on the fatigue behavior of fiber reinforced plastic (FRP) composites. Cyclic loading may propagate the initial damage, reduce the residual strength and decrease the fatigue life of a damaged laminate. However, the influence of the damage differs severely, depending on the initial damage type and the cyclic load case. The critical damage mode under compression load is typically delamination which permits sublaminates to buckle significantly reducing the laminate's capability to sustain load [1–3]. As delamination is located inside the laminate and not visible on the surface, the tolerance of delamination damage is particularly challenging to achieve a damage tolerant design. A typical cause of delamination damage is the low-velocity impact [4,5], as impact threat usually cannot be completely avoided on aerospace structures [6]. Hence, ways and means to design an airworthy structure with the expected damage have to be established.

Damage Tolerance (DT) is the state of the art design philosophy to maintain the structural integrity of an aircraft structure [7]. DT requires taking the expected damage into account during the design. Briefly summarized, there are three admissible concepts which can be applied to approve the DT of aerospace composite structures: slow growth, arrested growth and no-growth [8–10]. These concepts define, that an accidental damage may either grow stably (slow growth) or must not grow at all (no-growth) before it can be discovered and repaired during a scheduled inspection. Arrested growth is the combination of both variants.

Among these admissible DT concept, a no-growth policy currently seems to be the only viable method to cope with delamination [11,12]. Pascoe stated in a recent overview article [13], that even though a slow-growth approach possibly could offer weight benefits, insufficient information of impact damage evolving under fatigue loading prevents an application of slow-growth design approaches.

A damage tolerant design is driven by the load sustaining capability of the damaged structure throughout a design interval. The respective design interval of load cycles is either defined through the scheduled interval of an inspection procedure or throughout the service life of the structural component. The damage size to be considered for each interval depends on the respective visibility of the impact damage. Depending on the inspection procedure, a certain surface dent depth is considered as detectable. The largest damage which cannot be detected within this procedure is assumed to remain in the laminate until an inspection with higher sensitivity. Thus, the design is driven by the load sustaining capability of the largest damage which could possibly remain in the structure over the design interval. The determination of the respective damage size is based on the visibility criterion of the damage. While this criterion is the crack length for metallic structures [14], the metric to assess the damage visibility within a visual inspection procedure is the remaining dent depth on the surface. If the dent depth exceeds a certain threshold, it is considered as detectable. This surface dent accompanies each the impact damage and is caused by resin debris preventing crack closure after the impact [15].

According to the compression after impact (CAI) standards [16–18] the dent depth is measured directly after the impact event. However, for the practical application dent relaxation has to be taken into account as the decrease of the dent depth could result in an underestimation of the damage. In the CMH-17 Vol.3, a relaxation between 30% and 70% is reported [19], depending on the time, the environmental exposure and the mechanical loading. There are hardly any experimental investigations available which quantify the influence of these factors. The figure in the CMH derive from a study by Michèle Thomas from 1994, where the dent depth relaxation due to aging, environmental exposure, and mechanical loading were analyzed in coupon experiments [20]. The dent decrease through mechanical load was found to be up to 75% for 200,000 tension/compression load cycles with a load ratio of $R = -1$ and up to 32% for 200,000 compression/compression load cycles with a load ratio of $R = 10$. For load level under 50% of the static residual strength no relaxation was observed. Thomas proposed the relation in (1) for the maximum expectable relaxation, which considers the cumulative effects of time, aging and mechanical loading. The dent depth d_{relax} is the minimum achievable value through the mentioned relaxation effects. This minimum is assumed as a linear function of the initial dent depth after the impact d_{ini}. Further, Dubinskii et al. [21] studied the dent relaxation in stiffened panels, considering the influence of time and the environmental conditions. In their experimental analysis, a relaxation between 22% and 61% was reported for the investigated conditions.

$$d_{relax} = 0.45 \cdot d_{ini} - 0.1\,\text{mm} \qquad (1)$$

Due to the limited detectability of delamination within the standard inspection procedures and the dent relaxation, the considerable damage for such residual strength approval is comparably large. Only a special detailed inspection method [22] or a structural health monitoring system [23,24] offer possibilities to detect smaller damage which cannot be discovered through visual inspection procedures.

In practical application, the no-growth requirement is realized through strain allowables, which may never be exceeded [25,26]. As outlined by Dienel et al. [27], aircraft manufacturers determine strain allowables "based on extensive experimental campaigns". Such an experimental determination procedure for a DT strain allowable is based on residual strength tests after impact [28,29]. The operational load might further decrease the residual strength, due to fatigue and damage propagation effects. Hence, residual strength testing after cyclic loading is additionally required, as demanded in the AC 20-107B guidelines [8]. The corresponding behavior of impact-damaged composites under cyclic compression load is a topic which was studied over and over in the literature. An overview is presented in the review by Nettles et al. [30]. In 24 considered studies, hardly any delamination growth was observed before ultimate failure. If any growth was detected, it happened at high load levels and close to the ultimate collapse [12,31–33]. Only in few cases, a significant period of delamination growth before the collapse is reported

[11,25,34,35]. In another review article, Molent and Haddad investigate various experimental damage growth studies from the literature. They conclude, that there are sufficient examples in the literature where a slow and stable damage growth behavior was proven to occur systematically. The combination of this knowledge about stable damage growth with a suitable prediction method would enable a damage tolerant design based on the slow growth approach. Nonetheless, as outlined by Pascoe [13], there are still unresolved challenges before a reliable prediction method could be established. These challenges concern the driving phenomena of the damage propagation, the final failure, and even the characterization of a detected impact damage during an inspection. Further, in contrast to metallic materials, the growth of delamination under cyclic loading in a composites comprises barely one order of magnitude in the SERR [1]. Thus, a small uncertainty in the load might severely change how a delamination evolves. These challenges keep the no-growth concept the only currently applicable design concept.

In this work, the authors describe an experimental investigation of the damage tolerance behavior of carbon fiber reinforced plastic (CFRP) coupon specimens under cyclic compression load. The investigation includes the fatigue life, the damage growth behavior, the residual strength properties and the damage visibility including the relaxation. For that purpose, a set of 30 standard CAI specimens is tested in the undamaged and the damaged condition. For the latter case, quasi-static and cyclic compression-compression load are aspects of the inquiry. Through ultrasonic inspection, 3D-scanning, dent measurements and cross section images the DT-relevant metrics throughout the cyclic loading are investigated.

2. Experimental Methods and Procedures

The test campaign for the present study consists of compression after impact experiments on the coupon level. From the set of 30 coupons, three were used to determine the quasistatic, undamaged reference strength. The 27 remaining specimens were impacted according to the CAI standard AITM 1.0010 [17] to achieve a barely visible impact damage (BVID). Afterwards, quasi-static CAI tests determine the residual strength and cyclic compression after impact (CCAI) tests are used for damage propagation analysis. According to the CAI standard, the specimens have in-plane dimensions of 150×100 mm^2 and a nominal thickness of 4 mm which is achieved through the stacking sequence $[-45, 45, 0, 90]_{4s}$ with a total of 32 plies. All specimens were manufactured from an IMA/M21e prepreg material with a nominal ply thickness of 0.127 mm. The corresponding material properties are for example given in the article of Nezhad et al. [36] or Bogenfeld et al. [37].

2.1. Impact Tests

The primary subject of the test campaign is the damage evolution under cyclic compression load. For that aim, the results of the impact experiments are treated as initial conditions for the damage tolerance experiments after the impact. This section briefly describes the impact tests and the obtained results. All impact tests were conducted in a CEAST Fractovis impact drop tower as shown in Figure 1a. The impactor consisted of a hemispherical head with diameter of 16 mm and a total drop weight of 5.95 kg. The boundary conditions of the specimen fixture comply with the AITM standard, where the specimen is clamped on a frame with a 125×75 mm^2 cutout. A crucial part of the test procedure for this study is the measurement of the surface dent depth. Therefore, the impacted specimen is placed in a fixture and the surface is probed through a dial gauge. According to the standard, the maximum dent depth at the impact location is determined. Afterwards, the average dent depth at four reference points 20 mm removed from the impact point is subtracted from the maximum value. The sketch in Figure 1b shows these measurement points. This procedure permits an accuracy of less than 0.01 mm for repeated measuring of the same specimen.

The delamination damage is made visible through ultrasonic scanning. The ultrasonic scan was conducted in an immersion tank with an Olympus V309 ultrasonic transducer with a 5 MHz frequency and a focal length of 50 mm. The resulting D-scan provides detailed information about the delamination per interface and the C-Scan is the basis for the total damage envelope which is further referenced as the projected delamination area [38].

(a) Impact drop tower

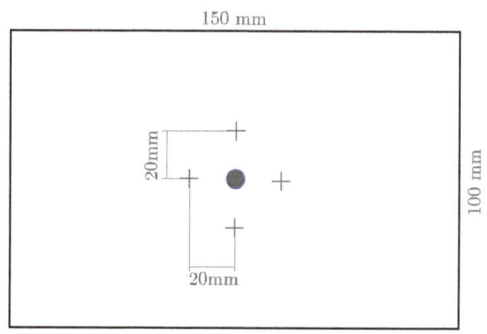

● Indented area at the impact location
+ Reference points for dent measurement

(b) CAI specimen sketch

Figure 1. CEAST Fractovis impact drop tower for the impact tests (a) and a sketch of the CAI standard specimen (b) according to AITM 1.0010 [17].

The impact energy level was determined through preliminary experiments in order to obtain a dent depth of 0.3 mm. This dent threshold marks a DT-critical damage for the service life of an aerospace structure, as the value of 0.3 mm defines the BVID limit for a damage which is likely to remain undetected in a detailed visual inspection [39,40]. An impact energy level of 35 J was found to correspond with the dent depth of 0.3 mm. This energy level was used for all impacts in this study.

All specified impact parameters and the characteristic measurement values of the impact tests are provided in Table 1. While the delamination area shows a scatter of around ±20%, which is typical for delamination damage [41,42]. Also, the dent depth reveals a significant scatter which even increases after the relaxation. This effect can also observed in the measured dent depth by Dubinskii et al. [21]. The other characteristic values which derive from the force history curves, reveals only slight scattering.

A metric of particular interest is the dent depth after the time-dependent relaxation. The value of 0.24 mm is 17% smaller than the value measured directly after the impact. This observation is in accordance with the 20% of dent relaxation reported by Thomas [20] and the relaxation of 22% reported by Dubinskii et al. [21]. Even though in both mentioned studies the relaxation was measured over a much longer time frame, in the present test series no further time-dependent relaxation was detected in measurements conducted after several days. Possible influence factors on the time-dependent relaxation velocity might be the damage size, the material and the stacking sequence of the laminate which we can deduce from the different relaxation behavior in reported in the studies by Dubinskii et al. [21] and Wagih et al. [43].

The Figure 2 depicts the set of curves for all 18 tests with 35 J impact energy. This set of curves confirms a relatively low scatter between the individual tests. Notably, the response curves of three specimens deviate from the set of curves after the peak force is reached. The major force drop which occurs in most cases around [12 kN] did not happen during these impacts. This might indicate that a certain damage, like a fiber fracture in a particular ply or a delamination in a particular interface did not occur in these tests.

Table 1. The characteristic values and the respective standard deviation of the coupon impact tests.

Parameter	Unit	Value
impact energy	[J]	35
impactor mass	[kg]	5.95
impactor diameter	[mm]	16
max. contact force	[kN]	13.85 ± 0.33
dent depth after impact	[mm]	0.29 ± 0.04
dent depth after 24 relaxation	[mm]	0.24 ± 0.07
proj. delamination	[mm^2]	889 ± 199
contact duration	[ms]	4.57 ± 0.07
energy absorption	[J]	17.0 ± 1.1

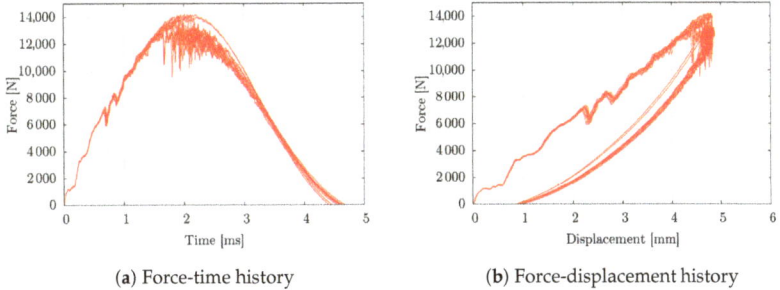

(a) Force-time history   (b) Force-displacement history

Figure 2. Set of force-time and force-displacement history curves of the 35 J CAI impacts.

The dent depth is the typical metric to assess the severity of a damage with regard to the residual strength reduction. However, this reduction is not driven by the dent but by the corresponding delamination area. Thus, the correlation of the projected delamination area and the dent depth shall be further investigated. This correlation is depicted in Figure 3a for the projected delamination area and the dent depth directly after the impact. The Figure 3b shows the correlation after a relaxation period of 24 h (Figure 3b). The damage area is measured through the area of the projected delamination in the ultrasonic C-scan.

Commonly, the impact energy correlates positively with the dent depth and this in turn with the delamination area [44,45]. However, for this coupon test series on the same impact energy level, the values of the dent depth scatter (cf. Figure 3). There is no significant correlation for the dent depth and delamination area on the same energy level. The dent depth after relaxation appears to be slightly negatively correlated (cf. Figure 3b), however the data does not allow the generalization of this observation, particularly due to the fact that the number data points with a same dent depth do not offer sufficient statistical certainty.

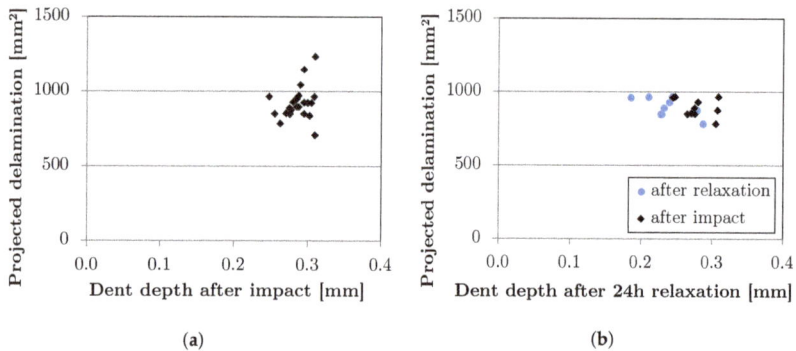

Figure 3. Data points of the measured dent depths and the projected delamination area of the 35 J impact experiments Not all specimens were subjected to a dent measurement after relaxation. Hence, not each data point in the diagram Figure 3a corresponds to a data point in the diagram Figure 3b. (a) Measurement after the impact, (b) Measurement after 24 h relaxation.

2.2. Compression after Impact Tests

The compression after impact tests were conducted on a 400 kN hydraulic testing machine equipped with compression plates. Figure 4 shows the experimental setup, where the CAI specimen fixture is placed in the test machine. To prevent lateral movement of the fixture, limit stops were mounted on the lower compression plate. The CAI specimen fixture prevents the global buckling of the specimen through vertical anti-buckling rails which are located around 10 mm from the specimen edges.

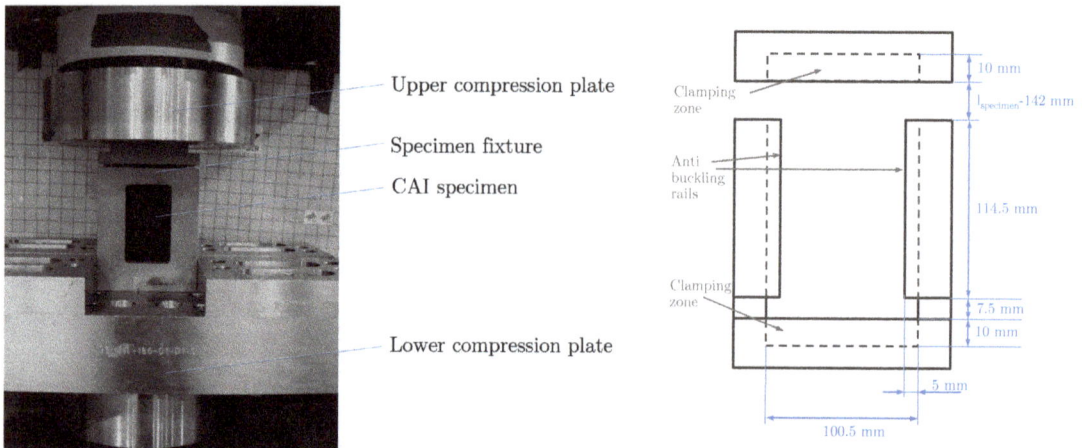

(a) Setup in the hydraulic test machine

(b) CAI specimen fixture

Figure 4. Test setup (a) and a sketch of the fixture (b) for the compression after impact tests.

A series of quasi-static compression tests with impacted and virgin specimens was conducted in order to determine the reference strengths. The average residual strength value determined in the quasi-static CAI tests defines the load level of 100% which causes instant failure of the specimen. For the cyclic experiments, the load level has to be reduced. Hence, the load levels for the tests were chosen relative to the quasi-static residual strength. High load levels are particularly interesting, inasmuch as these permit the highest probability of damage growth according to various studies [11,30]. The typical strain limit of a composite structure with a no-growth design [8] is commonly quantified between 2000 µε and 3000 µε as stated, for example, by Clark [25], Baker [46], or Riccio [3]. Baker even claims that no damage growth might occur for loads at as high as 80% of the limit strain. Hence, the test campaign in this study began with the load level of 90%. The further load levels were chosen during the test campaign, based on the observed behavior during the previous tests. The Table 2 gives an overview about all load level applied for the CCAI tests.

To conduct a cyclic test, additional load parameters have to be defined. For a given absolute maximum force—which is defined by the load level—the absolute minimum force of the cyclic load is defined by the load ratio. This load ratio was chosen as $R = 10$. This choice safely avoids the loosening of the contact between the compression plate of the testing machine and the specimen fixture. A permanent contact is necessary to prevent rattling. Hence, the absolute load minimum of the sinusoidal amplitude is exactly ten percent of the defined maximum. Furthermore, the control loop of the cyclic test has to be considered. It is not possible to keep the applied limit values for force and displacement constant throughout the load cycles because the specimen stiffness decreases under fatigue load. Constant displacement parameters are the safe approach to prevent sudden specimen failure. However, the stiffness reduction is likely to decrease the load to a level where delamination is unlikely to grow. In contrast to that, constant force parameters forward a progressive damage behavior. Hence, constant force parameters were chosen for this study, in order to forward the damage propagation.

A loading frequency of 3 Hz was experimentally determined as the maximum tolerable value for experiments on a load level of 90%. The limiting factor is the temperature of the specimen which rises due to internal friction during the cyclic loading. At a frequency of 3 Hz, the temperature increase measure on the surface of the specimen was less than 1 K. For long intervals of 50,000 and more cycles on the load levels 50% and 30% the frequency was increased to 5 Hz. The respective temperature increase on the specimen surface during the test was less than 1 K.

The detection and the measurement of the delamination growth and the development of the dent depth are vital to the objectives of the investigation. The respective information has to be obtained after a certain number of load cycles. For that purpose, the cyclic test was conducted in intervals. After each interval, the test was interrupted for ultrasonic testing and/or dent measurement. The interval length n_i was either a predefined value or arose from a required ultrasonic examination after acoustic evidence of damage propagation. Hence, the number of cycles in an interval varies for each cycle interval and for each specimen.

Table 2. Load levels for the CCAI tests and the distribution of the specimens.

Load Level	F_{max}	F_{min}	F_{amp}	ε_{max} (Nominal)	Tests Total [1]
100%	105.3 kN	10.5 kN	47.4 kN	6440 µε	5
90%	94.7 kN	9.5 kN	42.6 kN	5750 µε	5
80%	84.2 kN	8.4 kN	37.9 kN	5120 µε	5
75%	78.9 kN	7.9 kN	35.5 kN	4780 µε	4
70%	73.7 kN	7.4 kN	33.2 kN	4440 µε	5
50%	52.8 kN	5.3 kN	23.8 kN	3220 µε	2
30%	31.7 kN	3.2 kN	14.3 kN	1930 µε	1

[1] Not all specimens were subjected to cyclic load until failure. After a certain amount of cycles, some specimens were used for quasistatic residual strength tests, cyclic test with increased load or for cross section images.

3. Results and Discussion

3.1. Endurance and Residual Strength

A primary result of a CCAI experiment is the endurance of each specimen. The combined results from each load level result in the respective endurance limits, which permits us to derive an stress-life curve (S-N curve) as shown in Figure 5. The strong endurance increase for smaller load levels leads to a particularly flat S-N curve, which appears to be a typical phenomenon for compression/compression fatigue and was also observed by Post et al. [47]. Up to 300,000 load cycles were conducted. The specimens on the 50% and 30% load level as well as one specimen on the 75% load level did not fail after 300,000 cycles and were therefore excluded from the S-N-curve. For the specimen on the 75% load level a possible explanation can be found in the impact damage. The respective specimen corresponds to one of the impact history curves without the major force drop (cf. Section 2.1). This is due to fewer fiber breaks that occurred on the surface and as a consequence the endurance of the specimen increases.

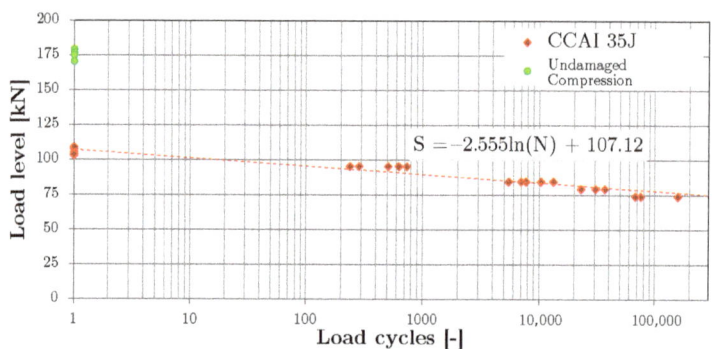

Figure 5. S-N-curve of the CCAI experiments with 35 J impact damage.

An issue of particular interest is the relation of the S-N curves with and without impact damage. However, the CAI specimens were found to be unsuitable for the cyclic test of pristine specimens. Friction between the specimen and the fixture initiated delamination at the loading edge. An entirely different failure mode occurred which does not permit to assess the fatigue behavior of the laminate. To overcome this deficiency, the fatigue behavior of the undamaged laminate could be determined on a standard compression specimen with tabs. Additional tests using standard compression specimens would have to be conducted to obtain the virgin S-N curve. As shown in the experimental work of Uda et al. [33], a linear superposition of the fatigue knock-down and the impact knock-down is a conservative estimate of the combined fatigue after impact knock-down. The advantage of this procedure is the reduction of the test effort. With a known S-N curve for the undamaged laminate, conservative estimates for the S-N curve corresponding to each each damage size could be obtained through the determination of the quasi-static residual strength.

Further, the influence of the cyclic loading on the residual strength was investigated. The force-strain curves for all quasi-static compression experiments are depicted in Figure 6. The first diagram in Figure 6a compares the compression tests with the CAI tests. The curves reveal a 1.0% stiffness decrease by the impact damage. Due to negligible scatter of the stiffness, this small value is statistically significant. The strength decrease is severe and constitutes 40%. Hence, the impact damage affects the strength more severely than the stiffness while the failure strain reduces proportionally to the strength. An observation which confirms that a sudden event—likely the sublaminate buckling—initiates the ultimate failure. The qualitative failure mode differs between the damaged and the undamaged specimens. While impact damaged specimens have their weakest zone where the damage is located, a virgin specimen has its weakest zone in the unsupported length where the anti-buckling rails end (due to the geometry of the CAI fixture, cf. AITM 1.0010 [17]). The failure emerges in this unsupported length. Due to this failure at a different weak point, the true undamaged reference strength of the CAI specimen is higher than this measured value.

The diagram in Figure 6b compares the CAI tests of the impacted specimens with those that were additionally subjected to cyclic loading. A significant stiffness decrease due to the cyclic load could not be measured in these tests. The experiments also didn't reveal a significant strength decrease. The ultimate failure load of most CCAI specimens is in the range of scatter of the CAI specimens. A similar behavior was observed for the cyclic tension after impact behavior in an earlier study [41]. Remarkably, there are two exceptions which revealed a higher failure load than the reference CAI tests. Whether these values shall be considered as a consequence of a natural scatter or whether there is indeed a cause for this is not clear. One possible explanation are the initial differences in the delamination size and shape before cyclic loading. In this case an increased sample size for the CAI specimens should reveal if the two specimens with higher values for the residual strength are consequence of the initial differences. Another possible explanation is that ply-debris shifted by cyclic loading into the interfaces might in certain cases impede sublaminate-buckling of the open delaminations and therefore slightly increase the residual strength.

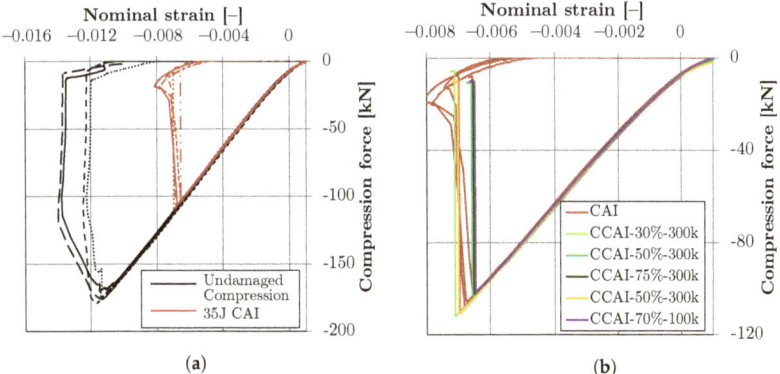

Figure 6. Force-strain history of the compression tests with the virgin specimens, the impacted CAI specimens, and the CCAI specimens with a cyclic load history. (**a**) Compression and CAI specimens. (**b**) CAI and CCAI specimens.

3.2. Damage Propagation

The information about the delamination growth is key to possibly establish a slow-growth approach in the damage tolerant design [13]. For that purpose, the damage was inspected and measured after certain intervals—either after a predefined interval of cycles or after acoustic evidence for damage propagation. After an interval, the specimen was removed from the testing machine for an ultrasonic inspection. However, the ultrasonic scanning revealed an additionally difficulty when assessing the damage size after cyclic loading. The scans exhibit an apparent decrease of the delamination area as depicted in Figure 7. A similar effect was reported by Feng et al. [48] in their study on the structural level.

This decrease could be observed in both scanning methods, the through-transmission scan (C-scan) and the time of flight evaluation of the pulse-echo scan (D-Scan). As an actual decrease of the damage is not a plausible explanation, this observation is probably due to a restrictions of both measurement methods: The C-Scan evaluates the echo from the bottom of the immersion tank. The amplitude of the sound wave which passed twice through the laminate is measured. A reduced delamination size in the C-Scan implies, a stronger echo signal. Hence, the damping of the laminate in the damaged region became less than directly after the impact. In the D-Scan, the echo produced by each delamination is evaluated. If a certain threshold value is exceeded, the respective location is considered as damaged. Thus, a decrease of the delamination area in the D-Scan has to be caused by a smaller echo from the delaminations themselves. In Section 3.4, we explain this effect through the mechanics of the cracked laminate.

For the damage evolution investigation in this campaign, the decrease of the delamination area has to be quantified. According to the conducted measurements, a maximum decrease of 15% and an average decrease of 9 ± 3% were identified. Depending on the load level, the minimum measured delamination area was found after less than one hundred cycles and or even after nearly 100,000 cycles, as the damage growth curves in Figure 8 show.

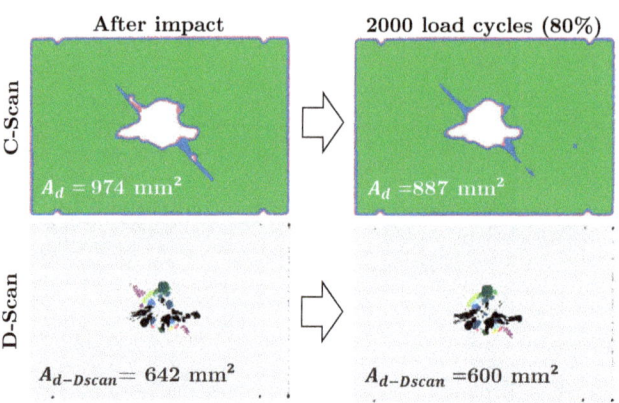

Figure 7. Example for the apparent shrinking of the delamination area A_d after 2000 load cycles at a load level of 80%.

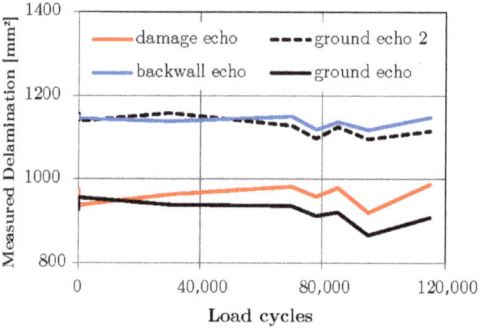

Figure 8. Evaluation of the different ultrasonic results for the sample specimen (70% load).

The quantitative delamination history for three specimens on each load level is shown in the four diagrams in Figure 9 (on the load levels 50% and 30%) no damage growth occurred). Further, the Figures A1–A3 depicts the D-scan images of all inspected specimens.

In accordance with the most results in the literature (cf. Nettles 2011 [30]), there was hardly any damage growth observed before the ultimate failure. The damage growth period was in most cases less than 1% of the total load cycles before failure. In many cases the damage growth could be perceived acoustically before ultimate failure. This indication was used to conduct an ultrasonic scan. In some cases, the test was afterwards continued in small cycling intervals. This procedure allowed to observe the CCAI failure as detailed as possible (specimens 13, 14, and 17 in Figure A1 and specimen 18 in Figure A2). These results show, that delamination begins to grow perpendicularly to the loading direction in the interfaces close to the specimen's back surface.

It is clearly visible that a considerable damage growth does not occur until a few cycles before the final specimen failure. On all four load levels, the specimens failed suddenly. In some cases, it was not possible to conduct an ultrasonic scan closely to the ultimate failure. The earliest detected damage growth is visible on specimen 21 in Figure A2. After 60,000 cycles—80% of the life until failure—a small increase of a delamination close to the back face was detected.

These observations do not provide any evidence for a possible slow-growth approach (in contrast to the tension after impact behavior [41]). Instead, the need for a no-growth design to account for the DT under compression load is confirmed.

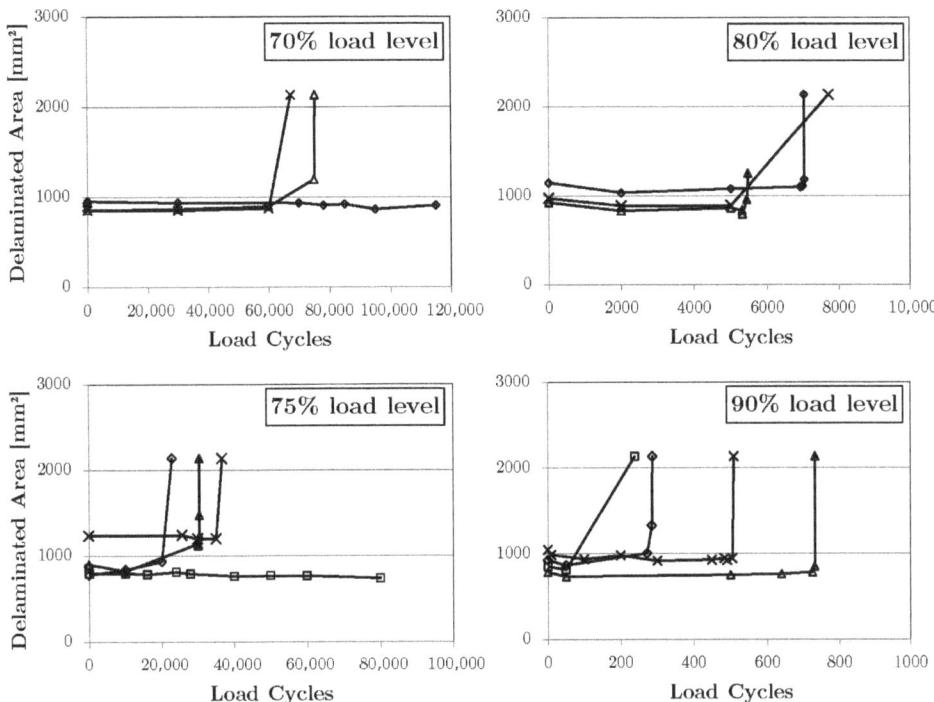

Figure 9. Delamination history of the specimens per load level obtained from the ultrasonic scanning between the load interval. The last data point of each curve represents the final failure.

3.3. Damage Visibility

The metric to assess the damage visibility in a damage tolerant design is the dent depth at the impact location. As shown by Thomas [20] or Dubinskii [21], significant relaxation of the dent depth after the impact has to be expected. The relevant dent depth to assess the visibility is the value after the relaxation, as stated in the Composite Materials Handbook [49]. Hence, the actual value of interest is the relaxed dent depth d_{relax}.

The diagram in Figure 10 depicts the measured relaxation throughout the cyclic tests. The curves approve a significant relaxation which increases with the load level. Both the relaxation velocity and the maximum achievable dent relaxation depend on the load level. While the loading with 90% and 80% caused the dent depth to reach nearly zero, the relaxation velocity of other specimens slowed down significantly after a certain limit value was reached. However, cycles with an increased load level, confirmed that this limit could be overcome and the relaxation continued. In contrast to the claim of Thomas [20], a lower limit for a load level where no load-driven relaxation occurs, was not found in this study. Nonetheless, in the experiments of Feng et al. [48] a relaxation of approx. 10% was reported on a load level of 40%.

According to the Equation (1) from Thomas [20], the average initial dent depth of 0.29 mm would result in a value of 0.03 mm after the load-driven relaxation. This value agrees with the observation in our study, that high load levels caused the dent to relax entirely.

The inversion of the equation Equation (1) suggests an initial dent depth of 0.9 mm to result in the limit value of 0.3 mm after the relaxation. However, the impact test on the CAI standard specimen does not permit an appropriate investigation of the respective damage size in certain cases. In the preliminary impact study to define the energy level for the test campaign, energies up to 50 J were tested. The damage width corresponding to the 50 J impact comprised already the entire supported specimen width, while the initial dent depth was still only 0.7 mm. Further increasing of the impact energy to achieve a larger dent is not possible, as the damage size would be limited by the supported width of 75 mm. Hence, the suitability of the CAI standard impact to assess the DT relevant BVID has to be questioned for the investigated laminate.

Further, the DT relevant damage size, which is associated with the visibility threshold of 0.3 mm, is in fact higher than the dent depth after relaxation apparently indicates. For the laminate under investigation, the respective BVID size would be at least 1400 mm^2 (result of the 50 J impact test with an initial dent depth of 0.7 mm). As the BVID is the smallest damage which can be detected in a detailed visual inspection, this large damage size indicates the need for the detection of smaller damage within the inspection. Improved damage detection methods beyond visual inspection have to be considered. Either through special detailed inspection procedures [22] or through structural health monitoring (cf. guided waves [23,50,51] or acoustic emission [52]) are possible options to detect damage independently from the dent depth.

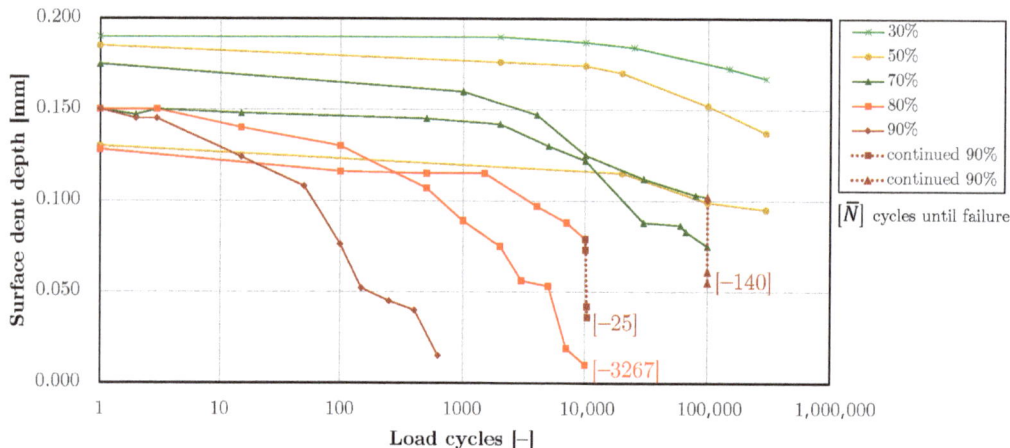

Figure 10. Relaxation of the dent depth at the impact spot depending on the cyclic load. $\bar{N}$ represents the number of cycles until ultimate failure under cyclic load for the specimens concerned.

3.4. Explanation of the Dent Relaxation

The causes of the dent relaxation and the seemingly decreasing delamination area were further investigated through polished cross sections of the impacted specimens. Two sample specimens were cut to gain insight on the damage mechanisms caused by the cyclic load. The first specimen, depicted in Figure 11a, was cut directly after the impact. The visible damage in this picture reveals open delaminations and inter-fiber fracture. This effect is already known by the work of Bouvet et al., who found that small resin debris prevents delamination cracks from closing [15].

The second image in Figure 11b corresponds to a specimen which was subjected to cyclic load which lead to a dent relaxation from 0.15 mm to 0.015 mm. Apparently, the cyclic load affects the crack pattern in the laminate. Through the continuous loading and unloading, the ply debris shifts enabling closure of previously blocked cracks and the cracked ply realigns. This further relaxation has direct influence on the surface dent depth as, according to Bouvet et al., the blocking of the crack closure is a primary cause of the surface dent accompanying an impact damage.

Also the seemingly reduced delamination area in the ultrasonic scan can be explained through this realignment. When a delamination crack entirely closes, the adjoining faces lie against each other. This direct contact of the separated plies reduces the damping of the ultrasonic sound wave in the through-transmission scan. In the same way, the delamination face produces a lower echo signal which is evaluated for the D-Scan. Thus, the measured delamination area decreases without an actual decrease of the damage.

(a) Directly after the impact   (b) After relaxation through 627 load cycles at 90%

Figure 11. Polished cross sections of 35 J impact damage of one specimen cut after the impact (**a**) and a second specimen cut after relaxation due to fatigue loading (**b**).

For the further investigation of the relaxation phenomena, the geometry of selected specimens was measured through 3D scanning by an ATOS digital image correlation test equipment (https://www.gom.com/en/products/3d-scanning/atos-compact-scan, accessed on 9 November 2021). The results of these measurement can be seen in the Figure 12. Two specimens were scanned directly after the impact, a third specimen was scanned after 10,000 cycles at a load level of 80%. Obviously, the deformation at the impact location is dominated by a significant dent (blue) on the impact side. This dent vanished entirely, in case of the third specimen which was subjected to cyclic loading. The deformed state on the back side is driven by a different phenomenon. A fiber bundle sticking out due to small fiber fracture combined with the ply splitting which occurs on the back side. This deformation on the back side is qualitatively similar for the specimens before and after the load-driven relaxation. In agreement with this discovery, no relaxation on the back side could be measured. The stick out height remained constant over the time and cyclic load. Eventually, the scans reveal a slight imperfection of the specimens, pertaining the deformation mode of the impact load. This deformation can be observed before and after the cyclic load. From these observations, it is concluded that the relaxation primarily affects the local dent at the impact location.

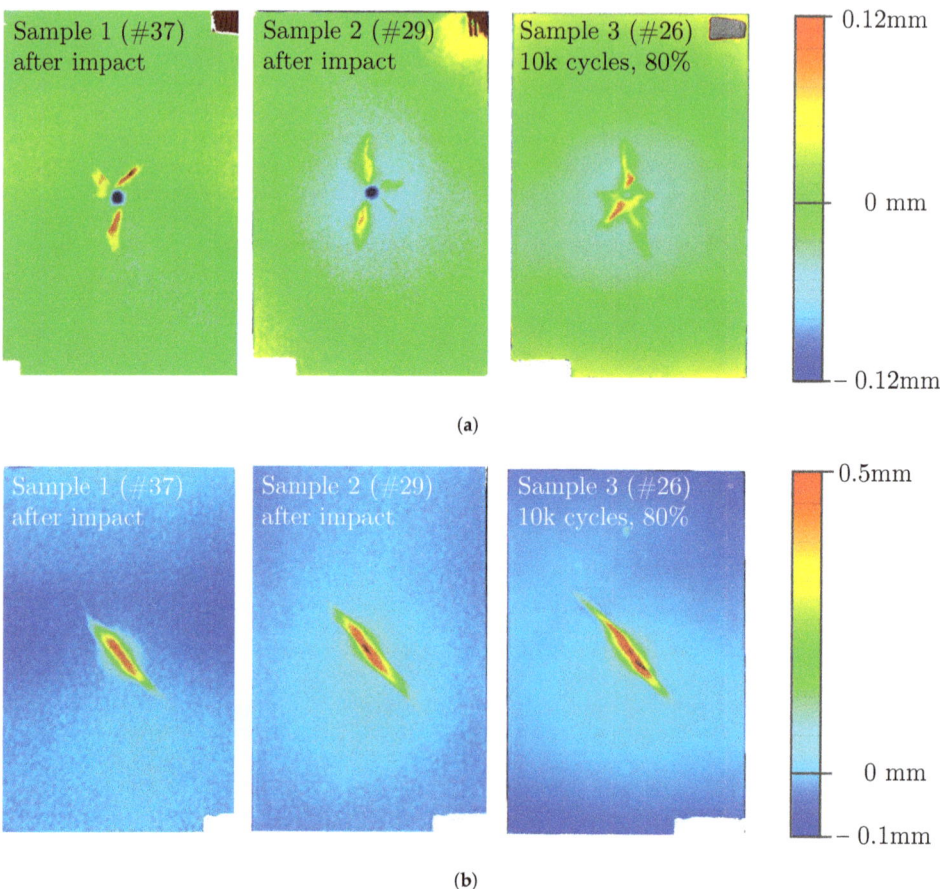

Figure 12. Surface geometry measured through 3D scanning directly after the impact (two specimens) and one specimen after the relaxation over 10k load cycles (one specimen). (**a**) Scans of the impact side, (**b**) Scans of the back side.

3.5. Damage Mode of the Ultimate Failure

To assess the influence of the cyclic load on the ultimate failure under compression, the observed damage pattern after a quasi-static test is compared with the damage pattern observed after failure through cyclic loading. The 35 J impact damage analyzed in the presented test campaign includes significant compression-driven fiber failure on the impact side. The impact back side shows hardly any fiber failure but ply splitting through inter-fiber failure. The typical damage pattern of the compression after impact experiments emerges through a fiber kink band which results from the propagation of the preexisting fiber cracks in the upper plies. As the images of a specimen after the quasi-static CAI test shows in Figure A4a, the kink band causes an actual crack in the uppermost ply over the entire specimen width. In contrast to that, no visible fiber cracking emerges on the backside of the same specimen. Instead, the lower plies buckle and sublaminates separate from the other plies. This crack pattern, which is referred as type I failure in the following, was observed on all five specimens subjected to quasi-static CAI tests.

The failure pattern of the specimen that were subjected to cyclic load was found to differ from the above-described failure type I. The Figure A4b,c show two specimens which failed under cyclic load on the load levels 70% and 90%, respectively. Instead of a fiber kink band with broken fibers, the uppermost ply buckles with the fibers remaining intact for the most part. This pattern is referred as failure type II. In some cases, broken fibers on the impact side were still observed, for instance on the left side on of the specimen in Figure A4c, which we interpret as a mixture both failure types. Also, on the coupons' back side differences between the quasi-static failure and the cyclic failure were found. Instead of the ply buckling with intact fibers, the cracking of fibers in the lowermost ply can be observed. In Figure A4c such cracking comprises the entire specimen width, in case of the specimen depicted in Figure A4b broken fibers emerge only in the edge regions. Failure of type II was not observed on any specimen after quasi-static failure in the CAI test. Nonetheless, in case of ultimate failure under cyclic load different types emerged. Finally, the residual strength tests with specimens previously subjected to cyclic loading revealed a behavior which resembles the CAI failure. The Table 3 provides an overview about the number of specimens for each test type and the respective failure type.

There are various possible reasons for the observed differences in the failure modes. First, the above-described dent relaxation under cyclic load affects the specimen geometry at the impact location (cf. Section 3.4). After the impact, the centric dent on the specimen (cf. Figure 12a) represents a geometric imperfection. With this dent relaxing through the cyclic loading, the imperfection gets reduced. Second, the cyclic compression causes fatigue effects to the entire laminate. Micro cracking in the matrix material can initiate independently from the impact damage [53,54]. Such cracking weakens the matrix and could facilitate the occurring of fiber kinking, as this failure mode depends on the capability of the matrix to stabilize the fibers [55]. This reduced stabilization could explain the cracked fibers on the back side under cyclic load. Third, it is possible that the relocation of the ply-debris out of blocked cracks into empty breakout-spots and the resulting realignment of the cracked plies might impede sublaminate buckling and facilitate fiber kinking resulting in a shift of failure modes. Eventually, the combination of all the phenomena can be the cause of the different failure patterns.

Table 3. Frequency of occurrence of the observed failure modes for specimen broken under quasi-static load (CAI), cyclic load (CCAI), and quasi-static load after cyclic loading (In total, 26 specimens were tested until failure. The missing 27th specimen in Table 2 was used for cross section images before the final failure (cf. Figure 11b).

Failure Type	CAI Failure	CCAI Failure	Quasi-Static Failure after CCAI
Type I	5/5	4/16	4/5
Type II	-	10/16	1/5
Between type I and II	-	2/16	-

4. Conclusions and Outlook

In this experimental investigation was studied how impact damage in CFRP coupons specimens evolves under cyclic compression load. The impact-damaged specimens were impacted and subjected to different cyclic load levels between 30 percent and 90 percent with a maximum of 300,000 cycles. Through ultrasonic inspection, 3D scanning, and dent measurement various effects were revealed concerning the damage propagation, the dent relaxation, and the influence to the residual strength:

- The most part of the specimen life under cyclic load the damage remains unchanged.
- Shortly before the final failure, the impact damage begins to propagate perpendicularly to the loading direction. In particular, delamination growth was found at the interfaces close to the back side.
- The dent at the impact location relaxes significantly under cyclic loading. Relaxation was observed on all load levels and can lead to the complete disappearance of the dent.
- The ultimate failure under cyclic loading reveals a different failure mode than the ultimate failure under quasi-static load. Under cyclic loading, fiber breakage is observed mainly on the back side while ply buckling occurs on the impact side.

The results confirm the known difficulties about the evolution of impact damage under cyclic compression loading, which require a damage tolerant design through the no-growth approach. Additionally, the dent relaxation decreases the damage detectability through visual inspection methods which indicates that improved detection methods can significantly reduce the DT relevant damage size.

Further, the investigation revealed effects of scatter concerning the impact damage and the resulting residual properties. To achieve a further understanding of this scattering, the occurring damage modes, their causes, and their correlation with particular residual properties have to be further investigated.

Author Contributions: Conceptualization, R.B.; Data curation, R.B.; Formal analysis, R.B. and C.G.; Investigation, R.B. and C.G.; Methodology, R.B.; Visualization, R.B.; Writing—original draft, R.B. and C.G.; Writing—review & editing, R.B. and C.G. All authors have read and agreed to the published version of the manuscript.

Funding: This research was funded through the DLR projects KonTeKst and oLAF.

Institutional Review Board Statement: Not applicable.

Informed Consent Statement: Not applicable.

Data Availability Statement: The data generated during the current study is currently not publicly available due to available from the corresponding author on reasonable request.

Acknowledgments: All research was accomplished within the DLR projects KonTeKst and oLAF. The authors acknowledge the DLR for funding these projects.

Conflicts of Interest: The authors declare no conflict of interest.

Appendix A. Result Figures of the Cyclic Coupon Tests

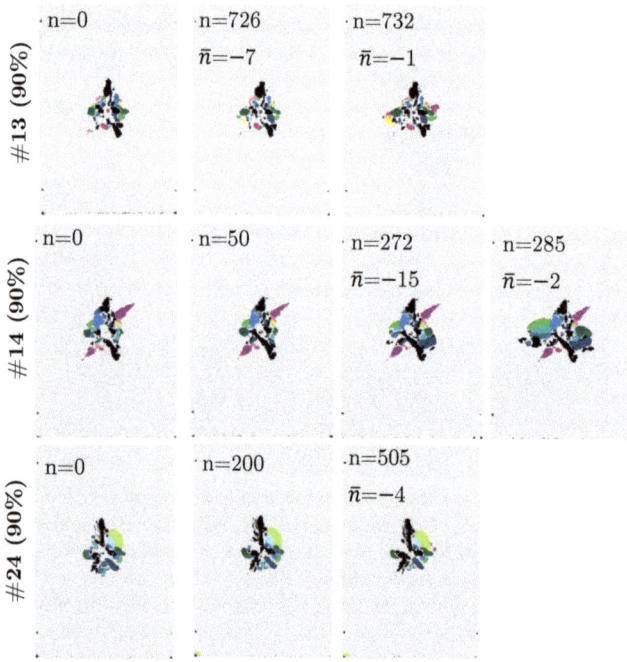

Figure A1. *Cont.*

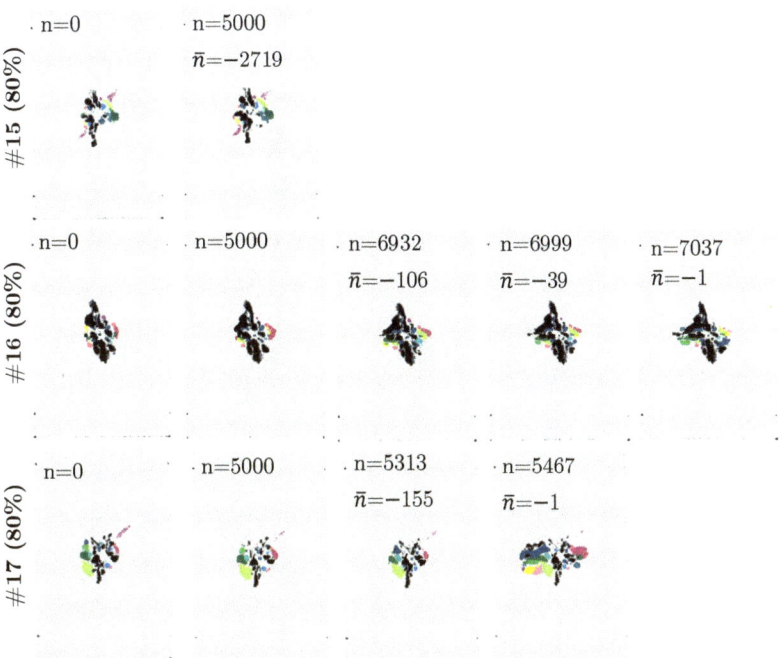

Figure A1. Observation of the delamination growth in the ultrasonic D-scan of selected specimens on the load levels 90% and 80%.

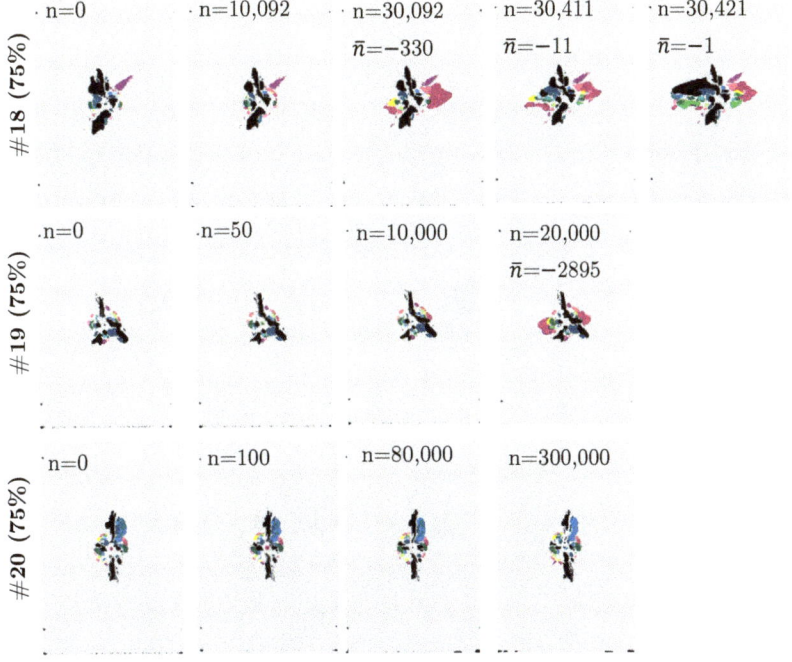

Figure A2. *Cont.*

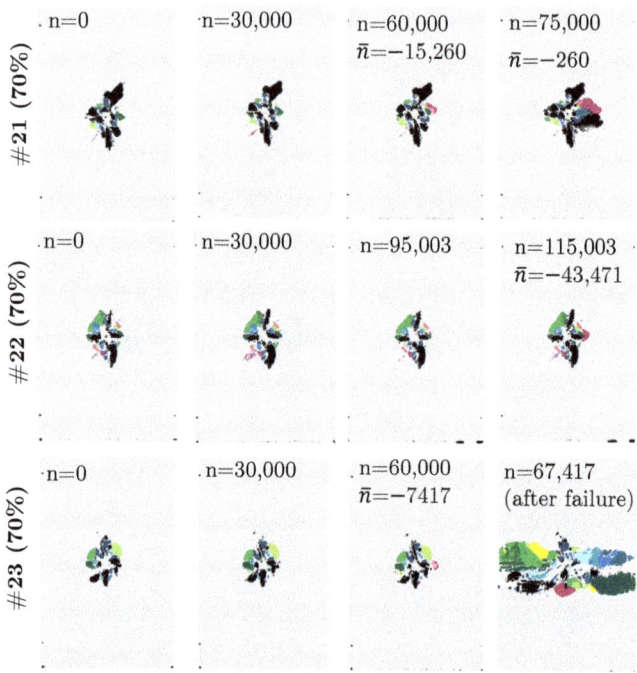

Figure A2. Observation of the delamination growth in the ultrasonic D-scan of selected specimens on the load levels 75% and 70%.

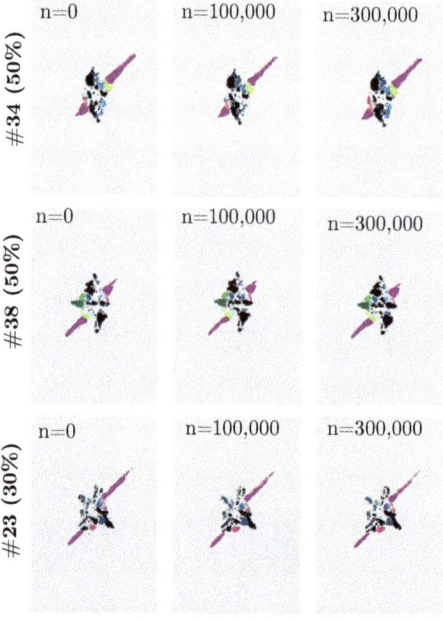

Figure A3. Observation of the delamination growth in the ultrasonic D-scan of selected specimens on the load levels 50% and 30%.

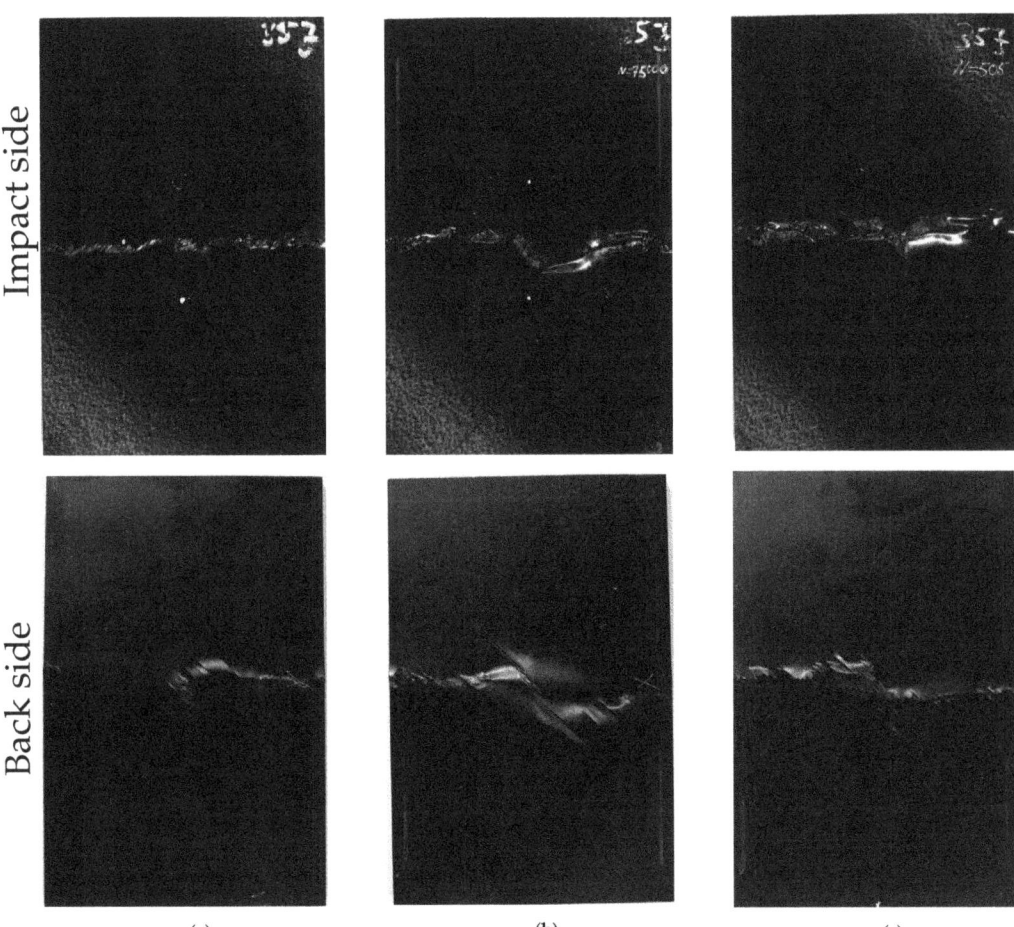

Figure A4. Exemplary images of three specimens after ultimate failure. The upper row shows the impact side, the lower row the back side. (**a**) Failure type I (Quasistatic failure, example: CAI failure), (**b**) Failure type II (Cyclic failure, example: 70% load level), (**c**) Mixed failure type (Cyclic failure, example: 90% load level).

References

1. O'Brien, T.K. *Towards a Damage Tolerance Philosophy for Composite Materials and Structures*; Technical Report March; NASA: Hampton, WV, USA, 1988.
2. Abir, M.; Tay, T.; Ridha, M.; Lee, H. On the relationship between failure mechanism and compression after impact (CAI) strength in composites. *Compos. Struct.* **2017**, *182*, 242–250. [CrossRef]
3. Riccio, A. *GARTEUR AG-32: Damage Growth in Composites Final Report TP-176*; Technical Report; Second University of Naples: Caserta, Italy, 2012.
4. Abrate, S. *Impact on Composite Structures*; Cambridge University Press: Cambridge, UK, 1998.
5. Richardson, M.; Wisheart, M. Review of low-velocity impact properties of composite materials. *Compos. Part A Appl. Sci. Manuf.* **1996**, *27*, 1123–1131. [CrossRef]
6. Dubinskii, S.; Feygenbaum, Y.; Senik, V.; Metelkin, E. A study of accidental impact scenarios for composite wing damage tolerance evaluation. *Aeronaut. J.* **2019**, *123*, 1724–1739. [CrossRef]
7. Federal Aviation Administration. *Advisory Circular 25.571-1D*; Federal Aviation Administration: Washington, DC, USA, 2011.
8. Federal Aviation Administration. *Advisory Circular: 20-107B—Composite Aircraft Structure*; Federal Aviation Administration: Washington, DC, USA, 2009.
9. EASA. *AMC 20-29 Composite Aircraft Structure—Annex II to ED Decision 2010/003/R of 19/07/2010*; EASA: Cologne, Germany, 2010; pp. 1–36.

10. Goudou, P. *Decision No 2010/003/R of the Executive Director of the European Aviation Safety Agency*; Technical Report; EASA: Cologne, Germany, 2010.
11. Mitrovic, M.; Hahn, H.T.; Carman, G.P.; Shyprykevich, P. Effect of loading parameters on the fatigure behaviour of impact damaged composite laminates. *Compos. Sci. Technol.* **1999**, *59*, 2059–2078. [CrossRef]
12. Davies, G.; Irving, P. *Impact, Post-Impact Strength and Post-Impact Fatigue Behaviour of Polymer Composites*; Elsevier Ltd.: Amsterdam, The Netherlands, 2014; pp. 231–259. [CrossRef]
13. Pascoe, J.A. Slow-growth damage tolerance for fatigue after impact in FRP composites: Why current research won't get us there. *Procedia Struct. Integr.* **2020**, *28*, 726–733. [CrossRef]
14. IASB. *Handbuch Struktur Berechnung*; Number 2016; Struktur-Berechnungsunterlagen, Industrie-Ausschuss: Ottobrunn, Germany, 1976; pp. 1–20.
15. Bouvet, C.; Rivallant, S.; Barrau, J.J. Low velocity impact modeling in composite laminates capturing permanent indentation. *Compos. Sci. Technol.* **2012**, *72*, 1977–1988. [CrossRef]
16. ASTM. *ASTM D7136—Standard Test Method for Measuring the Damage Resistance of a Fiber-Reinforced Polymer Matrix Composite to a Drop-Weight Impact Event*; ASTM: West Conshohocken, PA, USA, 2005. [CrossRef]
17. Airbus Industries. *AITM 1-0010 Determination of Compression Strength after Impact*; Airbus Industries: Leiden, The Netherlands, 2005.
18. Deutsches Institut für Normung. *DIN EN 6038—Bestimmung der Restdruckfestigkeit nach Schlagbeanspruchung*; Deutsches Institut für Normung: Berlin, Germany, 1996.
19. SAE International. Polymer Matrix Composites Materials Usage, Design, and Analysis. In *Composite Materials Hanbook (CMH-17)*; SAE International: Wichita, KS, USA, 2012; Chapter 12, Volume 3.
20. Thomas, M. Study on the Evolution of the Dent Depth due to an Impact on Carbon/Epoxy Laminates. Consequences on Impact Damage Visibility and on Service Inspection Requirements for Civil Aircraft Composite Structures. In Proceedings of the MIL-HDBK 17 Meeting, Monterey, CA, USA, 29–31 March 1994; pp. 126–129.
21. Dubinskii, S.; Fedulov, B.; Feygenbaum, Y.; Gvozdev, S.; Metelkin, E. Experimental evaluation of surface damage relaxation effect in carbon-fiber reinforced epoxy panels impacted into stringer. *Compos. Part B Eng.* **2019**, *176*, 107258. [CrossRef]
22. Papa, I.; Ricciardi, M.R.; Antonucci, V.; Langella, A.; Tirillò, J.; Sarasini, F.; Pagliarulo, V.; Ferraro, P.; Lopresto, V. Comparison between different non-destructive techniques methods to detect and characterize impact damage on composite laminates. *J. Compos. Mater.* **2019**, *54*, 617–631. [CrossRef]
23. Moix-Bonet, M.; Eckstein, B.; Wierach, P. Probability—Based Damage Assessment on a Composite Door Surrounding Structure. In Proceedings of the 8th European Workshop On Structural Health Monitoring (EWSHM), Bilbao, Spain, 5–8 July 2016.
24. Sikdar, S.; Banerjee, S. *Structural Health Monitoring of Advanced Composites Using Guided Waves*, 1st ed.; Lambert Academic Publishing: Beau Bassin, Mauritius, 2017.
25. Clark, G.; Van Blaricum, T.J. Load spectrum modification effects on fatigue of impact-damaged carbon fibre composite coupons. *Composites* **1987**, *18*, 243–251. [CrossRef]
26. Rhead, A.T.; Butler, R.; Hunt, G.W. Compressive strength of composite laminates with delamination- induced interaction of panel and sublaminate buckling modes. *Compos. Struct.* **2017**, *171*, 326–334. [CrossRef]
27. Dienel, C.P.; Meyer, H.; Werwer, M.; Willberg, C. Estimation of airframe weight reduction by integration of piezoelectric and guided wave–based structural health monitoring. *Struct. Health Monit.* **2018**, *18*, 1778–1788. [CrossRef]
28. Sebaey, T.A.; González, E.V.; Lopes, C.S.; Blanco, N.; Costa, J. Damage resistance and damage tolerance of dispersed CFRP laminates: Design and optimization. *Compos. Struct.* **2013**, *95*, 569–576. [CrossRef]
29. Dubary, N.; Bouvet, C.; Rivallant, S.; Ratsifandrihana, L. Damage tolerance of an impacted composite laminate. *Compos. Struct.* **2018**, *206*, 261–271. [CrossRef]
30. Nettles, A.; Hodge, A.; Jackson, J. An examination of the compressive cyclic loading aspects of damage tolerance for polymer matrix launch vehicle hardware. *J. Compos. Mater.* **2011**, *45*, 437–458. [CrossRef]
31. Melin, L.G.; Schön, J. Buckling behaviour and delamination growth in impacted composite specimens under fatigue load: An experimental study. *Compos. Sci. Technol.* **2001**, *61*, 1841–1852. [CrossRef]
32. Symons, D.D.; Davis, G. Fatigue testing of impact-damaged T300/914 carbon-fibre-reinforced plastic. *Compos. Sci. Technol.* **2000**, *60*, 379–389. [CrossRef]
33. Uda, N.; Ono, K.; Kunoo, K. Compression fatigue failure of CFRP laminates with impact damage. *Compos. Sci. Technol.* **2009**, *69*, 2308–2314. [CrossRef]
34. Saunders, D.S.; Van Blaricum, T.J. Effect of load duration on the fatigue behaviour of graphite/epoxy laminates containing delaminations. *Composites* **1988**, *19*, 217–228. [CrossRef]
35. Chen, A.S.; Almond, D.P.; Harris, B. In situ monitoring in real time of fatigue-induced damage growth in composite materials by acoustography. *Compos. Sci. Technol.* **2001**, *61*, 2437–2443. [CrossRef]
36. Nezhad, H.Y.; Merwick, F.; Frizzell, R.; McCarthy, C. Numerical analysis of low-velocity rigid-body impact response of composite panels. *Int. J. Crashworthiness* **2015**, *20*, 27–43. [CrossRef]
37. Bogenfeld, R.; Kreikemeier, J.; Wille, T. Review and Benchmark Study on the Analysis of Low-Velocity Impact on Composite Laminates. *Eng. Fail. Anal.* **2018**, *86*, 72–99. [CrossRef]

38. Bogenfeld, R. A Combined Analytical and Numerical Analysis Method for Low-Velocity Impact on Composite Structures. Ph.D. Thesis, Technische Universität Carola-Wilhelmina zu Braunschweig, Braunschweig, Germany, 2019.
39. Fualdes, C. Composites@Airbus Damage Tolerance Methodology. In *FAA Workshop for Composite Damage Tolerance and Maintenance*; National Institute for Aviation Research: Wichita, KS, USA, 2006.
40. Baaran, J.; Wölper, J.; Greef, M. *NSR Impact Damage Assessment*; German Aerospace Center: Braunschweig, Germany, 2009.
41. Bogenfeld, R.; Schmiedel, P.; Kuruvadi, N.; Wille, T.; Kreikemeier, J. An experimental study of the damage growth in composite laminates under tension–fatigue after impact. *Compos. Sci. Technol.* 2020, *191*, 108082. [CrossRef]
42. Schoeppner, G.; Abrate, S. Delamination threshold loads for low velocity impact on composite laminates. *Compos. Part A Appl. Sci. Manuf.* 2000, *31*, 903–915. [CrossRef]
43. Wagih, A.; Maimí, P.; González, E.V.; Blanco, N.; De Aja, J.R.; De La Escalera, F.M.; Olsson, R.; Alvarez, E. Damage sequence in thin-ply composite laminates under out-of-plane loading. *Compos. Part A Appl. Sci. Manuf.* 2016, *87*, 66–77. [CrossRef]
44. Caprino, G.; Lopresto, V.; Langella, A.; Leone, C. Damage and energy absorption in GFRP laminates impacted at low-velocity: Indentation model. *Procedia Eng.* 2011, *10*, 2298–2311. [CrossRef]
45. Rivallant, S.; Bouvet, C.; Hongkarnjanakul, N. Failure analysis of CFRP laminates subjected to compression after impact: FE simulation using discrete interface elements. *Compos. Part A Appl. Sci. Manuf.* 2013, *55*, 83–93. [CrossRef]
46. Baker, A.A. *Composite Materials for Aircraft Structures*; AIAA Education Series; American Institute of Aeronautics & Astronautics: Reston, WV, USA, 2004; p. 599.
47. Post, N.L.; Cain, J.J.; Lesko, J.J.; Case, S.W. Design knockdown factors for composites subjected to spectrum loading based on a residual strength model. In Proceedings of the ICCM International Conference on Composite Materials, Edinburgh, UK, 27–31 July 2009.
48. Feng, Y.; He, Y.; Zhang, H.; Tan, X.; An, T.; Zheng, J. Effect of fatigue loading on impact damage and buckling/post-buckling behaviors of stiffened composite panels under axial compression. *Compos. Struct.* 2017, *164*, 248–262. [CrossRef]
49. SAE International. *Composite Materials Handbook Volume 3—Polymer Matrix Composites: Materials, Usage, Design, and Analysis*; Wichita State University: Wichita, KS, USA, 1994; p. 85322.
50. Sikdar, S.; Kudela, P.; Radzieński, M.; Kundu, A.; Ostachowicz, W. Online detection of barely visible low-speed impact damage in 3D-core sandwich composite structure. *Compos. Struct.* 2018, *185*, 646–655. [CrossRef]
51. Willberg, C. Development of a New Isogeometric Finite Element and Its Application for Lamb Wave Based Structural Health Monitoring. Ph.D. Thesis, Otto von Guericke University Magdeburg, Magdeburg, Germany, 2012. [CrossRef]
52. Xu, J.; Wang, W.; Han, Q.; Liu, X. Damage pattern recognition and damage evolution analysis of unidirectional CFRP tendons under tensile loading using acoustic emission technology. *Compos. Struct.* 2020, *238*, 111948. [CrossRef]
53. Krause, D. A physically based micromechanical approach to model damage initiation and evolution of fiber reinforced polymers under fatigue loading conditions. *Compos. Part B Eng.* 2016, *87*, 176–195. [CrossRef]
54. Lüders, C. *Mehrskalige Betrachtung des Ermüdungsverhaltens Thermisch Zyklierter Faserkunststoffverbunde*; Technische Universität Carola-Wilhelmina zu Braunschweig: Braunschweig, Germany, 2020.
55. Pinho, S.T.; Darvizeh, R.; Robinson, P.; Schuecker, C.; Camanho, P.P. Material and structural response of polymer-matrix fibre-reinforced composites. *J. Compos. Mater.* 2012, *46*, 2313–2341. [CrossRef]

Article

Design of Low Cost Carbon Fiber Composites via Examining the Micromechanical Stress Distributions in A42 Bean-Shaped versus T650 Circular Fibers

Imad Hanhan and Michael D. Sangid *

School of Aeronautics and Astronautics, Purdue University, 701 W. Stadium Ave, West Lafayette, IN 47907, USA; ihanhan@purdue.edu
* Correspondence: msangid@purdue.edu

Abstract: Recent advancements have led to new polyacrylonitrile carbon fiber precursors which reduce production costs, yet lead to bean-shaped cross-sections. While these bean-shaped fibers have comparable stiffness and ultimate strength values to typical carbon fibers, their unique morphology results in varying in-plane orientations and different microstructural stress distributions under loading, which are not well understood and can limit failure strength under complex loading scenarios. Therefore, this work used finite element simulations to compare longitudinal stress distributions in A42 (bean-shaped) and T650 (circular) carbon fiber composite microstructures. Specifically, a microscopy image of an A42/P6300 microstructure was processed to instantiate a 3D model, while a Monte Carlo approach (which accounts for size and in-plane orientation distributions) was used to create statistically equivalent A42/P6300 and T650/P6300 microstructures. First, the results showed that the measured in-plane orientations of the A42 carbon fibers for the analyzed specimen had an orderly distribution with peaks at $|\phi| = 0°, 180°$. Additionally, the results showed that under 1.5% elongation, the A42/P6300 microstructure reached simulated failure at approximately 2108 MPa, while the T650/P6300 microstructure did not reach failure. A single fiber model showed that this was due to the curvature of A42 fibers which was 3.18 μm^{-1} higher at the inner corner, yielding a matrix stress that was 7 MPa higher compared to the T650/P6300 microstructure. Overall, this analysis is valuable to engineers designing new components using lower cost carbon fiber composites, based on the micromechanical stress distributions and unique packing abilities resulting from the A42 fiber morphologies.

Keywords: carbon fiber; low cost; finite element method; stress concentration

Citation: Hanhan, I.; Sangid, M.D. Design of Low Cost Carbon Fiber Composites via Examining the Micromechanical Stress Distributions in A42 Bean-Shaped versus T650 Circular Fibers. *J. Compos. Sci.* **2021**, *5*, 294. https://doi.org/10.3390/jcs5110294

Academic Editor: Jiadeng Zhu, Gouqing Li, Lixing Kang

Received: 7 September 2021
Accepted: 28 October 2021
Published: 7 November 2021

Publisher's Note: MDPI stays neutral with regard to jurisdictional claims in published maps and institutional affiliations.

Copyright: © 2021 by the authors. Licensee MDPI, Basel, Switzerland. This article is an open access article distributed under the terms and conditions of the Creative Commons Attribution (CC BY) license (https://creativecommons.org/licenses/by/4.0/).

1. Introduction

Adoption of polymer matrix carbon fiber composite materials in high volume applications (such as automotive applications) is not widespread and the rate of adoption has been slow. This is partly because of their higher costs compared to traditional materials (specifically the cost of the carbon fibers). Carbon fiber manufacturing can be split into 3 major steps: making the polyacrylonitrile (PAN) precursor, stabilization in air, and carbonization in an inert environment [1]. The PAN precursor is made from crude oil, which is refined and filtered (into dope), and coagulated in a specialized coagulation bath [2]. This precursor then undergoes stabilization in air by applying tension to the precursor at a temperature between 200 and 300 °C for about 2 h [1]. This is followed by carbonization in an inert environment, usually N_2, also under tension, but at an elevated temperature between 1000 and 1700 °C, followed by an additional carbonization step (also in an inert environment) at 2500–3000 °C [1]. The most expensive part of this process is the cost of the precursor fiber, which makes up 53% of the total cost of the carbon fiber, followed by the carbonization step, which makes up 24% of the total cost of the carbon fiber [3].

Work has been done to lower the cost of the precursor and the carbonization steps while retaining as much strength in the carbon fiber as possible. Decreasing the solvent content in the coagulation bath [4], lowering the pH level [5] and the temperature [6] of the coagulation bath, all lower the cost of the precursor production by increasing the mass transfer rate in the coagulation bath, resulting in faster precursor production. However, the increased mass transfer produces a fiber with an irregular cross-section, sometimes referred to as bean-shaped (an example can be seen in Figure 1A). This shape is a result of an instability at the surface of the precursor due to the high mass transfer rate, causing a collapse at one point (hence the bean-shape and the inner corner).

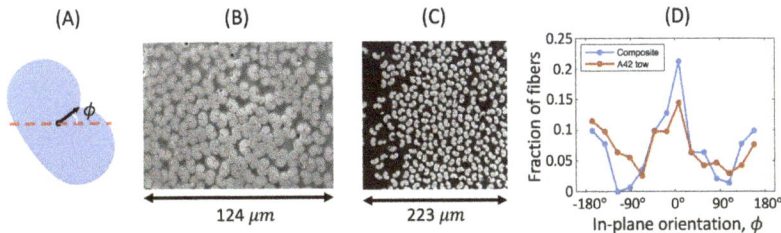

Figure 1. Shown in (**A**) is a schematic of an A42 carbon fiber with a definition of the in-plane angle, ϕ, (**B**) shows an SEM image of an A42/P6300 composite specimen (sectioned and polished), with vectors overlaid for the in-plane orientations, (**C**) shows an optical microscopy image of the A42 fiber tow prior to resin infusion also with vectors overlaid for the in-plane orientations, and (**D**) shows the computed in-plane orientation distributions.

Unfortunately lower cost precursors have higher traditional carbonization costs due to their molecular structure [2,7]. Researchers at Oak Ridge National Laboratory created novel techniques for reducing the carbonization costs of low cost precursors. Specifically, unique chemical baths and pre-treatments were used to lower the required temperature (thereby reducing operational costs) as well as developing a microwave assisted plasma carbonization technique [8]. Through these innovations, carbon fiber strengths in excess of 2.5 GPa and moduli of 220 GPa have been achieved using low cost precursors and carbonization processes [8].

Composites made with carbon fibers which have a bean-shaped cross-section have been used in the Ford Fusion B-pillar [9]. The composite material system, which used A42 carbon fibers and a P6300 epoxy matrix, was reported to have an ultimate stress (bulk) of 1568 MPa, with the P6300 epoxy system designed specifically to aid in high volume production (due to its ease in processing). As lower cost composite materials grow in their use, it becomes important to understand their micromechanical behaviors in order to predict their failure modes for component lifing. Computational tools have been shown to be very useful in this, such as the use of periodic representative volume elements to study the effect of fiber shape [10]. Additionally, researchers have shown that the extended finite element method can be useful in computing the transverse homogenized elastic constants [11]. The realm of virtual material design has enabled the computational analysis of highly complex fiber cross-sections [12].

While work has been done to explore the transverse properties of composites with irregular fibers, the effect of the fiber cross-section on the longitudinal micromechanical properties remains unknown, especially for bean-shaped fiber cross-sections. Therefore, this work analyzed the microstructure of an A42/P6300 composite material using finite element analysis (FEA) simulations, which were used to compare the longitudinal micromechanical behavior to a typical aerospace carbon fiber (T650) microstructure, which contains circular cross-section carbon fibers. Through the analysis of the microstructure, a definition of the in-plane angle for the bean-shaped fibers has been proposed, and shown to be effective in capturing the distribution of the in-plane orientation.

2. Material and Methods

The material studied in this work was a unidirectional A42/P6300 composite, which was manufactured using resin transfer molding at the MDLab in the Indiana Manufacturing Institute. The finished component was sectioned, polished, and inspected using scanning electron microscopy (SEM), where the SEM image can be seen in Figure 1B. Additionally, an A42 fiber bundle (prior to resin transfer) was inspected using SEM for comparison (Figure 1C). This was also conducted at the MDLab, and required the fibers to be loosely adhered together using a quick-setting adhesive, followed by polishing and SEM imaging to examine the fibers, as can be seen in Figure 1C.

The SEM image of the A42/P6300 composite (Figure 2A) was post-processed using a series of image processing steps. First, Weka segmentation was used, which is a machine learning image segmentation plug-in within Fiji [13]. Specifically, through point-and-click training, the Weka module was used to segment the pixels of the fibers and the matrix from the SEM image. Next, a watershed based method [14,15] (implemented in Matlab) was used to separate fiber features which were touching. Finally, ModLayer was used to validate the segmentation, and correct any instances of over- or under-segmentation [16], resulting in a segmented image, as can be seen in Figure 2B, where the fiber area fraction was found to be 65%.

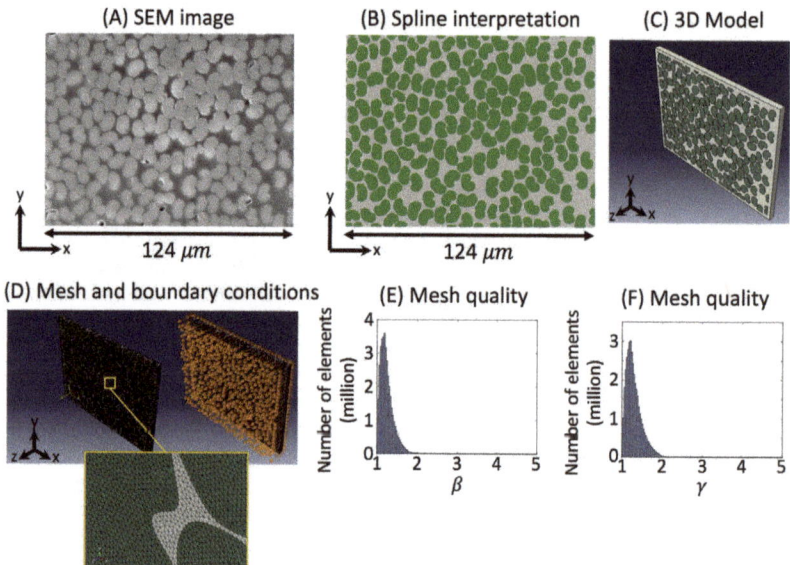

Figure 2. A SEM of the microstructure is shown in (**A**), where (**B**) shows the fiber detection and spline interpretation, (**C**) shows the 3D model with a 5 µm thickness used in this work, and (**D**) is the mesh and the boundary conditions, where the -X, -Y, and -Z surfaces had roller boundary conditions, and the +Z surface had a roller boundary condition with a displacement of +0.075 µm. The mesh quality is shown in (**E**) for β, and in (**F**) for γ.

Virtual microstructures (for the unidirectional fiber composites) were generated to explore varying distributions of fibers, as well as a statistically equivalent microstructures, where the fiber volume fraction and the fiber size distribution could be user-controlled to match across all microstructures. Specifically, a computer generated A42/P6300 microstructure (with random in-plane orientations) was created, as well as a statistically equivalent T650/P6300 microstructure (which had circular cross-section fibers). In order to accomplish this, a Matlab fiber packing algorithm was created that uses a template fiber (which can be either a bean-shaped A42 cross-section or a circular cross-section) to generate the microstructure of interest, where the template fiber could be iteratively processed and placed in the 2D model. This is similar to the procedure used by Gao et al. [17], however this work used a bean-shaped fiber as a template (where a representative fiber was chosen from the SEM image as the template fiber).

In the computer generated microstructures, the variations in fiber size (cross-sectional area) as well as the in-plane orientation was implemented by providing the algorithm a specific desired distribution (in the form of a histogram) of fiber size and in-plane orientation. The algorithm then iteratively used the distance transform and a Monte Carlo method to resize, rotate, and place each fiber, while achieving the correct size and orientation distributions and the desired fiber volume fraction (area fraction in 2D).

The microstructures were analyzed using FEA simulations, where each unidirectional fiber microstructure was padded with a 2.5 μm matrix buffer, and was extruded by 5 μm as can be seen in Figure 2C. The process involved an automated python script which extracted the boundary pixels of each fiber, then used Abaqus to create a 2D sketch of each fiber boundary using B-splines, followed by 3D extrusion of each fiber. The final 3D microstructure was meshed using tetrahedral elements, and the quality of the mesh was analyzed using β and γ, which are geometric parameters that use the radius of a circumscribed sphere for a tetrahedral element, CR, the radius of an inscribed sphere for a tetrahedral element, IR, the root mean square value of the lengths of an element's edges, S_{rms}, and the volume of the tetrahedral element, V, to compute β and γ [18,19]:

$$\beta = \frac{CR}{3 \times IR} \quad (1)$$

$$\gamma = \frac{S_{rms}^3}{8.48 \times V} \quad (2)$$

The mesh quality is shown in Figure 2E,F, where values of β and γ between 1 and 3 are considered good quality elements [18,19]. Elements for the fibers and the matrix were assigned isotropic linear elastic properties, with a user-defined material subroutine (UMAT) used for the ultimate strength of each constituent. The elements for the A42 fibers were assigned $E = 245$ GPa, $\nu = 0.28$, and $\sigma_{ult} = 4200$ MPa using an elastic-brittle failure model [17]. For simulations which used T650 fibers (with circular cross-sections), fiber elements were assigned $E = 255$ GPa, $\nu = 0.28$, and $\sigma_{ult} = 4280$ MPa using an elastic-brittle failure model [20]. The P6300 matrix elements (matrix used in all three simulations) were assigned $E = 3.8$ GPa, $\nu = 0.39$, and $\sigma_{ult} = 68$ MPa using an elastic-brittle failure model, which is representative of P6300 epoxy in tension (as was done in this work) [17]. The boundary conditions (Figure 2D) applied to the model were rollers on the -Z, -X, and -Y surfaces, with rollers and a displacement of +0.075 μm on the +Z surface (representing a 1.5% elongation, which is the expected elongation to failure) [9]. The UMAT allowed failure to be detected in either the fibers or the matrix, depending on which elements reached their failure criteria during the simulation.

3. Results and Discussion

Computing fiber orientation from 2D images, even for circular cross-sections, can be very challenging [15,21,22], and studies with bean-shaped fibers used an elliptical cross-sectional approximation [23]. In this work, the fibers are unidirectional, and therefore the out-of-plane orientation is 0°. For a circular cross-section fiber, the in-plane orientation would therefore be trivial. However, this is not the case for the bean-shaped cross-sections studied in this work. A definition is proposed in this work for the in-plane angle, ϕ, as the angle of the vector pointing from the centroid of the cross-section, to the collapsed surface point (referred to as the inner corner) as has been shown in Figure 1A. A Matlab algorithm which isolated the boundary pixels of each fiber cross-section was used to compute this vector for each fiber in the segmented and post-processed SEM image of the A42/P6300 composite, in order to compute the distribution of the in-plane orientations. It was found that the in-plane orientation of the fibers of the analyzed image was not random, and instead contained peaks at $|\phi| = 0°, 180°$, as can be seen in Figure 1D. A fiber bundle (prior to resin infusion) which was inspected using SEM (Figure 1C) and processed through the same procedure (to compute all in-plane angles) showed a matching distribution (Figure 1D). This implies that the in-plane orientation of the fibers is not a byproduct of the resin transfer molding, but is inherent to the fiber tow. Optimizing or altering this in-plane orientation distribution would therefore require altering the processes by which the fibers are rolled and packaged.

FEA analyses were performed on three microstructures: (i) A42 fibers instantiated from the SEM image (where ϕ showed peaks at $|\phi| = 0°, 180°$), (ii) statistically equivalent A42 fibers (with random ϕ, but equivalent volume fraction and fiber size distribution), and (iii) statistically equivalent T650 cylindrical fibers (with equivalent volume fraction and fiber size distribution). The stresses in the fibers are compared in Figure 3A, and the stresses in the polymer matrix are compared in Figure 3B, using a probability plot. Each probability plot shows a dashed line which represents a theoretical normal distribution. This can be useful for exploring the extremes of a distribution, which was important in this work which used a maximum stress failure criteria. It was found that the SEM to FEM A42 microstructural stress distributions aligned well with the computer generated A42 microstructural stresses. However, the random in-plane orientations of the computer generated A42 microstructure, as well as the use of a cross-sectional fiber template (where fiber curvature was more uniform across all fibers) resulted in higher extreme values of stress in the fibers, which can be seen in Figure 3A. This appears to have alleviated the matrix stresses in the computer generated A42/P6300 microstructure, and transferred more load to the fibers. There exists an opportunity to increase load transfer from the matrix to the fibers (and thereby load bearing capability) by optimizing fiber production processes to tailor the curvature of all fibers in a given microstructure and their in-plane orientations.

The SEM to FEM A42/P6300 microstructure was simulated to reach failure (at a local point within the matrix) at an average bulk stress of 2108 MPa. The location of failure, which was simulated and computed using the Abaqus UMAT, was found at the inner corner of a fiber near a resin-rich area, as can be seen in Figure 3C. In contrast, the location of failure in the computer generated A42/P6300 microstructure (with random ϕ) is shown in Figure 3D, also at a fiber's inner corner but at a region of closely packed fibers (fiber agglomeration). This shows that the morphology of the bean-shaped fibers will likely initiate microstructural fracture at a fiber's inner corner, however it is unclear whether resin rich areas or fiber agglomerations are more detrimental. Lastly, this can be compared to the computer generated T650/P6300 microstructure, which sustained an average bulk stress of 2260 MPa and failure was not established in the matrix or fiber elements. This is partly due to the slightly higher stiffness of a T650 fiber (which is 4% stiffer), as well as its uniform circular cross-section.

In order to study the specific effect of the A42 fiber curvature, a single fiber representative volume element was analyzed, as shown in Figure 4. A representative A42 fiber cross-section was used to generate the model in Figure 4A, and a circular cross-section T650 fiber was used in the statistically equivalent (fiber area and fiber volume fraction) model in Figure 4B. The models were loaded in the fiber direction with the same boundary conditions discussed in Section 2, and the stress in the matrix was probed at varying locations around the perimeter of the fiber (specifically focusing on the region of the collapsed surface point or inner corner). The curvature, κ, which is the inverse of the radius of curvature at the probed location, can be seen in Figure 4C, where as expected the A42 fiber had a much higher curvature (3.18 μm^{-1} higher) than the circular T650 carbon fiber. The corresponding matrix stress has been plotted in Figure 4D, where it can be seen that the higher curvature in the A42 fiber led to a matrix stress that was 7 MPa higher. This quantifies the contribution of the fiber curvatures leading to higher matrix stresses, and it shows that controlling the degree and uniformity of this curvature could lead to future optimizations in the desired microstructural stress distributions.

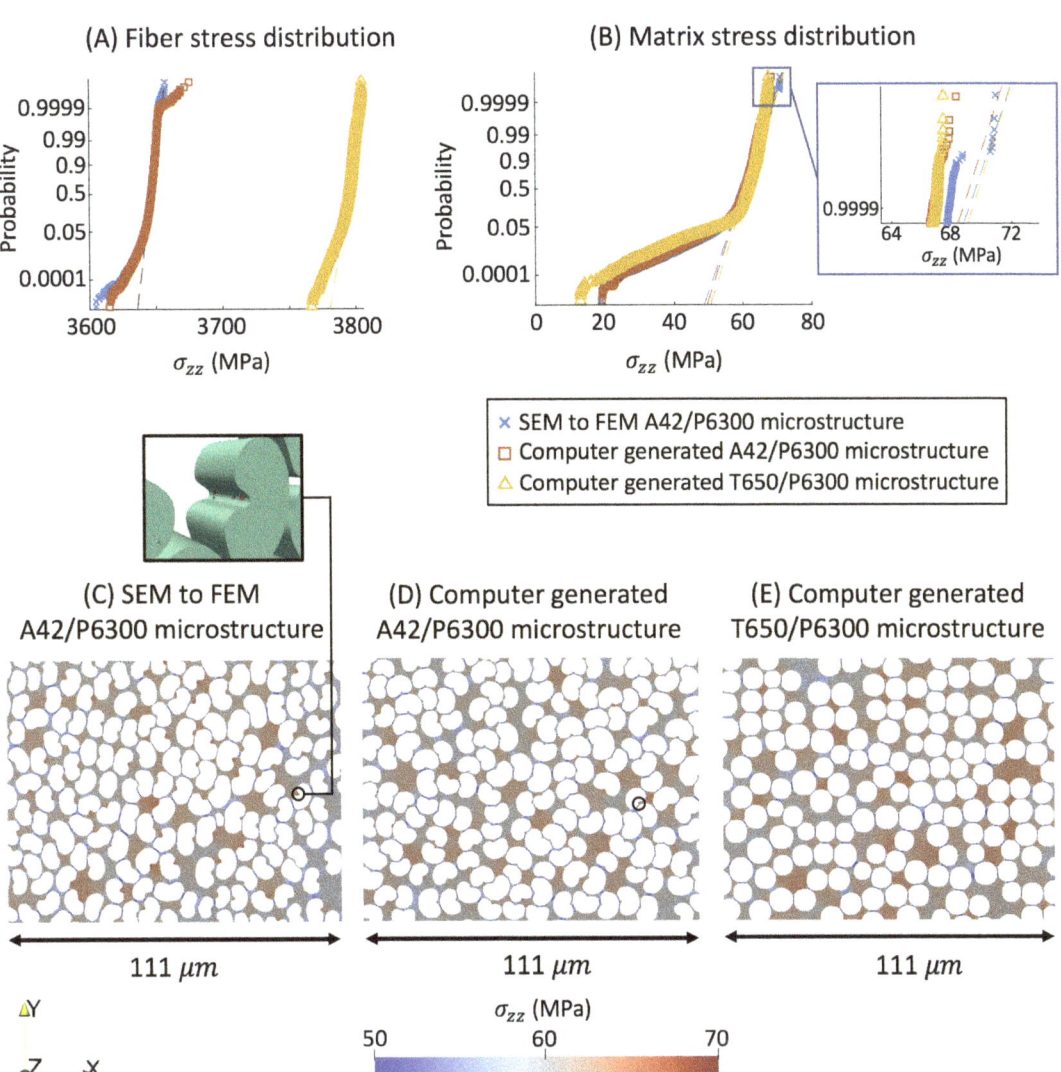

Figure 3. The probability plots of stress in the loading direction, σ_{zz}, for (**A**) the fibers, and (**B**) the matrix, for the real A42/P6300 microstructure, the computer generated A42/P6300 microstructure, and the computer generated T650/P6300 microstructure when elongated by 1.5% in the fiber direction. The stress results for σ_{zz} (cropped away from the boundary conditions) are also shown in (**C**) for the real A42/P6300 microstructure from the SEM image, (**D**) for the computer generated A42/P6300 microstructure with random in-plane orientations, and (**E**) for the computer generated T650/P6300 microstructure. Black circles in (**C**,**D**) point out the location of simulated failure determined by a simulation using a UMAT.

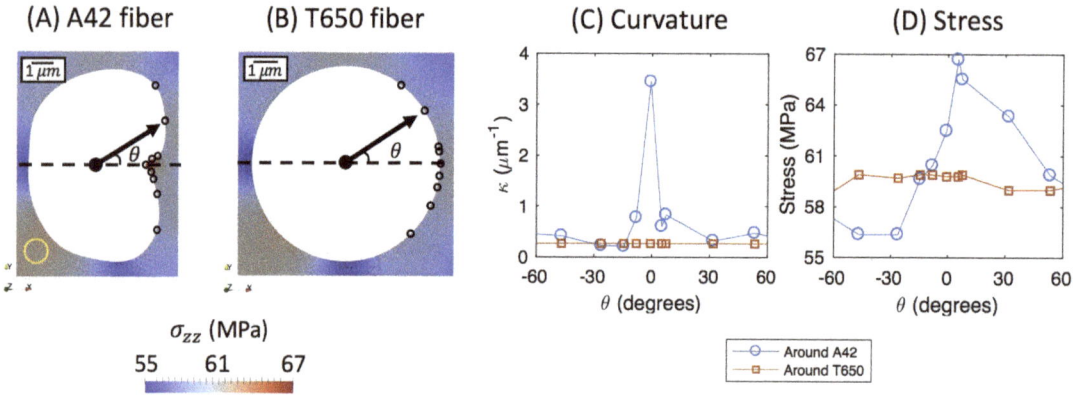

Figure 4. The curvature and stress comparisons of a single fiber model, where (**A**) is the matrix stress around a representative A42 fiber taken from the SEM image, (**B**) is the matrix stress around an ideal T650 fiber, (**C**) is the curvature compared for both fiber models, and (**D**) the stress compared for both fiber models. The yellow circle in (**A**) shows the effects of the roller boundary conditions at the −X and −Y surfaces.

It is important to note that this work has examined only the longitudinal (fiber direction) properties. This is partly because other works have examined transverse properties in detail of other irregularly shaped fiber cross-sections [10–12]. Additionally, similar types of behavior (in terms of stress concentration) under more complex stress states would be expected, including laminates made of these materials. Any fine-tuning of these properties would hinge on the potential to optimize the microstructure, including the curvature of the fiber cross-sections. Tailoring the curvature of the collapsed surface point of the bean-shaped fibers may require adjustments to the coagulation bath during the precursor production, including possibly developing new chemical treatments that help control the curvature of the collapsed surface point to ensure it is consistent across all precursor fibers. On the other hand, tailoring the in-plane orientation may require mechanical adjustments to the production process, such as modifications to the fiber rolling process, which may prove to be very challenging, since it would require quality inspections of the in-plane orientation of fibers during production. The present study could potentially aid in the design of composites with improved performance, which needs to be balanced with the associated increased production costs. Without any optimizations to the processing, the use of these bean-shaped composites may contribute potential enhancements or benefits, such as the increased surface area contact which could improve fiber-matrix bonding. There remains a need to examine such potential enhancements in the fiber-matrix interface and its impact on potential fiber-matrix debonding.

4. Conclusions

The microstructure of an A42/P6300 composite was analyzed using an SEM image of the sectioned and polished surface of a manufactured composite component. The SEM image was post-processed through a number of steps in order to segment the fiber and matrix pixels and was used to generate an FEA model. Additionally, a fiber packing algorithm was created, which used a fiber template to create a computer generated microstructure which can be statistically equivalent (in fiber volume fraction and fiber size distribution). The fiber packing algorithm allows for variations in the in-plane orientation, which was used to study an A42 microstructure with random in-plane orientations. The observations for the bean-shaped A42 carbon fibers were compared to the microstructure of a T650 carbon fiber composite, which has circular fiber cross-sections. Overall:

1. A definition for the in-plane angle of a bean-shaped low cost carbon fiber was defined as the vector pointing from the centroid of the fiber to the collapsed surface point (or inner corner).
2. The in-plane orientation distribution for the SEM image analyzed was not random and was found to persist in the carbon fiber tow (prior to resin transfer) with peaks at $|\phi| = 0°, 180°$.
3. A FEA simulation analysis showed that a computer generated A42 microstructure (with equivalent fiber volume fraction and fiber size distribution) that had random in-plane fiber orientations and uniform curvatures of each fiber, showed slightly better load transfer to the fibers, with a slight alleviation in the matrix stress.
4. A single fiber model showed that an A42 carbon fiber had a curvature that was 3.18 μm^{-1} higher than a circular T650 fiber, resulting in a matrix stress that was 7 MPa higher.

Author Contributions: Conceptualization, M.D.S.; methodology, I.H.; software, I.H.; formal analysis, I.H.; resources, M.D.S.; data curation, I.H.; writing—original draft preparation, I.H.; writing—review and editing, M.D.S.; visualization, I.H.; supervision, M.D.S; project administration, M.D.S.; funding acquisition, M.D.S. All authors have read and agreed to the published version of the manuscript.

Funding: This research was funded by the Indiana Manufacturing Competitiveness Center (IN-MaC) Faculty Fellows program and the National Science Foundation CMMI MoM, Award No. 1662554 (Program Manager Dr. Siddiq Qidwai).

Institutional Review Board Statement: Not applicable; this study did not involve humans or animals subjects.

Informed Consent Statement: Not applicable; this study did not involve humans or animals subjects.

Data Availability Statement: The authors will make the data available upon request.

Acknowledgments: The authors would like to thank Alex Reichanadter and Jan-Anders Mansson at the MDLab for providing the SEM images used in this work.

Conflicts of Interest: The authors declare no conflict of interest.

References

1. Newcomb, B.A. Processing, structure, and properties of carbon fibers. *Compos. Part A Appl. Sci. Manuf.* **2016**, *91*, 262–282. [CrossRef]
2. Khayyam, H.; Jazar, R.N.; Nunna, S.; Golkarnarenji, G.; Badii, K.; Fakhrhoseini, S.M.; Kumar, S.; Naebe, M. PAN precursor fabrication, applications and thermal stabilization process in carbon fiber production: Experimental and mathematical modelling. *Prog. Mater. Sci.* **2020**, *107*, 100575. [CrossRef]
3. Choi, D.; Kil, H.S.; Lee, S. Fabrication of low-cost carbon fibers using economical precursors and advanced processing technologies. *Carbon* **2019**, *142*, 610–649. [CrossRef]
4. Peng, G.Q.; Zhang, X.H.; Wen, Y.F.; Yang, Y.G.; Liu, L. Effect of coagulation bath DMSO concentration on the structure and properties of polyacrylonitrile (PAN) nascent fibers during wet-spinning. *J. Macromol. Sci. Part B Phys.* **2008**, *47*, 1130–1141. [CrossRef]
5. Walczak, Z.K. *Processes of Fiber Formation*; Elsevier: Amsterdam, The Netherlands, 2002.
6. Dong, R.; Keuser, M.; Zeng, X.; Zhao, J.; Zhang, Y.; Wu, C.; Pan, D. Viscometric measurement of the thermodynamics of PAN terpolymer/DMSO/water system and effect of fiber-forming conditions on the morphology of PAN precursor. *J. Polym. Sci. Part B Polym. Phys.* **2008**, *46*, 1997–2011. [CrossRef]
7. Tsai, J.S.; Lin, C.H. The effect of molecular weight on the cross section and properties of polyacrylonitrile precursor and resulting carbon fiber. *J. Appl. Polym. Sci.* **1991**, *42*, 3045–3050. [CrossRef]
8. Paulauskas, F.L.; Warren, C.D.; Eberle, C.C.; Naskar, A.K.; Ozcan, S.; Da Costa Mendes Fagundes, A.P.; Barata Dias, R.M.; De Magalhães Correia, P.F. Novel precursor materials and approaches for producing lower cost carbon fiber for high volume industries. In Proceedings of the ICCM International Conferences on Composite Materials, Edinburgh, UK, 27–31 July 2009.
9. Balijepalli, B.; Bank, D.H.; Baumer, R.E.; Lowe, M.; Ma, L.; James, A. High quality carbon fiber epoxy prepregs for a wide range of reinforcement architectures. In Proceedings of the SPE ACCE: Opportunities & Challenges with Carbon Composites + Carbon Composites, Novi, MI, USA, 6–8 September 2017.
10. Herráez, M.; González, C.; Lopes, C.S.; de Villoria, R.G.; LLorca, J.; Varela, T.; Sánchez, J. Computational micromechanics evaluation of the effect of fibre shape on the transverse strength of unidirectional composites: An approach to virtual materials design. *Compos. Part A Appl. Sci. Manuf.* **2016**, *91*, 484–492. [CrossRef]

11. Higuchi, R.; Yokozeki, T.; Nagashima, T.; Aoki, T. Evaluation of mechanical properties of noncircular carbon fiber reinforced plastics by using XFEM-based computational micromechanics. *Compos. Part A Appl. Sci. Manuf.* **2019**, *126*, 105556. [CrossRef]
12. Yang, L.; Li, Z.; Sun, T.; Wu, Z. Effects of Gear-Shape Fibre on the Transverse Mechanical Properties of Unidirectional Composites: Virtual Material Design by Computational Micromechanics. *Appl. Compos. Mater.* **2017**, *24*, 1165–1178. [CrossRef]
13. Arganda-Carreras, I.; Kaynig, V.; Rueden, C.; Eliceiri, K.W.; Schindelin, J.; Cardona, A.; Seung, H.S. Trainable Weka Segmentation: A machine learning tool for microscopy pixel classification. *Bioinformatics* **2017**, *33*, 2424–2426. [CrossRef] [PubMed]
14. Malmberg, F.; Lindblad, J.; Östlund, C.; Almgren, K.M.; Gamstedt, E.K. Measurement of fibre-fibre contact in three-dimensional images of fibrous materials obtained from X-ray synchrotron microtomography. *Nucl. Instruments Methods Phys. Res. Sect. A Accel. Spectrometers Detect. Assoc. Equip.* **2011**, *637*, 143–148. [CrossRef]
15. Sharma, B.N.; Naragani, D.; Nguyen, B.N.; Tucker, C.L.; Sangid, M.D. Uncertainty quantification of fiber orientation distribution measurements for long-fiber-reinforced thermoplastic composites. *J. Compos. Mater.* **2017**, *52*, 1781–1797. [CrossRef]
16. Hanhan, I.; Sangid, M.D. ModLayer: A MATLAB GUI Drawing Segmentation Tool for Visualizing and Classifying 3D Data. *Integr. Mater. Manuf. Innov.* **2019**, *8*, 468–475. [CrossRef]
17. Gao, J.; Shakoor, M.; Domel, G.; Merzkirch, M.; Zhou, G.; Zeng, D.; Su, X.; Liu, W.K. Predictive multiscale modeling for Unidirectional Carbon Fiber Reinforced Polymers. *Compos. Sci. Technol.* **2020**, *186*, 107922. [CrossRef]
18. Prithivirajan, V.; Sangid, M.D. Examining metrics for fatigue life predictions of additively manufactured IN718 via crystal plasticity modeling including the role of simulation volume and microstructural constraints. *Mater. Sci. Eng. A* **2020**, *783*, 139312. [CrossRef]
19. Parthasarathy, V.N.; Graichen, C.M.; Hathaway, A.F. A comparison of tetrahedron quality measures. *Finite Elem. Anal. Des.* **1994**, *15*, 255–261. [CrossRef]
20. Solvay. Thornel T650. 2020. Available online: https://www.solvay.com/en/product/thornel-t650 (accessed on 27 October 2021)
21. Bay, R.S.; Tucker, C.L. Stereological measurement and error estimates for three-dimensional fiber orientation. *Polym. Eng. Sci.* **1992**, *32*, 240–253. [CrossRef]
22. Hanhan, I.; Agyei, R.; Xiao, X.; Sangid, M.D. Comparing non-destructive 3D X-ray computed tomography with destructive optical microscopy for microstructural characterization of fiber reinforced composites. *Compos. Sci. Technol.* **2019**, *184*, 107843. [CrossRef]
23. Sharp, N.; Goodsell, J.; Favaloro, A. Measuring Fiber Orientation of Elliptical Fibers from Optical Microscopy. *J. Compos. Sci.* **2019**, *3*, 23. [CrossRef]

Article

Simplified Approach for Parameter Selection and Analysis of Carbon and Glass Fiber Reinforced Composite Beams

Reza Moazed *, Mohammad Amir Khozeimeh and Reza Fotouhi

Department of Mechanical Engineering, University of Saskatchewan, Saskatoon, SK S7N 5A9, Canada; mok101@mail.usask.ca (M.A.K.); reza.fotouhi@usask.ca (R.F.)
* Correspondence: reza.moazed@usask.ca

Abstract: In this study, a simplified approach that can be used for the selection of the design parameters of carbon and glass fiber reinforced composite beams is presented. Important design parameters including fiber angle orientation, laminate thickness, materials of construction, cross-sectional shape, and mass are considered. To allow for the integrated selection of these parameters, structural indices and efficiency metrics are developed and plotted in design charts. As the design parameters depend on mode of loading, normalized structural metrics are defined for axial, bending, torsional, and combined bending-torsional loading conditions. The design charts provide designers with an accurate and efficient approach for the determination of stiffness parameters and mass of laminated composite beams. Using the design charts, designers can readily determine optimum fiber direction, number of layers in a laminate, cross-sectional shape, and materials that will provide the desired mass and stiffness. The laminated composite beams were also analyzed through a detailed finite element analysis study. Three-dimensional solid elements were used for the finite element modelling of the beams. To confirm design accuracy, numerical results were compared with close-form solutions and results obtained from the design charts. To show the effectiveness of the design charts, the simplified method was utilized for increasing the bending and torsional stiffness of a laminated composite robotic arm. The results show that the proposed approach can be used to accurately and efficiently analyze composite beams that fall within the boundaries of the design charts.

Keywords: composite beams; finite element analysis; glass fiber; carbon fiber; polymer matrix

Citation: Moazed, R.; Khozeimeh, M.A.; Fotouhi, R. Simplified Approach for Parameter Selection and Analysis of Carbon and Glass Fiber Reinforced Composite Beams. *J. Compos. Sci.* **2021**, *5*, 220. https://doi.org/10.3390/jcs5080220

Academic Editors: Jiadeng Zhu, Gouqing Li and Lixing Kang

Received: 24 July 2021
Accepted: 12 August 2021
Published: 18 August 2021

Publisher's Note: MDPI stays neutral with regard to jurisdictional claims in published maps and institutional affiliations.

Copyright: © 2021 by the authors. Licensee MDPI, Basel, Switzerland. This article is an open access article distributed under the terms and conditions of the Creative Commons Attribution (CC BY) license (https://creativecommons.org/licenses/by/4.0/).

1. Introduction

Numerous researchers have spent a significant amount of effort to study the behavior of composite beams and structures when subjected to various loading conditions [1–10]. These studies have included a wide range of composite materials ranging from laminated polymer matrix composites [1–4] to steel reinforced grout composite-concrete structures [9,10]. The studies outlined in [5–8] have considered the static and dynamic behavior, vibration of composite structures, damping characteristics of composite materials, thermal effects, and effects of boundary conditions such as bolted joints. As part of the extensive research effort regarding structures constructed using composite materials, several approaches and methodologies have been proposed for the selection of the parameters and variables that are important in their design. These parameters include fiber angle orientation, layup sequence, layer thickness, number of layers, geometrical dimensions, and constituent material properties. Various approaches including two and three-dimensional finite element analysis (FEA), genetic algorithm (GA), and particle swarm optimization (PSO) have been utilized for the selection of the optimal parameters of composite beams and structures [11–13]. It should be noted that, even though these methods can be used for determining the optimal design variables and parameters, the methods are time-consuming, computationally expensive, and can be difficult to implement. The development of a simplified approach and design charts will allow engineers and scientists to more readily observe the impact of the numerous parameters that govern the static and dynamic performance of composite structures.

A simplified approach using beam elements and finite element (FE) modelling for the analysis and design of thin-walled T-joint connections has been proposed by Moazed et al. [14,15]. In their studies, the authors considered the behavior of the T-joints subjected to in-plane and out-of-plane loading conditions. It was shown that, with the use of certain normalized correction factors, one-dimensional beam elements can be used to accurately predict all the important information regarding the structural performance of these connections. Important information of the structural analysis included the stresses, deformations, reaction forces, and natural frequencies. Without the proposed correction factors, scientists and engineers need to develop complicated and computationally expensive FE models using two-dimensional plate and shell elements and three-dimensional solid elements. Other studies that have considered proposing simplified approaches for structural analysis include Pasini [16] and Ashby [17,18]. In Ashby's study, it is shown that structural properties can be mapped, displaying the ranges of static and dynamic performance that they offer. In each map, some spaces are filled, and some gaps exist. These gaps can be further investigated for developing materials with new properties [18]. It should be noted that the proposed approaches in the above-mentioned studies have been limited to structures made of traditional isotropic materials.

Studies and research work by Kollar et al. [19] provide an extensive evaluation of the mechanics of composite structures and the parameters that impact the structural performance of laminated composite beams. These parameters include the layup sequence, number of plies, cross-sectional shape, and materials of construction. Based on the works of [19], An et al. [20] and Pasini [21] provide structural efficiency metrics for the integrated selection of layup, materials, and cross-sectional shape of laminated composite beam structures. These authors have developed selection graphs for laminated composite beams subjected to bending and torsional loading. The performance selections graphs were used to study the strength of the structures using the Tsai–Wu failure criterion [20]. Buckney et al. [22] studied the behavior of members subjected to asymmetric bending and showed that shape factors are a useful tool to assess the structural efficiency of various structural shapes. Wanner [23] developed a systematic materials selection procedure for spatially limited, light-weight structural components. This approach led to a specific class of objective equations and performance indices. Pasini [24] introduced a material and shape selection method to reduce the challenges and effort involved in the selection of materials and cross-sectional shapes in applications involving flexural vibrations. The author developed certain parameters such as shape transformers and performance indices to map the beams' mechanical properties. This method was used to assess flexural vibrations in multilayered resonators (MRs). The developed shape transformers can be used to predict the impact of cross-sectional shape, symmetry, number of layers, and material properties on the vibration of MRs [24]. In future studies, the authors devised shape transformers for different loading scenarios such as bending, torsion, shear, and combined loading [25]. The methods and charts were generalized for single and multi-criteria selection of light-weight shafts subjected to a combination of bending, shear, and torsional loading [26].

Laminated composite beams have been used in structures as well as mechanical systems. Application of composite beams in the construction of mechanisms includes linkages of robotic manipulators. Robotic manipulators can perform repetitive work at much higher speeds and efficiency than human operators [27]. The steel and aluminum manipulator designed by Zhang et al. [28] has a maximum reach of 3 m and can carry weights of up to 15 kg. Even though manipulators constructed using conventional materials (e.g., aluminum and steel) benefit from a high, axial, bending, and torsional stiffness, the mass of the manipulator is also typically high. The mass of the robot linkages affects the mechanism's dynamic performance and a higher mass results in larger motor power consumption. Utilizing thin-walled members and lighter materials of construction (with high specific strength and stiffness ratios) for the robot's structure can improve the robot's dynamic performance and lower the power consumption requirements. Materials with a high specific strength and specific stiffness include polymer matrix composites (PMCs) such as glass fiber reinforced plastic (GFRP) and carbon fiber reinforced plastic (CFRP). These materials have been used for the construction of robotic manipulators to successfully reduce the weight by up to 70% compared with steel materials [29,30]. Other researchers such as Yin et al. [31] have designed hybrid structures for manipulators using composites and aluminum beams. The manufactured prototype of the manipulator in [31] showed that the hybrid structure has a higher operational speed and lower settling time in start-stop operations. Lee et al. [32] designed a composite wrist to replace the aluminum version. The intent of the study was to design the composite wrist such that the dynamic performance is improved, and deflection and mass are minimized.

In this study, the recent research works performed by An et al. [20] have been further extended to include design charts for two different envelopes. The first envelope has a square shape, and the second envelope has a rectangular shape. Compared with the envelopes used in the studies performed by [20], these envelopes allow for the assessment of a larger number of cross-sectional types and sizes. In this paper, the behavior of laminated composite beams subjected to axial and bending loadings (about the two principal directions Y and Z) and torsional loadings are considered. The works performed by [20] were also limited to single cell beams. In this paper, in additional to single open and close-loop cross sections, the applicability of the developed design charts for multi-cell beam design is investigated. The results of the study show that the proposed axial and bending design charts can also be used for the parameter selection and design of multi-cell laminated composite beams. The effects of beam thin-walledness are also studied in this paper. For this purpose, the bending and shear deformation effects are considered, and the effects of different boundary conditions are investigated. A detailed three-dimensional FEA using solid elements is performed. The results of the numerical simulations are compared to the analytical results and those predicted by the design charts for various loading conditions. Finally, the design charts are used to design the linkages of a robotic manipulator using thin-walled laminated composite beams. It is shown that the simplified approach and design charts can be an effective tool for the design of laminated composite beam structures. The proposed charts help in reducing the inherent complexity involved in the selection of the numerous design variables of laminated composite beams. The simplified approach also allows the designer to accurately predict the deflections and rotations of the aforementioned beams without the need to develop detailed and computationally expensive FE models.

2. Laminated Composite Beams—Structural Efficiency Metrics and Performance Indices

The static and dynamic performance of a given laminated composite prismatic beam (e.g., shown in Figure 1a) depends on the number of layers, fiber angle orientation, layup sequence, materials of construction (e.g., CFRP, GFRP), the cross-section shape (open or closed section), and the geometrical dimensions. By assigning the elastic material properties to the composite beam, normalized shape transformer parameters φ_G can be defined for a given geometric quantity (G). These geometric quantities can include parameters such as area (A), volume (V), and the second moment of area (I). For example, $\varphi_G = \frac{I}{I_{ref}}$ is the second moment of area transformer, where I is the second moment of area of the cross-section and I_{ref} is the second moment of area of a defined reference envelope. Equations (1) to (4) present the normalized axial stiffness (φ_{EA}), bending stiffness (φ_{EI}), torsional stiffness (φ_{GJ}), and mass ratio index (φ_m).

$$\varphi_{EA} = \frac{EA_\theta}{(EA)_{Ref}} \quad (1)$$

$$\varphi_{EI} = \frac{EI_\theta}{(EI)_{Ref}} \quad (2)$$

$$\varphi_{GJ} = \frac{GJ_\theta}{(GJ)_{Ref}} \tag{3}$$

$$\varphi_m = \frac{(\rho A)_\theta}{(\rho A)_{Ref}} \tag{4}$$

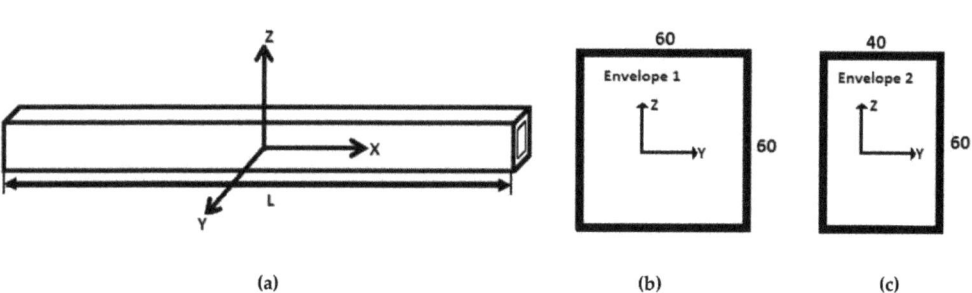

Figure 1. (a) Prismatic Beam, (b) square envelope, and (c) rectangular envelope. Note: all dimensions are in mm.

In Equations (1)–(4), subscript (Ref) indicates the reference beam properties and subscript (θ) indicates the beam with a certain fiber angle direction. That is, $(EA)_{Ref}$ is the reference beam's equivalent axial stiffness, (EI_{Ref}) is the reference beam's equivalent bending stiffness, $(GJ)_{Ref}$ is the reference beam's equivalent torsional stiffness, and $(\rho A)_{Ref}$ is the reference beam's mass. In turn, $(EA)_\theta$, $(EI)_\theta$, $(GJ)_\theta$, and $(\rho A)_\theta$ are the axial, bending, torsional stiffnesses, and mass of the cross-section under consideration, respectively. In this study, the fiber angle direction is assumed to be constant for all laminae. The square and rectangular envelopes utilized in this study are presented in Figure 1b,c. For the square envelope, the bending stiffness in the Y-direction and, for rectangular envelope, the bending stiffnesses in the Y and Z directions are analyzed. All of the parameters are normalized to a reference cross section; for the square envelope, the reference beam is a CFRP hollow square beam made of 16 plies with fibers oriented in the 0° direction (measured from the X-axis). Similarly, for the rectangular envelope, the reference cross-section is a CFRP hollow rectangular beam with 16 plies in the 0° direction (measured from the X-axis). The equivalent axial, bending, and torsional stiffnesses in Equations (1)–(4) are calculated using the close-form equations provided in Table A1 in the appendices. This applies to the reference composite beam cross sections as well as sections with various ply angles. Interested readers are referred to the works of [19,20] for more details and derivation of these equations. Several performance parameters corresponding to different loading conditions are defined for the purpose of developing design charts and comparing the structural and mechanical performance of the composite beams. These performance indices are outlined in Equations (5)–(8) as follows:

The performance index for axial stiffness:

$$P_a = \frac{\varphi_{EA}}{\varphi_m} = \frac{EA_\theta}{(EA)_{Ref}} \frac{(\rho A)_{Ref}}{(\rho A)_\theta} \tag{5}$$

The performance index for bending stiffness:

$$P_b = \frac{\varphi_{EI}}{\varphi_m} = \frac{EI_\theta}{(EI)_{Ref}} \frac{(\rho A)_{Ref}}{(\rho A)_\theta} \tag{6}$$

The performance index for torsional stiffness:

$$P_t = \frac{\varphi_{GI}}{\varphi_m} = \frac{GJ_\theta}{(GJ)_{Ref}} \frac{(\rho A)_{Ref}}{(\rho A)_\theta} \tag{7}$$

The performance index for combined torsional-bending stiffness:

$$P_{tb} = \frac{\varphi_{GI}}{\varphi_{EI}} = \frac{GJ_\theta}{(GJ)_{Ref}} \frac{(EI)_{Ref}}{EI_\theta} \qquad (8)$$

Equations (5)–(8) are used in the next section to map the structural and mechanical performance indices of six types and sizes of laminated composite beams. The corresponding beam cross sections are shown in Figure 2. As mentioned above, two envelopes were considered for mapping the performance indices. The envelope sizes are 60 mm × 60 mm for the square envelope, which is larger than the 50 mm × 50 mm envelope size proposed by An et al. [20]. This will allow for the application of the design charts to beams with larger cross sections. To study beams of various cross sections and sizes, a second envelope was also introduced in this study. This envelope has a rectangular geometry with dimensions of 40 mm × 60 mm. The dimensions of this envelope were chosen to correspond with the dimensions of the aluminum beams used in the robotic manipulator presented in [28].

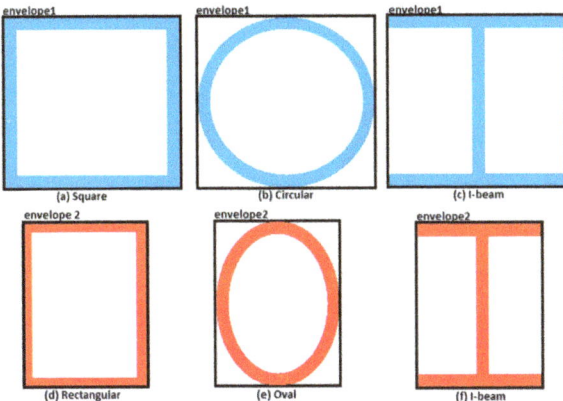

Figure 2. Beam cross sections used to fill the square and rectangular envelopes.

In Section 3, design charts are developed for laminated composite beams with ply numbers ranging from 4 to 24 with a symmetric and balanced layup.

3. Thin-Walled Laminated Composite Beam Design Charts

The in-plane and out-plane behavior of beams, more specifically their static and dynamic response, strongly depends on their stiffness and mass. In beams constructed using traditional isotropic materials (e.g., steel, aluminum, and other metals), the relevant stiffnesses can readily be determined using standard engineering formulas. However, for beams constructed using laminated composite materials, the designer must select several parameters that govern the final stiffnesses of the beams. This can be a challenging and cumbersome engineering effort. The design charts presented in this section were developed to provide engineers and scientists with a means to easily observe the effect of these parameters on the stiffness and mass of composite beams. For this purpose, design charts for axial stiffness, bending stiffness, and torsional stiffness were developed and the results are presented in Figures 3–6. Each point on these charts corresponds to a laminated composite beam with a defined number of layers, fiber orientation, mass, and a specific loading condition. For ease of reference to the design points, the following labeling format is used throughout this paper: "X_M^{N-FD}" where "X" refers to the cross-sectional shape; "FD" refers to the fiber direction; and "N" and "M" present the number of layers and materials, respectively. Moreover, for labeling the cross-sectional shapes, the letter "R" refers to a rectangular beam, "I" to an I-beam, "O" to an oval beam, "S" to a square beam, and "C" to the circular beam. All material properties (e.g., CFRP and GFRP) are provided in Table A2 in the appendices. It should be noted that the developed design charts apply to thin-walled beams (for which shear deformation effects are small), assuming linear elastic material models and small deformations (i.e., geometric and material nonlinearities are not considered).

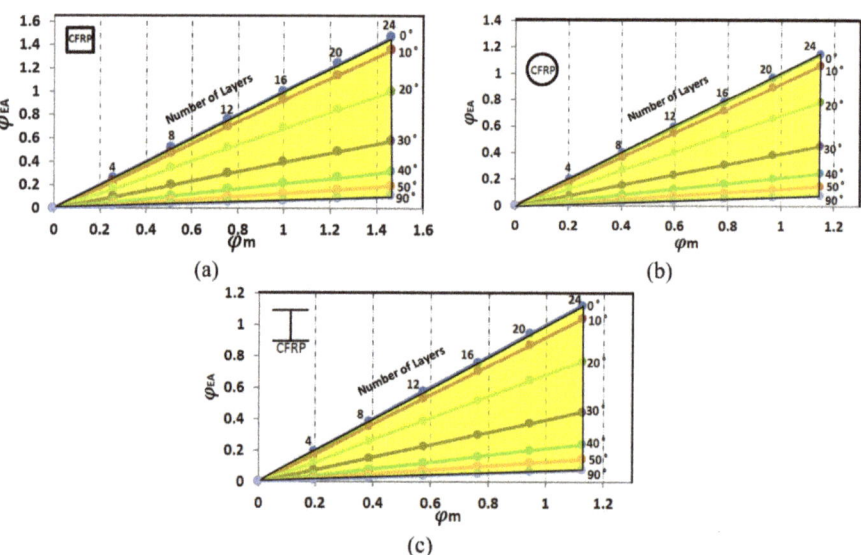

Figure 3. Normalized axial stiffness versus mass ratio index for CFRP materials of construction: (a) square beam, (b) circular tube, and (c) I-beam.

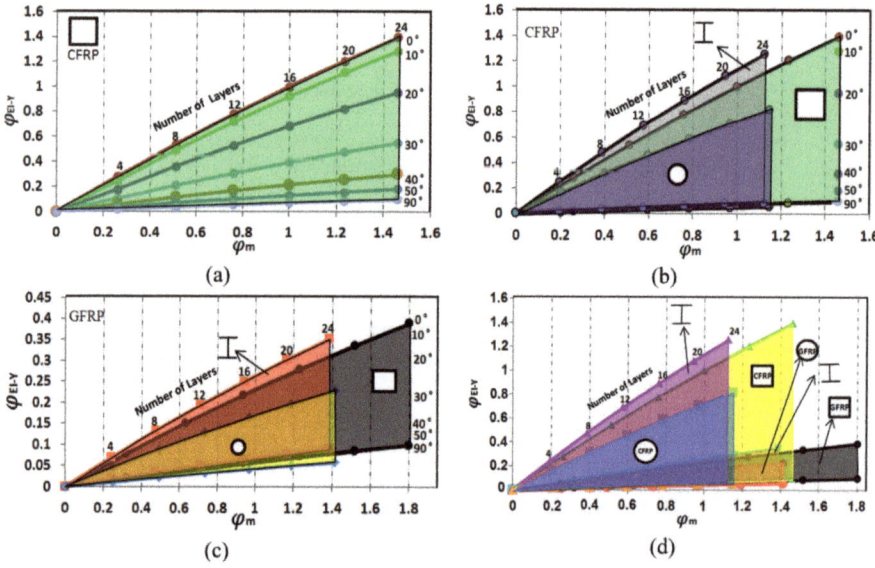

Figure 4. Normalized bending performance index: (a) square CFRP cross section, (b) performance comparison for different cross-sectional shapes using CFRP materials, (c) performance comparison for different cross-sectional shapes using GFRP materials, and (d) performance comparison for various cross-sections constructed from CFRP and GFRP materials.

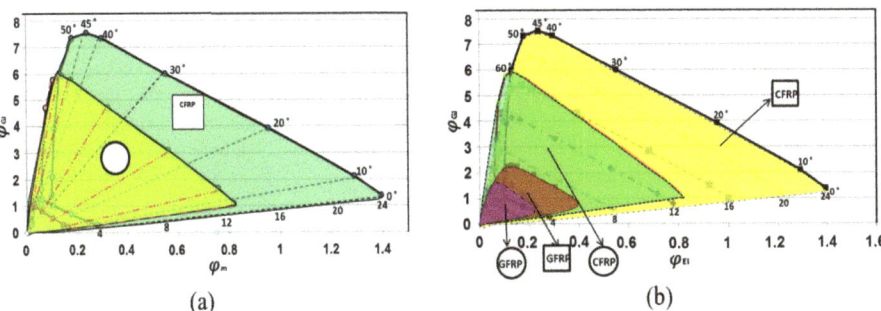

Figure 5. Normalized torsional performance index: (**a**) comparison of square and circular cross-sections using CFRP and (**b**) comparison of square and circular cross sections with CFRP and GFRP materials.

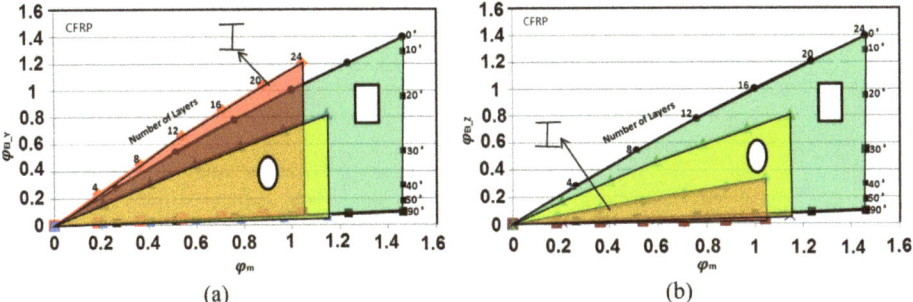

Figure 6. Normalized bending stiffness versus mass ratio: (**a**) in the Y-direction and (**b**) in the Z-direction.

Figure 3a–c depict the normalized axial stiffness versus the mass ratio index for the square, circular, and I-beam cross sections. The results in these figures show that the square cross section has a higher axial efficiency ($\varphi_{EA} = 1.46$) when compared with the circular tube ($\varphi_{EA} = 1.14$), and the I-beam ($\varphi_{EA} = 1.12$) is the least efficient beam when subjected to axial loading. As expected, the axial stiffness performance index (and off-axis strength) decreases by increasing the fiber direction from 0° to 90°, and a rapid decrease in axial stiffness is observed between the 10° to 40° fiber orientation. A further increase in fiber angle orientation (i.e., between 40° and 90°) does not result in a significant decrease is the axial stiffness. This behavior shows that the optimum values for designing a laminated composite beam subjected to axial loading conditions is for a fiber angle orientation between 0° and 10°, where a significant decrease in axial stiffness is not observed.

Figure 4a shows the normalized bending ratio versus the mass ratio index for the square cross section. A similar trend to that observed for the axial loading conditions and axial stiffness can be seen. By increasing the fiber angle from 0° to 50°, the normalized bending stiffness changes from 1.39 to 0.18, which is a decrease of approximately 90%.

For ease of comparison and owing to space limitations and constraints, except for the results shown in Figure 4a for the square cross section, the results for the other cross sections are grouped into the same design charts. For example, Figure 4b provides the results for the I-beam, square, and circular cross sections with CFRP material properties. Plotting the results for the different cross sections on the same charts also allows the designer to select the cross-sectional shape of the beam while simultaneously considering parameters such as number of layers, fiber orientation, materials of construction, and mass. The results in Figure 4b show that S_{CFRP}^{24-0} has the highest normalized bending stiffness ratio in comparison with I_{CFRP}^{24-0} and C_{CFRP}^{24-0}. More specifically, S_{CFRP}^{24-0} is 1.11 times stiffer and 1.30 times heavier than I_{CFRP}^{24-0} and 1.71 times stiffer and 1.28 times heavier than C_{CFRP}^{24-0}. Depending on the beam selection criteria (e.g., lower deflection vs. lower mass), the designer can readily choose the most appropriate beam for the application. For example, the S_{CFRP}^{24-0} beam will have a higher stiffness and hence lower deflection, but at the same time, will have a larger mass. Using the results from Figure 4b, the designer can also determine the "family" of cross sections and parameters that provide the desired normalized bending stiffness. For example, if a normalized bending stiffness of 0.6 is desired, the following beams can readily be identified: I_{CFRP}^{10-0}, I_{CFRP}^{16-20}, S_{CFRP}^{14-20}, S_{CFRP}^{10-10}, and C_{CFRP}^{16-0}. Similarly, if the designer is concerned about the mass of the beams, the mass ratio index (φ_m) can be specified. For example, if a mass ratio index of 0.8 is desired, the following beams have the closest values: I_{CFRP}^{16-0}, I_{CFRP}^{16-20}, S_{CFRP}^{12-0}, C_{CFRP}^{16-0}, and C_{CFRP}^{16-20}, and all satisfy the design criteria.

Figure 4c,d show the effect of materials of construction (GFRP vs. CFRP) on the normalized bending stiffness and mass ratios. In Figure 4c, the same parameters as that of Figure 4b are used, but the CFRP material properties are replaced with GFRP material properties. The figures show that, for identical parameters, beams with GFRP material properties are significantly heavier and less stiff than the corresponding beams fabricated using CFRP materials. For instance, the normalized bending stiffness and mass ratios for S_{CFRP}^{24-0} are equal to 1.39 and 1.46, respectively. The corresponding values for S_{GFRP}^{24-0} are 0.38 and 1.8, respectively. This shows that S_{CFRP}^{24-0} is 3.66 times stiffer and 1.23 lighter than S_{GFRP}^{24-0}. Considering that lower deflection and weight are important factors in material selection, the developed design charts can serve as a valuable tool in the design of laminated composite beams.

The normalized torsional stiffness (φ_{GI}) versus mass index ratio (φ_m) and the performance index for combined torsional-bending stiffness (P_{tb}) are shown in Figure 5a,b, respectively. The design charts were developed for the closed cross sections using the square envelope with CFRP and GFRP materials properties. A general observation from the design charts is that square cross sections have a better torsional performance than circular cross sections when subjected to pure torsion as well as combined torsional-bending loading. As expected, the highest torsional efficiency is achieved when the fiber direction is 45°. For design points with a constant fiber direction, increasing the number of layers from 0 to 24 will result in an increase in P_{tb}. Another observation is that, as the fiber orientation increases from 0° to 45°, the normalized torsional stiffness and torsional-bending performance index also increase. As the fiber orientation is further increased from 45° to 90°, there is a rapid reduction of the normalized torsional stiffness and P_{tb}. Consequently, for design purposes, these design charts can be separated into two sections, namely, the section to the right of 45° and the section to left. In the case that the composite beam will be subjected to torsional loading, design points in the left section of these charts should be avoided owing to the rapid decrease in the values of φ_{GI} and P_{tb}.

A further observation that can be made based on the results presented in Figures 4 and 5 is that S_{CFRP}^{24-45} and C_{CFRP}^{24-45} have the highest torsional efficiency, while S_{GFRP}^{24-0} and C_{GFRP}^{24-0} have the highest bending efficiency. The torsional stiffness for S_{CFRP}^{24-45} is 7.54 higher than the reference cross section (i.e., S_{CFRP}^{16-0}). Regarding the performance of CFRP and GFRP composite beams subjected to combined torsional and bending loads, Figure 5b shows that the torsional efficiency of S_{CFRP}^{24-45} is 3.3 times higher than that of S_{GFRP}^{24-45}.

The rectangular envelope of 40 mm × 60 mm was used to develop design charts for rectangular, oval, and I-beams with unequal flange and web dimensions. For these beams, the bending stiffness is different in the cross-section's principal directions (i.e., Y and Z directions). Figure 6a,b show the normalized bending stiffness in the Y and Z directions versus the mass index ratio.

4. Three-Dimensional Finite Element Analysis and Numerical Simulations

In this section, the results obtained analytically using the equations provided in Table A1 [19] (and plotted in the design charts) are compared with the FEA results. It should be noted that the analytical solutions and design charts do not take into account shear and warping deformations, while the developed FE models include the effects of shear and warping deformations due to torsion.

The numerical modelling of the composite beams is performed using the commercial finite element analysis package, ANSYS [33]. The developed FE model geometry, applied loading, and boundary conditions are shown in Figure 7. The cantilever beam shown in this figure is 1200 mm long and has a square cross section with dimensions of 60 mm × 60 mm (corresponding to the square envelope). A concentrated load of 120 N in the Z-direction and a torque moment of 3600 N.mm were applied at the free end of the cantilever. The developed FE models are based on linear elastic material models, static loading conditions, and small deformations (i.e., geometric and material nonlinearities were not considered). Quadratic and linear solid elements were used for the meshing of the composite beams. More specifically, Solid185 and Solid186 from the FEA software element library were selected. These elements are three-dimensional elements and have been shown [14,15] to be effective for the three-dimensional modeling of structures. The elements have three degrees of freedom at each node: translations in the nodal x, y, and z directions, and are either linear and defined by eight nodes (Solid 185) or are higher order (i.e., quadratic) defined by 20 nodes (Solid 186).

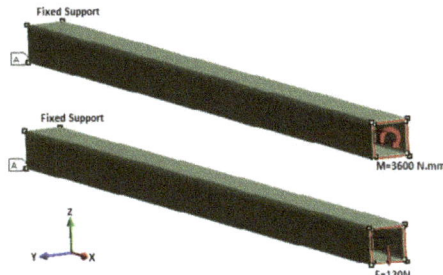

Figure 7. Finite element model geometry, applied loading, and boundary conditions (developed FE models consist of up to ~72,000 nodes and ~68,000 elements depending on laminate thickness).

For the purposes of comparison, the maximum deflection in the Z-direction (δ_{max}) and angle of twist around the X-direction (θ_{max}) were obtained from the FEA. The normalized bending and torsional numerical stiffness ratios obtained from the FEA are defined as $(\frac{EI}{EI_{Ref}})_{FEA}$ and $(\frac{GJ}{GJ_{Ref}})_{FEA}$ and are calculated as follows:

$$\left(\frac{EI}{EI_{Ref}}\right)_{FEA} = \left(\frac{\delta_{Ref}}{\delta_{max}}\right)_{FEA} \tag{9}$$

$$\left(\frac{GJ}{GJ_{Ref}}\right)_{FEA} = \left(\frac{\theta_{Ref}}{\theta_{max}}\right)_{FEA} \tag{10}$$

The analytical and finite element analysis results and normalized stiffness ratios that are defined in Equations (9) and (10) are summarized in Tables 1–3. The results are presented for fiber directions of 0°, 10°, and 20° with the number of layers ranging from 8 to 24. The results for other fiber directions were also obtained but are not presented here. In the tables, the following equations are used to analytically determine the deflections and rotations for further confirmation of the FEA results:

$$\delta_{max\ (analytical)} = \frac{PL^3}{3EI} \tag{11}$$

Table 1. Bending and torsional results obtained using analytical and FEA approaches (FD = 0°).

No. of Layers	θ_{max} FEA (rad)	δ_{max} FEA (mm)	θ_{max} Analytical (rad)	δ_{max} Analytical (mm)	$\left(\frac{\theta_{ref}}{\theta_{max}}\right)$ FEA	$\left(\frac{GJ_\theta}{GJ_{Ref}}\right)$ Design Charts	$\left(\frac{\delta_{ref}}{\delta_{max}}\right)$ DEA	$\left(\frac{EI_\theta}{EI_{Ref}}\right)$ Design Charts
8	2.17 × 10⁻³	2.80	1.99 × 10⁻³	2.43	0.5	0.53	0.50	0.53
12	1.43 × 10⁻³	1.87	1.37 × 10⁻³	1.68	0.76	0.77	0.75	0.77
16	1.09 × 10⁻³	1.41	1.07 × 10⁻³	1.31	1	1	1	1
20	9.04 × 10⁻⁴	1.16	8.92 × 10⁻⁴	1.08	1.21	1.20	1.21	1.21
24	7.76 × 10⁻⁴	0.99	7.72 × 10⁻⁴	0.94	1.4	1.38	1.42	1.39

Table 2. Bending and torsional results obtained using analytical and FEA approaches (FD = 10°).

No. of Layers	θ_{max} FEA (rad)	δ_{max} FEA (mm)	θ_{max} Analytical (rad)	δ_{max} Analytical (mm)	$\left(\frac{\theta_{ref}}{\theta_{max}}\right)$ FEA	$\left(\frac{GJ_\theta}{GJ_{Ref}}\right)$ Design Charts	$\left(\frac{\delta_{ref}}{\delta_{max}}\right)$ FEA	$\left(\frac{EI_\theta}{EI_{Ref}}\right)$ Design Charts
8	1.44 × 10⁻³	2.94	1.35 × 10⁻³	2.66	0.75	0.80	0.48	0.49
12	9.81 × 10⁻⁴	1.95	9.27 × 10⁻⁴	1.83	1.11	1.17	0.72	0.71
16	7.44 × 10⁻⁴	1.49	7.20 × 10⁻⁴	1.42	1.47	1.51	0.94	0.92
20	6.10 × 10⁻⁴	1.23	5.97 × 10⁻⁴	1.17	1.79	1.82	1.14	1.11
24	5.23 × 10⁻⁴	1.06	5.17 × 10⁻⁴	1.01	2.08	2.10	1.33	1.28

Table 3. Bending and torsional results obtained using analytical and FEA approaches (FD = 20°).

No. of Layers	θ_{max} FEA (rad)	δ_{max} FEA (mm)	θ_{max} Analytical (rad)	δ_{max} Analytical (mm)	$\left(\frac{\theta_{ref}}{\theta_{max}}\right)$ FEA	$\left(\frac{GJ_\theta}{GJ_{Ref}}\right)$ Design Charts	$\left(\frac{\delta_{ref}}{\delta_{max}}\right)$ FEA	$\left(\frac{EI_\theta}{EI_{Ref}}\right)$ Design Charts
8	7.80 × 10⁻⁴	3.77	7.48 × 10⁻⁴	3.63	1.40	1.48	0.37	0.36
12	5.32 × 10⁻⁴	2.54	5.09 × 10⁻⁴	2.48	2.05	2.18	0.55	0.52
16	4.10 × 10⁻⁴	1.96	3.94 × 10⁻⁴	1.92	2.66	2.81	0.71	0.68
20	3.37 × 10⁻⁴	1.61	3.27 × 10⁻⁴	1.59	3.23	3.39	0.87	0.82
24	2.88 × 10⁻⁴	1.39	2.82 × 10⁻⁴	1.37	3.78	3.92	1.01	0.95

The deflections calculated using Equation (11) include only the deformation due to bending. In this equation, P is the applied load, L is the length, and EI is the equivalent bending stiffness of the composite beam.

$$\theta_{max \, (analytical)} = \frac{T}{GJ}\left(\frac{\sinh(\mu L)}{\mu \times \cosh(\mu L)}(\cosh(\mu L) - 1) + X - \frac{\sinh(\mu L)}{\mu}\right) \quad (12)$$

Equation (12) is used to determine the rotations (i.e., angle of twist) and includes the effects of warping [34]. In this equation, T is the applied torque; GJ is the equivalent torsional stiffness; L is the length of the beam; X is the distance from the fixed support boundary condition; and, for the case of a square cross-section, $\mu^2 = \frac{12G}{Ea^2}$, where E is the equivalent elastic modulus and a is the dimension of the side length of the square section [34].

The results presented in Tables 1–3 show that the FEA results are in close agreement with the results obtained from closed form analytical solutions that take into account the effects of warping. The close agreement between the analytical and FEA results validate the developed FE model.

The difference between the design charts and FEA results is summarized in Table 4. The % difference values were determined as follows:

$$\% \, difference = \left|\frac{Design \, chart \, result - FEA \, result}{Design \, chart \, result}\right| \quad (13)$$

In Table 4, D_{GJ}^0, D_{GJ}^{10}, and D_{GJ}^{20} correspond to the percent differences between the normalized torsional stiffness values for the 0°, 10°, and 20° fiber orientations obtained using the FEA and analytical approaches. In turn, D_{EI}^0, D_{EI}^{10}, and D_{EI}^{20} correspond to the percent differences between the normalized bending stiffness values for the 0°, 10°, and 20° fiber orientations, respectively, obtained using the FEA and analytical approaches.

Table 4. Difference between the FEA and analytical method.

Layers	D_{GJ}^{0} (% Difference)	D_{GJ}^{10} (% Difference)	D_{GJ}^{20} (% Difference)	D_{EI}^{0} (% Difference)	D_{EI}^{10} (% Difference)	D_{EI}^{20} (% Difference)
8	5.66	6.25	5.40	5.66	2.04	2.77
12	1.29	5.12	5.96	5.60	1.40	5.76
16	0	2.64	5.33	0	2.17	5.88
20	0.83	1.64	4.72	0	2.70	6.10
24	1.45	0.95	3.57	2.15	3.9	6.31

The values for the % difference provided in Table 4 range from nearly zero up to 6.31%. These differences are due to some of the assumptions that are inherently included in the design charts. For instance, the design charts do not take into account the effects of shear and warping deformations. In using the design charts, designers should remind themselves that the results and graphs are provided for thin-walled beams, and the underlying assumption is that the layer thickness is small. For reference, the individual lamina thicknesses used in this study were taken as 0.18 mm.

To investigate the effects of shear deformation on the design charts, the shear deformation (δ_s) and bending deformation (δ_b) were determined using the equations provided in Table A3 in Appendix B. As can be seen in the table, in addition to the geometric parameters of the beam, shear deformation effects (and their contribution to the total deformation) also depend on the beam end boundary conditions. The results reported here correspond to the laminated composite hollow square beam and I-beam with 24 layers and fiber orientation ranging from 0° to 90°. Figure 7 and Figures S1–S3 refer to the square beam with boundary conditions corresponding to cases 1, 2, and 3 in Table A3, respectively. In turn, I1, I2, and I3 refer to an I-beam with boundary conditions corresponding to cases, 1, 2, and 3 in Table A3, respectively.

From Figure 8, it is observed that, for beams with angle ply layup and fiber angle direction ranging between approximately 30° and 60°, the effect of shear deformation is lower than that of fiber angle orientations between 0° and 30°. This is expected as the shear stiffness has higher values when the fiber orientation is between 30° and 60°. On the other hand, for angle ply layups with fibers oriented between approximately 0° and 30°, the shear deformation effect is significantly higher. The same observations can be made for both the square and I-beam cross sections. Furthermore, observations from Figure 8 show that the effects of shear deformation are significant for boundary conditions other than cantilever beams (Case 1) and lower fiber angle orientations. That is, the effect of shear deformation contribution for fiber orientations between approximately 0° and 30° and simply-simply supported boundary conditions (case 2) and clamped-clamped boundary conditions (case 3) can be close to 60%.

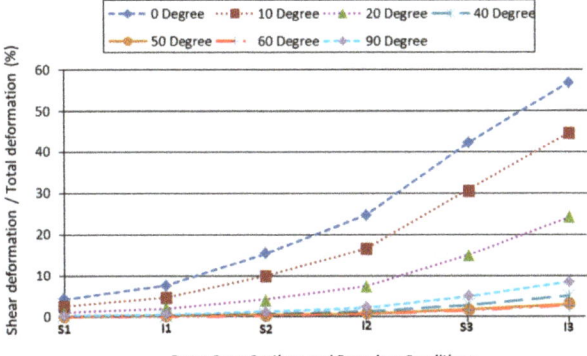

Figure 8. Ratio of shear deformation to total deformation.

The results presented in this section show that the developed design charts can be used for the design of laminated composite beams if the effects of shear deformation are not significant (i.e., less than approximately 10%). The designers can use the results of Figure 8 (and similar plots) to determine the percent contribution of shear deformation. In the case that the shear deformation contribution is larger than 10%, the design charts can have significant inaccuracies and should not be used. The use of the design charts for the design of laminated composite beams is investigated in Sections 7–9 of this paper.

5. Application of Design Charts for the Analysis of Multi-Cell Beams

Laminated composite beams with closed cross-sections that have more than one cell occur in many engineering applications, including aircrafts, ships, and oil rigs, as well as robotic manipulators. Two composite beams with rectangular and I cross sections that fit within the rectangular envelope presented in Section 2 are shown in Figure 9. The cross sections were used to create a three-cell composite beam. These thin-walled multi-cell beams can be analyzed by an extension of the single-cell analysis approach and design charts described in the previous sections. For a multi-cell composite beam, the equivalent normalized bending ratio can be calculated using Equation (14) as follows:

$$\varphi EI_{multicell} = \frac{\sum_{i=1}^{N} EI_{Yi}}{\sum_{i=1}^{N} EI_{Refi}} \qquad (14)$$

where EI_{Yi} is the equivalent bending stiffness for each cell with respect to the Y-axis and EI_{Refi} is the equivalent bending stiffness of the reference beam. It is assumed that all cells have the same geometrical sizes, number of laminae, thicknesses, and material properties, that is, $EI_{Y1} = EI_{Y2} = \ldots = EI_{YN}$. Therefore, $\sum_{i=1}^{N} EI_{Yi} = N \cdot EI_{Yi}$. Similarly, we can write the denominator of Equation (14) as $\sum_{i=1}^{N} EI_{Refi} = N \cdot EI_{Refi}$. Substituting these expressions back into Equation (14), the following equation is obtained:

$$\varphi EI_{multicell} = \varphi EI \qquad (15)$$

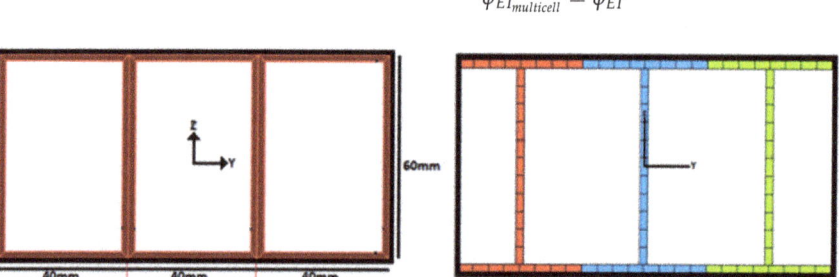

Figure 9. Multi-cell composite beams with rectangular and I-beam cross sections. Each of the single cells has the dimensions of the rectangular envelope (40 mm by 60 mm).

Equation (15) shows that a multi-cell beam's bending stiffness ratio is equal to the beam bending efficiency ratio with respect to the Y-axis (defined in Section 2). Therefore, the bending efficiency design charts can be applied to multi-cell beams with the same limitations and constraints that govern their use for single cell sections.

Similarly, the mass ratio index for single cell beams can be extended to multi-cell beams as follows:

$$\varphi m_{multicell} = \frac{\sum_{i=1}^{N} \rho A_i}{\sum_{i=1}^{N} \rho A_{Di}} \rightarrow \varphi m_{multicell} = \varphi m \qquad (16)$$

The same conclusions do not apply to the cross-sectional properties with respect to the Z-axis (e.g., Z-direction bending stiffness), because the centroidal Z-axis of each cell does not coincide with the centroidal Z-axis of the multi-cell cross section. As a result, in the Z-direction, the design charts for a single cell beam cannot be used for the design of multi-cell beams. This also applies to the torsional stiffness as the shear center for each cell does not coincide with the shear center of the multi-cell beam. However, the axial stiffness plots can be used for the analysis of the multi-cell beams subjected to axial loading.

An example of multi-cell beams used in robotic applications is the plant phenotyping robot, which is partially shown in Figure 10 [28]. The robot is only partially shown as it is currently being patented. The robot is a long-reach 5-DOF manipulator and is designed and manufactured at the University of Saskatchewan. The robot has one prismatic joint that extends its reach to approximately three (3) meters. Each of the two top links are made of three aluminum beams with $L = 1200$ mm and with a rectangular cross section of 40×60 mm. Therefore, the resulting multi-cell beam has dimensions of 120×60 mm. The robot carries several sensors with a total mass of 15 kg at its end-effector. The manipulator's tip displacement at the end effector affects the sensors' performance; therefore, the designer is concerned with designing a robot with minimum tip deflections. As mentioned previously, the robot in [28] was fabricated using aluminum and steel materials. To reduce the mass of the structure and beam deflections (i.e., manipulator's tip deflections), in this study, the authors have examined the effectiveness of using laminated composite beams for the materials of the manipulator. The design charts presented in the previous sections were utilized for this purpose. Results of this work are presented in Sections 7 and 9.

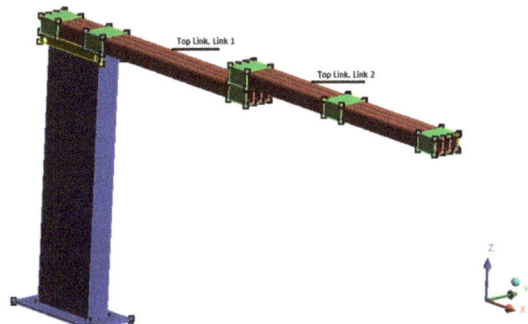

Figure 10. 5-DOF robot manipulator used in plant phenotyping.

6. Simplified Approach and Methodology for Utilizing the Design Charts

This section outlines the simplified approach and methodology for using the design charts. The methodology can be itemized as follows:

1. Select appropriate design chart—As there are different design charts corresponding to the different modes of loading, the designer must first determine the type of loading so that the appropriate design chart(s) can be utilized. The design charts were developed for axial, bending, torsional, and bending-torsional loadings.
2. Calculate mass, deflections, and rotations of reference beam—Once the type of loading is identified, the designer must then calculate the mass and deflections/rotations when the reference beam is subjected to the applied loadings. The reference beam(s) utilized for the development of the design charts are presented and discussed in Section 2. It should be noted that beams other than the reference beam can be used for these calculations. That is, any beam that corresponds to a design point on the design charts can be used for this purpose.
3. Specify desired decrease in mass, deflections, and rotations—Once the deflections and rotations of the reference beam have been determined, the designer must then specify

the desired decrease in deformations and rotations (i.e., increase in axial, bending, or torsional stiffness) or reduction in mass relative to the reference beam(s). That is, the designer specifies the desired normalized axial stiffness (φ_{EA}), bending stiffness (φ_{EI}), torsional stiffness (φ_{GJ}), and mass ratio index (φ_m) as appropriate.

4. Select a "family" of design points that satisfy requirements—Based on the desired normalized parameters in step 3, a "family" of cross sections with the number of layers, fiber angle orientation, mass, and materials of construction that satisfy the design criteria can readily be determined from the design charts.

Figure 11 is a flow chart depicting the required steps for utilizing the design charts.

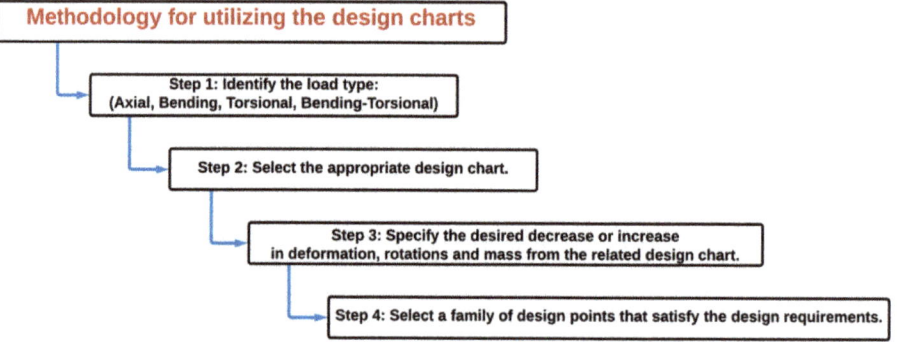

Figure 11. Methodology for utilizing the design charts.

To determine the number of layers, the designer can select any arbitrary lamina thickness for the individual plies. That is, there are no constraints regarding the individual lamina thicknesses as long as the laminated composite beam remains thin-walled and the design chart assumptions discussed in Section 4 are not violated (e.g., shear deflection contribution remains lower than approximately 10%).

In the next sections, the above design methodology is utilized to determine the appropriate cross-sectional shape, layup, and fiber angle orientations for laminated composite beams. Three general case studies were completed. More specifically, in the first case study, the design charts corresponding to the rectangular envelope are used for the design of multi-cell beams for the robotic arm shown in Figure 10. In the second case study, the design charts corresponding to the square envelope are examined. The laminated composite beams are subjected to bending loads in both case studies. In the third case study, tortional loads are applied to the laminated composite beam made up of a single cell rectangular cross section.

It should be noted that the aluminum robotic arm presented in Figure 10 has a maximum deflection of 2 mm when subjected to a 150 N force in the negative Z-direction at its tip [28]. To improve the performance of the robot and accuracy of sensor measurements, designers can consider increasing the beam stiffness while decreasing the mass using the design charts provided in this paper.

7. Case Study 1

In the first case study, a concentrated load of 150 N is applied to the free end of a multi-cell laminated composite beam. As the beam is subjected to in-plane bending, the bending efficiency design charts will have to be utilized. The design aims to decrease the beam tip deflection by increasing the laminated composite beam's Y-direction equivalent stiffness. The notation used to identify the multi-cell beams will be as follows: three rectangular beams (RRR), three oval beams (OOO), and three I-beams (III). The following design point RRR_{CFRP}^{12-20} was selected as the starting design beam and used to calculate the relevant information. It should be noted that any other point on the design charts could have been chosen for this purpose. From Figure 6a, this beam has a bending efficiency of 0.53. To utilize the design charts, two design scenarios are considered. In the first scenario, φ_{EI} is increased from 0.53 to 0.8, and in the second scenario, it is increased from 0.53 to 1.2. Using the design charts provided in Figure 6a, the design points for scenario one can be III_{CFRP}^{16-10}, RRR_{CFRP}^{12-0}, RRR_{CFRP}^{14-10}, and OOO_{CFRP}^{24-0}, and those for scenario two can be III_{CFRP}^{24-0}, RRR_{CFRP}^{20-0}, and RRR_{CFRP}^{22-10}. Therefore, using the design charts provided in Figure 6a, the designer can estimate the decrease in deflection without performing a detailed and computationally expensive FEA as follows:

$$\left(\frac{\varphi_{EI-desired}}{\varphi_{EI-reference\ beam}}\right) = \left(\frac{\delta_{max-reference\ beam}}{\delta_{desired}}\right) \rightarrow \frac{0.8}{0.53} = \frac{2\ mm}{\delta_{desired}} \rightarrow \delta_{desired} \approx 1.32\ mm \tag{17}$$

$$\left(\frac{\varphi_{EI-desired}}{\varphi_{EI-reference\ beam}}\right) = \left(\frac{\delta_{max-reference\ beam}}{\delta_{desired}}\right) \rightarrow \frac{1.2}{0.53} = \frac{2\ mm}{\delta_{desired}} \rightarrow \delta_{desired} \approx 0.88\ mm \tag{18}$$

For purposes of verification and comparison, a detailed FEA of the beams corresponding to the selected design points from the design charts was completed in Ansys. The beams were meshed using the Solid 185 (8-noded linear) element from the FEA software element library. For the purposes of verifying the convergence of the FE results, the beams were also meshed using Solid 186 (20-noded quadratic) element. Figure 12 shows the loading and boundary conditions on these multi-cell beams. The FE models consist of up to ~168,000 nodes and ~158,500 elements depending on the number of layers and laminate thickness.

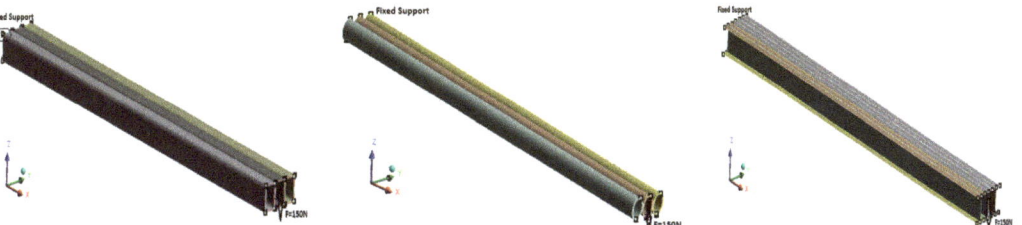

Figure 12. Proposed multi-cell beams (based on the design charts) subjected to bending loads. Cross sections are three rectangular beams (RRR), three oval beams (OOO), and three I-beams (III).

The numerical results and corresponding percent difference with the results obtained using the design charts are summarized in Table 5. As can be observed, there is a very close agreement between the design charts and the deflections predicted by the detailed three-dimensional FEA. It can be seen that, for all of the proposed multi-cell cross sections, the difference between the deflections predicted by the design charts and FEA is less than 8%, which is reasonable. The differences between the results obtained using the design charts and FEA are due to some of the assumptions and limitations of the design charts. For instance, as discussed previously, the design charts do not take into account the effects of shear deformations. In using the design charts, designers should keep in mind that the results and performance indices are provided for thin-walled beams, and the underlying assumption is that shear deformations are negligible.

Table 5. FEA results—maximum deflections under bending loading for the proposed multi-cell beams (using rectangular envelope).

$\varphi_{EI}=0.53 \rightarrow 0.8$		$\varphi_{EI}=0.53 \rightarrow 1.2$	
$\delta_{III^{16,10}_{CFRP}} = 1.30$ mm	1.51% difference	$\delta_{III^{24,0}_{CFRP}} = 0.81$ mm	7.95% difference
$\delta_{RRR^{12,0}_{CFRP}} = 1.37$ mm	3.79% difference	$\delta_{RRR^{20,0}_{CFRP}} = 0.86$ mm	2.27% difference
$\delta_{RRR^{14,10}_{CFRP}} = 1.27$ mm	3.79% difference	$\delta_{RRR^{22,10}_{CFRP}} = 0.85$ mm	3.41% difference
$\delta_{OOO^{24,0}_{CFRP}} = 1.31$ mm	0.76% difference		

8. Case Study 2

In this case study, the effectiveness of the design charts developed using the square envelope are examined. The same loading and applied boundary conditions as that of case study 1 are considered. However, laminated composite beams with square and circular cross-sections that fit the square envelope are examined. The same approach as that of case study 1 was followed to utilize the design charts shown in Figure 4. The notation used to identify the multi-cell beams is as follows: three square beams (SSS), three circular beams (CCC), and three I-beams (III). The following design point SSS^{20-40}_{CFRP} was arbitrarily selected with a bending efficiency of 0.25. To utilize the design charts, two design scenarios are considered. In the first scenario, φ_{EI} is increased from 0.25 to 0.4, and in the second scenario, it is increased from 0.25 to 0.8. Using the design charts provided in Figure 4a, the design points for scenario one can be III^{8-10}_{CFRP}, SSS^{16-30}_{CFRP}, and CCC^{12-10}_{CFRP}, and those for scenario two can be III^{16-10}_{CFRP} and CCC^{24-0}_{CFRP}. Similar to the first case study, the designer can estimate the decrease in deflection without performing a detailed and time-consuming FEA as follows (using design charts in Figure 4):

$$\left(\frac{\varphi_{EI-desired}}{\varphi_{EI-reference\ beam}}\right) = \left(\frac{\delta_{max-reference\ beam}}{\delta_{desired}}\right) \rightarrow \frac{0.4}{0.25} = \frac{2.1\ mm}{\delta_{desired}} \rightarrow \delta_{desired} \approx 1.31\ mm \quad (19)$$

$$\left(\frac{\varphi_{EI-desired}}{\varphi_{EI-reference\ beam}}\right) = \left(\frac{\delta_{max-reference\ beam}}{\delta_{desired}}\right) \rightarrow \frac{0.8}{0.25} = \frac{2.1\ mm}{\delta_{desired}} \rightarrow \delta_{desired} \approx 0.65\ mm \quad (20)$$

A detailed FEA of the beams corresponding to the selected design points from the design charts was also completed. Table 6 shows the FEA results. It can be seen that, for all of the proposed square and circular multi-cell cross sections, there is a good agreement between the deflections predicted by the design charts and FEA.

Table 6. FEA results—maximum deflections under bending loading for the proposed multi-cell beams (using square envelope).

$\varphi_{EI}=0.25 \rightarrow 0.4$		$\varphi_{EI}=0.25 \rightarrow 0.8$	
$\delta_{III^{8-10}_{CFRP}} = 1.30$ mm	0.76% difference	$\delta_{III^{16-10}_{CFRP}} = 0.67$ mm	3.08% difference
$\delta_{CCC^{12-10}_{CFRP}} = 1.32$ mm	0.76% difference	$\delta_{CCC^{24-0}_{CFRP}} = 0.69$ mm	6.15% difference
$\delta_{SSS^{16-30}_{CFRP}} = 1.38$ mm	5.34% difference		

9. Case Study 3

In addition to bending loads, the laminated composite beams in structures and mechanisms can be subjected to torsional loads. For instance, if the sensors at the tip of the beam are installed at a location that is not in-line with the shear center of the cross section of the beam, the beam will be subjected to torsional loading. In this section, the torsional efficiency design charts for the square envelope were used to increase the torsional stiffness of the beam, and hence decrease the angle of twist of the beam. For this purpose, a torsional moment of 36,000 N.mm. was applied at the free end of a cantilever beam with a length of 1200 mm. The cross-sectional dimensions of this beam are such that it fills the square envelope defined in Section 2. A detailed FEA was also completed on the beam to determine the angle of twist. The developed FE model, meshing, applied boundary conditions, and loading are shown in Figure 13.

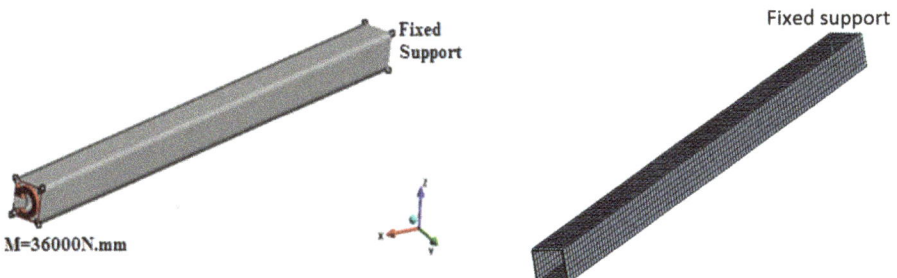

Figure 13. Applied boundary conditions and loading of beam subjected to torsional loading (FE models consist of up to ~61,000 nodes and ~57,400 elements depending on laminate thickness).

As the developed torsional efficiency design charts cannot be applied to multi-cell beams, only single cell beams are considered here. To utilize the design charts, several design points are considered. The beam corresponding to the design point S_{CFRP}^{20-0} with φ_{GJ} of 1.2 was arbitrarily selected as the starting beam. The aim is to choose a laminated composite beam with a certain fiber angle orientation, mass, and cross section that can provide the required stiffness and normalized mass index ratio φ_M, without performing a detailed and time-consuming FEA on numerous laminated composite beams. The design charts provided in Figure 5 are utilized using the same approach as that in case studies 1 and 2 to determine the angle of twist for various beams with higher stiffness than the starting reference beam. These beam cross sections are outlined in Table 7. In practice, the desired angle of twist (and mass) will be specified, and the designer can readily choose a laminated composite beam from the design charts that will provide the require stiffness (and mass) to meet the specified criteria.

Table 7. Maximum angle of twist for single cell beam subjected to torsion.

Beam Cross-Section and Layup	φ_{GJ}	φ_M	Angle of Twist—Design Chart (rad)	Angle of Twist—FEA (rad)	% Difference
S_{CFRP}^{24-40}	7.35	1.46	1.50×10^{-3}	1.60×10^{-3}	6.67
C_{CFRP}^{20-40}	5.00	0.96	2.10×10^{-3}	2.30×10^{-3}	9.52
S_{CFRP}^{16-30}	4.29	1	2.50×10^{-3}	2.70×10^{-3}	8.00
C_{CFRP}^{12-30}	2.61	0.59	4.10×10^{-3}	4.30×10^{-3}	4.88
S_{CFRP}^{20-0}	1.20	1.23	8.90×10^{-3}	9.00×10^{-3}	1.12

The results obtained from the designed charts and FEA are shown in Table 7. Note that the difference between the angles of twist calculated using the design charts versus FEA is less than 10% for all cases.

As can be observed, the design charts can be a very useful tool for determining a "family" of possible beams with the desired stiffness that meet the deflection or rotation criteria. The final selection of an appropriate beam from the range of possible options depends on application, ease of fabrication, costs, and available manufacturing machines and technologies. Regarding the robotic arm, it can be observed how the design charts can aid in selecting a laminated composite beam with similar or better structural performance than the aluminum beam (e.g., higher stiffness and lower mass).

10. Conclusions

In this study, structural efficiency metrics and performance indices were defined for the integrated selection of the design parameters that govern the static and dynamic performance of laminated thin-walled composite beams. Important parameters in the design of laminated composite beams include the number of layers, thickness of laminate, fiber angle orientation, materials of construction, and cross-sectional shape. Based on these design parameters, performance indices such as normalized axial stiffness, bending stiffness, torsional stiffness, combined torsional-bending stiffness, and the mass index ratio were defined and are summarized in design charts. The developed design charts are a useful tool for comparing the structural performance of laminated composite beams of various cross-sections, fiber angle orientation, and materials. To analyze cross sections of various types and sizes, results are generated for two envelopes. The proposed charts assist the designer in the design variable selection process and analysis of beams with square, rectangular, circular, oval, and I-beam cross sections. In addition to single cell beams, the application of the design charts was extended to multi-cell beams for certain loading conditions.

Three-dimensional finite element models using linear and quadratic solid elements were developed and the numerical simulation results were compared to the closed form solution and design chart results. It was observed that the design charts can be used to assist engineers and researchers to accurately and efficiently predict the structural behavior (e.g., stiffnesses, deflections, and rotations) of laminated composite beams that fall within the boundaries of the design charts. It was observed that the two main factors that impact the accuracy of the results predicted by the design charts are shear deformation and warping effects. The results show that, for cases where the shear deformation contribution to total deformation is below 10%, the design charts provide an accurate prediction of the behavior of the composite beam.

The simplified approach for parameter selection and analysis of laminated composite beams was summarized in a series of steps. The design charts and simplified approach were then used as a tool for selecting laminated composite beam(s) with the required stiffness and mass to replace an aluminum robotic arm. It was observed that the proposed approach can be used in the design of laminated composite beams without the need to create detailed FE models that are time-consuming to develop and computationally expensive.

Author Contributions: The Phenotyping robot (hardware and software parts) was developed by the Robotics Lab at the University of Saskatchewan over a period of four years; Conceptualization, M.A.K. and R.M.; methodology, R.M., M.A.K. and R.F.; software, M.A.K.; validation, R.M., M.A.K. and R.F.; formal analysis, M.A.K. and R.M.; investigation, M.A.K., R.M. and R.F.; resources, R.F.; data curation, R.M. and M.A.K.; writing—original draft preparation, M.A.K.; writing—review and editing, R.M. and R.F.; visualization, M.A.K. and R.M.; supervision, R.M. and R.F.; project administration, R.M. and R.F.; funding acquisition, R.F. All authors have read and agreed to the published version of the manuscript.

Funding: This work is supported by the Canada First Research Excellence Fund (CFREF) through the Global Institute for Food Security (GIFS), University of Saskatchewan, Canada. Also, funding was provided by an NSERC-CRD grant of R.F.

Data Availability Statement: Not applicable.

Conflicts of Interest: The authors declare no conflict of interest.

Appendix A. Equivalent Axial, Bending, and Torsional Stiffness Equations and Material Properties

Table A1. Equivalent axial, bending, and torsional stiffnesses [19].

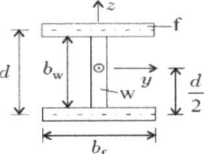

$$EA = \frac{2b_f}{(a_{11})_f} + \frac{2b_w}{(a_{11})_w}$$

$$EI_{yy} = \frac{b_f}{(a_{11})_f}\frac{d^2}{2} + \frac{2b_f}{(d_{11})_f} + \frac{2b_w^3}{12(a_{11})_w}$$

$$EI_{zz} = \frac{b_w}{(a_{11})_f}\frac{d_f^2}{2} + \frac{2b_w}{(d_{11})_w} + \frac{2b_f^3}{12(a_{11})_f}$$

$$GI_t = \frac{2d_f^2 d^2}{(a_{66})_f d_f + (a_{66})_w d}$$

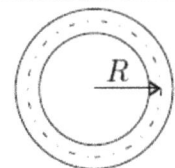

$$EA = \frac{2b_f}{(a_{11})_f} + \frac{b_w}{(a_{11})_w}$$

$$EI_{yy} = \frac{b_f}{(a_{11})_f}\frac{d^2}{2} + \frac{2b_f}{(d_{11})_f} + \frac{2b_w^3}{12(a_{11})_w}$$

$$EI_{zz} = \frac{b_w}{(d_{11})_w} + \frac{2b_f^3}{12(a_{11})_f}$$

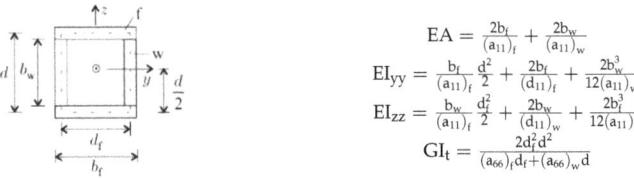

$$EA = \left(\frac{2R\tau}{a_{11}}\right)$$

$$EI_{yy} = EI_{zz} = \tau\left(\frac{R^3}{a_{11}} + \frac{R}{d_{11}}\right)$$

$$GI_t = \frac{2R^3\pi}{a_{66}}$$

$$EA = \frac{2b_f}{(a_{11})_f} + \frac{b_w}{(a_{11})_w}$$

$$EI_{yy} = \frac{b_f}{(a_{11})_f}\frac{d^2}{2} + \frac{2b_f}{(d_{11})_f} + \frac{b_w^3}{12(a_{11})_w}$$

$$EI_{zz} = \frac{b_w}{(a_{11})_w}(d_f - y_c)^2 + \frac{b_w}{(d_{11})_w} + \frac{2}{(a_{11})_f}\left(\frac{y_c^3}{3} + \frac{(b_f - y_c)^3}{3}\right)$$

$$y_c = \frac{1}{EA}\left(\frac{2b_f}{(a_{11})_f}\frac{b_f}{2} + \frac{b_w}{(a_{11})_w}d_f\right)$$

Table A2. Properties for the CFRP and GFRP lamina, values adopted from [33].

Property	CFRP	GFRP
Density, p, g/cm^3	1.60	1.97
Longitudinal modulus, E1, GPa	147.0	41.0
Transverse in-plane modulus, E2, GPa	10.3	10.4
Transverse out-of-plane modulus, E3, GPa	10.3	10.4
In-plane shear modulus, G12, GPa	7.00	4.30
Out-of-plane shear modulus, G23, GPa	3.70	3.50
Out-of-plane shear modulus, G13, GPa	7.00	4.30
Major in-plane Poisson's ratio, v12	0.27	0.28
Out-of-plane Poisson's ratio, v23	0.54	0.50
Out-of-plane Poisson's ratio, v13	0.27	0.28

Appendix B. Comparison of FEA Results and Shear to Bending Deformation Ratios

Table A3. Bending and shear deformation equations, formula adopted from [19].

Boundary Conditions	Bending Deformation δ_b	Shear Deformation δ_s	$\frac{\delta_s}{\delta_b}$
Case 1	$\frac{PL^3}{3EI}$	$\frac{PL}{S}$	$\frac{3EI}{L^2 S}$
Case 2	$\frac{PL^3}{48EI}$	$\frac{PL}{4S}$	$\frac{12EI}{L^2 S}$
Case 3	$\frac{PL^3}{192EI}$	$\frac{PL}{4S}$	$\frac{48EI}{L^2 S}$

References

1. Carrera, E.; Filippi, M.; Mahato, P.K.; Pagani, A. Advanced models for free vibration analysis of laminated beams with compact and thin-walled open/closed sections. *J. Compos. Mater.* **2015**, *49*, 2085–2101. [CrossRef]
2. Jun, L.; Jin, X. Coupled Bending-Torsional Dynamic Response of Axially Loaded Slender Composite Thin-Walled Beam with Closed Cross-Section. *J. Compos. Mater.* **2004**, *38*, 515–534. [CrossRef]
3. Infante, V.; Madeira, J.; Ruben, R.B.; Moleiro, F.; de Freitas, S.T. Characterization and optimization of hybrid carbon–glass epoxy composites under combined loading. *J. Compos. Mater.* **2019**, *53*, 2593–2605. [CrossRef]
4. Rozylo, P.; Teter, A.; Debski, H.; Wysmulski1, P.; Falkowicz, K. Experimental and Numerical Study of the Buckling of Composite Profiles with Open Cross Section under Axial Compression. *Appl. Compos. Mater.* **2017**, *24*, 1251–1264. [CrossRef]
5. Pan, S.; Dai, Q.; Safaei, B.; Qin, Z.; Chu, F. Damping characteristics of carbon nanotube reinforced epoxy nanocomposite beams. *Thin-Walled Struct.* **2021**, *166*, 108127. [CrossRef]
6. Dai, Q.; Qin, Z.; Chu, F. Parametric study of damping characteristics of rotating laminated composite cylindrical shells using Haar wavelets. *Thin-Walled Struct.* **2021**, *161*, 107500. [CrossRef]
7. Li, H.; Lv, H.; Gu, J.; Xiong, J.; Han, Q.; Liu, J.; Qin, Z. Nonlinear vibration characteristics of fibre reinforced composite cylindrical shells in thermal environment. *Mech. Syst. Signal Process.* **2021**, *156*, 107665. [CrossRef]
8. Li, H.; Lv, H.; Sun, H.; Qin, Z.; Xiong, J.; Han, Q.; Liu, J.; Wang, X. Nonlinear vibrations of fiber-reinforced composite cylindrical shells with bolt loosening boundary conditions. *J. Sound Vib.* **2021**, *496*, 115935. [CrossRef]
9. D'Antino, T.; Sneed, L.H.; Carloni, C.; Pellegrino, C. Effect of the inherent eccentricity in single-lap direct-shear tests of PBO FRCM-concrete joints. *Compos. Struct.* **2016**, *142*, 117–129. [CrossRef]
10. Ombres, L.; Verre, S. Experimental and numerical investigation on the steel reinforced grout (SRG) composite-to-concrete bond. *J. Compos. Sci.* **2020**, *4*, 182. [CrossRef]
11. Gillet, A.; Francescato, P.; Saffre, P. Single- and Multi-objective Optimization of Composite Structures: The Influence of Design Variables. *J. Compos. Mater.* **2010**, *44*, 457–480. [CrossRef]
12. Walker, M.; Smith, R.E. A Technique for the Multi-objective Optimization of Laminated Composite Structures using Genetic Algorithms and Finite Element Analysis. *Compos. Struct.* **2003**, *62*, 123–128. [CrossRef]
13. Feil, R.; Pflumm, T.; Bortolotti, P.; Morandini, M. A Cross-Sectional Aeroelastic Analysis and Structural Optimization Tool for Slender Composite Structures. *Compos. Struct.* **2020**, *253*, 112755. [CrossRef]
14. Moazed, R.; Fotouhi, R.; Szyszkowski, W. Out-of-plane Behaviour and FE Modeling of a T-Joint Connection of Thin-Walled Square Tubes. *Thin-Walled Struct.* **2012**, *51*, 87–98. [CrossRef]
15. Moazed, R.; Fotouhi, R.; Szyszkowski, W. The In-Plane Behaviour and FE Modeling of a T-Joint Connection of Thin-Walled Square Tubes. *Thin-Walled Struct.* **2009**, *47*, 816–825. [CrossRef]
16. Pasini, D. Shape transformers for material and shape selection of lightweight beams. *Mater. Des.* **2007**, *28*, 2071–2079. [CrossRef]
17. Ashby, M.F. Overview No. 92: Materials and Shape. *Acta Metall. Mater.* **1991**, *39*, 1025–1039. [CrossRef]
18. Ashby, M.F.; Bréchet, Y.J.M. Designing Hybrid Materials. *Acta Mater.* **2003**, *51*, 5801–5821. [CrossRef]
19. Kollár, L.; Springer, G. *Mechanics of Composite Structures*; Cambridge University Press: Cambridge, UK, 2003. [CrossRef]
20. An, H.; Singh, J.; Pasini, D. Structural Efficiency Metrics for Integrated Selection of Layup, Material, and Cross-Section Shape in Laminated Composite Structures. *Compos. Struct.* **2017**, *170*, 53–68. [CrossRef]
21. Pasini, D. A New Theory for Modeling the Mass-Efficiency of Material, Shape and Form. Ph.D. Thesis, University of Bristol, Bristol, UK, 2003.
22. Buckney, N.; Pirrera, A.; Weaver, P.M. Structural Efficiency Measures for Sections Under Asymmetric Bending. *ASME. J. Mech. Des.* **2015**, *137*, 011405. [CrossRef]
23. Wanner, A. Minimum-Weight Materials Selection for Limited available Space. *Mater. Des.* **2010**, *31*, 2834–2839. [CrossRef]

24. Pasini, D. Shape and Material Selection for Optimizing Flexural Vibrations in Multilayered Resonators. *J. Microelectromech. Syst.* **2006**, *15*, 1745–1758. [CrossRef]
25. Amany, A.; Pasini, D. Material and Shape Selection for Stiff Beams Under Non-Uniform Flexure. *Mater. Des.* **2009**, *30*, 1110–1117. [CrossRef]
26. Singh, J.; Mirjalili, V.; Pasini, D. Integrated shape and material selection for single and multi-performance criteria. *Mater. Des.* **2011**, *32*, 2909–2922. [CrossRef]
27. Lewis, F.L.; Dawson, D.M.; Abdallah, C.T. *Robot Manipulator Control: Theory and Practice*; CRC Press: Boca Raton, FL, USA, 2003. [CrossRef]
28. Zhang, Q.-W.; Fotouhi, R.; Cote, J.; Pour, M.K. Lightweight Long-Reach 5-DOF Robot Arm for Farm Application. In Proceedings of the ASME 2019 International Design Engineering Technical Conferences and Computers and Information in Engineering Conference, Anaheim, CA, USA, 18–21 August 2019. IDETC2019-98366.
29. Ghazavi, A.; Gordaninejad, F.; Chalhoub, N.G. Dynamic Analysis of a Composite-Material Flexible Robot Arm. *Comput. Struct.* **1993**, *49*, 315–327. [CrossRef]
30. Caprino, G.; Langella, A. Optimization of robotic arms made of composite materials for maximum fundamental frequency. *Compos. Struct.* **1995**, *31*, 1–8. [CrossRef]
31. Yin, H.; Liu, J.; Yang, F. Hybrid Structure Design of Lightweight Robotic Arms Based on Carbon Fiber Reinforced Plastic and Aluminum Alloy. *IEEE Access.* **2019**, *7*, 64932–64945. [CrossRef]
32. Lee, C.S.; Lee, D.G.; Oh, J.H.; Kim, H.S. Composite Wrist Blocks for Double Arm Type Robots for Handling Large LCD Glass Panels. *Compos. Struct.* **2002**, *57*, 345–355. [CrossRef]
33. *Ansys Version 18.1 Standard User's Manual*. 2018. Available online: www.ansys.com (accessed on 17 August 2021).
34. Pluzsik, A.; Kollár, L.P. Effects of Shear Deformation and Restrained Warping on the Displacements of Composite Beams. *J. Reinf. Plast. Compos.* **2002**, *21*, 1517–1541. [CrossRef]

Article

Influences on Textile and Mechanical Properties of Recycled Carbon Fiber Nonwovens Produced by Carding

Frank Manis [1,*], Georg Stegschuster [2], Jakob Wölling [1] and Stefan Schlichter [2]

[1] Fraunhofer Institute for Casting, Composite and Processing Technology IGCV, 86159 Augsburg, Germany; jakob.woelling@igcv.fraunhofer.de
[2] Institut Für Textiltechnik Augsburg gGmbH, 86159 Augsburg, Germany; georg.stegschuster@ita-augsburg.de (G.S.); stefan.schlichter@ita-augsburg.de (S.S.)
* Correspondence: frank.manis@igcv.fraunhofer.de

Abstract: Nonwovens made of recycled carbon fibers (rCF) and thermoplastic (TP) fibers have excellent economic and ecological potential. In contrast to new fibers, recycled carbon fibers are significantly cheaper, and the CO_2 footprint is mostly compensated by energy savings in the first product life cycle. The next step for this promising material is its industrial serial use. Therefore, we analyzed the process chain from fiber to composite material. Initially, the rCF length at different positions during the carding process was measured. Thereafter, we evaluated the influence of the TP fibers on the processing, fiber shortening, and mechanical properties. Finally, several nonwovens with different TP fibers and fiber volume contents between 15 vol% and 30 vol% were produced, consolidated by hot-pressing, and tested by four-point bending to determine the mechanical values. The fiber length reduction ranged from 20.6% to 28.4%. TP fibers cushioned the rCF against mechanical stress but held rCF fragments back due to their crimp. The resulting bending strength varied from 301 to 405 MPa, and the stiffness ranged from 16.3 to 30.1 GPa. Design recommendations for reduced fiber shortening are derived as well as material mixtures that offer better homogeneity and higher mechanical properties.

Citation: Manis, F.; Stegschuster, G.; Wölling, J.; Schlichter, S. Influences on Textile and Mechanical Properties of Recycled Carbon Fiber Nonwovens Produced by Carding. *J. Compos. Sci.* **2021**, *5*, 209. https://doi.org/10.3390/jcs5080209

Keywords: carbon fiber; recycling; nonwoven; carding; hot pressing; polyamide 6; polyethylene terephthalate

Academic Editors: Jiadeng Zhu, Gouqing Li and Lixing Kang

Received: 25 May 2021
Accepted: 30 July 2021
Published: 6 August 2021

Publisher's Note: MDPI stays neutral with regard to jurisdictional claims in published maps and institutional affiliations.

Copyright: © 2021 by the authors. Licensee MDPI, Basel, Switzerland. This article is an open access article distributed under the terms and conditions of the Creative Commons Attribution (CC BY) license (https://creativecommons.org/licenses/by/4.0/).

1. Introduction

Composite materials have shown a strong increase in their versatility in recent years as well as a strong increase in demand. In particular, the wind energy and aerospace markets cannot be imagined without fiber reinforced polymers (FRP), such as glass fiber reinforced (GFRP) or carbon fibers reinforced plastics (CFRP). Carbon fibers are widely used because of their high specific strength and stiffness, which are superior to those of conventional metals and glass fiber composites. Therefore, carbon fiber reinforced plastics showed an increase in demand of 8.84% (Compound Annual Growth Rate 2010–2020) [1]. New production sites have been built, and new applications have been found and marketed [1].

Negative impacts from usage of this lightweight material are due to its limited recyclability and limited possibilities for circularity. Both pre-consumer waste as well as post-consumer waste are increasing, and the availability of secondary fibers is higher than the demand for recycled carbon fibers (rCF). Some of the reasons include low or unknown mechanical properties, inhomogeneous material behavior, limited possibility of simulation, and a high material price.

For this reason, the technologies of processing and evaluating recycle carbon fibers have been under steady investigation for many years starting with the work of Pickering et al. and Pimenta et al. [2–5]. The first scientific works were mostly related to the fiber-matrix separation by pyrolysis and oxidation [6–8]. In recent years, more papers regarding solvolysis were published [9–12], and lab and medium scale devices were built [13,14].

After the fibers are chopped and reclaimed, they can be used for textile processes, such as carding and wet-laid technologies. The carding process offers a higher technology readiness level (TRL) as well as productivity, and many companies have started to rebuild and change their carding design to improve carbon fiber handling [15–20]. Fiber breakage and unstable textile properties, including the coefficient of variation (CV value) of the areal weight or varying fiber volume contents, still pose great challenges for the market access of this material.

The main advantages of the nonwoven process are excellent economic efficiency due to the high production output and high flexibility regarding properties, like the areal weight, isotropy, and degree of compaction. To be able to successfully process recycled carbon fibers on a nonwoven line, the line is modified to minimize fiber damage. The compact carding process combines several conversion processes to produce a homogeneous textile surface from a fiber blend. A major goal of recycling is to keep the fiber length as long as possible in order to enable multiple recycling cycles and, thus, contribute to a complete circular economy. Furthermore, longer CFs can contribute to better mechanical properties of the composite [21,22]. The minimum fiber length for carding is around 30 mm; therefore, in this paper, this threshold was chosen for the evaluation of the process.

The fibers are successively separated along the process into single fibers, which are parallelized. In technical jargon, those processes are called opening and carding. These processes are carried out by rollers, which are covered with a saw-tooth-wire—the clothing. The clothing enhances the grip between fibers and rollers [23]. The carded fibers are finally accumulated on the doffer roller and a coherent, uniform fiber mat forms: the carding web.

The carding web builds the basis for the nonwoven fabric. As the areal weight of the carding web ranges from only 10 to 40 g/m^2, it is layered and stacked to achieve the desired areal weight and improve the handling. The stacked webs are bonded together by needling with a needle loom. The fibers are intertwined by the needles, and the resulting frictional connection with the surrounding fibers creates an irreversible bond [24].

The production of needle-punched nonwovens has been developed and optimized for thermoplastic and/or staple fibers of natural origin. The use of fiber blends with carbon and thermoplastic fibers offers two significant advantages in the production process. First, the thermoplastic fibers act as carriers for the carbon fibers, which reduces the fiber length shortening [25]. Second, the thorough mixing during the nonwoven formation results in minimal flow paths for the matrix [26]. Usually, high-viscosity thermoplastics are impregnated into textile surfaces by high pressure and temperature [27]. Nonwovens from carbon thermoplastic fiber blends do not require this step in the same way.

Nonwovens from rCF can be processed in many ways. If thermoplastic fibers are added to the carding or wet-laid process, the hybrid nonwovens can be consolidated directly by variothermal hotpressing or isothermal pressing. Pure CF nonwovens are suitable for infiltration processes, such as resin transfer molding or wet compression molding [28]. As hybrid nonwovens already have their polymer included, process time and costs can be saved. Furthermore, the thermoplastic components are formable and weldable and can be further functionalized by injection molding.

In this study, we investigate possible improvements for dry-laid nonwovens from recycled carbon fibers on two different levels: carbon fiber and composite. At the carbon fiber level, the degree of carbon fiber breakage is studied depending on three factors. The factors are the web formation process in the carding machine, the number of worker–stripper pairs of the carding machine, and the blend ratio of thermoplastic fibers in a blend with carbon fibers. The aims on the composite level are the investigation of the mechanical influence of the carbon fiber volume content as well as the influence of different polymer types.

2. Materials and Methods

2.1. Materials and Methods at the Carbon Fiber Level

Two fundamentally different kinds of carbon fibers are used for the trials, which are representative types of recycled carbon fibers. On the one hand, carbon fiber cut-offs from textile production are employed, whose mechanical and rheological properties correspond to those of virgin carbon fibers. On the other hand, pyrolized carbon fibers are utilized, which undergo a heat treatment process during the reclamation to separate the matrix from the carbon fibers.

The pyrolysis removes the friction-reducing sizing on the surface of the carbon fibers, thus, leading to higher friction and stress on the carbon fibers during subsequent processing. In addition, the heat treatment can cause degradation of the carbon fibers, which increases the probability of breakage [2]. The carbon fiber of type CarboNXT chopped 60,000 NP5 R (VCF) produced by CarboNXT GmbH, Stade, is a carbon fiber that originates from automotive cut-off waste and still has an intact sizing attached from the carbon fiber production.. Two length classes of VCF are used, which differ in the number of cutting procedures:

- VCF3x: three-fold cutting by guillotine cutting (see Figure 1) with a 60-mm blade distance.
- VCF1x: one-fold cutting by guillotine cutting with a 60-mm blade distance.

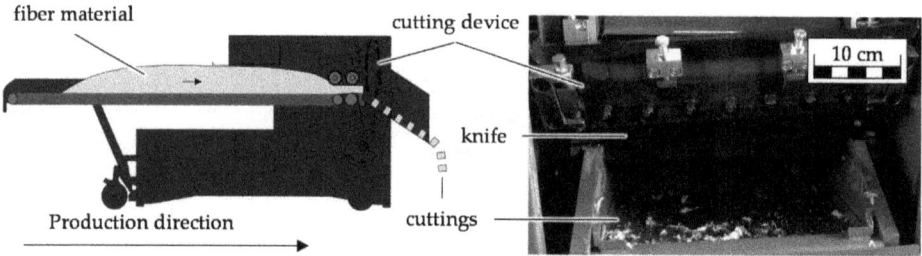

Figure 1. Schematic representation of the guillotine cutting [29].

The pyrolized carbon fiber used was a Carbiso C SM45R 60/90 (PCF) from ELG Carbon Fibre Ltd., Coseley, UK. According to the manufacturer, the fibers are between 60-mm and 90-mm long and do not have any sizing left on the fibers. This carbon fiber originates from automotive waste but is additionally thermally treated by ELG Carbon Fibre.

Three thermoplastic (TP) fibers were used in this study as shown in Table 1. The types P300 and P301 are polyamide 6 (PA6) fibers with different textile properties, which were produced by EMS-CHEMIE HOLDING AG, Dormat/Ems, Switzerland. In addition, a polyethylene terephthalate (PET) fiber type TREVIRA® 290 from Trevira GmbH, Bobingen, Germany, was employed. The TP fiber properties differ in length, fineness, and crimp and are summarized in Table 1.

Table 1. Thermoplastic fibers and their textile properties.

Type	Length [mm]	Fineness [dtex]	Crimp [B/cm]	Material Code
P300	40	1.7	8	PA6-40-1.7-8
P301	60	6.7	6	PA6-60-6.7-6
TREVIRA® 290	60	6.7	4	PET-60-6.7-4

Polyamid 6 is commonly used in the automotive industry and was, therefore, chosen in this study as a benchmark material. PET is widely used in textiles but is not a typical matrix material. The fibers, however, are cheaper than polyamid 6 fibers and additionally offer higher stiffness and strength values. For those reasons, this material was also investigated as a matrix material in this study. Fineness is a measure for the weight of textile fibers regarding their length. The unit dtex describes the weight in gram per 10,000 m of fiber length. The crimp is a unit for the number of curls per centimeter that a textile fiber possesses. Those curls lead to a better grip between fibers and the wires of the rolls of the carding machine.

All fibers and blends were processed at ITA Augsburg on the nonwoven line KC11 2–4 SD/MEK 11 from Dilo Systems GmbH, Eberbach, at ITA Augsburg. Three separate trial series were conducted to investigate the influence on the carbon fiber length, which is explained in detail in the following subsections.

2.1.1. Materials and Methods for Investigating the Influence of Web Formation on Carbon Fiber Length

In the first trial series, three different fiber blends were processed with the nonwoven line using a fixed set of machine parameters to determine the influence of the web formation process on the carbon fiber length. Therefore, the carbon fiber VCF3x was processed pure and as a blend with 60 wt% of the polyamid6 fiber PA6-40-1.7. Preliminary trials showed that a thermoplastic fiber content of 60 wt% was sufficient for a secure consolidation and production of composites, which are used for mechanical testing in the later stages of the study. The pyrolized carbon fiber PCF was only processed as a blend with 60 wt% of the PA6-40-1.7 since the high fiber-metal friction was expected to result in severe fiber length losses in the pure processing of pyrolized carbon fibers. The materials and the respective ratio of mixture are shown in Table 2.

Table 2. Processed fiber blends for fiber length measurement regarding web formation.

Carbon Fiber	wt%	TP Fiber	wt%	Worker–Stripper Pair	Material Code
VCF3x	100	-	-	3	VCF3x
VCF3x	40	PA6-40-1.7-8	60	3	VCF3x_PA6
PCF	40	PA6-40-1.7-8	60	3	PCF_PA6

The influence of the web formation by the carding machine was analyzed by measuring the fiber length at four positions along the carding machine. The fiber length measuring positions as well as the machine setup used are shown in Figure 2.

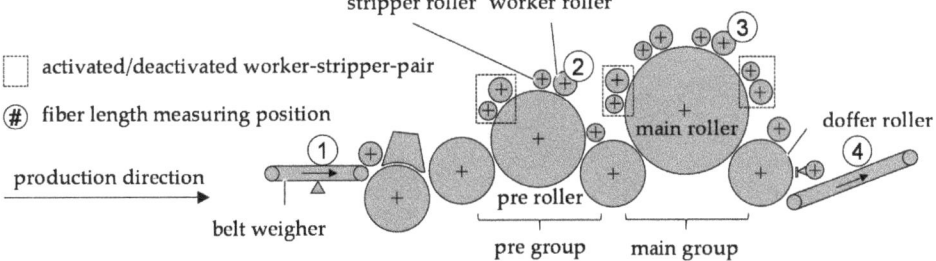

Figure 2. Schematic representation of the carding machine. The numbers indicate the positions of sampling for the fiber length measurements [30].

The first fiber length measuring position (pos. 1) is located on the belt weigher. The results at this position are the reference. The second fiber length measuring position (pos. 2) is located at the second worker roll of the pre group giving the length reduction caused by the material intake and the pre group. The third fiber length measuring position (pos. 3) is located at the third worker roll of the main group. The results show the length reduction caused by the main group where the most intense carding takes place. The fourth fiber length measurement position (pos. 4) is located after the carding machine. The samples are taken directly from the web, and the results are used to evaluate the shortening effect of the web formation on the doffer roller.

2.1.2. Materials and Methods for Investigating the Influence of Worker–Stripper Pair Numbers on Carbon Fiber Length

In the second trial series two different machine setups were investigated where three and six active worker–stripper pairs were employed as shown in Figure 2. Between the worker rollers and the main rollers, the fibers were carded. Carding is a trade-off between fiber orientation and fiber damage. The degree of fiber orientation in the web is directly related to the number of worker–stripper pairs.

Six worker–stripper pairs were the maximum number possible in the current machine setup resulting in a high degree of orientation, which is desired in the web. Three worker–strippers were chosen as the minimum number as this number still provides acceptable fiber orientation while significantly reducing stress on the fibers. To evaluate the influence of the machine setup, the carbon fiber VCF1x was processed using three and six active worker–stripper pairs (Table 3(a)) measuring the carbon fiber length on the belt weigher and on the produced web.

Table 3. Processed fiber blends for fiber length measurement regarding the machine setup.

	Carbon Fiber	wt%	wt% of Thermoplastic Fiber	Worker–Stripper Pair	Material Code
(a)	VCF1x	100	-	3	VCF1x_100wt%_3
	VCF1x	100	-	6	VCF1x_100wt%_6
(b)	VCF1x	40	60	3	VCF1x_40wt%_3
(c)	VCF1x	90	10	6	VCF1x_90wt%_6

2.1.3. Materials and Methods for Investigating the Influence of Blend Ratio of Thermoplastic Fibers and Carbon Fibers on Carbon Fiber Length

In the third trial series, the influence of different blend ratios between thermoplastic fibers and carbon fibers is investigated using three and six worker–stripper pairs, respectively. The trials using three active worker–stripper pairs are designed to study the lowest possible damaging of carbon fibers with the nonwoven line. Therefore 60 wt% of thermoplastic fibers are used as shown in Table 3(b).

The tests with six active worker–stripper pairs serve to investigate the influence of a small quantity of thermoplastic fibers when the carbon fibers are subjected to the highest possible stress. Therefore, 10 wt% of thermoplastic fibers are used as shown in Table 3(c). Although the addition of 10 wt% of thermoplastic fibers will not suffice to a complete wet out of a composite, the effect on fiber length reduction is significant and is thus investigated. Small amounts of thermoplastic fibers could be used to greatly enhance the fiber length of carbon fiber nonwovens used solely for resin impregnation. Carbon fiber samples are taken for measurement on the belt weigher and the web (fiber length measuring positions 2 and 4, respectively, shown in Figure 2).

2.1.4. Carbon Fiber Length Measurement

The length of the carbon fibers from each position was measured by the two-tweezers method in accordance to DIN 53808-1. The sample size included 200 single carbon fiber measurements. The results were sorted into length classes of 5- or 1-mm width. Thus, the mean fiber length and the fiber length distribution were determined as shown in fiber length diagrams. Due to the manual procedure and the subjective selection of carbon fibers by hand as well as the fracture behavior of the recycled carbon fibers under mechanical load, this procedure is prone to errors.

Therefore, the fiber length measurement was carried out by a designated person to minimize the subjective influence during testing. Since individual carbon fibers break during handling with tweezers, the measuring concentrated on rovings—carbon fiber bundles consisting of several thousand individual fibers—instead (Figure 3). The rovings account for 15.6 wt% of the carbon fibers [30]. Although the fiber length measurements, thus, only cover a small portion of the carbon fibers there is, at present, no more sophisticated measuring method available.

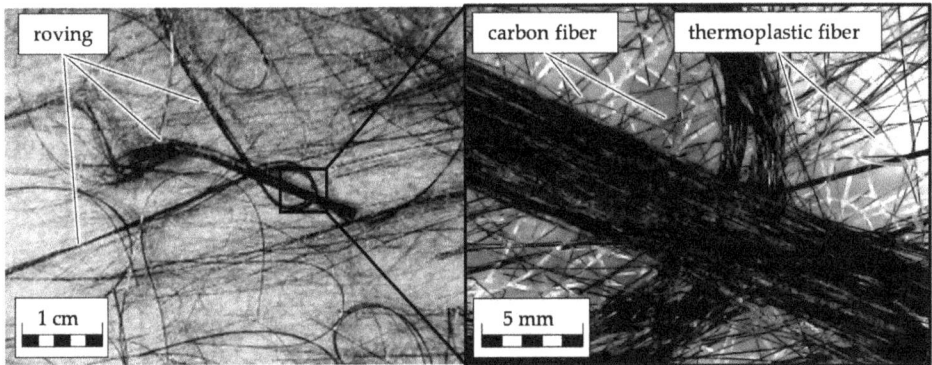

Figure 3. Schematic representation of the carding machine and sampling positions [30].

To assess the significance between the fiber lengths at different positions, statistical methods were employed. The Kolmogorov–Smirnov test was used to determine whether the data is normally distributed, which does not apply to the fiber length measurement. Therefore, the Mann–Whitney test was used to calculate the probability of the null hypothesis [31]. The null hypothesis states that the mean values of fiber length at different positions do not show any significant difference between them. The alternative hypothesis states that the calculated mean values do show a significant difference between them. The significance level is 5%. If the probability of the occurrence of the null hypothesis is less than 5%, the significance level is fulfilled, and the null hypothesis is rejected.

2.2. Material and Methods at the Composite Level-Bending Properties of rCF Nonwoven Panels

The second part of our investigation focuses on the composite properties of carbon fiber nonwoven reinforced plastics. The nonwovens are processed into panels at Fraunhofer IGCV in Augsburg. A complete list of the used materials is shown in Table 4. The variable parameters are carbon fibers from cut VCF and pyrolyzed carbon fibers PCF, the amount of worker–stripper pairs, the properties of the thermoplastic fibers and the proportion of thermoplastic fibers of the blends.

Table 4. Parameters of hybrid nonwovens for the hot-pressing process.

Carbon Fiber	Vol% (rCF)	TP Fiber	Worker–Stripper Pairs	Material Code
PCF	32	P300	6	PCF_P300_WS6
PCF	33	P300	3	PCF_P300_WS3
PCF	32	P301	3	PCF_P301_WS3
PCF	30	TREVIRA® 290	3	PCF_T290_WS3
VCF1x	15	TREVIRA® 290	3	15%_VCF1x_85%_PET
VCF1x	18	TREVIRA® 290	3	18%_VCF1x_82%_PET
VCF1x	25	TREVIRA® 290	3	25%_VCF1x_75%_PET
VCF1x	29	TREVIRA® 290	3	29%_VCF1x_71%_PET

An LZT-OK-130-L press from the company Langzauner GmbH, Lambrechten (AUT), with a maximum force of 1370 kN was used to manufacture 2-mm thick samples, which were mechanically characterized afterwards. Therefore, in dependence to the areal weight, 10–11 layers of the nonwoven were stacked above each other without rotating them to enable a clear evaluation of the machine and the cross direction properties. The TK 350 × 350 mm² tooling was used for the production of the samples because, here, the material only has limited contact with the air due to the optimized gab geometry.

For PA6 and PET hybrid nonwovens, a pressing temperature of 290 °C was set. The temperature was controlled at the top and bottom part of the tooling as well as inside the sample. The maximum deviation of temperature was about 10 K. For the pressing, a variothermal process was used starting at room temperature and compressing the material with 50 kN (4 bar). Afterward, the material was heated with 10 K/min until it reached the maximum temperature.

Then, the maximum pressure up to 50 bar was applied to the material without dwell time. The material was cooled down by 10 K/min with an isobar pressure of 50 bar. In dependence of the fiber volume content, homogeneity, and polymer viscosity, those parameters were slightly changed to achieve the optimal result. After the consolidation, the thickness of each panel was measured, and the surface quality was evaluated visually. If the panel met the required quality criteria, it was used for mechanical testing.

After the consolidation, the samples were cut from the panel by water jet cutting. Ten samples in 0°, +45°, −45°, and 90° were cut from the panel. In respect to the previous measurements, six samples were used for the bending test because this allows a statical evaluation in an appropriate quality. The others were used for the fiber volume content determination according to a wet-chemical solvolysis (DIN EN 2564).

The bending tests were carried out according to DIN EN ISO 14125 and in compliance with the four-point bending set-up. Material class II was assumed, and therefore samples of 40 × 15 × 2 mm³ were manufactured. After the testing, the average values as well as the standard deviation were calculated. Broken samples were used for cross section images to gather more information regarding the quality of the impregnation.

3. Results

The results are discussed in two parts. First, the results of the fiber length measurements regarding the web formation, machine setup, and blend ratio of thermoplastic fibers are presented. The second part shows the results for the investigation of the mechanic bending properties of nonwoven reinforced plastics regarding the carbon fiber volume content and the use of different thermoplastic fibers.

3.1. Results of the Fiber Length Measurements Regarding the Web Formation, Machine Setup, and Blend Ratio of Thermoplastic Fibers with Carbon Fibers

3.1.1. Results of the Fiber Length Measurements Regarding Web Formation

The results of the fiber length measurements conducted during the web formation are shown in Table 5. The average carbon fiber length prior to the carding process ranges from 59 to 62.3 mm with more than 72.6% of fibers being over 30-mm long. During the process, the fibers lost between 20.6% up to 28.4% of length. Pure carbon fibers and pyrolized carbon fibers showed a higher reduction in length. The addition of thermoplastic fibers reduced the fiber shortening by almost 28% for VCF.

Table 5. Fiber length of VCF, VCF_PA6, and PCF_PA6 at pos. 1 and pos. 4.

Material Code	Avg. Length at Pos. 1	Avg. Length at Pos. 4	Length Reduction	Fibers Over 30 mm (Pos. 1)	Fibers Over 30 mm (Pos. 4)
VCF3x	62.3 mm	44.6 mm	28.4%	78.6%	71.6%
VCF3x_PA6	62.3 mm	46.0 mm	20.6%	72.6%	69.2%
PCF_PA6	59.0 mm	42.7 mm	27.6%	75.6%	63.2%

The quantity of fibers over 30 mm decreased for all tested blends during the process. The VCF_PA6 showed the least reduction (3.4%), and PCF_PA6 showed the highest reduction (12.4%). The relatively lower starting point of 72.6% of fibers over 30 mm for VCF3x_PA6 was caused by the fiber preparation, which was passed by all materials prior to the measuring location at pos. 1. The fiber preparation contained a suction system, and it mixed and dosed the materials leading to different starting values.

Figure 4a–c represent carbon fiber length diagrams of the three processed materials. The optimal length for the used nonwoven process was between 30 and 90 mm. The carbon fiber length curves of VCF3x, VCF3x_PA6, and PCF_PA6 show a shift to the left as the web formation process progresses, indicating fiber length loss due to mechanical stress on the fibers from the carding process.

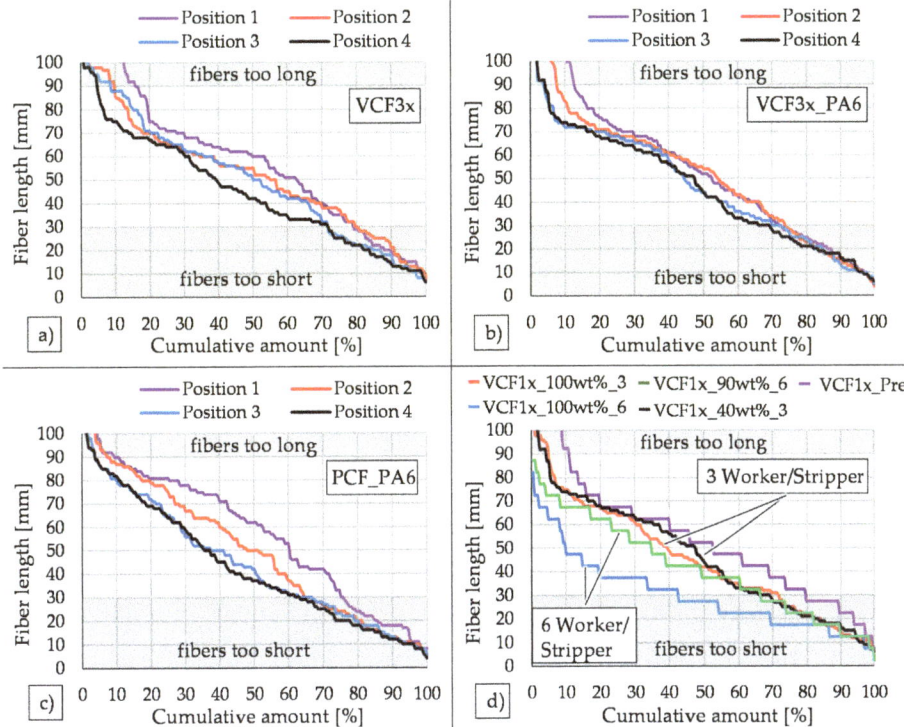

Figure 4. Carbon fiber length diagrams for VCF (**a**), VCF_PA6 (**b**), PCF_PA6 (**c**) and carbon fibers processed by different machine setups (**d**).

In Figure 4a, a clear shift to the left is observed for fibers above 40 mm from pos. 1 to pos. 2. Below the 70-mm carbon fiber length, the graphs of pos. 2 and pos. 3 are almost identical. Above this threshold, the quantity of longer fibers is reduced. From pos. 3 to pos. 4, a distinct decrease in carbon fiber length is observed between 35 and 60 mm, and the amount of longer fibers decreases even further. All changes are significant as stated in Table 6 except for pos. 2 to pos. 3, which are almost identical.

Table 6. Statistical significance between the fiber lengths of the positions (significant difference indicated as "+", no difference indicated as "−").

Parameter	Pos. 1 to Pos. 2	Pos. 1 to Pos. 3	Pos. 1 to Pos. 4	Pos. 2 to Pos. 3	Pos. 2 to Pos. 4	Pos. 3 to Pos. 4
VCF3x	+	+	+	−	+	+
VCF3x_PA6x	−	+	+	−	−	−
PCF_PA6	+	+	+	−	+	−

The length reduction is greatly reduced, and the doffing between pos. 3 and pos. 4 shows almost no shortening of the fibers. At the same time, more fibers are below 30 mm. One reason is the crimp of the thermoplastic fibers. Short carbon fibers are retained by the thermoplastic fibers in the web, and thus the suction system of the card draws fewer short carbon fibers from the mixture [30].

The carbon fibers of the PCF_PA6 blend were longer than the carbon fibers of VCF. The reduction from pos. 1 to pos. 2 is visible by a prominent left shift for all fibers shorter than 80 mm. Fibers above were not affected. A shortening between pos. 2 and pos. 3 is visible for fibers above 30 mm. Nevertheless, no significant difference can be proven. Pos. 3 and pos. 4 are almost identical.

The results of Table 6 were used to distinguish between parts of the carding machine with high and low carbon fiber length reduction. The carding machine exhibited a significant carbon fiber length reduction when processing VCF3x, which is supported with statistical significance. Only the main group located between pos. 2 and pos. 3 showed no significant fiber length reduction. The reduction in carbon fiber length for VCF3x_PA6 between two positions next to each other—pos. 1 to pos. 2, pos. 2 to pos. 3, and pos. 3 to pos. 4—is too small to be distinguished with statistical significance.

Between pos. 1 and pos. 3 as well as between pos. 1 and pos. 4, which contains the complete carding machine, the carbon fiber length was reduced with statistical significance. The length of the carbon fibers of the blend PCF_PA6 were reduced significantly at the intake and the pre group. The shortening in carbon fiber length caused by the main group (pos. 2 to pos. 3) and the doffing (pos. 3 to pos. 4) was not statistically significant. The pyrolized carbon fiber length was, however, statistically significantly reduced if the main group and the doffing are considered together as shown by the validation of significance between pos. 2 to pos. 4.

3.1.2. Results of the Fiber Length Measurements Regarding Machine Setup

The diagram in Figure 4d presents the fiber lengths before and after the nonwoven process using different machine setups. The processing of the pure carbon fiber VCF1x_100wt%_6 with six active pairs of workers and strippers exhibits an average fiber length reduction of 49%. The processing with three worker–stripper pairs led to a length reduction of 28.4%.

3.1.3. Results of the Fiber Length Measurements Regarding the Blend Ratio of Thermoplastic Fibers and Carbon Fibers

Processing a thermoplastic carbon fiber blend with 60 wt% thermoplastic fibers using three active worker–stripper pairs led to a reduction of carbon fiber length of 20.6%. This corresponds to a decrease of 27.5% in comparison to the processing of pure carbon fibers with the same machine setup. When six active worker–stripper pairs were used to process a blend of 10 wt% thermoplastic fibers and 90 wt% carbon fibers, the carbon fiber shortening was reduced from 49% to 21.5%, which is a reduction of 56%.

3.2. Mechanical Evaluation rCF Nonwoven with Different TP-Fibers and Fiber Volume Content

The produced panels were tested by four-point bending. The results of each test were used to calculate an average mechanical value in each of the four directions and its standard deviation. For the investigation of the orientation and isotropy, different approaches can be found in literature. In this paper, the orientation ratio of a nonwoven material is described as the ratio of the highest average stiffness divided by the lowest. For carded material, that usually means bending strength at 90° in direction $\sigma_{90°}$ divided by the bending strength at 0° in direction $\sigma_{0°}$ because of the higher orientation in 90° due to the cross-lapping process.

For the investigation, the average values were calculated by the mean values of all four testing directions (0°, +45°, −45°, and 90°) and compared to the ratio of the 90° value divided by the 0° value. Those values show very similar numbers. Therefore, the 90° to 0° average value was used in this study. The influence of the fiber volume content was investigated. The mechanical properties of the samples were not normalized to a fixed fiber volume content, which is usually done with a linear recalculation to be able to directly compare samples of different fiber content.

The investigated PET materials (Figure 5) had fiber volume contents of 15%, 18%, 25%, and 30%. The properties of the 15% and 18% material were very similar. This similarity highlights that the deviation of the material was higher than the theoretical increase of performance by adding 3% carbon fiber. With a higher fiber volume content of 25%, the stiffness and strength increased by 32–33%. The material with 30% content of fibers did not, however, show any further increase of its strength and showed a slight increase in stiffness.

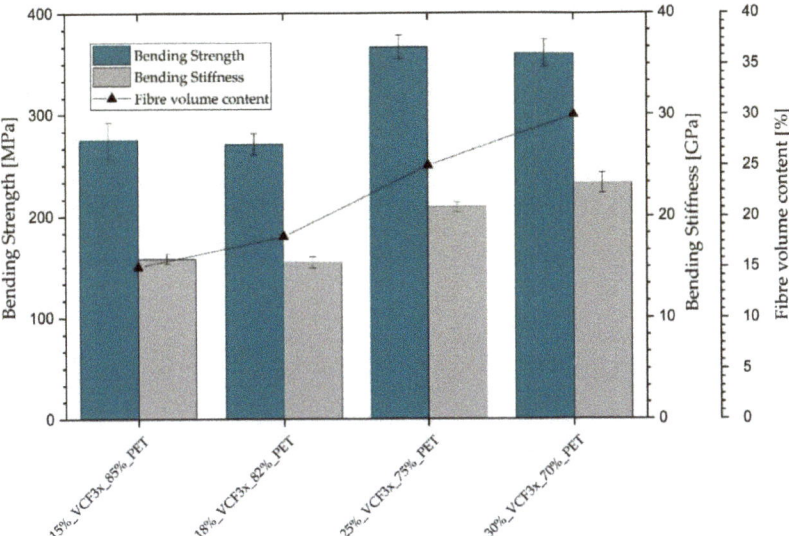

Figure 5. The bending strength and bending stiffness of four PET nonwovens with 15%, 18%, 25%, and 30% fiber volume contents.

Overall, a stiffness of 15.8 GPa for the 15% material and 23.3 GPa for the 30% FVC material are reported. The bending strength showed a minimum of 275 MPa and a maximum value of 366 MPa for the 18% and 30% material, respectively. All those values are additionally shown in Table 7. In addition to the mechanical properties, cross section images are made to determine the quality of impregnation and fiber breakage (Figure 6).

Table 7. The 0° and 90° properties of nonwovens by using different carbon fiber volume contents.

Material Code	Bending Strength 0° [MPa]	Bending Strength 90° [MPa]	Average Strength [MPa]	Bending Stiffness 0° [GPa]	Bending Stiffness 90° [GPa]	Average Stiffness [GPa]
15%_VCF3x_85%PET	270.6	279.3	274.9	16.0	15.6	15.8
18%_VCF3x_82%PET	265.2	277.7	271.4	15.1	15.8	15.4
25%_VCF3x_75%PET	353.0	380.2	366.6	19.2	22.6	20.9
30%_VCF3x_70%PET	335.6	385.6	360.6	21.2	25.4	23.3

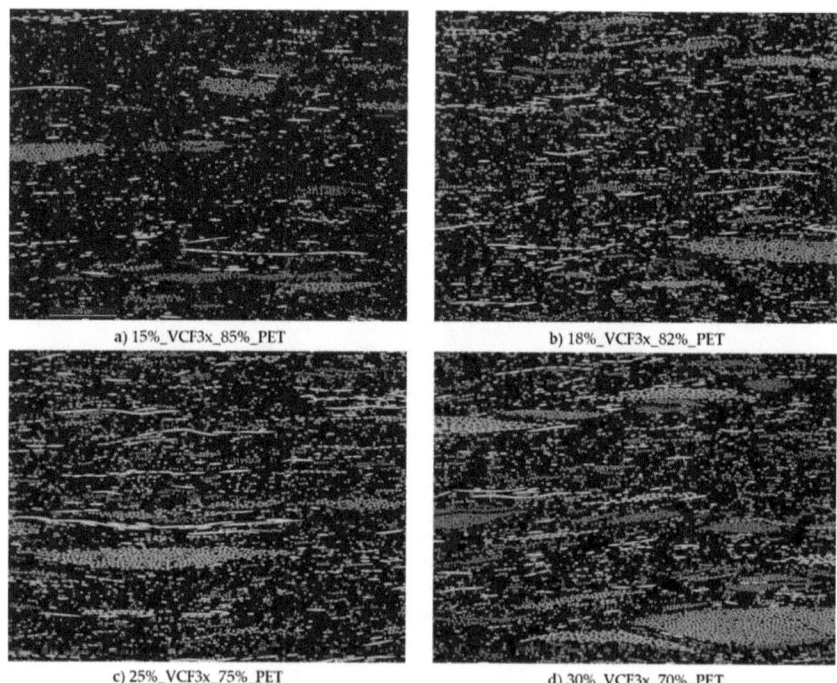

Figure 6. Cross section images of PET nonwoven with different fiber volume contents. (**a**) 15%, (**b**) 18%, (**c**) 25%, and (**d**) 30% fiber volume content. All with the same resolution. Red bar in picture (**a**) indicates 200 µm.

In addition to the investigation of the FVC influence, the effects of different textile properties of the fibers as well as different machine setups on the mechanical properties of nonwoven reinforced composites were evaluated. Four different polymer fibers blended with the same fiber volume content of PCF were used to manufacture nonwovens. The blends are shown in Table 8. The materials were used for sample production by hot-pressing and mechanical characterization by four-point-bending. Figure 7 shows the bending strength and bending stiffness of those materials.

Table 8. The 0° and 90° properties of nonwovens using different polymer types.

Material Code	Bending Strength 0° [MPa]	Bending Strength 90° [MPa]	Average Strength [MPa]	Bending Stiffness 0° [GPa]	Bending Stiffness 90° [GPa]	Average Stiffness [GPa]
PCF_P300_WS6	328	344	336	18.3	20.7	19.5
PCF_P300_WS3	341	360	351	19.2	20.4	19.8
PCF_P301_WS3	301	415	358	16.3	25.4	20.9
PCF_T290_WS3	305	405	355	21.2	30.1	25.6

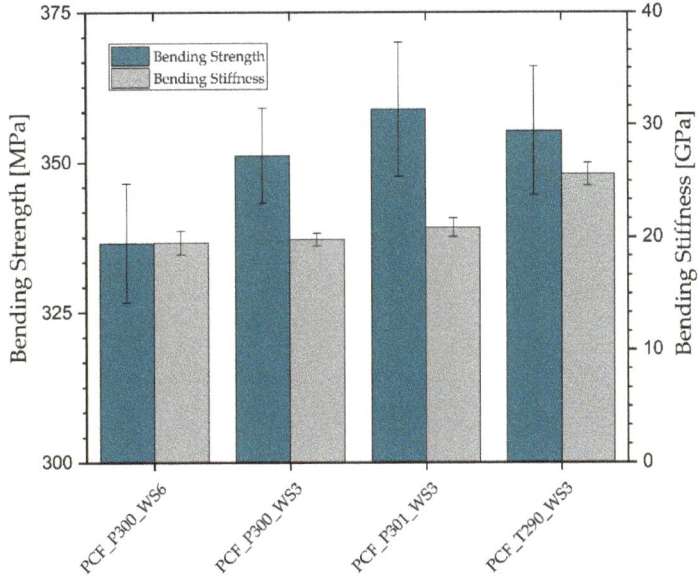

Figure 7. The bending strength and bending stiffness of different machine set-up and different TP fibers. (P300 and P301 are PA6 fibers, and T290 is the PET fiber). WS3 and WS6 equal the number of worker and stripper pairs within the carding line. The FVC of the PA6 materials was scattered between 31% and 33%, and the PET T290 material had an FVC of 30%.

The axis of the strength is zoomed in to highlight the effect. The illustrated values are average values of the 0° and 90° directions. The PCF_P300_WS6 fiber blend showed the overall lowest properties with a 336 MPa bending strength and 19.5 GPa bending stiffness. The reduction of worker–stripper pairs from six to three led, on average, to an increase of bending strength by 4.4% to 351 MPa and to an increase of bending stiffness by 1.5% to 19.8 GPa. By changing the textile properties of the thermoplastic fibers—the fiber length, fineness, and crimp—an increase to 358 MPa (+6.5%) for strength and an increase to 20.9 GPa (7.2%) for stiffness were achieved.

The PA6 materials had a similar fiber volume content (31–33%), and thus it can be assumed that there should be no influence on the mechanical properties according to the results in the previous trials. Compared to PA6, the use of PET strongly increased the average bending stiffness to 25.6 GPa. This material also had a slightly lower fiber volume content (30%), and therefore the potential of the use of PET for stiffness driven applications is even higher. The strength could not be increased using PET fibers, however, and the average value at 355 MPa was even lower than for PCF_P300.

In the cross direction (90°), the average tested value was 30.1 GPa, and, in the machine direction, 21.2 GPa. The orientation ratio calculated to 1.42 for the PET/rCF nonwoven. In comparison, the PCF_P300_WS6 showed an almost isotropic behavior with an orientation ratio of 1.13. The orientation ratio was increased to 1.56 by changing the PA6 properties using fiber type P301, which is longer, less fine, and less crimped. Longer, thicker, and less crimped fibers not only increased the overall properties but also the orientation degree of the web and, therefore, influenced the overall nonwoven orientation and orientation ratio.

4. Discussion

A study by Dauner et al. investigated the length reduction by the position of the card [25]. Although the setup of the nonwoven machine was not explicitly stated, the carding was similar, and the roller cards were constructed in similar ways. The results are compared to the current findings in Table 9.

Table 9. Length reduction by position in percent.

Origin of the Data	Pos. 1 to Pos. 2 Intake and Pre Group	Pos. 2 to Pos. 3 Main Group	Pos. 3 to Pos. 4 Web Formation on the Doffer Roller	Total Fiber Length Reduction Pos. 1 to Pos. 4
Dauner [25]	33.3%	16.7%	8.0%	48.9%
VCF3x	15.9%	4.6%	10.8%	28.4%
VCF3x_PA6	11.1%	9.1%	1.7%	20.6%
PCF_PA6	14.2%	13.0%	3.0%	27.6%

The length reduction was strongest at the beginning of the carding machine for all measured fibers and blends. The length reduction in comparison to previous works showed a lower overall shortening. Possible reasons for the shorting at the intake and the pre group are the clamping of the fibers at the intake where fibers are torn leading to high mechanical stress and fiber breakage. The carbon fibers were, furthermore, unscathed at the intake, and thus the decrease in carbon fiber length due to the first contact with rollers and wires is most significant. Concerning fiber blends, the subsequent positions were each reduced in their effective shortening. Pure VCFs, however, were shortened by only 4.6% from pos. 2 to pos. 3 but 10.8% to pos. 4, reversing the effect. The reduction of the carbon fiber length between pos. 2 and pos. 3 was caused by the main group, consisting of the main roller and four pairs of worker–strippers. In this part of the machine, the highest degree of fiber separation, parallelization, and blending takes place. These operations result in increased mechanical stress on the carbon fibers, which ultimately leads to fiber breakage. More brittle fibers, such as PCF, which may be degraded by the pyrolysis process, showed a higher degree of carbon fiber length reduction. The VCFs used in this study, on the other hand, are likely already too short to be further damaged by the main group.

The web formation on the doffing roller is located between pos. 3 to pos. 4. The VCF and PCF blends did not show any significant reduction in carbon fiber length at this location. The thermoplastic fibers presumably ensured a dampening effect for the carbon fibers during the web formation. The fibers were slowed down significantly, and thermoplastic fibers might dampen the extreme forces caused by the deceleration. Pure VCFs, however, were not cushioned and exhibited a significant decrease in carbon fiber length due to the high mechanical stress as the results show.

Pure carbon fibers and pyrolized carbon fibers showed a higher reduction in length. The pyrolized fibers were more brittle after the reclamation process and the friction reducing sizing. The thermoplastic fibers in the blends had two opposite effects. On the one hand, the fibers functioned as a cushion for the carbon fibers and reduced the fiber damage especially for longer carbon fibers. The effect was shown by the addition of 10 wt% TP fibers in different machine setups, which led to a reduced fiber shortening up to 58%.

On the other hand, the crimped thermoplastic fibers kept shorter carbon fibers in the web. Therefore, this increased the amount of short fibers that are otherwise removed by the suction system of the carding machine. The decreasing quantity of fibers over 30 mm proved this effect, which was highest for VCF and significantly lower for the blends with thermoplastic fibers. This could also be the reason for the reversed amount of shortening by position observed for the processing of VCF.

Another key finding is the positive influence of less worker–stripper-pairs on the fiber length. With three worker–stripper pairs, the length reduction decreased substantially, while the mechanical results did not show lower orientation values. In fact, the mechanical values became higher as shown in Figure 6.

The comparison of the measured bending properties in Figure 5 indicates that an increase of the bending strength as well as the bending stiffness occurred due to the rising fiber volume content. This fits the Voigh–Reuss–Hill assumption of an isotropic material [32,33]. It was shown that a rising fiber volume content has an influence on the mechanical properties especially in terms of bending load. However, the observed effect is not always linear and, therefore, does not follow the rule of mixtures for fiber-reinforced polymers.

One reason is that the deviation of mechanical properties is strongly influenced by the carbon and thermoplastic fiber properties as well as the carding adjustments and is not based only on the principles of load distribution onto the reinforcement fiber. By increasing the FVC from 15 vol% to 30 vol% of carbon fibers, the strength can be increased by 33% of the original strength and the stiffness by 50% of its original value. Still, the materials with 25 vol% showed a similar mechanical behavior to the 30 vol% material, which proves that the strength and stiffness were limited.

A similar effect was shown by Pickering et al. [3]. He also demonstrated that the stiffness of a nonwoven web can be increased by fiber volume content in a linear correlation in certain limits. On the other hand, the strength began to drop at a critical fiber volume content. His measurements were applied to a thermoset material and are, however, not completely transferable to this thermoplastic nonwoven study. One reason for the non-linearity of correlations could be due to the infiltration quality by the thermoplastic fibers within the hot-pressing route.

Textile TP fibers have relatively high viscosities due to the spinning process. In the pressing process, the polymer chains need to be melted, and the viscosity should be reduced as much as possible to achieve a good impregnation behavior. With a higher fiber volume content, the isotropic nonwoven material has a lower permeability and is, therefore, harder to infiltrate. This leads to dry spots and fiber shortening by pressing onto dry fibers that are not embedded by the resin. This effect can be seen by the cross-section images in Figure 6. The 15%, 18%, and 25% material did not show many pores, but the 30% materials showed a higher density of fibers as well as a higher amount and size of voids within the material cross section.

By increasing the orientation ratio, the in-plane permeability can be increased, and higher fiber volume contents could be possible. Within the project MAI CC4 Carina [19,20], the project partners were investigating the correlation of the orientation ratio, the fiber volume content, and the mechanical performance. It was shown that up to 38 vol% fiber volume content was possible, and the mechanical properties could be further increased. However, there are other complications in the thermoplastic processing route for such high fiber volume contents. By using a thermoset polymer or TP powder polymer, the infiltration can be improved, and higher fiber volume contents are achievable.

The second mechanical investigation in this study was the influence of different TP fiber properties on the bending properties. By the reduction of the worker and stripper pairs, the composite strength was increased by 4.5%. In changing the textile properties to thicker, longer, and less crimped fibers, the orientation ratio of the web and the nonwoven were increased as well. The average strength was increased slightly by 1.9% and the stiffness by 5.5%. Although those increases were not statistically significant and, therefore, only represent a trend.

By using PET instead of PA6, the stiffness was improved by 22.4%, while the strength remained on the same level. PET exhibits a higher stiffness in comparison to PA6, which also transfers to the improved mechanical properties of PET nonwoven panels compared to PA6 nonwoven panels. The high availability of recycled PET and circularity could suggest PET over PA6 for the material choice of future sustainable materials. In conclusion, we demonstrated that the choice of polymer, as well as the textile properties and the machine setup, have a strong influence on the mechanical properties of nonwoven reinforced composites.

5. Conclusions

Carbon fibers always undergo shortening in the carding process. The amount of length reduction depends heavily on the carbon fiber type, if the carbon fibers are blended with thermoplastic fibers, and the number of worker–stripper-pairs employed. Our recommendation for the processing of carbon fibers by the carding process is the reduction of the worker–stripper pairs to a suitable minimum as well as the addition of thermoplastic fibers whenever possible.

This study of mechanical properties showed that the thermoplastic polymer, their fiber properties, as well as the fiber volume content had a dominant influence on the bending properties of the composite material. The linearity of the carbon fiber volume content and bending properties were observed in a limited way.

Knowledge of the carding and fiber properties leads to the production of enhanced nonwovens. One major issue that must be solved in future research projects is the high degree of variation in the areal weight as well as the material distribution within blends. Looking forward, industrial serial application and production must be proven, and the processing parameters need to be optimized even further.

Author Contributions: Conceptualization, G.S. and F.M.; methodology, G.S. and F.M.; validation, G.S. and F.M.; formal analysis, G.S. and F.M.; investigation, G.S. and F.M.; resources, G.S. and F.M.; writing—original draft preparation, G.S. and F.M.; writing—review and editing, G.S., F.M., S.S., and J.W.; visualization, G.S. and F.M.; project administration, F.M. (Project leader) and G.S. (Work package textile); funding acquisition, G.S. and F.M. All authors have read and agreed to the published version of the manuscript.

Funding: This research originates from the project MAI CC4 CaRinA (Carbonfaser Recyclingwerkstoffe für industrielle Anwendungen), which was kindly funded by Bayerisches Staatsministerium für Wirtschaft, Landesentwicklung und Energie under grant number 2-NW-1707.

Acknowledgments: We like to thank Christina Aust and Petra Amann for their great support with all their experience in testing of composites. Furthermore, we like to thank Werner Münch, Dajan Kheder, Matthias Abbt, and Christoph Klement for their support in producing nonwovens and counting vast numbers of fibers. They greatly enhanced the quality of this study with all their support. Our gratitude to the company Dilo Machines GmbH, Eberbach for the generous provision of a complete nonwoven line as well as their unwavering technical support.

Conflicts of Interest: The authors declare no conflict of interest.

References

1. Sauer, M. *Composites-Marktbericht 2020*; Composite United e.V.: Berlin, Germany, 2021.
2. Pickering, S. Recycling technologies for thermoset composite materials—Current status. *Compos. Part. A* **2006**, *37*, 1206–1215. [CrossRef]
3. Pickering, S.J.; Liu, Z.; Turner, T.A.; Wong, K.H. Applications for carbon fibre recovered from composites. *IOP Conf. Ser. Mater. Sci. Eng.* **2016**, *139*, 12005. [CrossRef]
4. Pimenta, S.; Pinho, S.T. The effect of recycling on the mechanical response of carbon fibres and their composites. *Compos. Struct.* **2012**, *94*, 3669–3684. [CrossRef]
5. Pimenta, S. Toughness and strength of recycled composites and their virgin precursors. *Diss. Imp. Coll. Lond.* **2013**. [CrossRef]
6. Park, J.M.; Kwon, D.J.; Wang, Z.J.; Gu, G.Y.; DeVries, K.L. Effect of thermal treatment temperatures on the reinforcing and interfacial properties of recycled carbon fiber–phenolic composites. *Compos. Part A Appl. Sci. Manuf.* **2013**, *47*, 156–164. [CrossRef]
7. Nahil, M.A.; Williams, P.T. Recycling of carbon fibre reinforced polymeric waste for the production of activated carbon fibres. *J. Anal. Appl. Pyrolysis* **2011**, *91*, 67–75. [CrossRef]
8. López, F.A.; Rodríguez, O.; Alguacil, F.J.; García-Díaz, I.; Centeno, T.A.; García-Fierro, J.L. Recovery of carbon fibres by the thermolysis and gasification of waste prepreg. *J. Anal. Appl. Pyrolysis* **2013**, *104*, 675–683. [CrossRef]

9. Greco, A.; Maffezzoli, A.; Buccoliero, G.; Caretto, F.; Cornacchia, G. Thermal and chemical treatments of recycled carbon fibres for improved adhesion to polymeric matrix. *J. Compos. Mater.* **2013**, *47*, 369–377. [CrossRef]
10. Iwaya, T.; Tokuno, S.; Sasaki, M.; Goto, M.; Shibata, K. Recycling of fiber reinforced plastics using depolymerization by solvothermal reaction with catalyst. *J. Mater. Sci.* **2008**, *43*, 2452–2456. [CrossRef]
11. Xu, P.; Li, J.; Ding, J. Chemical recycling of carbon fibre/epoxy composites in a mixed solution of peroxide hydrogen and N,N-dimethylformamide. *Compos. Sci. Technol.* **2013**, *82*, 54–59. [CrossRef]
12. Morin, C.; Loppinet-Serani, A.; Cansell, F.; Aymonier, C. Near- and supercritical solvolysis of carbon fibre reinforced polymers (CFRPs) for recycling carbon fibers as a valuable resource: State of the art. *J. Supercrit. Fluids* **2012**, *66*, 232–240. [CrossRef]
13. From Laboratory to Industrial Scale. Available online: https://www.phyre-recycling.com/project (accessed on 27 January 2021).
14. The Future in Carbon Fibre Recycling. Available online: http://catack-h.com/technology-2/?lang=en (accessed on 27 January 2021).
15. Hofmann, M.; Nestler, A. CarboLace-from Recycled Carbon Fibres to Spunlace Nonwovens. In Proceedings of the Aachen-Dresden-Denkendorf International Textile Conference 2017, Stuttgart, Germany, 30 November–1 December 2017. [CrossRef]
16. Manis, F.; Schmieg, M.; Sauer, M.; Drechsler, K. Properties of second life carbon fibre reinforced polymers. *Key Eng. Mater.* **2017**, *742*, 562–567. [CrossRef]
17. Shah, D.U.; Schubel, P.J. On recycled carbon fibre composites manufactured through a liquid composite moulding process. *J. Reinf. Plast. Compos.* **2015**, *35*, 533–540. [CrossRef]
18. Viale, S. Recycled Carbon Fibers for High Added-Value Product, CU e.V. Themenwoche Nachhaltigkeit. 6 May 2020. Available online: https://www.youtube.com/watch?v=Ze6bO82J3J0 (accessed on 13 May 2020).
19. CaRinA | Carbon Fiber Recycled Materials for Industrial Applications. Available online: https://www.igcv.fraunhofer.de/en/about_us/fraunhofer_igcv/composites/reference_projects/CaRinA.html (accessed on 27 January 2021).
20. Stegschuster, G.; Schlichter, S. Perspectives of web based composites from RCF material. *IOP Conf. Ser. Mater. Sci. Eng.* **2018**, *406*, 012022. [CrossRef]
21. Thomas, G. Thermoplastische formmassen. In *Handbuch Faserverbundkunststoffe/Composites: Grundlagen, Verarbeitung, Anwendungen*, 4th ed.; AVK-Industrievereinigung Verstärkter Kunststoffe e.V., Ed.; Springer Vieweg: Wiesbaden, Germany, 2014; pp. 278–290. [CrossRef]
22. Harbers, T. Beitrag Zum Materialverständnis kohlenstofffaserbasierter Nassvliese. Ph.D. Thesis, Technische Universität München, München, Germany, 2016. Available online: http://mediatum.ub.tum.de/?id=1319156 (accessed on 8 August 2016).
23. Brydon, A.G.; Pourmohammadi, A. Card clothing. In *Handbook of Nonwovens*; Russell, S.J., Ed.; Woodhead Publishing Limited: Cambridge, UK; CRC Press LLC: Abington, PA, USA, 2007; pp. 44–53.
24. Dilo, J.P. Vernadelungsverfahren. In *Vliesstoffe: Rohstoffe, Herstellung, Anwendung, Eigenschaften, Prüfung*, 2nd ed.; Fuchs, H., Albrecht, W., Eds.; Wiley-VCH: Weinheim, Germany, 2012; pp. 255–311.
25. Dauner, M.; Baz, S.; Geier, M.; Gresser, G.T. *Rahmenbedingungen zur Verarbeitung von Carbonfasern durch Krempeltechnik. 31;* Hofer Vliesstofftage: Hof, Germany, 2016; Available online: https://www.hofer-vliesstofftage.de/vortraege/2016/2016-05-d.pdf (accessed on 3 August 2021).
26. Schlichter, S.; Stegschuster, G. Web based composites: Potenziale vliesbasierter verbundbauteile. *Tech. Text.* 60 **2017**, *4*, S271–S272.
27. Schlichter, S.; Stegschuster, G. *Web Based Composites-Potenziale Vliesbasierter Verbundbauteile. 31. Hofer Vliesstofftage, Hof, Germany, 9–10 November 2016*; Bildungswerk der Bayerischen Wirtschaft (bbw) Gemeinnützige GmbH Hof: Hof, Germany, 2016; Available online: https://www.hofer-vliesstofftage.de/vortraege/2016/2016-06-d.pdf (accessed on 3 August 2021).
28. Albrecht, F.; Rosenberg, P.; Heilos, K.; Hoffmann, M.; Henning, F. Vliesrtm–Reuse of carbon fiber waste in composite structures. In Proceedings of the SAMPE Europe Conference 2019, Nantes, France, 17–19 September 2019.
29. Pierret GmbH: Schneidemaschinen. Available online: https://www.pierret.com/de/nos-produits/coupeuses/ (accessed on 29 January 2021).
30. Stegschuster, G. Analyse des Kardierverfahrens zur Herstellung von Carbonfaservliesstoff als Verstärkungstextil für Faserverbundwerkstoffe. Ph.D. Thesis, Universität Augsburg, Augsburg, Germany, 2021.
31. Lohninger, H. Welch-Test. Available online: http://www.statistics4u.info/fundstat_germ/ee_welch_test.html (accessed on 11 December 2020).
32. Voigt, W. Ueber die Beziehung zwischen den beiden Elasticitätsconstanten isotroper Körper. *Ann. Phys.* **1889**, *274*, 573–587. [CrossRef]
33. Reuss, A. Berechnung der Fließgrenze von Mischkristallen auf Grund der Plastizitätsbedingung für Einkristalle. *Z. Angew. Math. Mech.* **1929**, *9*, 49–58. [CrossRef]

Article

Fabrication of Porous Carbon Films and Their Impact on Carbon/Polypropylene Interfacial Bonding

Yucheng Peng [1,2,*], Ruslan Burtovyy [2], Rajendra Bordia [2] and Igor Luzinov [2,*]

[1] School of Forestry and Wildlife Sciences, Auburn University, 602 Duncan Drive, Auburn, AL 36849, USA
[2] Department of Materials Science and Engineering, Clemson University, 161 Sirrine Hall, Clemson, SC 29634, USA; rburtov@g.clemson.edu (R.B.); rbordia@clemson.edu (R.B.)
* Correspondence: yzp0027@auburn.edu (Y.P.); luzinov@clemson.edu (I.L.)

Abstract: Porous carbon films were generated by thermal treatment of polymer films made from poly(acrylonitrile-co-methyl acrylate)/polyethylene terephthalate (PAN/PET) blend. The precursor films were fabricated by a dip-coating process using PAN/PET solutions in hexafluoro-2-propanol (HFIP). A two-step process, including stabilization and carbonization, was employed to produce the carbon films. PET functioned as a pore former. Specifically, porous carbon films with thicknesses from 0.38–1.83 µm and pore diameters between 0.1–10 µm were obtained. The higher concentrations of PET in the PAN/PET mixture and the higher withdrawal speed during dip-coating caused the formation of larger pores. The thickness of the carbon films can be regulated using the withdrawal speed used in the dip-coating deposition. We determined that the deposition of the porous carbon film on graphite substrate significantly increases the value of the interfacial shear strength between graphite plates and thermoplastic PP. This study has shown the feasibility of fabrication of 3D porous carbon structure on the surface of carbon materials for increasing the interfacial strength. We expect that this approach can be employed for the fabrication of high-performance carbon fiber-thermoplastic composites.

Keywords: carbon-polymer adhesion; mechanical interlocking; PAN; PET; polypropylene; composites

1. Introduction

The increasing utilization of carbon fiber reinforced polymer composites (CFRPC), especially thermosetting polymers, has raised environmental and economic awareness of the need to recycle the composites [1–7]. To this end, employment of polyolefins for fabrication of advanced and recyclable carbon fiber reinforced thermoplastic composites (CFRTCs) possessing high solvent/environmental resistance, high modulus, strength, and toughness, can relieve the economic, environmental, and political pressure [8–19]. Lightweight CFRTCs already demonstrate great potential for automobile, aerospace, defense sectors, civil infrastructures, sport/leisure goods, and energy sector [1,5,6,12,20–22]. However, the key challenge remaining in the fabrication of advanced CFRTCs is poor interfacial bonding between fiber surface and polyolefins (exemplified by the most extensively explored polypropylene). The low adhesion level originates from low surface energy and the inert surface of carbon fibers (CFs) made from highly crystallized graphitic basal planes [20–42].

Several chemical and physical strategies are being explored to increase the interfacial adhesion between the CFs and polypropylene (PP). This includes the functionalization of PP with polar/reactive functional moieties and/or surface modification of CF such as oxidation, plasma and ozone treatment, thermal treatments, sizing, coupling agent treatments, and micro/nanofiber deposition, including whiskerization [23–32,43]. The latter methods of CF surface modification, where a change of surface topography/morphology of the fiber is targeted, employ mechanical interlocking to considerably increase the interlaminar shear strength (ILSS) of the resulting carbon fiber composites [31,32,43,44]. The mechanical anchoring diminishes the relative sliding between the CF and polymer matrix and improves stress transfer between the fiber and matrix, leading to higher modulus and strength. Thus, the properties of CFRTCs can be significantly improved via manipulation of the interfacial geometry. The interlocking approach appears to be quite universal in respect to polymer matrices used, as long as polymer material can impregnate the deposited fibrous surface structures. The main challenges remaining for the approach are complexity/sophistication of the morphological modification of CF exterior, mechanical robustness and uniformity of the fibrous layer deposited, effective impregnation of the layer with thermoplastic (e.g., PP) melt of high viscosity, and decrease of mechanical strength of CF as a result of the modification.

To this end, we have been developing an original surface modification method for carbon graphitic surfaces (such as the ones of carbon fibers) to create a non-fibrous mechanically anchoring robust carbon-based boundary. Specifically, our focus is on the fabrication of three-dimensional (3D) micrometric porous carbon structures attached to the carbon substrate surface. A 3D micromechanical interlocking interphase between carbon materials and a thermoplastic polymer is built during the composite manufacturing process through the infiltration of polymer melt into the pores created on carbon's surface, leading to improvement of interfacial shear strength. The 3D micromechanical interlocking interphase has been designed to equate the stress at the carbon/polymer interface to the stress in the continuous phase of the polymer component when external shear/tension forces are applied to the composites. In this case, the maximum interfacial shear stress that can be sustained is determined by the strength of the polymer matrix.

The porous anchoring layer formation is conducted in just two straightforward steps: deposition of polymeric precursor on carbon surface and carbonization of the precursor layer. The deposition of carbon precursor is conducted from a solvent using a dip-coating procedure that can be realized in a conventional industrial setting. The carbonization is conducted using the same protocols as those typically used to obtain CF from polyacrylonitrile (PAN) fibers. In essence, to obtain the porous carbon coatings, we followed here procedures previously reported by us elsewhere [45]. In brief, it was found that crack-free carbon films with nanoscale roughness and controllable thickness can be produced by high-temperature treatment of the PAN-based film fabricated through dip-coating from 1,1,1,3,3,3-hexafluoro-2-propanol (HFIP) solution. The same basic procedure was used to form porous carbon films with tunable porous structure when PAN and polyethylene oxide (PEO) blend was used instead of PAN. PEO functioned as the pore-forming component in the precursor and carbon films, whose pore diameter could be varied from hundreds of nanometers to micrometers. However, PEO was not the ideal pore-forming polymer since it has relatively low glass transition and melting temperatures and is in a viscous molten state at the temperature of PAN stabilization. Thus, while carbon film obtained had significant internal porosity, surface porosity was relatively low [45].

In the current work, we replaced PEO with polyethylene terephthalate (PET), which has a melting temperature of about 265 °C [46] and does not flow at the temperature used for PAN stabilization. The PET's employment allowed us to produce carbon films with the open-pore structure suitable to serve as the porous anchoring layer, connecting the polymer matrix and CF via the interlocking mechanism. The initial 3D micrometric porous carbon structure development and characterization were performed on silicon wafers. The porous carbon structure was then fabricated on the surfaces of graphite plates using the protocols developed for depositing porous carbon structure on the silicon wafers. Finally, the inter-laminar shear strength for PP/carbon interface was measured for graphite substrates using a lap-shear test. It was determined that ILSS increased ~70% when the porous coating was deposited on graphite plates.

2. Materials and Methods

2.1. Materials

Poly(acrylonitrile-co-methyl acrylate) (PAN, M_w = 100,000 g/mol) with 94 wt.% of acrylonitrile and polyethylene terephthalate (PET, M_V = 18,000 g/mol) were purchased from Sigma-Aldrich (St. Louis, MO, USA). 1,1,1,3,3,3-hexafluoro-2-propanol (HFIP) was obtained from Oakwood Chemical (West Columbia, SC, USA). Silicon wafers with one highly polished side were obtained from WRS Materials (Spring City, PA, USA). Homopolymer polypropylene (PP) grafted with 1 wt.% maleic anhydride with a melt flow index of 115 g/10 min at 2.16 kg/190 °C was purchased from Addivant (Danbury, CT, USA). The melting point of the PP is reported as 160–170 °C. High-density graphite plates (1.91 g/cm^3) were provided by Toyo Tanso USA, Inc. (Durham, CT, USA).

2.2. Polymer Solution Preparation

PAN solution in HFIP at the concentration of 3 wt.% was prepared by dissolving PAN in HFIP at room temperature. PAN/PET solutions were prepared by dissolving both polymers in HFIP at room temperature. The concentration of PAN/PET solutions represents the total weight of PAN and PET in the solution. Six solutions having 3 wt.% concentration with PAN/PET weight ratios of 60:40, 65:35, 70:30, 75:25, 80:20, and 90:10 were prepared.

2.3. Precursor Polymer Film Preparation

A dip-coating process was employed to deposit a polymer film from the solution on the silicon wafers using dip-coater D-3400 (Mayer Feintechnik, Göttingen, Germany). Before the dip-coating, silicon wafers were first cleaned in D.I. (deionized) water for one hour using an ultrasonic bath, then placed in a piranha solution (3:1 concentrated sulfuric acid/30% hydrogen peroxide) at 80 °C and sonicated for one hour, and finally rinsed several times with D.I. water. In the dip-coating process, two limiting withdrawal speeds of the equipment (25 and 250 mm/min) were used. All depositions were performed after ultrasonic treatment of the polymer solutions for about 10 min to remove air bubbles. After the dip-coating, the samples were placed under a hood in an ambient environment for three days to dry. The typical dimensions of the polymer films produced were 1 cm × 2 cm.

The following sample nomenclature used in this study is shown in Table 1. For PAN films, the first number after "PAN" shows the solution concentration and the second number indicates the dip-coating withdrawal speed. For example, PAN-3-25 is used to denote the sample of PAN film produced from a 3 wt.% solution at a dip-coating speed of 25 mm/min. The sample name for polymer films of PAN/PET prepared from 3 wt.% solution starts with PAN/PET followed by the weight percentage of PET in the total polymer weight and then the dip-coating withdrawal speed. For instance, PAN/PET film obtained at a dip-coating withdrawal speed of 25 mm/min with the weight ratio of PAN to PET of 90:10 was labeled as PAN/PET-10-25.

Table 1. Sample nomenclature.

Sample Name	Polymer Concentration in Solution (wt.%)	PAN/PET (by Weight)	Dip-Coating Withdrawal Speed (mm/min)
PAN-3-25	3	100/0	25
PAN-3-250	3	100/0	250
PAN/PET-10-25	3	90/10	25
PAN/PET-20-25	3	80/20	25
PAN/PET-25-25	3	75/25	25
PAN/PET-30-25	3	70/30	25
PAN/PET-35-25	3	65/35	25
PAN/PET-40-25	3	60/40	25
PAN/PET-10-250	3	90/10	250
PAN/PET-20-250	3	80/20	250
PAN/PET-25-250	3	75/25	250
PAN/PET-30-250	3	70/30	250
PAN/PET-35-250	3	65/35	250
PAN/PET-40-250	3	60/40	250

2.4. Carbon Film Fabrication

The two-step carbonization procedure previously reported by us was employed here [45]. The films of PAN and PAN/PET were first thermally stabilized under air in an oven. The oven temperature was raised from 30–230 °C at a heating rate of 3 °C/min, held at 230 °C for 6 h, and then cooled down to room temperature. A high-temperature carbonization treatment was applied to the thermally stabilized films under a stream of ultrahigh purity nitrogen. The temperature was raised from 30 °C at a heating rate of 3 °C/min to 600 °C and then was subsequently heated to 1000 °C at a heating rate of 10 °C/min. The temperature was kept at 1000 °C for 30 min. The films were finally cooled down to room temperature in nitrogen.

2.5. Materials Characterization

Characterization of the films followed the procedures used in our previous study reported elsewhere [45]. The surface morphologies of the polymer films, stabilized films, and carbon films were characterized by atomic force microscopy (AFM) and scanning electron microscopy (SEM). AFM Dimension™ 3100 (Digital Instruments Inc., Tonawanda, NY, USA) in tapping mode was used to image the surface topography. NSC16 type silicon probes (MikroMasch) with a resonance frequency ~170 kHz, a spring constant ~45 N/m, and a tip radius of 8–10 nm were used. Imaging was performed at a scan rate of 1 Hz for pure PAN films and 0.5 Hz for PAN/PET films. A high-resolution scanning electron microscope (S4800 Hitachi, Tokyo, Japan) was used for the surface and fracture morphologies characterization. Before the SEM examination, the samples were coated with platinum. SEM images were obtained at an accelerating voltage of 10 kV at various magnifications. The thicknesses of the polymer films, stabilized films, and carbon films were determined by AFM cross-sectional measurements using the scratch method [47]. Scratches were formed on the polymer films by a razor blade. The same scratches were used to measure the thickness of the stabilized and carbonized films.

2.6. Thermogravimetric Analysis

Thermogravimetric analysis (TGA) of PAN and PET was conducted on an analyzer of Q500 from TA Instruments (New Castle, DE, USA). The TGA analysis was performed using the conditions simulating the two-step thermal treatment process. The TGA steps were specified as follows: (1) raise the temperature from 30–230 °C in air at 3 °C/min, (2) stay at 230 °C for 6 h in air, (3) cooldown to 30 °C in air, (4) raise the temperature from 30–600 °C at 3 °C/min in nitrogen, (5) raise the temperature from 600–1000 °C at 10 °C/min in nitrogen, and (6) stay at 1000 °C for 30 min.

2.7. Interlaminar Shear Strength Measurement

Graphite plates coated with solid and porous carbon films were used in this part of our study. The dip-coating and carbonization procedures were the same as the ones used for the model silicon wafer substrates. Before the dip-coating, graphite plates (2.5 cm × 2 cm) were cleaned using the HFIP solvent. PAN and PAN/PET solutions with the PAN to PET ratio of 65:35 were used in the dip-coating process at the withdrawal speed of 250 mm/min. Initial bonding of the graphite plates coated with the carbon films was performed in a vacuum oven. Inside the oven, PP pellets were placed between the two graphite plates with a vacuum applied at the beginning of the consolidation process. The temperature was raised to 200 °C to melt the PP, and the assembly was kept at 200 °C for 30 min. Then, the two graphite plates with molten PP in between were transferred to a laboratory hot press set at 200 °C and low pressure (less than 0.7 MPa) was applied to fabricate a uniform PP layer between the plates. The bonding area was 2 cm × 2 cm, and 5 specimens for each sample group were prepared. The single lap-joint test was performed using an Instron machine (Model 5582) in compression mode at a loading head motion rate of 1 mm/min. The interlaminar shear strength is reported. The fracture interface between the two graphite plates was then characterized using a high-resolution scanning electron microscope (S4800 Hitachi). Before the SEM examination, the fracture surfaces were coated with platinum. SEM images were obtained at an accelerating voltage of 10 kV at various magnifications.

3. Result and Discussion

3.1. Fabrication and Morphology of Polymer Films

To obtain films from a binary polymer blend via a controllable and reproducible solvent-based process (such as dip-coating employed in this work), both polymers must be soluble in a solvent. It is also required that no phase separation occurs when the polymers are dissolved in the solvent simultaneously. One-component PAN and PET solutions in HFIP are visually clear, and the solid films obtained from the solution are optically transparent and uniform. The result indicated significant PAN and PET solubility in the solvent. The binary PAN/PET solutions were also transparent and homogeneous at the concentrations used here (PAN to PET ratio from 90:10 to 60:40), indicating good solubility of the polymer mixture in the solvent at 3 wt.%. Thus, PET could be added to the PAN solution to generate pores in the carbon films in the course of the high-temperature carbonization process.

Due to drying, the PAN/PET films at the PAN to PET ratios of 75:25, 70:30, 65:35, and 60:40 lost their transparency. The opaqueness of films suggested that phase separation occurred in these films. At the same time, the films with the PAN to PET ratios of 90:10 and 80:20 remained visually transparent. To further characterize the morphology of the PET/PAN films, their surface was visualized with AFM. Figure 1 shows the images for the films obtained by the dip-coating at the withdrawal speed of 25 mm/min followed by drying. One can see that, for all PET concentrations, circular pores formed within the films. The porous structure formation is associated with the different affinity of phases separating PAN and PET to the solvent [45]. Specifically, the pore formation indicates that the PAN matrix is more rapidly depleted of the solvent and solidifies before the solidification of the PET phase, which has a higher affinity to the solvent [48,49]. When a withdrawal speed of 250 mm/min is used, the films' morphology was practically the same at the qualitative level (images are not shown). However, at higher PET content, the pores' size is somewhat larger for the PET/PAN films obtained at a withdrawal speed of 250 mm/min. We suggest that the formation of larger pores can be caused by the deposition of a thicker swollen polymer film on the surface of the substrate at the higher dip-coating withdrawal speed (as discussed in the sections below). For the thicker film, it takes longer for the solvent to evaporate, and, therefore, larger PET domains can form during the phase separation process.

For both withdrawal speeds, the pore size increased with PET concentration. The diameter of the pores was estimated from AFM images (Figure 2a). At PAN/PET ratios of 90:10 and 80:20, the pores at both dip-coating withdrawal speeds were smaller than 220 nm in diameter, and, therefore, the films appeared transparent after drying. With the increase of PET concentration, the pore size increased to about 242 nm, 563 nm, 2 µm, to 3 µm for films having the PAN to PET ratios of 75:25, 70:30, 65:35, and 60:40, respectively, fabricated at the withdrawal speed of 25 mm/min. For a withdrawal speed of 250 mm/min, the pore sizes were 340 nm, 755 nm, 2.74 µm, and 7 µm for films with the PAN to PET ratios of 75:25, 70:30, 65:35, and 60:40, respectively. Our results indicate that the porous structure's parameters can be varied via PET concentration and the withdrawal speed in a certain range.

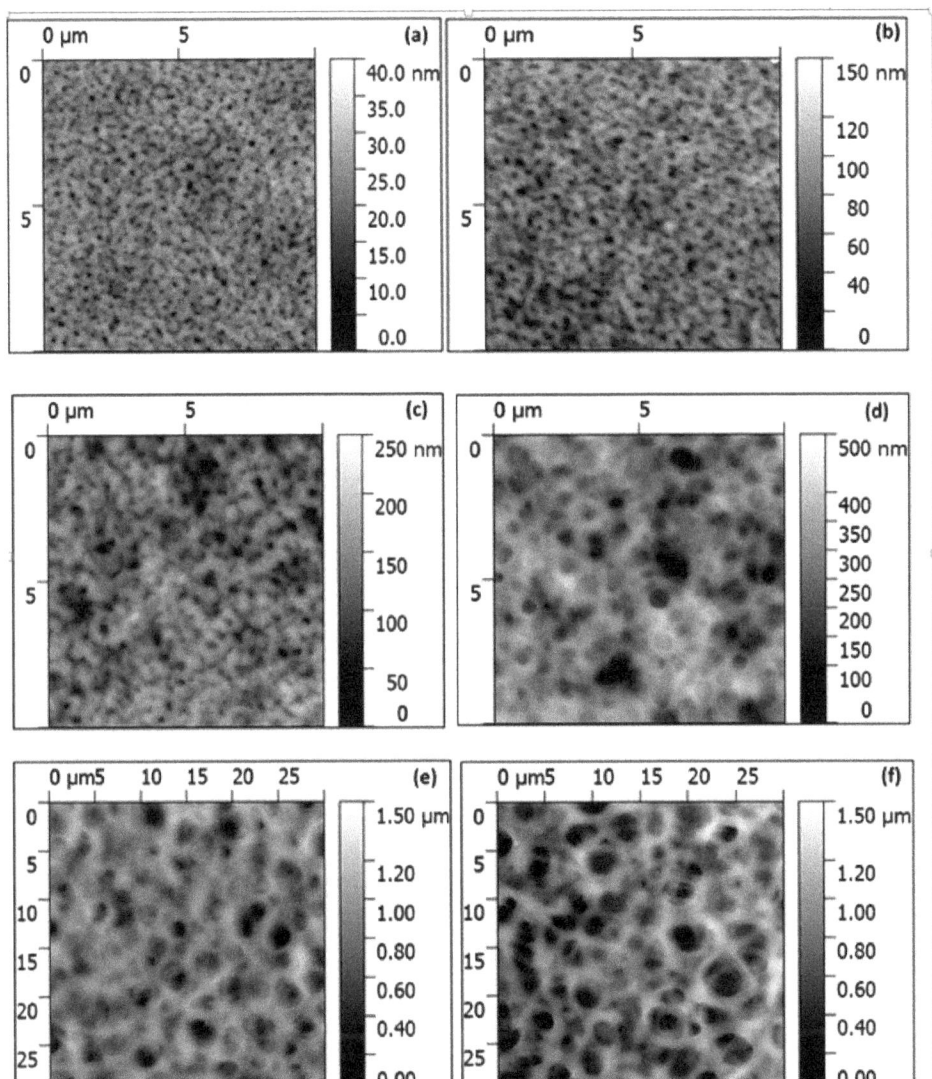

Figure 1. AFM topographical images of polymer films as deposited on silicon wafer by dip-coating: (**a**) PAN/PET-10-25, (**b**) PAN/PET-20-25, (**c**) PAN/PET-25-25, (**d**) PAN/PET-30-25, (**e**) PAN/PET-35-25, and (**f**) PAN/PET-40-25. The scale bar on the right side of the AFM images depicts the height scale of the image, where light and dark shades correspond to high and low relative heights, respectively.

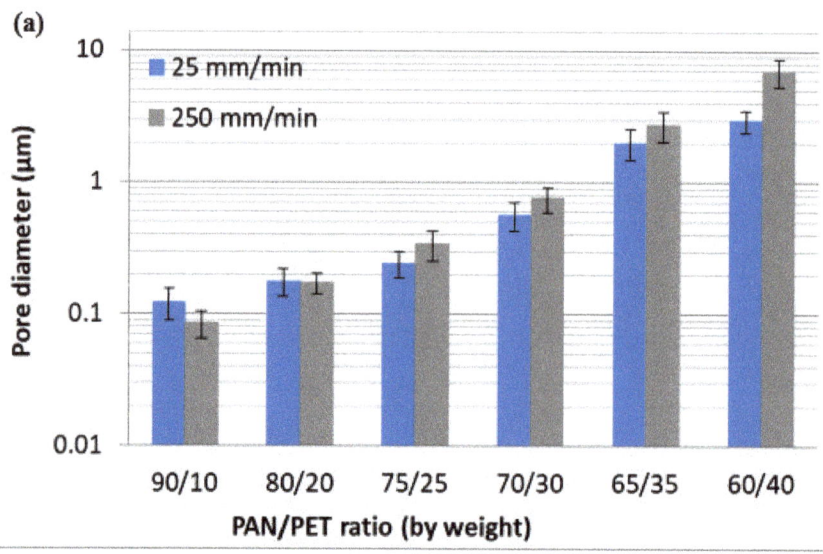

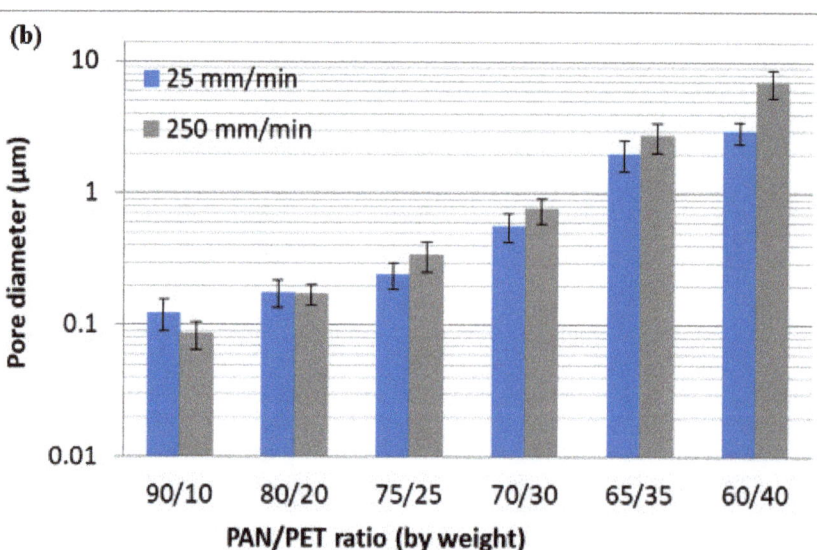

Figure 2. Pore diameters for porous polymer (**a**) and carbon (**b**) films obtained from AFM images.

The thicknesses of polymer films obtained were measured using AFM (Figure 3). For the films fabricated using the speed of 250 mm/min, all the polymer films had thicknesses higher than the vertical limit of our AFM (6.42 µm), and, thus, the thicknesses were not determined. Therefore, the data is reported for the films obtained using the lower withdrawal speed. The pure PAN film had a thickness of 2.73 µm. The PAN/PET films produced in this study had significantly lower thicknesses than that of the PAN film. The blended films' thicknesses with 10, 20, 25, 30, 35, and 40 wt.% of PET were 1.72, 1, 1.25, 1.03, 1.43, and 1.28 µm, respectively. The decrease of the thickness is associated with the decrease of the solution viscosity as part of PAN is replaced with PET. Indeed, according to the classical Landau–Levich theory, the thickness of a film (h) produced by dip-coating on a flat substrate is related to the viscosity of the solution, dip-coating withdrawal speed, solution surface tension, and density [45,50,51]:

$$h \propto \frac{(\eta \cdot v)^{2/3}}{\sigma^{1/6} \cdot (\rho \cdot g)^{1/2}} \qquad (1)$$

where η, σ, and ρ are the viscosity, the surface tension, and the density of the fluid, respectively, while v and g are the dip-coating withdrawal speed and the standard gravitational acceleration. In our case, the effect of the viscosity and dip-coating speed dominated the process. At the higher speed, significantly thicker films were fabricated. At the constant speed of 25 mm/min, the PAN/PET films were significantly thinner than the PAN film. We suggest that the decrease of solution viscosity was the main cause in this case. Indeed, the viscosity (η) of the PET solution was significantly lower than that of PAN at the same concentration. It is associated with the significant molecular weight (M) difference between the two polymers, as indicated by the relationship shown below [46]:

$$\eta \propto (c^\gamma M)^\beta \qquad (2)$$

where c is the solution concentration, parameter β changes with molecular weight and concentration, generally from 2 to 3.4, and γ varies from 1.85 to 1.4 as solvent affinity to polymer increases. From Equation (2), it is evident that replacing part of the PAN (Mw = 100,000 g/mol) with PET (Mw = 18,000 g/mol) would decrease the solution viscosity, leading to a thinner film. The effect of PET concentration in the solution on the film thickness is not yet clear, since many factors, such as surface tension, density, and polymer-solvent affinity influence the polymer film thickness. At the same time, it is obvious that the thickness of the films can be tuned significantly by variation of the withdrawal speed used for the dip-coating.

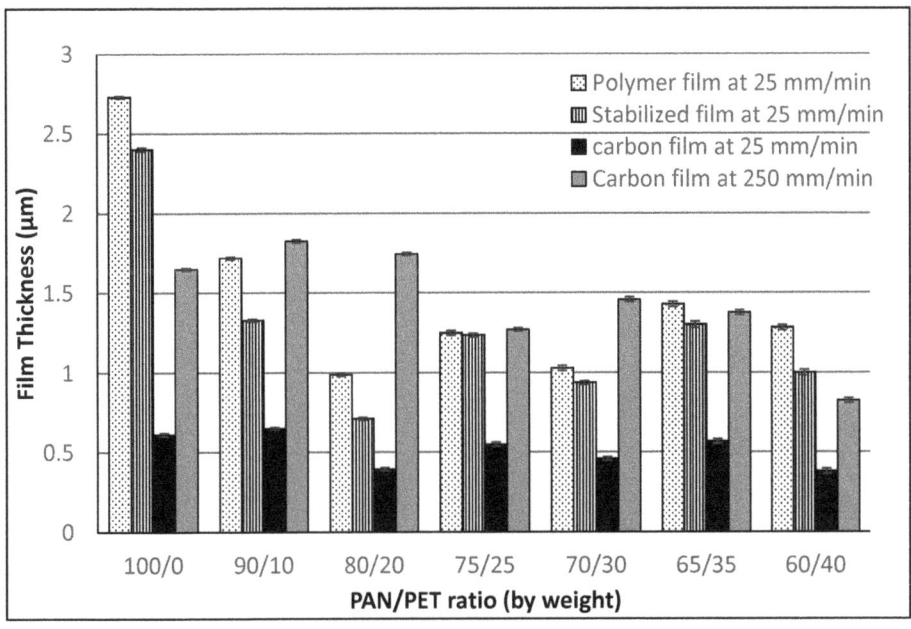

Figure 3. The thicknesses of the deposited polymer films, stabilized films, and carbon films.

3.2. Fabrication and Morphology of Carbon Films

The general procedures used here to fabricate the solid/porous carbon films were reported in detail in our preceding publication [45]. Before the fabrication, TGA measurements were conducted to determine the behavior of the polymers during the PAN stabilization (at 230 °C for 6 h) and the carbonization process conducted by heating the material to 1000 °C. PAN's stabilization is necessary to prepare the PAN-based material for further high-temperature carbonization [22,45]. The results are shown in Figure 4. It was observed that during the stabilization process, PET and PAN had 3 wt.% and 7 wt.% mass loss, respectively. Thus, no significant thermal degradation of PET occurred during the stabilization process. We associate the mass loss with water absorbed by the polymer. The mass loss of PAN (beside absorbed water) was mainly caused by the chemical reactions of macromolecular cyclization to form a ladder-type polymer in addition to the dehydrogenation and oxidation [22]. During the initial phase of carbonization (230–600 °C), significant thermal degradation of PET was observed. Specifically, the major mass change of the material (>80 wt.%) takes place between 370–500 °C, and the mass of PET residue was around 12 wt.% at the end of the first phase of carbonization. In this initial carbonization stage, PAN lost about 25 wt.% of its mass. During the next step of carbonization (600–1000 °C), PET lost ~2 wt.% of its original mass. Simultaneously, the mass of PAN residue decreased from 68 wt.% to 52 wt.%. In the course of the final annealing at 1000 °C, PET lost an additional 4 wt.%, and, therefore, the carbon residue of the polymer is ~6 wt.%. PAN lost approximately 6 wt.% during the final annealing leaving the carbon residue of about 46 wt.%.

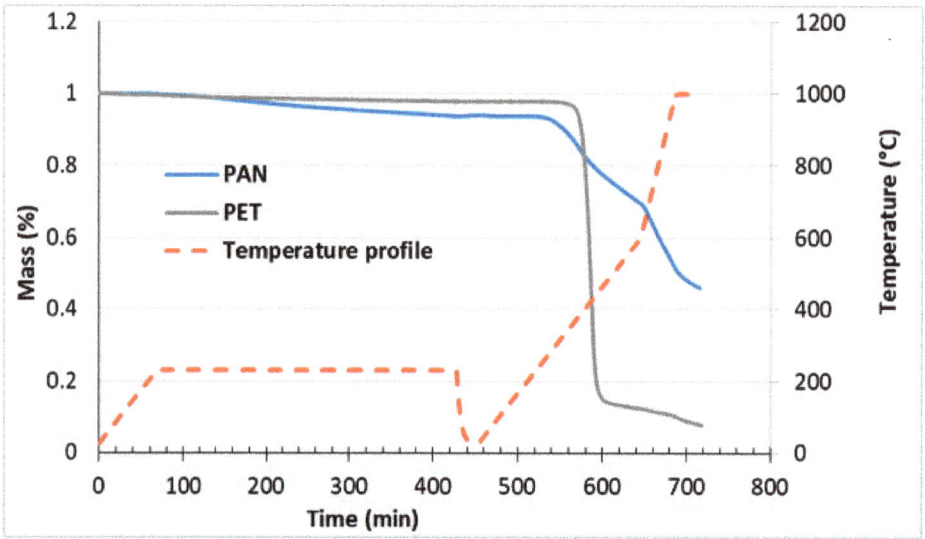

Figure 4. Temperature profile for the stabilization and carbonization processes used and corresponding TGA traces for pure PAN and PET.

The temperature profile used in the TGA analysis Figure 4 was employed for the fabrication of carbon films. First, the polymer films after air drying were stabilized under air at 230 °C for 6 h. After the stabilization, the film thickness decreased in the range of 10–25% for PAN and PAN/PET films (Figure 3). Only for the film with a 75/25 PAN/PET ratio, there was no measurable change of thickness. The reason for this was not clear at the time of this writing. The color of stabilized polymer films was dark brown. After carbonization, the films became black. The carbon film thickness was measured using AFM (Figure 3). The thickness decreased significantly in comparison to that of stabilized films.

The thickness of solid carbon film made from PAN (at withdrawal speed of 25 mm/min) is around 0.61 µm, which is a 75% decrease from that of the initial polymer film. The reduction of thickness for the carbon films obtained for PAN/PET mixtures was about 40–60%. The porous carbon films with the PET percentages of 10, 20, 25, 30, 35, and 40 had thicknesses of 0.65, 0.39, 0.55, 0.46, 0.57, and 0.38 µm, respectively. The thicknesses of carbon films obtained from polymer films dip-coated at the withdrawal speed of 250 mm/min were also measured using AFM Figure 3. The solid carbon film was 1.65 µm. The porous carbon film thicknesses were on the similar level 1.83, 1.75, 1.27, 1.46, 1.38, and 0.82 µm for 10, 20, 25, 30, 35, and 40% PET in the PAN/PET mixtures, respectively. We did not find a significant correlation between the composition of the PAN/PET mixture and the thickness of the carbon films obtained. However, there is a clear dependence of the thickness on the withdrawal speed used during the polymer film fabrication. Thus, the speed can be used to fabricate the porous carbon film of different thicknesses.

The morphology of carbon films was observed with AFM. The images of the porous films generated from polymer films dip-coated at the withdrawal speed of 25 mm/min are shown in Figure 5. Continuous porous carbon structures were observed, which indicated that a continuous PAN structure was formed during polymer film formation from the binary PAN/PET solutions. The continuous network of the porous PAN structure survived the stabilization and carbonization process without forming cracks. The same observation was made for the films obtained using the higher withdrawal speed (images are not shown). The correlation between PET content and pore size and the one between the withdrawal speed and pore size are maintained through the carbonization process (Figure 2). As the percentage of PET and withdrawal speed increases, the pore size in the carbon films increases. For the carbon films generated from polymer films dip-coated at 25 mm/min, the pore diameters are 0.11, 0.19, 0.32, 0.70, 2.67, and 3.16 µm as the PET percentage in the original polymer films increased from 10, 20, 25, 30, 35, to 40. When the polymer films were formed at the dip-coating withdrawal speed of 250 mm/min, the corresponding pore diameters were 0.11, 0.23, 0.40, 0.72, 3.81, and 10.16 µm. The carbon structure's pore sizes were also larger than the corresponding pore sizes in the polymer films Figure 2. The mass loss of PAN and PET contributed to the larger pore size in the carbon structure. In general, we determined that the PET content and withdrawal speed could be employed to vary the diameter of pores in the carbon films.

AFM provided accurate information only on the topmost morphology of the films. To complement AFM imaging, the cross-section of carbon films, fabricated from PAN and PAN/PET, were examined using SEM (Figure 6). Figure 6a,b show that the films made from pure PAN do not have a porous structure. Figure 6c–h display the cross-section of the films obtained from the PAN/PET mixture. These films have a significant number of pores in their structure. The pores were found to be elliptical in shape because of the film shrinkage in the vertical direction during stabilization and carbonization processes. The shrinkage in the horizontal direction was constrained by the bonding of the film to the substrates. The film's thickness and the pores' size were dependent on PET content. At the PET concentration of 10 wt.%, the carbon films obtained have the internal pore of the smallest size (Figure 6c) that appear not to be interconnected. As the concentration of PET in the polymer film increased, the pores formed in the carbon films became larger, and the number of interconnected pores increased (Figure 6d–h). At the PET concentration of 40% (Figure 6h), the carbon film included a significant fraction of pores in the structure. It is necessary to note that the thickness of the carbon films measured with AFM and the one estimated from SEM were within 10–20% of each other.

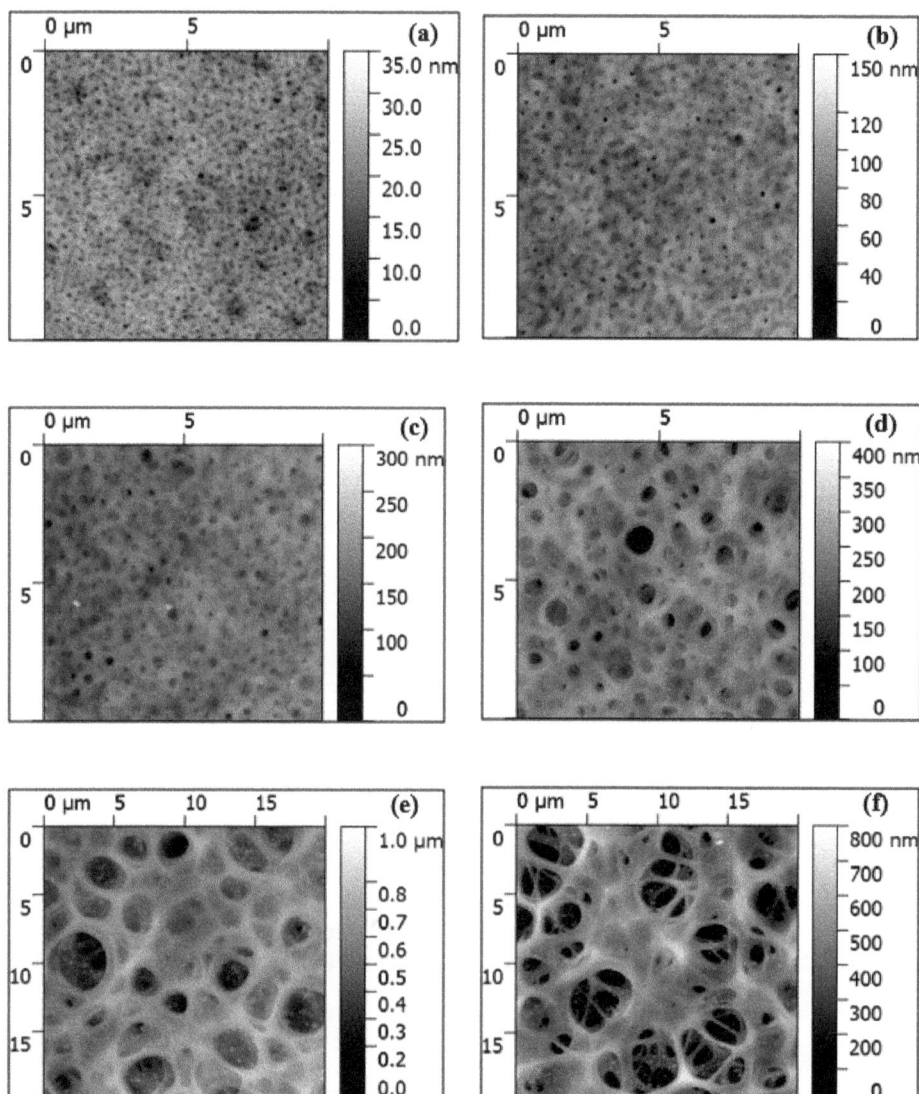

Figure 5. AFM topographical images of the carbonized polymer films obtained from (**a**) PAN/PET-10-25, (**b**) PAN/PET-20-25, (**c**) PAN/PET-25-25, (**d**) PAN/PET-30-25, (**e**) PAN/PET-35-25, and (**f**) PAN/PET-40-25. The scale bar on the right side of the AFM images depicts the height scale of the image, where light and dark shades correspond to high and low relative heights, respectively.

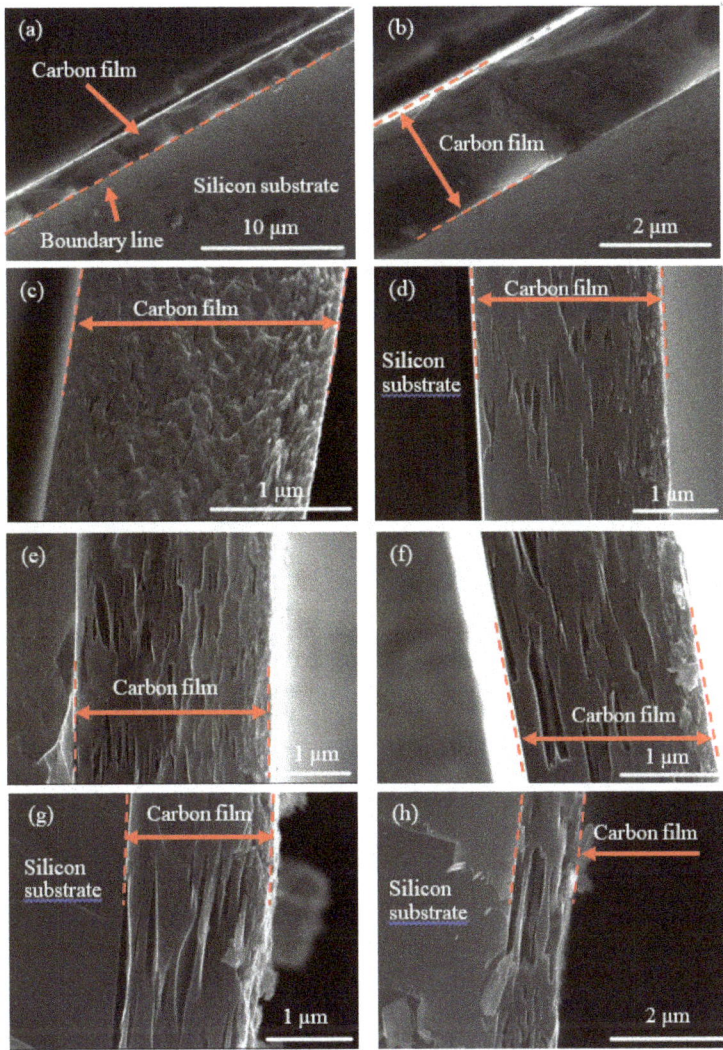

Figure 6. The SEM micrographs showing the cross-section of carbon films prepared from polymer films: (**a**,**b**) pure PAN, (**c**) PAN/PET-10-250, (**d**) PAN/PET-20-250, (**e**) PAN/PET-25-250, (**f**) PAN/PET-30-250, (**g**) PAN/PET-35-250, and (**h**) PAN/PET-40-250.

3.3. Interlaminar Shear Testing

We hypothesized that porous carbon structure will demonstrate higher adhesion to thermoplastic PP in comparison with non-porous carbon substrate. To test this, we evaluated the effect of the porous coating on interfacial bonding strength between PP and graphite plates. To this end, the porous carbon films were fabricated on the surfaces of the plates using a PAN/PET ratio of 65/35 and withdrawal speed of 250 mm/min. PP grafted with 1 wt.% of maleic anhydride was used in the adhesion test, since it was previously reported that maleic anhydride units improve carbon-PP adhesion [26,28,30]. The schematic of the single lap-joint specimen used in our measurements is shown in Figure 7a. The solid carbon film fabricated from pure PAN film was also deposited on the graphite plates to be used as the control samples. The shear strength results calculated from the tests are shown in Figure 7b. The graphite plates coated with the porous carbon film had a significantly higher shear strength (~8.2 MPa) than those coated with the solid carbon film (~4.9 MPa). Therefore, the modification of carbon materials with the porous surface structure developed in this study improved the interfacial shear strength by ~70%.

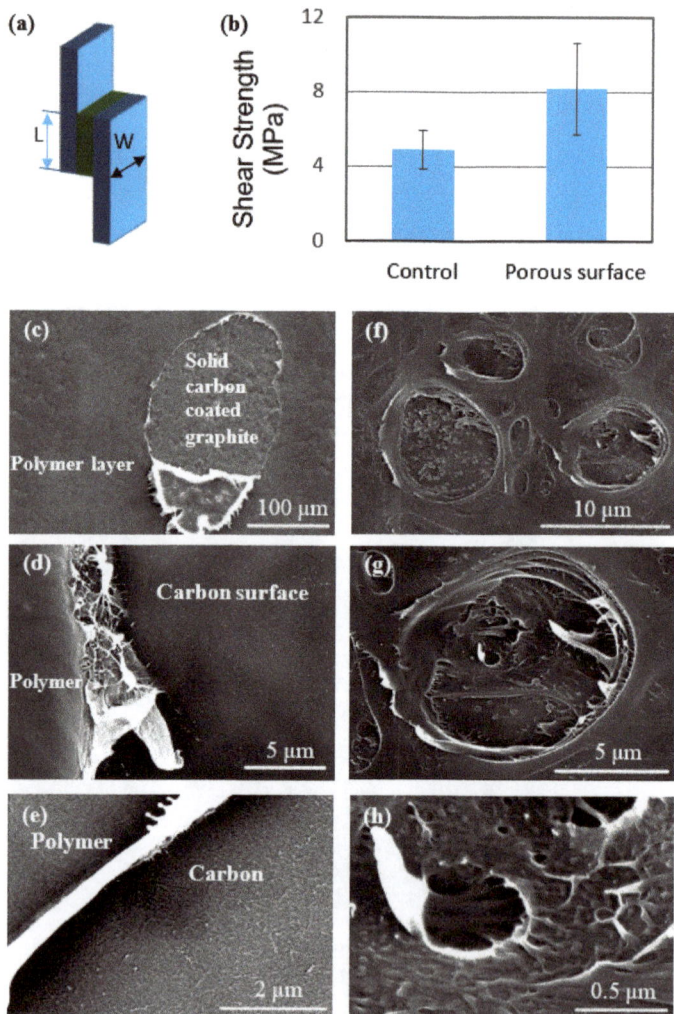

Figure 7. The schematic of the single lap-joint specimen (a), the shear strength results (b), and the SEM micrographs showing the fracture surfaces of solid (c–e) and porous (f–h) carbon-coated graphite plates.

The morphology of the fractured interface was observed with SEM (Figure 7c–h). The adhesive failure for the graphite plates coated with the solid carbon film occurred at the polymer/carbon interface (Figure 7c–e). For the plates coated with the porous carbon films, penetration/infiltration of polymer into the pores can be identified from the fracture interface (Figure 7f–h), which also shows the predominantly cohesive failure of the PP matrix. In this failure mode, the interfacial stress causes shear deformation and polymer matrix breakage, indicating that the carbon/PP interfacial bonding is significantly enhanced. The increase of the inter-laminar shear strength is also associated with the increased contact area for the porous structure due to polymer penetration into the internal pores. Indeed, the polymer matrix penetrated in the surface (Figure 7f) and internal pores (Figure 7g,h), and interlocking interphase between PP and the porous carbon structure was formed. The polymer that penetrated the internal pores was stretched during the failure. This conclusion is supported by the extensive fibrillation of the polymer phase (Figure 7f–h). We also note that the carbon porous coating was not damaged during the mechanical test, indicating a strong porous carbon structure was formed using the technique reported here.

4. Conclusions

We demonstrated that porous carbon films with controllable thickness and surface/internal structures could be fabricated by the carbonization of PAN/PET films made through a dip-coating process. PET functioned as a pore former in the polymer film. After carbonization of the polymer, porous carbon films with thicknesses from 0.38–1.83 μm and pore diameters between 0.1–10 μm were obtained. The higher concentrations of PET in the PAN/PET mixture and the higher withdrawal speed during dip-coating caused the formation of larger pores. The thickness of the carbon films can be controlled by the withdrawal speed during dip-coating. We determined that deposition of the porous carbon film on graphite substrate significantly increases the value of the interfacial shear strength between graphite plates and thermoplastic PP. This study has shown the feasibility of fabrication of 3D porous carbon structure on the surface of carbon materials for increasing the interfacial strength. This approach can be used for the fabrication of high-performance carbon fiber-thermoplastic composites.

Author Contributions: Investigation, Y.P., R.B. (Ruslan Burtovyy), R.B. (Rajendra Bordia) and I.L.; Writing—original draft, Y.P. and I.L.; Writing—review & editing, Y.P., R.B. (Rajendra Bordia) and I.L. All authors have read and agreed to the published version of the manuscript.

Funding: This work was supported, in part, by the National Science Foundation EPSCoR Program under NSF Award # OIA-1655740.

Acknowledgments: Any opinions, findings, and conclusions, or recommendations expressed in this material are those of the author(s) and do not necessarily reflect those of the National Science Foundation. Y.P. also acknowledges a postdoctoral research fellowship funded by Clemson University.

Conflicts of Interest: The authors declare no conflict of interest

References

1. Bledzki, A.K.; Seidlitz, H.; Krenz, J.; Goracy, K.; Urbaniak, M.; Rosch, J.J. Recycling of Carbon Fiber Reinforced Composite Polymers-Review-Part 2: Recovery and Application of Recycled Carbon Fibers. *Polymers* **2020**, *12*, 3003. [CrossRef] [PubMed]
2. Deng, J.Y.; Xu, L.; Liu, J.H.; Peng, J.H.; Han, Z.H.; Shen, Z.G.; Guo, S.H. Efficient Method of Recycling Carbon Fiber from the Waste of Carbon Fiber Reinforced Polymer Composites. *Polym. Degrad. Stabil.* **2020**, *182*. [CrossRef]
3. Gopalraj, S.K.; Karki, T. A Review on the Recycling of Waste Carbon Fibre/Glass Fibre-Reinforced Composites: Fibre Recovery, Properties and Life-Cycle Analysis. *Sn. Appl. Sci.* **2020**, *2*, s42452-s020.
4. Kiss, P.; Stadlbauer, W.; Burgstaller, C.; Stadler, H.; Fehringer, S.; Haeuserer, F.; Archodoulaki, V.M. In-House Recycling of Carbon- and Glass Fibre-Reinforced Thermoplastic Composite Laminate Waste into High-Performance Sheet Materials. *Compos. Part A Appl. Sci. Manuf.* **2020**, *139*. [CrossRef]
5. Kumar, S.; Krishnan, S. Recycling of Carbon Fiber with Epoxy Composites by Chemical Recycling for Future Perspective: A Review. *Chem. Pap.* **2020**, *74*, 3785–3807. [CrossRef]
6. Utekar, S.; Suriya, V.K.; More, N.; Rao, A. Comprehensive Study of Recycling of Thermosetting Polymer Composites - Driving Force, Challenges and Methods. *Compos. Part B Eng.* **2021**, *207*. [CrossRef]
7. Zhang, J.; Chevali, V.S.; Wang, H.; Wang, C.H. Current Status of Carbon Fibre and Carbon Fibre Composites Recycling. *Compos. Part. B Eng.* **2020**, *193*. [CrossRef]
8. Pimenta, S.; Pinho, S.T. Recycling Carbon Fibre Reinforced Polymers for Structural Applications: Technology Review and Market Outlook. *Waste Manag.* **2011**, *31*, 378–392. [CrossRef]
9. Li, P.D.; Zhao, Y.; Long, X.; Zhou, Y.W.; Chen, Z.Y. Ductility Evaluation of Damaged Recycled Aggregate Concrete Columns Repaired With Carbon Fiber-Reinforced Polymer and Large Rupture Strain FRP. *Front. Mater.* **2020**, *7*. [CrossRef]

10. Tapper, R.J.; Longana, M.L.; Norton, A.; Potter, K.D.; Hamerton, I. An Evaluation of Life Cycle Assessment and its Application to the Closed-Loop Recycling of Carbon Fibre Reinforced Polymers. *Compos. Part. B Eng.* **2020**, *184*. [CrossRef]
11. Giorgini, L.; Benelli, T.; Brancolini, G.; Mazzocchetti, L. Recycling of Carbon Fiber Reinforced Composite Waste to Close Their Life Cycle in a Cradle-to-Cradle Approach. *Curr. Opin. Green Sust.* **2020**, *26*. [CrossRef]
12. Liu, Z.; Turner, T.A.; Wong, K.H.; Pickering, S.J. Development of High Performance Recycled Carbon Fibre Composites with an Advanced Hydrodynamic Fibre Alignment Process. *J. Clean Prod.* **2021**, *278*. [CrossRef]
13. Wellekotter, J.; Resch, J.; Baz, S.; Gresser, G.T.; Bonten, C. Insights into the Processing of Recycled Carbon Fibers via Injection Molding Compounding. *J. Compos. Sci.* **2020**, *4*, 161. [CrossRef]
14. Ota, H.; Jespersen, K.M.; Saito, K.; Wada, K.; Okamoto, K.; Hosoi, A.; Kawada, H. Effect of the interfacial nanostructure on the Interlaminar Fracture Toughness and Damage Mechanisms of Directly Bonded Carbon Fiber Reinforced Thermoplastics and Aluminum. *Compos. Part A Appl. Sci. Manuf.* **2020**, *139*. [CrossRef]
15. Khurshid, M.F.; Hengstermann, M.; Hasan, M.M.B.; Abdkader, A.; Cherif, C. Recent Developments in the Processing of Waste Carbon Fibre for Thermoplastic Composites—A Review. *J. Compos. Mater.* **2020**, *54*, 1925–1944. [CrossRef]
16. Roux, M.; Eguemann, N.; Dransfeld, C.; Thiebaud, F.; Perreux, D. Thermoplastic Carbon Fibre-Reinforced Polymer Recycling with Electrodynamical Fragmentation: From Cradle to Cradle. *J. Compos.* **2017**, *30*, 381–403. [CrossRef]
17. Hirayama, D.; Saron, C.; Botelho, E.C.; Costa, M.L.; Ancelotti, A.C. Polypropylene Composites Manufactured from Recycled Carbon Fibers from Aeronautic Materials Waste. *Mater. Res.-Ibero-Am. J.* **2017**, *20*, 519–525. [CrossRef]
18. Friedrich, K. Carbon Fiber Reinforced Thermoplastic Composites for Future Automotive Applications. *AIP Conf. Proc.* **2016**, *1736*. [CrossRef]
19. Stoeffler, K.; Andjelic, S.; Legros, N.; Roberge, J.; Schougaard, S.B. Polyphenylene Sulfide (PPS) Composites Reinforced with Recycled Carbon Fiber. *Compos. Sci. Technol.* **2013**, *84*, 65–71. [CrossRef]
20. Sharma, M.; Gao, S.L.; Mader, E.; Sharma, H.; Wei, L.Y.; Bijwe, J. Carbon Fiber Surfaces and Composite Interphases. *Compos. Sci. Technol.* **2014**, *102*, 35–50. [CrossRef]
21. Huang, X.S. Fabrication and Properties of Carbon Fibers. *Materials* **2009**, *2*, 2369–2403. [CrossRef]
22. Donnet, J.-B.; Wang, T.K.; Peng, J.C.M.; Rebouillat, S. *Carbon Fibers*, 3rd ed.; Marcel Dekker Inc.: New York, NY, USA, 1998.
23. Fu, S.Y.; Lauke, B.; Mäder, E.; Yue, C.Y.; Hu, X. Tensile Properties of Short-Glass-Fiber- and Short-Carbon-Fiber-Reinforced Polypropylene Composites. *Compos. Part A Appl. Sci. Manuf.* **2000**, *31*, 1117–1125. [CrossRef]
24. Gamze Karsli, N.; Aytac, A.; Akbulut, M.; Deniz, V.; Güven, O. Effects of Irradiated Polypropylene Compatibilizer on the Properties of Short Carbon Fiber Reinforced Polypropylene Composites. *Radiat. Phys. Chem.* **2013**, *84*, 74–78. [CrossRef]
25. Han, S.H.; Oh, H.J.; Kim, S.S. Evaluation of Fiber Surface Treatment on the Interfacial Behavior of Carbon Fiber-Reinforced Polypropylene Composites. *Compos. Part. B Eng.* **2014**, *60*, 98–105. [CrossRef]
26. Karsli, N.G.; Aytac, A. Effects of Maleated Polypropylene on the Morphology, Thermal and Mechanical Properties of Short Carbon Fiber Reinforced Polypropylene Composites. *Mater. Des.* **2011**, *32*, 4069–4073. [CrossRef]
27. Rezaei, F.; Yunus, R.; Ibrahim, N.A. Effect of Fiber Length on Thermomechanical Properties of Short Carbon Fiber Reinforced Polypropylene Composites. *Mater. Des.* **2009**, *30*, 260–263. [CrossRef]
28. Tian, H.; Yao, Y.; Liu, D.; Li, Y.; Jv, R.; Xiang, G.; Xiang, A. Enhanced Interfacial Adhesion and Properties of Polypropylene/Carbon Fiber Composites by Fiber Surface Oxidation in Presence of a Compatibilizer. *Polym. Compos.* **2019**, *40*, E654–E662. [CrossRef]
29. Tian, H.F.; Yao, Y.Y.; Wang, C.Y.; Jv, R.; Ge, X.; Xiang, A.M. Essential Work of Fracture Analysis for Surface Modified Carbon Fiber/Polypropylene Composites with Different Interfacial Adhesion. *Polym. Compos.* **2020**, *41*, 3541. [CrossRef]
30. Unterweger, C.; Duchoslav, J.; Stifter, D.; Fürst, C. Characterization of Carbon Fiber Surfaces and Their Impact on the Mechanical Properties of Short Carbon Fiber Reinforced Polypropylene Composites. *Compos. Sci. Technol.* **2015**, *108*, 41–47. [CrossRef]
31. Vishkaei, M.S.; Salleh, M.A.M.; Yunus, R.; Biak, D.R.A.; Danafar, F.; Mirjalili, F. Effect of Short Carbon Fiber Surface Treatment on Composite Properties. *J. Compos. Mater.* **2010**, *45*, 1885–1891. [CrossRef]
32. Zhang, K.; Li, Y.; He, X.; Nie, M.; Wang, Q. Mechanical Interlock Effect Between Polypropylene/Carbon Fiber Composite Generated by Interfacial Branched Fibers. *Compos. Sci. Technol.* **2018**, *167*, 1–6. [CrossRef]
33. Huang, Y.L.; Young, R.J. Interfacial Micromechanics in Thermoplastic and Thermosetting Matrix Carbon Fibre Composites. *Compos. Part A Appl. Sci. Manuf.* **1996**, *27*, 973–980. [CrossRef]
34. Tang, L.G.; Kardos, J.L. A Review of Methods for Improving the Interfacial Adhesion between Carbon Fiber and Polymer Matrix. *Polym. Compos.* **1997**, *18*, 100–113. [CrossRef]
35. Gravis, D.; Moisan, S.; Poncin-Epaillard, F. Characterization of surface physico-chemistry and morphology of plasma-sized carbon fiber. *Thin Solid Films* **2021**, *721*, 11. [CrossRef]
36. Donnet, J.B.; Brendle, M.; Dhami, T.L.; Bahl, O.P. Plasma Treatment Effect on the Surface-Energy of Carbon and Carbon-Fibers. *Carbon* **1986**, *24*, 757–770. [CrossRef]
37. Donnet, J.B.; Cazeneuve, C.; Schultz, J.; Shanahan, M.E.R. Surface-Energy of Carbon-Fibers. *Carbon* **1980**, *18*, 61.
38. Donnet, J.B.; Dhami, T.L.; Dong, S.; Brendle, M. Microwave Plasma Treatment Effect on the Surface-Energy of Carbon-Fibers. *J. Phys. D Appl. Phys.* **1987**, *20*, 269–275. [CrossRef]
39. Donnet, J.B.; Ehrburger, P. Carbon-Fiber in Polymer Reinforcement. *Carbon* **1977**, *15*, 143–152. [CrossRef]
40. Donnet, J.B.; Guilpain, G. Surface Treatments and Properties of Carbon-Fibers. *Carbon* **1989**, *27*, 749–757. [CrossRef]
41. Donnet, J.B.; Guilpain, G. Surface Characterization of Carbon-Fibers. *Composites* **1991**, *22*, 59–62. [CrossRef]

42. Ehrburger, P.; Herque, J.J.; Donnet, J.P. Recent Developments in Carbon-Fiber Treatments. *Acs. Sym. Ser.* **1976**, 324–334.
43. Sadeghvishakei, M.; Yunus, R.; Salleh, M.A.M.; Pignolet, A. Mechanical Investigation in Whiskerized Carbon Fiber/Polypropylene Composite. In Proceedings of the Smart Materials, Structures & NDTin Aerospace, Montreal, QC Canada, November 2011.
44. Zhang, Q.H.; Liu, J.W.; Sager, R.; Dai, L.M.; Baur, J. Hierarchical Composites of Carbon Nanotubes on Carbon Fiber: Influence of Growth Condition on Fiber Tensile Properties. *Compos. Sci. Technol.* **2009**, *69*, 594–601. [CrossRef]
45. Peng, Y.C.; Burtovyy, R.; Yang, Y.; Urban, M.W.; Kennedy, M.S.; Kornev, K.G.; Bordia, R.; Luzinov, I. Towards scalable Fabrication of Ultrasmooth and Porous thin Carbon Films. *Carbon* **2016**, *96*, 184–195. [CrossRef]
46. Fried, J. *Polymer Science and Technology*, 3rd ed.; Pearson: London, UK, 2014.
47. Seeber, M.; Zdyrko, B.; Burtovvy, R.; Andrukh, T.; Tsai, C.-C.; Owens, J.R.; Kornev, K.G.; Luzinov, I. Surface Grafting of Thermoresponsive Microgel Nanoparticles. *Soft Matter.* **2011**, *7*, 9962–9971. [CrossRef]
48. Walheim, S.; Böltau, M.; Mlynek, J.; Krausch, G.; Steiner, U. Structure Formation via Polymer Demixing in Spin-Cast Films. *Macromolecules* **1997**, *30*, 4995–5003. [CrossRef]
49. Elbs, H.; Fukunaga, K.; Stadler, R.; Sauer, G.; Magerle, R.; Krausch, G. Microdomain Morphology of Thin ABC Triblock Copolymer Films. *Macromolecules* **1999**, *32*, 1204–1211. [CrossRef]
50. Landau, L.; Levich, B. Dragging of a Liquid by a Moving Plate. *Acta. Phys. URSS* **1942**, *17*, 42–54.
51. Mayer, H.C.; Krechetnikov, R. Landau-Levich Flow Visualization: Revealing the Flow Topology Responsible for the Film Thickening Phenomena. *Phys. Fluids* **2012**, *24*. [CrossRef]

Article

Impact Damage Ascertainment in Composite Plates Using In-Situ Acoustic Emission Signal Signature Identification

Robin James *, Roshan Prakash Joseph and Victor Giurgiutiu

Department of Mechanical Engineering, University of South Carolina, Columbia, SC 29208, USA; rjoseph@email.sc.edu (R.P.J.); victorg@sc.edu (V.G.)
* Correspondence: rj11@email.sc.edu

Abstract: Barely visible impact damage (BVID) due to low velocity impact events in composite aircraft structures are becoming prevalent. BVID can have an adverse effect on the strength and safety of the structure. During aircraft inspections it can be extremely difficult to visually detect BVID. Moreover, it is also a challenge to ascertain if the BVID has in-fact caused internal damage to the structure or not. This paper describes a method to ascertain whether or not internal damage happened during the impact event by analyzing the high-frequency information contained in the recorded acoustic emission signal signature. Multiple 2 mm quasi-isotropic carbon fiber reinforced polymer (CFRP) composite coupons were impacted using the ASTM D7136 standard in a drop weight impact testing machine to determine the mass, height and energy parameters to obtain approximately 1″ impact damage size in the coupons iteratively. For subsequent impact tests, four piezoelectric wafer active sensors (PWAS) were bonded at specific locations on each coupon to record the acoustic emission (AE) signals during the impact event using the MISTRAS micro-II digital AE system. Impact tests were conducted on these instrumented 2 mm coupons using previously calculated energies that would create either no damage or 1″ impact damage in the coupons. The obtained AE waveforms and their frequency spectrums were analyzed to distinguish between different AE signatures. From the analysis of the recorded AE signals, it was verified if the structure had indeed been damaged due to the impact event or not. Using our proposed structural health monitoring technique, it could be possible to rapidly identify impact events that cause damage to the structure in real-time and distinguish them from impact events that do not cause damage to the structure. An invention disclosure describing our acoustic emission structural health monitoring technique has been filed and is in the process of becoming a provisional patent.

Keywords: barely visible impact damage (BVID); composite structures; damage detection; carbon fiber reinforced polymer (CFRP); acoustic emission; structural health monitoring; piezoelectric wafer active sensors (PWAS)

Citation: James, R.; Joseph, R.P.; Giurgiutiu, V. Impact Damage Ascertainment in Composite Plates Using In-Situ Acoustic Emission Signal Signature Identification. J. Compos. Sci. 2021, 5, 79. https://doi.org/10.3390/jcs5030079

Academic Editors: Jiadeng Zhu, Gouqing Li and Lixing Kang

Received: 11 November 2020
Accepted: 9 March 2021
Published: 12 March 2021

Publisher's Note: MDPI stays neutral with regard to jurisdictional claims in published maps and institutional affiliations.

Copyright: © 2021 by the authors. Licensee MDPI, Basel, Switzerland. This article is an open access article distributed under the terms and conditions of the Creative Commons Attribution (CC BY) license (https://creativecommons.org/licenses/by/4.0/).

1. Introduction

1.1. Background and Motivation

Recent advances in manufacturing technologies have led to the increasing usage of composite materials being used in aerospace primary and secondary structures due to their high strength to weight ratio and lightweight. Structures manufactured using composite materials, whether thermosets or thermoplastics, must be made in a nearly perfect state such that they do not introduce any dangerous risks during the operational lifetime of the aerospace structure. The manufacturing process of composite structures can introduce significant manufacturing flaws and operational damage during its service life. These types of defects may lead to catastrophic failures if they are not detected at the earliest stages of development using efficient structural health monitoring techniques.

Barely visible impact damage (BVID) is a type of damage that occurs most often in composite structures. It can occur during the manufacturing stages or during the operational lifetime of the composite structure. During BVID causing impact event, the debris (impactor) may lead to internal damage within the composite structure such as complex delaminations, matrix cracks, fiber fracture and a combination of all three. BVID comprises of surface indentations, which are not clearly visible due to the coating of paint during aircraft inspections done visually. If gone unnoticed, the internal damage can grow and propagate leading to catastrophic failures.

The damage tolerance concepts introduced around 20–30 years ago paved the way for understanding BVID and how it led to complicated damage in composites [1,2]. Following these developments, inspection standards needed to be developed for the inspection of composite structures where BVID became an important aspect and needed to be distinguished from visible impact damage (VID). A damage is characterized as BVID if it is visible at a distance of less than 1.5 m using regular vision. Similarly, a damage is characterized as VID if it is visible at a distance of 1.5 m or greater [3]. Depending on how a damage is characterized (as BVID or VID), important decisions regarding repairs to be conducted on composite in the areas where impact events occur, are taken. VID's need to be repaired immediately based on this understanding. However, there may be a situation when the damage is characterized as BVID based on visually conducted inspections but may in fact have an impact damage size of 1" (25 mm) or greater which could significantly aggravate the strength of the composite part (see Figure 1). Since this damage is now characterized as BVID, it may get ignored from being repaired. Figure 1 clearly demonstrates that 1" impact damage diameter can seriously reduce the strength of a composite structure compared to any other damage type (delamination, porosity, open hole), having the same size. This clearly demonstrates the seriousness and significance detecting and monitoring impact damage having a diameter of 1" or greater [4,5].

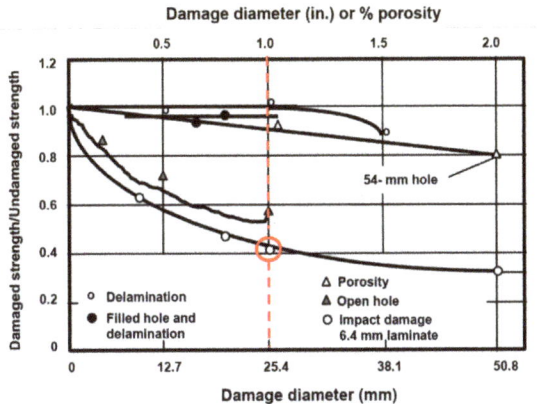

Figure 1. Remaining composite strength as a function of impact damage diameter [3].

Nondestructive evaluation (NDE) and structural health monitoring (SHM) methodologies needed to be developed to effectively detect and monitor impact damage due to the increasing occurrences of BVID in composite structures. Ultrasonic NDE was one of the first few methods to be used for impact damage detection [6–8]. Ultrasonic guided wave (GW) propagation methodologies in composite laminates have been extensively used to observe how different wave modes interact with impact damage in composites [9–12]. In recent years, Innovative eddy current testing (ECT) methods have been explored extensively by researchers to detect different types of manufacturing flaws in CFRP composites [13–15]. These methods can be extended to detect impact damage in conductive fiber reinforced composite materials. Microwave nondestructive evaluation (MNDE) techniques have also been investigated by researchers to detect low velocity and high velocity impact damage in composites due to environmental effects such as hail stone impact and bird hits [16–19].

Advanced NDE methods based on heat dissipation such as Infrared thermography are also being explored by scientists as a viable option of detecting impact damage in a non-contact, rapid manner [20–22]. X-ray micro computed tomography is also being developed to give a 3D visualization of impact damage in composite structures through multiple B-scans and C-scans that can be observed at different orientations [23,24]. The authors of this paper are also exploring advanced guided wave propagation methods [25–28] for and long-distance propagation of the guided waves in the composite structure which will enable large area examination of the composite structures subjected to controlled impact damage creation.

In recent years extensive work has been done to understand effective acoustic emission methods for structural health monitoring of impact damage in composite materials. Prosser et al. [29] analyzed AE signals created by impact sources in thin aluminum structures and graphite/epoxy composites subjected to low and high-velocity impacts. Rosa et al. [30,31] have primarily focused on the post-impact behavior of natural fiber composites and hybrid composites using acoustic emission methods. Other researchers [32,33] used acoustic emission sensor networks to reconstruct the force-time history to better understand the loading phenomena from the impact event and compare it to the experimental force-time history. The uniqueness in our research is to use existing PWAS sensors to record AE signals in real-time during impact events and ascertain if a sizable damage has occurred or not. This will greatly reduce system downtime and ensure that necessary composite repairs are conducted.

1.2. Objectives of This Paper

In this paper, the authors have described an AE based structural health monitoring method [34] that can analyze the AE signal signatures obtained from an impact event and can ascertain if the impact event has indeed caused an extensive damage inside the composite structure or not. To do this, preliminary drop weight impact tests were conducted on various 2 mm thick quasi-isotropic CFRP composite coupons conforming to the ASTM D7136 standard for drop weight impact testing [35]. These preliminary experiments were useful in estimating the mass, height and energy combination to obtain a certain size (approximately 1" damage diameter) of impact damage in the composite coupon iteratively.

After estimating the mass, height and energy combination for creating approximately 1" impact damage diameter in a 2 mm thick composite coupon, subsequent impact tests were conducted on AE instrumented composite coupons on which four PWAS were bonded at specific locations based on the fiber orientation angles in the composite coupons. The drop weight impact testing system along with the AE signal capture using the MISTRAS AE system is displayed in Figure 2. Two sets of experiments were conducted–one experiment with low energy (1 J) impact that created no damage in an instrumented composite coupon and the second test with a higher energy (16 J) impact which created approximately 1" impact damage size. AE signal analysis and mode separation study were performed to understand both the impact events and clearly differentiate between a catastrophic impact that creates sizable damage and a benign impact that creates no damage.

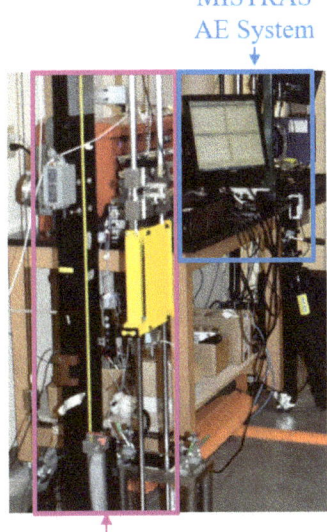

Figure 2. Drop weight impact testing with AE signal capture.

2. Manufacturing Process and Experimental Setup

2.1. Manufacturing of CFRP Composite Laminates

CFRP Composite laminates were fabricated using the CYCOM® 5320-1 epoxy resin system with the Hexcel IM7 12K fiber in a Wabash hot press using the cure cycle provided by the manufacturer of the prepreg. To manufacture quasi-isotropic composite plates with the correct thicknesses, a stacking sequence with the appropriate number of layers had to be chosen [36–38]. A $[-45/90/+45/0]_{2S}$ stack up was chosen for fabricating the composite laminate with 16 layers having a nominal thickness of approximately 2 mm. From the cured composite laminate, 6" × 4" coupons were cut out for conducting standardized impact tests [35]. The Wabash hot press machine, cure cycle and the cured composite plate with the 6" × 4" cut-outs is displayed in Figure 3.

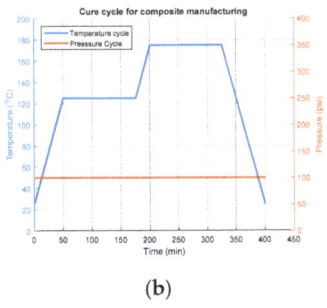

(a)　　　　　　　　　　(b)　　　　　　　　　　(c)

Figure 3. Manufacturing of composite coupons for impact testing: (**a**) Wabash hot press machine; (**b**) Manufacturing cure-cycle; (**c**) Cured composite laminate with 6" × 4" cut-outs.

2.2. Experimental Setup for Acoustic Emission Recording of Impacted Composite Coupon

Preliminary drop weight impact experiments were conducted on numerous 2 mm 6" × 4" quasi-isotropic CFRP composite coupons [5] to determine the mass, height and energy combination to obtain a certain size of impact damage. These impact tests were conducted on a drop weight impact tower conforming to the ASTM D7136 standard as displayed in Figure 4. After this, real-time acoustic emission experiments were supposed to be carried out on more 2 mm 6″ × 4″ quasi-isotropic CFRP composite coupons. In order to do this, four piezoelectric wafer active sensors (PWAS), 7-mm in diameter and 0.5-mm in thickness, were bonded on each composite coupon at different locations corresponding to fiber orientation angles in the stacking sequence of the composite. PWAS 1 was bonded 45-mm from the impact location in the 90-degree fiber direction. PWAS 2 was bonded 75-mm away from the impact location was installed in the −45-degree fiber direction. PWAS 3 was bonded 75-mm away from the impact location in the 0-degree fiber direction. PWAS 4 was bonded 75-mm away from the impact location in the 45-degree fiber direction as can be observed in Figure 5. In this way the impact coupons were instrumented to carry out real-time acoustic emission recording of impact tests to be conducted on them.

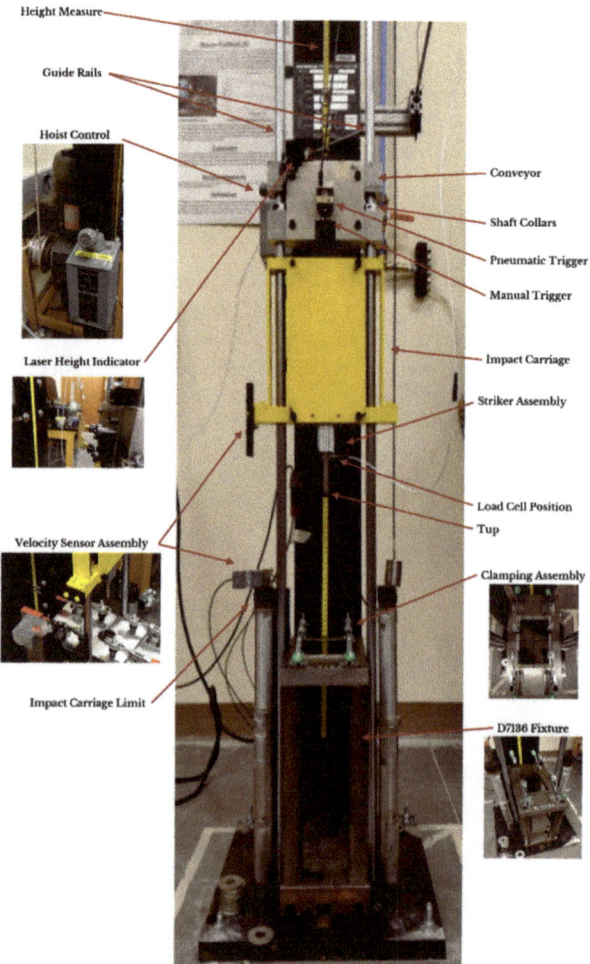

Figure 4. Dynatup 8200 drop weight impact testing machine instrumented with load cell and velocity sensor.

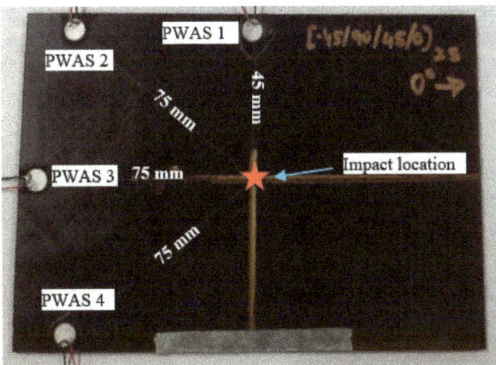

Figure 5. Location of four PWAS with respect to impact location on composite coupon.

To conduct the real-time acoustic emission experiment, the instrumented coupon with the four PWAS was clamped on the ASTM D7136 fixture on the drop weight impact testing machine. The wires from the four PWAS were connected to a pre-amplifier and the connections from the pre-amplifier were connected to the MISTRAS AE system for capturing the AE signals during the drop weight impact testing experiment so that all the signals associated with the impact event using the four PWAS bonded in the different fiber orientation angles could be analyzed. The acoustic preamplifier is a bandpass filter, which can filter out signals between 30 kHz to 700 kHz. Provided with 20/40/60 dB gain (can be selected using a switch), this preamplifier operates with either a single-ended or differential sensor. In the present experiment, 40 dB gain was selected. The preamplifier was connected to the MISTRAS AE system. A sampling frequency of 10 MHz was chosen to capture any high-frequency AE signals. The timing parameters set for the MISTRAS system were: peak definition time (PDT) = 200 µs, hit definition time (HDT) = 800 µs, and hit lockout time (HLT) = 1000 µs. This complete experimental setup with the AE instrumentation used is displayed in Figure 6.

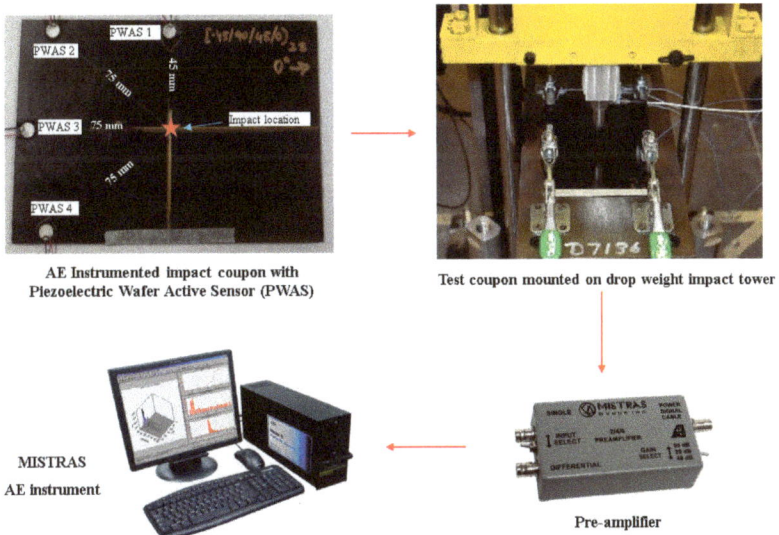

Figure 6. Experimental setup of ASTM drop weight impact test on AE instrumented coupon.

3. AE Signal Analysis from Instrumented Impact Tests

3.1. 1 J Impact Test on AE Instrumented 2 mm Composite Coupon–No Damage

The first instrumented impact test conducted on a 2 mm composite coupon is a low energy impact i.e., about 1 J impact that produces no damage in the composite coupon. To conduct this impact test, the instrumented coupon displayed in Figure 5 was clamped on the ASTM D7136 fixture and the real-time AE signal hit were acquired by all the four PWAS using the MISTRAS AE system as displayed in the experimental setup given in Figure 6. Since the impact energy is only 1 J, the height from which the impactor is dropped on the composite coupon is only a few centimeters. In such a scenario, it becomes very difficult to avoid a rebounding or secondary impact on the composite coupon after the first impact. The AE hits acquired at all the four PWAS for this 1 J impact event is displayed in Figure 7.

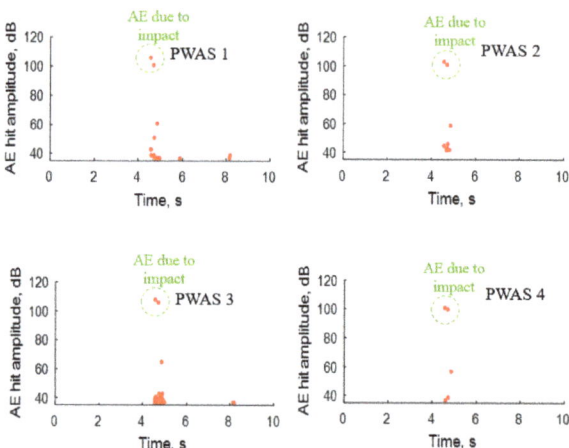

Figure 7. AE hits observed at the four PWAS due to 1 J impact event.

In Figure 7 there are two successive AE impact hits due to the rebound of the impactor on the composite coupon. These two hits are obtained by all the four PWAS and are clearly separated from other low amplitude hits which could consist of background noise or boundary reflections from the edges of the composite coupon, since we are assuming that this low energy of approximately 1 J did not create any damage in the composite coupon.

After this, the waveforms of the impact hits were extracted from the MISTRAS AE system. The waveforms of the 1st and 2nd impact hits and their FFT's are presented in Figures 8 and 9. We can clearly observe that the signals from these two successive impact hits had a major frequency content in the low-frequency range below 200 kHz which indicates low-frequency flexural modes in the composite coupon. We can also observe that the signal amplitude for the 1st impact hit was higher at PWAS 1 which was in the 90-degree direction and PWAS 3, which was in the 0-degree direction.

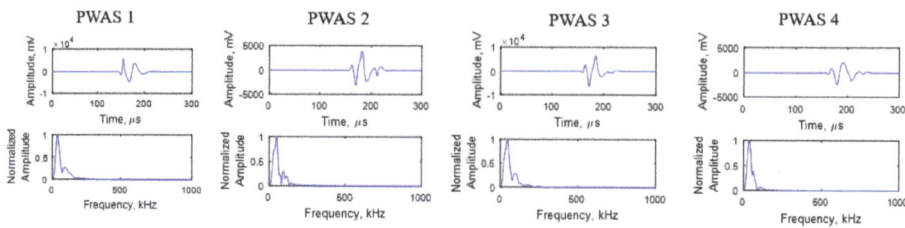

Figure 8. Signal correspondence at all four PWAS due to the 1st impact hit.

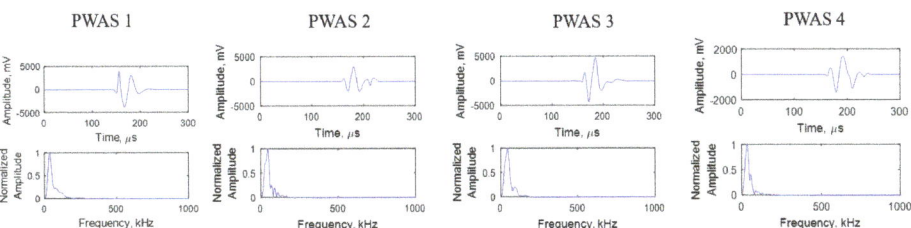

Figure 9. Signal correspondence at all four PWAS due to the 2nd impact hit.

3.2. 16 J Impact Test on AE Instrumented 2 mm Composite Coupon–1" Impact Damage

The second instrumented impact test conducted on a 2 mm composite coupon is a 16 J energy impact based on the preliminary impact tests conducted on various 2 mm composite coupons, as described in a previous work [5]. The energy of 16 J was chosen such that it produces an impact damage size of approximately 1" in the 2 mm composite coupon. To conduct this impact test, the instrumented coupon labeled AE1-Q2A similar to the coupon displayed in Figure 5 was clamped on the ASTM D7136 fixture and the real-time AE signals during the impact event were acquired by all the four PWAS using the MISTRAS AE system as displayed in the experimental setup given in Figure 6. Since the impact energy for this impact event was 16 J, the height from which the impactor is dropped on the composite coupon is higher than the previous impact test and it was easily possible to catch the impact cart with weights after the 1st impact to avoid a secondary or rebound impact on the composite coupon. The AE hits were acquired at all the four PWAS for this 16 J impact event. The force-time history for this impact event was acquired by the dynamic load cell attached to the impactor and the energy-time was deduced using the force-time history data and the impact velocity measured by the velocity sensor.

Figure 10 shows four plots related to coupon AE1-Q2A, displaying the force-time plot, the energy-time plot, the B-scan and C-scan from ultrasonic testing (UT). The force-time plot is parabolic in shape and shows a peak at a certain maximum load of approximately 4.48 kN. Anomalies in the parabolic shape of the force-time plot indicate that the coupon has undergone extensive damage when undergoing impact.

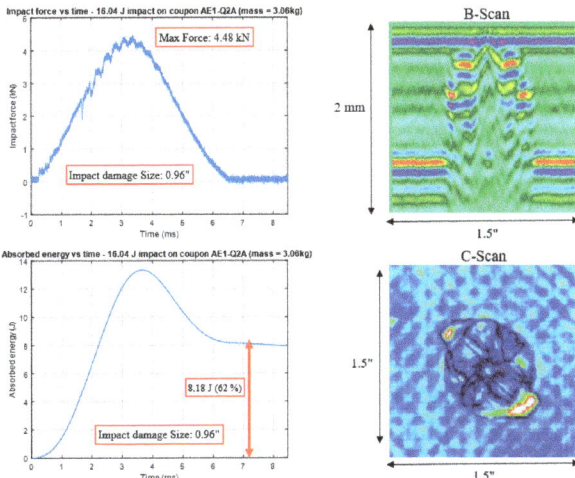

Figure 10. Impact results of 2 mm coupon AE1-Q2A consisting of force-time plot, energy-time plot, B-scan and C-scan from UT.

The energy-time plot clearly demonstrates the energy absorbed (62%) by the coupon during the impact event to create the irreversible process of a 1" impact damage diameter in coupon AE1-Q2A. From the B-scan, it can be seen, that although the center of the damaged area undergoes permanent deformation similar to a dent, it does not have a delamination, since a clear back wall reflection from the center of the damage can be seen in the B-scan. From the C-scan, we can clearly see the fiber break and push out in the −45° fiber direction and this can be physically seen by looking at the rear surface of the impacted coupon as well. It can be observed that bonding the four PWAS on the AE1-Q2A composite coupon had little to no change in its impact characteristics.

Next, we analyze the AE signals received at all four PWAS. We can clearly observe in Figure 11 that the impact hit i.e., the hit which is received at the four PWAS when the first contact is made between the impactor and the composite coupon, can be clearly separated from the remaining hits received by the four PWAS. The other low amplitude hits consist of hits obtained due to the damage propagation within the composite coupon mixed with background noise and boundary reflections from the edges of the composite coupon. It is also important to note that at the PWAS 1, only the impact hit was received and after that no more hits were received by PWAS 1. This issue occurred at PWAS 1 because at the moment of impact, one of the cables connected to the PWAS 1 got unintentionally or accidentally detached from the PWAS 1 after the high amplitude flexural wave was experienced at the location where PWAS 1 was bonded to the composite coupon. Due to the detachment of the cable from PWAS 1 it was only able to capture the impact hit and was not able to capture any of the other low amplitude hits which could have valuable information about the impact damage propagation. In future experiments, all the cables will be properly reinforced so that signals at all PWAS can be received in an uninterrupted manner.

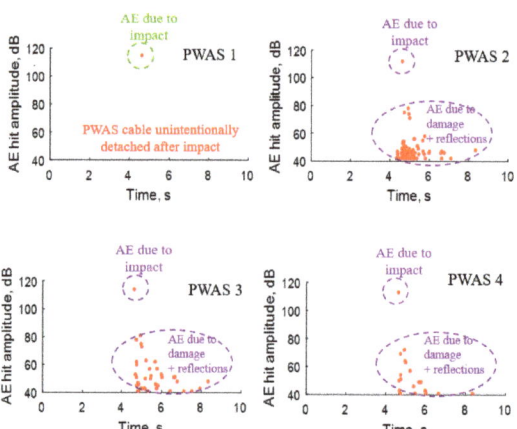

Figure 11. AE hits observed at the four PWAS due to 16 J impact event.

If we separate the time domain signals and their FFT's, received at all four PWAS from the impact hit as observed in Figure 12, we can clearly observe that the signals from the impact hit has a major frequency content in the low-frequency range below 200 kHz with a large amplitude which indicates low-frequency flexural modes in the composite coupon.

If we separate the time domain signals and their FFT's, received at all four PWAS from a hit that corresponds to damage propagation in the composite as observed in Figure 13, we can clearly observe that the signals from this hit at all the PWAS has a major frequency content in the frequency range between 300 and 500 kHz with a much lower amplitude in comparison to the impact hit. It is also important to note that there is no signal correspondence at PWAS 1 for a hit that corresponds to damage growth since no AE hits were received by PWAS 1 other than the impact hit, as stated earlier.

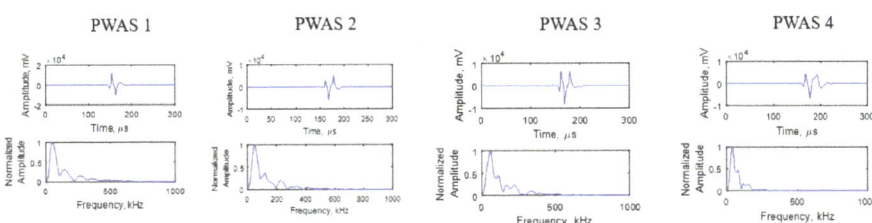

Figure 12. Signal correspondence at all four PWAS due to the 1st impact hit.

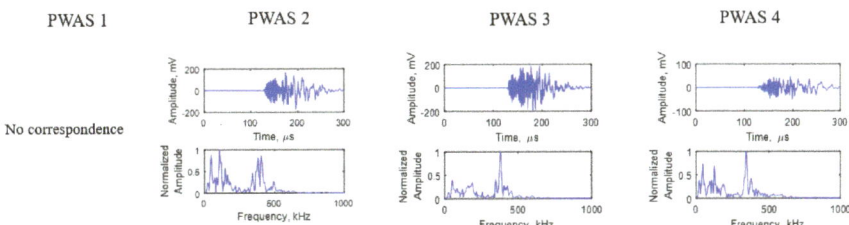

Figure 13. Signal correspondence at all four PWAS due to a hit corresponding to damage formation.

As observed from the C-scan in the quad plot displayed in Figure 10, we can clearly see that the maximum extent of damage due to the impact event occurs at the −45 degree direction. Therefore, we take a closer look at the signals obtained from some of the hits at PWAS 2 which is bonded in the −45-degree direction in Figure 14. We can clearly separate the high amplitude, low-frequency impact hit and its signal from some other hits and their signals that correspond to damage propagation. Within the class of hits and their signals that correspond to damage, there are subtle differences in the signals because they may represent different types of damage such as matrix cracking, fiber break, and delamination growth. One of the goals in future experiments will be to separate the damage signals from different types of damage experienced by the composite coupon upon impact.

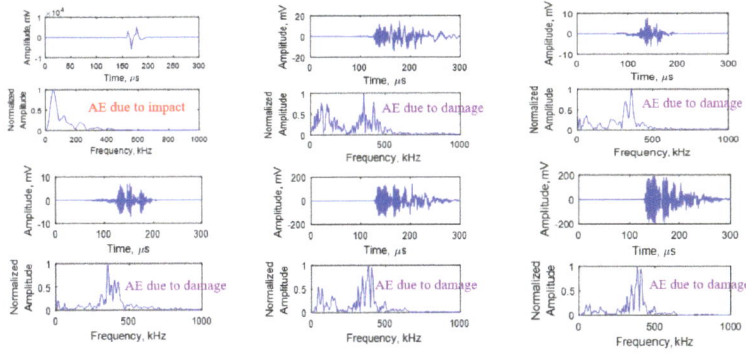

Figure 14. Various signals observed at PWAS2 due to the 16 J impact event.

3.3. Mode Separation Study of AE Signals Due to Impact Event

After acquiring all the AE hits and performing the signal analysis from the AE hits, time-frequency analysis of the AE signals was also performed. The analysis aimed to study the Lamb wave mode content in the AE signals recorded. To do this, we first use the Semi-Analytical Finite Element (SAFE) method to obtain the group velocity dispersion curve for the 2 mm composite coupon with a stacking sequence of $[-45/90/+45/0]_{2S}$ as displayed in Figure 15.

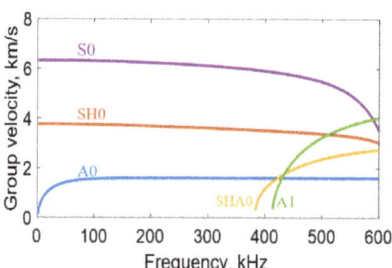

Figure 15. Group velocity dispersion curve for 2 mm composite coupon having a stacking sequence of $[-45/90/+45/0]_{2S}$.

To perform the mode separation study for the AE due to the impact event, we first analyze the impact hits from the 1 J impact hit that caused no damage in a 2 mm composite coupon, and the 16 J impact hit that caused a 1" impact damage in a composite coupon. We conduct the time-frequency analysis for both the impact hits and superimpose it with the group velocity dispersion curve of the 2 mm composite coupon. These plots can be observed in Figure 16a,b. If we compare these two plots, we can clearly observe that the strong A0 mode can be observed due to the impact hit in both the plots. We can also observe that 16 J impact hit has a stronger A0 content. We can also see the signals obtained at PWAS 2 for both impact hits in Figure 16c,d. Upon comparing these two plots we can observe that the 16 J impact hit has an additional higher frequency content between 200 kHz and 400 kHz due to a higher energy impact of 16 J compared to a lower energy impact of 1 J.

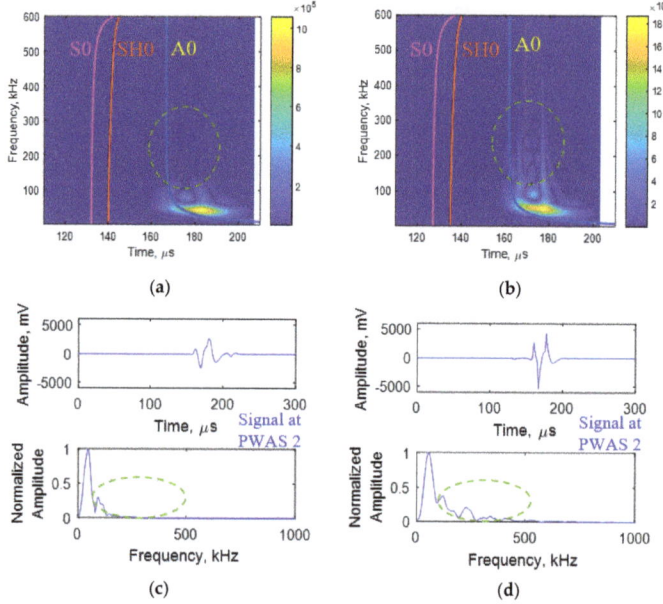

Figure 16. (a) Time-frequency plot from 1 J impact hit signal (b) Time-frequency plot from 16 J impact hit signal (c) AE signal at PWAS 2 due to 1 J impact hit and frequency spectrum (d) AE signal at PWAS 2 due to 16 J impact hit and frequency spectrum.

To perform the mode separation study for the AE due to damage growth, we analyze an AE hit that corresponds to damage growth from the 16 J impact event that caused a 1" impact damage in the composite coupon. We conduct the time-frequency analysis of the signal and superimpose it with the group velocity dispersion curve of the 2 mm composite coupon. This plot can be observed in Figure 17a. We can also observe the signal due to the damage growth obtained at PWAS 2 displayed in Figure 17b. From these plots we can clearly observe that the damage growth has a strong S0 and SH0 mode between 300 kHz and 500 kHz. We can also see that the damage growth has weak A0 mode along with many boundary reflections. If we were do conduct a preliminary inspection, we can see that SH0 mode is found stronger than the S0 mode. Previous work [25–28] has also indicated that SH0 mode is very sensitive to impact damage and can be used to detect impact damage.

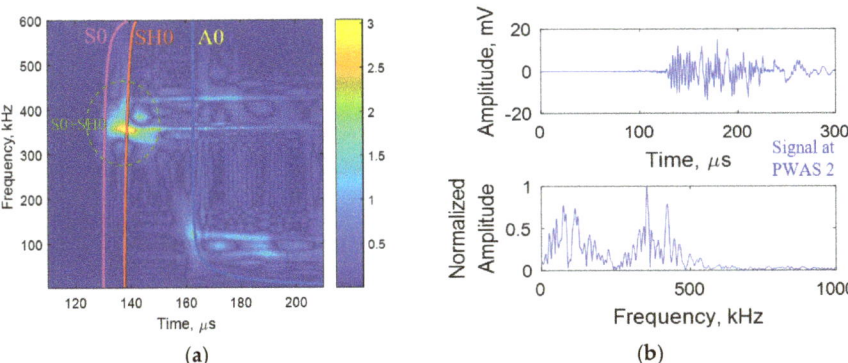

Figure 17. (a) Time-frequency plot from 16 J damage hit (b) Signal at PWAS 2 and the frequency spectrum due to 16 J damage hit.

4. Summary, Conclusions and Future Work

4.1. Summary

In this paper, the AE signal signature identification was used to ascertain if an impact event creates a sizable damage in a composite coupon or not. This was done by modifying the existing standardized test method for drop weight impact testing by introducing an instrumented composite coupon to acquire real-time AE signals.

Using the mass, height and energy combinations from the preliminary impact tests [5], four PWAS were bonded on two composite coupons at locations corresponding to fiber orientation angles and then drop weight impact tests conforming to ASTM D7136 standard on these instrumented composite coupons was conducted. On the first instrumented coupon a 1 J impact that creates no damage, was conducted and on the second instrumented coupon a 16 J impact that creates 1" impact damage diameter was conducted. We found that we could separate the impact AE hit from an AE hit corresponding to damage growth and perform a mode separation study.

4.2. Conclusions

Preliminary impact tests conducted on 2 mm quasi-isotropic coupons were used to estimate the mass, height and energy combinations to obtain approximately 1" impact damage size using incremental energy impacts on various test coupons and post-impact data analysis to estimate force-time histories and energy-time histories. UT scans enabled us to characterize the impact damage size, shape and location.

Impact tests conducted on AE instrumented 2 mm composite coupons showed similar impact characteristics despite bonding four PWAS to acquire real-time AE signals. AE signals corresponding to the impact hits were identified clearly and separated from the AE signals that corresponded to internal damage growth in the composite coupons. It was observed that the AE due to impact hit has a stronger low-frequency content with high amplitude at a region below 200 kHz. It was also observed that the AE signals due to the irreversible process of damage has a stronger high –frequency content in the range of 300 to 500 kHz.

Upon performing the mode separation study on the impact hits, it was observed that the impact hit has a strong A0 mode content depending on the energy of the impact. The mode separation study on the AE hit corresponding to damage growth indicated that it has a strong S0 mode and SH0 mode content where the SH0 mode seems to be the dominant mode and more sensitive to the impact damage.

An invention disclosure [39] covering our novel findings has been prepared and is in the process of becoming a provisional patent.

4.3. Future Work

Further controlled impact tests will be conducted on AE instrumented 2 mm composite coupons using the mass, height and energy combinations estimated from the preliminary impact experiments to obtain multiple impact damage sizes for a comparative study. A deviation from the ASTM D7136 standard for drop weight impact testing will be employed to use larger size coupons (12" × 6") to use-non reflective boundary and receive clean signals from the impact tests which are free from boundary reflections. AE signal analysis will be used to investigate the separation of AE signals from different types of damages processes (matrix crack, fiber break and delamination) that occur during an impact event.

Further work could be performed towards the practical application of the research results presented in this paper by exploring the possibility of using PWAS for real-time AE structural health monitoring of impact events in composites to make sure if the impact has indeed created damage inside the composite. Estimating the size, location, shape and extent of the impact damage by analyzing the AE signals received by a network of PWAS will be of paramount interest. Computer simulations and equipment development could be conducted independently or in collaboration with an industrial partner.

Author Contributions: In this article, the experiments, calculations, and analysis were done by R.J., and R.P.J. with advice and inspiration from V.G. The details methodology was provided by R.J., R.P.J., and V.G. The original draft was prepared by R.J. The review and editing were done by R.J., R.P.J., and V.G. All authors have read and agreed to the published version of the manuscript.

Funding: This work was supported by the Air Force Office of Scientific Research (AFOSR) grant number FA9550-16-1-0401 and the Office of Naval Research (ONR) grant number N00014-17-1-2829.

Acknowledgments: The authors would like to thank the Ronald E. McNair Center for Aerospace Innovation and Research for providing them the manufacturing facilities to manufacture the composite coupons. The authors would also like to thank Xinyu Huang for providing the drop-weight impact tower for conducting the ASTM D7136 impact tests.

Conflicts of Interest: The authors declare no conflict of interest.

References

1. Davis, M.J.; Jones, R.T. Damage Tolerance of Fibre Composite Laminates. In *Fracture Mechanics Technology Applied to Material Evaluation and Structure Design*; Springer: Dordrecht, The Netherlands, 1983; pp. 635–655.
2. Anderson, B. Factors affecting the design of military aircraft structures in carbon fibre reinforced composites. In *Fracture 84*; Pergamon: New Delhi, India, 1984; pp. 607–622.
3. MIL-HDBK-17-3F. *Composite Materials Handbook*; Chapter 7; US Department of Defense: Arlington, VA, USA, 2002; Volume 3, pp. 7–29.
4. James, R.; Giurgiutiu, V.; Flores, M.; Mei, H.; Haider, M.F. Challenges of generating controlled one-inch impact damage in thick CFRP composites. In Proceedings of the AIAA Scitech 2020 Forum, Orlando, FL, USA, 6–10 January 2020; p. 0723.
5. James, R.; Giurgiutiu, V. Towards the generation of controlled one-inch impact damage in thick CFRP composites for SHM and NDE validation. *Compos. Part B Eng.* **2020**, *203*, 108463. [CrossRef]
6. Cantwell, W.; Morton, J. Detection of impact damage in CFRP laminates. *Compos. Struct.* **1985**, *3*, 241–257. [CrossRef]
7. Bar-Cohen, Y.; Crane, R.L. Washington, DC: U.S. Patent and Trademark Office. U.S. Patent No. 4,457,174, 3 July 1984.

8. Aymerich, F.; Meili, S. Ultrasonic evaluation of matrix damage in impacted composite laminates. *Compos. Part B Eng.* **2000**, *31*, 1–6. [CrossRef]
9. Gresil, M.; Giurgiutiu, V. Guided wave propagation in composite laminates using piezoelectric wafer active sensors. *Aeronaut. J.* **2013**, *117*, 971–995. [CrossRef]
10. Santos, M.; Santos, J.; Amaro, A.; Neto, M.A. Low velocity impact damage evaluation in fiber glass composite plates using PZT sensors. *Compos. Part B Eng.* **2013**, *55*, 269–276. [CrossRef]
11. Rogge, M.D.; Leckey, C.A. Characterization of impact damage in composite laminates using guided wavefield imaging and local wavenumber domain analysis. *Ultrasonics* **2013**, *53*, 1217–1226. [CrossRef]
12. Tai, S.; Kotobuki, F.; Wang, L.; Mal, A. Modeling Ultrasonic Elastic Waves in Fiber-Metal Laminate Structures in Presence of Sources and Defects. *J. Nondestruct. Eval. Diagn. Progn. Eng. Syst.* **2020**, *3*, 1–42. [CrossRef]
13. James, R.; Haider, M.F.; Giurgiutiu, V.; Lilienthal, D. A simulative and experimental approach towards eddy current non-destructive evaluation of manufacturing flaws and operational damage in CFRP composites. *J. Nondestruct. Eval. Diagn. Progn. Eng. Syst.* **2019**, *3*, 1–14. [CrossRef]
14. Liang, T.; Ren, W.; Tian, G.Y.; Elradi, M.; Gao, Y. Low energy impact damage detection in CFRP using eddy current pulsed thermography. *Compos. Struct.* **2016**, *143*, 352–361. [CrossRef]
15. He, Y.; Tian, G.; Pan, M.; Chen, D. Impact evaluation in carbon fiber reinforced plastic (CFRP) laminates using eddy current pulsed thermography. *Compos. Struct.* **2014**, *109*, 1–7. [CrossRef]
16. Narayanan, R.M.; James, R. Microwave nondestructive testing of galvanic corrosion and impact damage in carbon fiber reinforced polymer composites. *Int. J. Microw. Appl.* **2018**, *7*, 1–15. [CrossRef]
17. Greenawald, E.C.; Levenberry, L.J.; Qaddoumi, N.; McHardy, A.; Zoughi, R.; Poranski, C.F., Jr. Microwave NDE of impact damaged fiberglass and elastomer layered composites. In *AIP Conference Proceedings*; American Institute of Physics: Melville, NY, USA, 2000; Volume 509, pp. 1263–1268.
18. Li, Z.; Haigh, A.D.; Soutis, C.; Gibson, A.A.P. Simulation for the impact damage detection in composites by using the near-field microwave waveguide imaging. In *NDT 2014—53rd Annual Conference of the British Institute of Non-Destructive Testing*; British Institute of Non-Destructive Testing: Northampton, UK, 2014.
19. Li, Z.; Soutis, C.; Haigh, A.; Sloan, R.; Gibson, A. Application of an electromagnetic sensor for detection of impact damage in aircraft composites. In Proceedings of the 2016 21st International Conference on Microwave, Radar and Wireless Communications, MIKON 2016, Krakow, Poland, 9–11 May 2016.
20. Meola, C.; Carlomagno, G.M. Impact damage in GFRP: New insights with infrared thermography. *Compos. Part A Appl. Sci. Manuf.* **2010**, *41*, 1839–1847. [CrossRef]
21. Tan, K.; Watanabe, N.; Iwahori, Y. X-ray radiography and micro-computed tomography examination of damage characteristics in stitched composites subjected to impact loading. *Compos. Part B Eng.* **2011**, *42*, 874–884. [CrossRef]
22. Meola, C.; Carlomagno, G.M. Infrared thermography to evaluate impact damage in glass/epoxy with manufacturing defects. *Int. J. Impact Eng.* **2014**, *67*, 1–11. [CrossRef]
23. Bull, D.; Helfen, L.; Sinclair, I.; Spearing, S.; Baumbach, T. A comparison of multi-scale 3D X-ray tomographic inspection techniques for assessing carbon fibre composite impact damage. *Compos. Sci. Technol.* **2013**, *75*, 55–61. [CrossRef]
24. Schilling, P.J.; Karedla, B.R.; Tatiparthi, A.K.; Verges, M.A.; Herrington, P.D. X-ray computed microtomography of internal damage in fiber reinforced polymer matrix composites. *Compos. Sci. Technol.* **2005**, *65*, 2071–2078. [CrossRef]
25. Mei, H.; Haider, M.F.; James, R.; Giurgiutiu, V. Pure S0 and SH0 detections of various damage types in aerospace composites. *Compos. Part B Eng.* **2020**, *189*, 107906. [CrossRef]
26. James, R.; Mei, H.; Giurgiutiu, V. SH-mode guided-wave impact damage detection in thick quasi-isotropic composites. In Proceedings of the Health Monitoring of Structural and Biological Systems IX, Houston, TX, USA, 10–13 February 2020; Volume 11381, p. 113810R.
27. Mei, H.; James, R.; Giurgiutiu, V. Damage detection in laminated composites using pure SH guided wave excited by angle beam transducer. In Proceedings of the Health Monitoring of Structural and Biological Systems IX, Houston, TX, USA, 10–13 February 2020; Volume 11381, p. 113810P.
28. Mei, H.; James, R.; Haider, M.F.; Giurgiutiu, V. Multimode Guided Wave Detection for Various Composite Damage Types. *Appl. Sci.* **2020**, *10*, 484. [CrossRef]
29. Prosser, W.H.; Gorman, M.R.; Humes, D.H. *Acoustic Emission Signals in Thin Plates Produced by Impact Damage*; Journal of Acoustic Emission, Acoustic Emission Working Group: Houston TX, USA, 1999.
30. De Rosa, I.M.; Santulli, C.; Sarasini, F.; Valente, M. Post-impact damage characterization of hybrid configurations of jute/glass polyester laminates using acoustic emission and IR thermography. *Compos. Sci. Technol.* **2009**, *69*, 1142–1150. [CrossRef]
31. De Rosa, I.M.; Santulli, C.; Sarasini, F. Acoustic emission for monitoring the mechanical behaviour of natural fibre composites: A literature review. *Compos. Part A Appl. Sci. Manuf.* **2009**, *40*, 1456–1469. [CrossRef]
32. Choi, I.; Hong, C. Low-velocity impact response of composite laminates considering higher-order shear deformation and large deflection. *Mech. Compos. Mater. Struct.* **1994**, *1*, 157–170. [CrossRef]
33. Park, J.-M.; Kong, J.-W.; Kim, J.-W.; Yoon, D.-J. Interfacial evaluation of electrodeposited single carbon fiber/epoxy composites by fiber fracture source location using fragmentation test and acoustic emission. *Compos. Sci. Technol.* **2004**, *64*, 983–999. [CrossRef]

34. James, R.; Joseph, R.; Giurgiutiu, V. Impact damage detection in composite plates using acoustic emission signal signature identification (Conference Presentation). In Proceedings of the Active and Passive Smart Structures and Integrated Systems IX, Houston, TX, USA, 10–13 February 2020; Volume 11376, p. 113760K.
35. ASTM D7136/D7136M-15. Standard Test Method for Measuring the Damage Resistance of a Fiber-Reinforced Polymer Matrix Composite to a Drop-Weight. *Impact Event ASTM Int.* **2015**, *15*, 1–7.
36. Caputo, F.; De Luca, A.; Lamanna, G.; Borrelli, R.; Mercurio, U. Numerical study for the structural analysis of composite laminates subjected to low velocity impact. *Compos. Part B Eng.* **2014**, *67*, 296–302. [CrossRef]
37. Flores, M.; Mollenhauer, D.; Runatunga, V.; Beberniss, T.; Rapking, D.; Pankow, M. High-speed 3D digital image correlation of low-velocity impacts on composite plates. *Compos. Part B Eng.* **2017**, *131*, 153–164. [CrossRef]
38. Wallentine, S.M.; Uchic, M.D. A study on ground truth data for impact damaged polymer matrix composites. In *AIP Conference Proceedings*; AIP Publishing LLC: Melville, NY, USA, 2018; Volume 1949, p. 120002.
39. Giurgiutiu, V.; James, R.; Joseph, R. Acoustic emission method to ascertain damage occurrence in impacted composites. *USC-IPMO* **2019**, 1448.

Article

Analysis of Composite Structures in Curing Process for Shape Deformations and Shear Stress: Basis for Advanced Optimization

Niraj Kumbhare [1], Reza Moheimani [1,2] and Hamid Dalir [1,*]

[1] Advanced Composite Structures Engineering Laboratory, Department of Mechanical and Energy Engineering, Purdue School of Engineering and Technology, Indianapolis, IN 46202, USA; nirkumbh@iupui.edu (N.K.); rezam@purdue.edu (R.M.)
[2] School of Mechanical Engineering, Purdue University, West Lafayette, IN 47907, USA
* Correspondence: hdalir@purdue.edu

Citation: Kumbhare, N.; Moheimani, R.; Dalir, H. Analysis of Composite Structures in Curing Process for Shape Deformations and Shear Stress: Basis for Advanced Optimization. *J. Compos. Sci.* **2021**, *5*, 63. https://doi.org/10.3390/jcs5020063

Academic Editor: Jiadeng Zhu

Received: 24 December 2020
Accepted: 18 February 2021
Published: 23 February 2021

Publisher's Note: MDPI stays neutral with regard to jurisdictional claims in published maps and institutional affiliations.

Copyright: © 2021 by the authors. Licensee MDPI, Basel, Switzerland. This article is an open access article distributed under the terms and conditions of the Creative Commons Attribution (CC BY) license (https://creativecommons.org/licenses/by/4.0/).

Abstract: Identifying residual stresses and the distortions in composite structures during the curing process plays a vital role in coming up with necessary compensations in the dimensions of mold or prototypes and having precise and optimized parts for the manufacturing and assembly of composite structures. This paper presents an investigation into process-induced shape deformations in composite parts and structures, as well as a comparison of the analysis results to finalize design parameters with a minimum of deformation. A Latin hypercube sampling (LHS) method was used to generate the required random points of the input variables. These variables were then executed with the Ansys Composite Cure Simulation (ACCS) tool, which is an advanced tool used to find stress and distortion values using a three-step analysis, including Ansys Composite PrepPost, transient thermal analysis, and static structural analysis. The deformation results were further utilized to find an optimum design to manufacture a complex composite structure with the compensated dimensions. The simulation results of the ACCS tool are expected to be used by common optimization techniques to finalize a prototype design so that it can reduce common manufacturing errors like warpage, spring-in, and distortion.

Keywords: advanced composite cure simulation; carbon fiber-reinforced composites; optimization study; process-induced shape deformation; design of experiments; Latin hypercube sampling

1. Introduction

Composite structures have a vast range of applications in the aerospace, automobile, and energy industries. As time has passed, the manufacturing of composites has grown exponentially. The necessity of lightweight design and higher performance in the aerospace and motorsports industries has increased the demand for carbon fiber-reinforced polymer composite materials. Residual stresses and shape deformations are the main obstacles for high performance in a composite structure. The use of deformed components in assembly can cause higher internal stresses, which then hamper a final product's performance [1–6].

The thermomechanical analysis of composites is a hot topic for all researchers in the composite world. An ample amount of data have recently been collected from innovative research on composite structures' behavior. However, there is still a lack of standards to measure deformation values and residual stresses during the curing process of composite manufacturing [7,8]. In the automobile industry, tool designers decide the parameters based on trial results and their experienced guesses to estimate distortion and warpage. The most frequent problem found in the approximation method is that the results are not precise for intricate designs [9,10]. Finding optimum design parameters for tool development becomes a tedious task because of these deformations and residual stresses. The need for a reliable approach to predict these shape distortions has driven researchers to develop new methods to find deformation values [7,11,12].

A variety of research has been done on the process-induced shape deformations and stresses of nonplanar parts. However, thin planar composite plates show complex deformations that cannot be verified with old methods. Ersoy et al. [11] implemented a two-step finite element analysis method to calculate curved parts' deformations. Wisnom and M.R. [12] found that the spring-in angle of curved composites is proportional to the laminate's through-thickness. Zhang, G. and J. Wang [13] represented the results of an investigation on the process-induced stress and deformation of variable-stiffness composite cylinders. Mezeix et al. [14] presented a method to predict composite flight structures' deformation using ABAQUS software

Ansys composite cure simulation is a novel method developed using the Ansys tool, which was used as a part of the work presented in this paper. Curing process analysis is the building block for obtaining accurate results for predicting the final shape of a composite part [15]. G. Fernlund [16] explained the significance of the cure cycle in the dimensional fidelity of autoclave-processed composite parts. The final laminate quality is dependent on the heating rate, initial cure temperature, and dwelling time [17,18]. Temperature has been proven to leave a significant effect on the body of an alloy while using finite element analysis (FEA) [19]. The coefficient of thermal expansion (CTE) is different in different directions for composite materials. Previous experimental studies have shown that the CTE of resin-dominated directions is much higher than that of fiber-dominated directions [20,21]. The stresses get built up on the fiber-matrix, lamina–laminate, and structural levels. When a composite part is cooled, the generated residual stresses are compensated for by the tool dimensions when they are in contact with the tool. The piece gets distorted to its equilibrium state to balance the internal residual stresses when the support gets removed [22]. In a literature survey, it was found that some of the factors responsible for creating deformations are the layup sequence, resin cure shrinkage, tool–part interaction, part angles, curing time and cycle, ply thickness, and layup angles.

The next step is to analyze simulation results to compare and find optimized parameters with an objective of minimum deformation. Efficient design is obtained by sizing a composite part's geometry, altering the manufacturing process, and tailoring design variables that control mechanical properties such as fiber orientation, the number of plies, and the stacking sequence. The manufacturability of a virtually optimized composite structure is a crucial precondition for the usability of the best results. In the current work, the consideration of manufacturing constraints ensured that all compared solutions are producible. This paper shows a study of the fundamentals of the design optimization of composite parts by considering manufacturing limitations. The optimization of composite parts that combine finite element (FE) models takes a has-run time for each FE simulation [23]. Generally, composite design optimization is a non-convex, multimodal optimization problem that involves continuous and discrete variables. In such cases, population-based algorithms like genetic algorithms (GAs) are preferred because they use several design responses in each iteration to find the optimal solution instead of gradient information. In the current research work, the simulation results were analyzed to prepare the base of an optimization algorithm [24–26]. Amir Ehsani [27] demonstrated the GA technique used to optimize composite angle grid plates.

This paper presents a reliable engineering approach to find process-induced shape deformations using the Ansys Composite Cure Simulation (ACCS) tool, utilizing the results of its analysis to find an optimal design with the least deformation using the fundamentals of design optimization. A novel FEA method is proposed to address the challenge of predicting residual stresses, distorted values, and the importance of the study of composites' thermal behavior. The global optimization gave optimal model results, which helped to finalize the design.

2. Materials and Methods

An optimal design can be found by combining ACCS simulation results with a standard optimization process. ACCS has been used to analyze deformations like spring-in, warpage, and residual stresses in composite parts. Deformation results vary with different inputs, including the layup sequence, angles, support constraint, and cure cycle. These variations were studied to find the best design to manufacture a complex composite structure. Figure 1 illustrates the full methodology starting from the selection of design samples to the final optimized design selection. The process started with the design optimization method, which was further divided into sub-processes, such as the objective function selection, constraints, and design of experiments (DOEs). Executing FEA was the second important step, which included the simulation process to find deformation results. These results were further analyzed and compared to find the optimized design. The methods explained in this section include all the details about the simulation and optimization process.

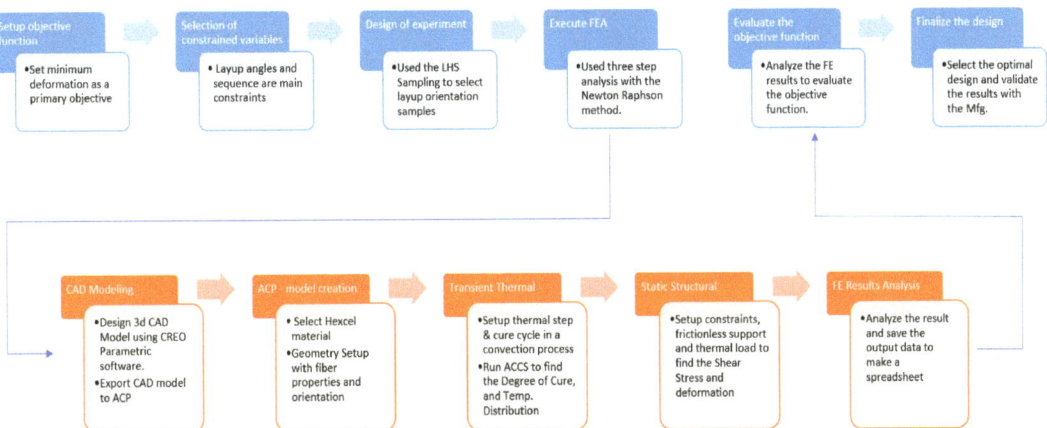

Figure 1. Process flow chart to find the final design. ACP: ANSYS Composite PrepPost; LHS: Latin hypercube sampling; FE: finite element; ACCS: Ansys Composite Cure Simulation.

2.1. Materials

The automotive and aerospace industry uses various composite materials, including natural composites, carbon–carbon composites, metal matrix composites (MMCs), and polymer matrix composites (PMCs). They provide benefits such as weight reduction, durability, high strength, and energy absorption. In this project, a Hexcel AS4-8552 prepreg—a unidirectional prepreg with a high-performance polymer matrix—was used for the simulation process. Hexcel material can make it possible to achieve weight reductions by maintaining a component's high structural performance. Using Hexcel AS4-8552 was beneficial because of its properties like high impact resistance, reasonable translation of fiber properties, and high strength. The mechanical properties of the Hexcel AS4 material provided by the manufacturer are mentioned in Table 1 [15]. The X and Y directions indicate the fiber directions as parallel (0°) and transverse (90°), respectively, to the matrix. Ansys has the feature of defining customized material properties in the engineering data section. These properties were applied to the L plate of dimensions $50 \times 86 \times 1.6$ mm with a flange height of 50 mm.

Table 1. Material properties of Hexcel AS4-8552.

Material Property	Value
Density	1580 kg/m^3
Coefficient of Thermal Expansion	
i. X-Direction	1×10^{-20}/°C
ii. Y/Z-Direction	3.261×10^{-5}/°C
Young's Modulus	
i. X-Direction	135 GPa
ii. Y/Z-Direction	9.5 GPa
Poisson's Ratio	
i. XY	0.3
ii. ZY	0.45
iii. XZ	0.3
Shear Modulus	
i. XY	4.90 GPa
ii. ZY	3.27 GPa
iii. XZ	4.90 GPa
Orthotropic Thermal Conductivity:	
i. X-Direction	5.5 W/(m°C)
ii. Y-Direction	0.489 W/(m°C)
iii. Z-Direction	0.658 W/(m°C)
Specific Heat, C_p	1300 W/(m°C)
Fiber Volume Fraction	0.5742
Resin Properties:	
Initial Degree of Cure	0.0001
Maximum Degree of Cure	0.9999
Gelation Degree of Cure	0.33
Total Heat of Reaction	540 KJ
Glass Transition Temperature:	
Initial Value	2.670 °C
Final Value	218.27 °C
λ	0.4708 °C
Orthotropic Cure Shrinkage:	
i. X-Direction	1×10^{-20}/mm
ii. Y-Direction	0.0073/mm
iii. Z-Direction	0.0073/mm
Orthotropic Liquid Pseudo Elasticity:	
i. X-Direction	132 GPa
ii. Y-Direction	165 GPa
iii. Z-Direction	165 GPa

2.2. Simulation Process

The FEA simulation process was developed using the ACCS package to measure the deformations and residual stresses in the complex composite structure. The ACCS package is a combination of the ANSYS Composite PrepPost (ACP), transient thermal, and static structural analysis modules. The ACCS solver is a crucial tool within the transient thermal module to develop polymerization and find the internal heat generated due to conducted exothermic reactions. It was connected to the structural analysis in this research to study the deformations and shear stresses using the thermal results. ACCS utilizes a fast three-step simulation approach for comparatively thin laminates (<5 mm thick), where an even temperature distribution is assumed [15].

2.2.1. Composite Model Creation

The simulation process starts with the Computer Aided Design (CAD) modeling. In this research, the presented solid model was an L-shaped plate of dimensions 50 × 86 × 1.6 mm and a flange height of 50 mm, as shown in Figure 2a. The analysis was done on a solid composite model to obtain entire thickness cure properties. The design part included 3D solid modeling and .step file conversion in the Creo 4.0 parametric software.

The step file was imported to the ACP module for the composite fiber setup as necessary. The imported solid model's surface was extracted with the ANSYS Spaceclaim platform to create the shell model. A solid model of composite structure with the desired fiber ply thickness, layup angles, and sequence was created using the ACP module. Figure 2b shows the L-plate model generated in the ACP module with the given fiber properties such as 6 layers of Hexcel material oriented by $(0/45/90)_s$. The thicknesses of the fibers used in the model were 0.2 and 0.4 mm.

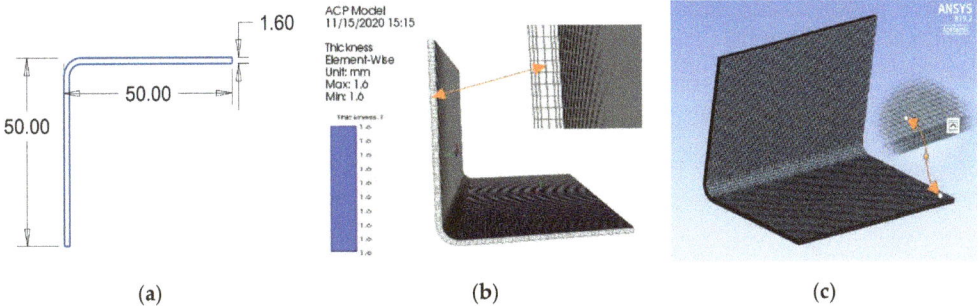

Figure 2. Fiber orientation and mesh details in the Advanced Composite PrePost: (**a**) layup details, (**b**) dimensions, and (**c**) mesh details.

Mesh generation was the next significant step in the FEA simulation. Any stress analysis is subject to several types of errors including user error, error due to assumptions and simplifications in the model, and errors due to insufficient mesh discretization. An analysis gives precise results if critical stresses converge to a reasonable level of accuracy. Hemesh Patil [28] presented a mesh convergence study on cylindrical parts. Mesh convergence in a finite elemental study defines the relationship between the number of elements and the accuracy of results. It is necessary to use a desired mesh that is acceptable to the shape and size of the used elements [15]. In the current work, element size was selected to be 1 mm by considering the simulation time and accuracy of results. The sum of observed nodes and elements were 8900 and 8712, respectively. Figure 2c shows the meshing detail for a design structure with a layup of $(0/45/90)_s$. During the simulation, the meshing was generated with smooth quality, including an element size from 0.5 to 2 mm. A significant difference could be observed in the shear stress accuracy and deformation of the results. A comparison of the results with different element sizes is presented in Table 2.

Table 2. Significance of mesh size in the deformation.

Element Size	Nodes	Elements	Deformation	Shear Stress
0.5	35,575	35,199	1.14	96.676
1	8900	8712	4.6191	76.893
1.5	4080	3953	27.968	75.213
2	2295	2200	22.332	75.773

2.2.2. Transient Thermal Analysis

This module's primary purpose is to obtain thermal properties, such as the degree of cure, glass transition temperature, and heat of reaction. This is a three-step non-linear analysis that uses the Newton–Raphson method. A convection condition is given to a composite plate as a thermal input load. The relationship of heat transfer by convection is like the conduction process, which is proportional to the surface area, temperature, and heat transfer coefficient. The rate of reaction in the curing process is proportional to the rate of heat flow [29]. The measured heat generated by the resin during the cure can be calculated using Equation (1), while the degree-of-cure of the resin can be obtained by integrating the area under the curve of cure rate vs. time, as shown in Equation (2)

$$\frac{d\alpha}{dt} = \frac{1}{H_T}\frac{dH}{dt} \tag{1}$$

$$\alpha = \frac{1}{H_T}\int_0^1 \frac{dH}{dt}dt \tag{2}$$

The convective load resembles the heating process of a composite part. The convection coefficient was taken as 25 W/mm^2, and the time step that defined the cure cycle was 240 s, which was kept constant for the full thermal analysis. Two types of cure cycles were used to analyze the thermal behavior of a composite structure. The types of cure cycles and layup sequences have major effects on shape deformation [17,30–32]. Figure 3 shows the double-hold cure cycle in which the temperature rose from 20 to 120 °C and was maintained for 1800 s in the first hold; then, it rose to 180 degrees with a dwell of 1800 s for the second hold; and in the last step, it cooled down to the normal temperature. However, in the single cure cycle as shown in the Figure 4, temperature rose from 20 to 180 degrees and was held for 3600 s [15].

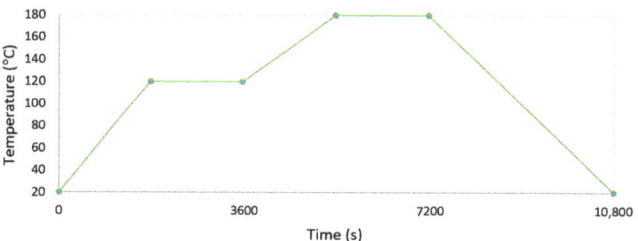

Figure 3. Double hold cure cycle.

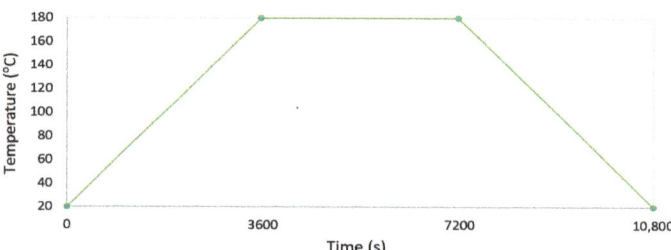

Figure 4. Single hold cure cycle.

2.2.3. Static Structural Analysis

Structural analysis is a three-step non-linear analysis used to study mechanical behavior like stresses developed in a composite structure. Figure 5 shows a design model with the constraint of given thermal load as an input to get results. The model was constrained at the endpoint of inside curve edges. A fully cured model could provide initial boundary conditions, while frictionless support and remote displacements were the applied load conditions for the structural analysis. The development of residual stresses can also be a function of time [30,31].

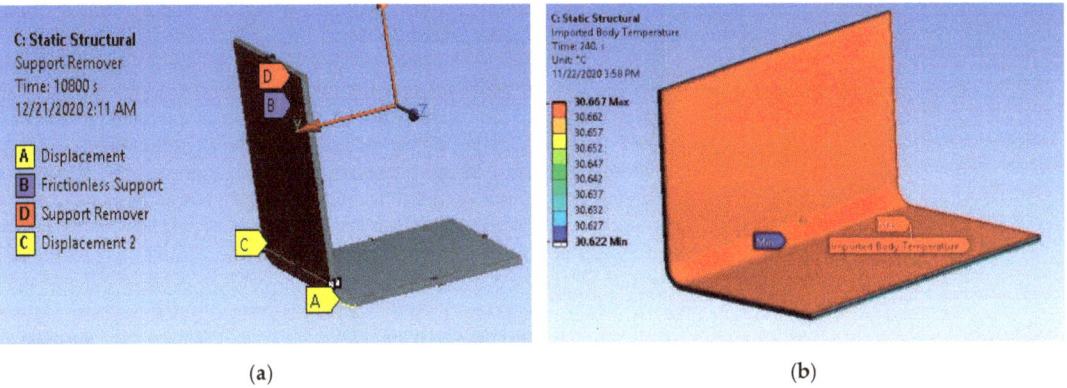

(a) (b)

Figure 5. Static structural analysis model setup: (a) Constraints in static structural analysis and (b) imported load in static structural analysis.

The stresses in laminate or ply can be calculated from classical laminate theory. Figure 6 explains classical laminate theory in detail. Equation (3) mentions the formula to calculate ply stress [15,32].

$$\begin{pmatrix} \sigma_{xx} \\ \sigma_{yy} \\ \tau_{xy} \end{pmatrix} = \begin{pmatrix} Q_{11} & Q_{12} & Q_{16} \\ Q_{12} & Q_{22} & Q_{26} \\ Q_{16} & Q_{26} & Q_{66} \end{pmatrix} \begin{pmatrix} \varepsilon_{xx} \\ \varepsilon_{yy} \\ \varepsilon_{xy} \end{pmatrix} \quad (3)$$

where, σ, τ, ε, and Q stand for ply stress, shear stress, strain in a single direction, and the transformed reduced stiffness matrix, respectively [33].

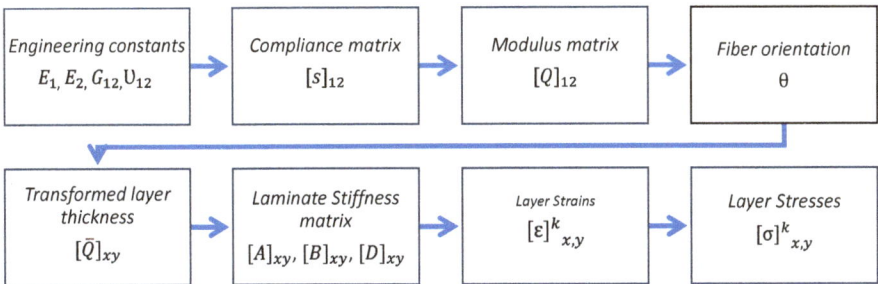

Figure 6. Flowchart to find the shear stresses.

2.3. Multi-Objective Design Optimization

Design engineers use the optimization technique most of the time to find the parameters that give the best system performance with the desired objectives. This research work presents the basis for a composite structure's optimization by considering limitations in the manufacturing processes and materials. A comprehensive geometrical model is considered for a sample structure to find a precise and useful design with deformation and shear stress. The optimization algorithm is based on the Design and Analysis of Computer Experiments (DACE), in which smart sampling and objectives evaluation drive the design towards a global optimum [23].

Gradient-based and population-based methods can be used in the maximization of the objective function. Gradient-based methods are computationally efficient but local in nature. Population-based methods like GAs and particle swarm optimization (PSO) have more chances to find global solutions for non-linear functions, but they cannot be guaranteed. The randomly generated population is updated based on the fitness values and random methods until the optima is found.

2.3.1. Artificial Neural Network (ANN)

An optimization problem needs realistic boundaries, such as constraints, to keep the result in the desired limit set. Depending on several parameters, such as the number of variables, the solution method, and the used theories, an optimization problem may take a long time. There are several accurate and fast numerical approximation methods to find a given variable function's comparable value. An artificial neural network (ANN) is one of the most accepted ways that is implemented by researchers as a perfect tool to provide fast and reasonable results [27].

An ANN is a biologically inspired computer program formed from hundreds of single units, artificial neurons, or processing elements (PEs) [33]. It comprises several component layers, neurons, and connections. ANNs are trained by detecting patterns and are capable of processing data and making accurate predictions. Figure 7 illustrates the layout of a neural network. 'i' represent the input layer with n parameters specified by the design variables, h_1–h_n are hidden layers, and 'o' represents the network's output layer with n objectives. A multi-layer perceptron (MLP) network with three hidden layers and one output layer with two neurons could be defined for the given problem [34–37]. Two neurons were the objectives with the minimum deformation and shear stress.

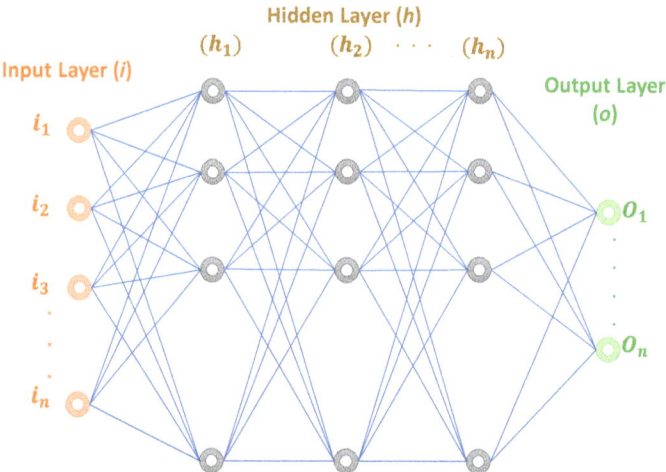

Figure 7. The artificial neural network (ANN) structure.

2.3.2. Latin Hypercube Sampling

Many design optimization processes start with the selection of design variables. DACE arises when selecting the sample points to be simulated to generate quality results in order to satisfy the objective. It is essential to define the sampling technique that avoids irrelevant simulations. The most frequently used sampling strategies with excellent filling properties are random sampling, stratified sampling, and Latin hypercube sampling (LHS) [34,35]. In the present work, the primary design variable was the layup angle, and the LHS sampling plan was used to generate a sample design variable set. This method ensures that all design space portions are represented in a stratified manner [36]. LHS was used her to generate the initial sample for the layup angles. These variables were then combined with the categorical data [37,38]. The sampling plan used for the project work was as shown in Table 3.

Table 3. Latin hypercube sampling plan.

α_1	α_2	α_3
0	45	90
0	−45	90
0	0	45
0	0	−45
0	0	90
0	45	0
0	−45	0
0	0	0
90	90	90
90	0	45

2.3.3. Multi-Objective Optimization Formulation

Multi-objective optimization works as a trade-off analysis because improving one objective implies the worsening of others [2,3,39–42]. In contrary to single-objective optimization problems, generally, there is no unique solution for multi-objective optimization. Therefore, the results of these problems are typically presented by a Pareto frontier curve, which is a set of optimal solutions [27]. A multi-objective optimization problem is defined as:

$$\text{find } x \in \mathbb{R}^{n_{dv}}$$
$$\text{minimize} \quad f(x)$$
$$\text{Subject to } g_i(x) \leq 0, i = 1, 2 \ldots, n_c$$

where $x = [x_1, \ldots, x_{n_{dv}}]$ is the design vector, n_{dv} is the number of design variables, and $f(x)$ is the vector of the objective functions such as $f(x) = [f_1(x), \ldots, f_k(x)]$, where k is the number of objective functions (output vector). Design constraints can be given by $g_i(x)$, where n_c is the number of constraints.

Many factors affect the amount of residual stress generated in a composite structure during the manufacturing process. A few main ones were considered in this project as design variables. The main input parameters were layup angles and sequences, which were further categorized with the type of cure cycle used and how the material was constrained. A sample number of layup angles considered by using the LHS method is shown in Table 3. In the given design, a total of six layers were taken into consideration. For example, $(0/45/90)_s$ is the structure of the symmetric sequence that was defined as $(0/45/90/90/45/0)$; however, an asymmetric sequence could be defined as $(0/45/90/90/-45/0)$. The present work considered minimum deformation as a primary objective function, and the least value of maximum shear stress generated in the structure was also considered. The design constraints used for the layup angles were limited from −90 to 90. Therefore, the multi-objective optimization statement for the current work is formulated as follows:

$$\text{find } x = [x_1, x_2, x_3, x_4] \in \mathbb{R}^4$$
$$\text{minimize} \begin{bmatrix} f_1(x_1, x_2) \\ f_2(x_1, x_2) \end{bmatrix}$$
$$\text{Subject to } g_i(x) = \alpha(x_1, x_2) \leq 0$$
$$-90 \leq x_1, x_2 \leq 90$$

3. Results and Discussion

3.1. Thermal Analysis Results

The convection method implemented during the cure process caused major changes in the resin. At the start, resin was in a viscous flow state. The second stage, which was the resin transition phase, occurred between 3600 and 7200 s and caused the resin to be violently cured. Its elastic modulus significantly increased, and the resin volume shrank. The structural analysis was performed on the layup with the (0/45/90)$_s$ stacking sequence. The last stage showed that the resin was wholly cured and no chemical reaction took place. Figure 8 shows thermal analysis results for the changes that occurred in the resin transformation phase. It can be seen in the Figure 8 that glass transition temperature changed during the second phase from 3600 to 7200 s, where the heat of reaction (HOR) was at its maximum. Figure 9 shows the temperature distribution in the transient thermal analysis. The temperature was at its maximum at the center and gradually reduced towards the outer ply. Figure 9 also shows the final temperature of the structure after the curing process. The temperature of the structure reduces from 180 to approximately 20 °C, with a final temperature of 20.044 °C. The glass transition temperature T_g significantly affected the resin mechanical properties, changing it from a rubbery state to a glassy state [15].

$$\frac{T_g - T_{g0}}{T_{g\infty} - T_{g0}} = \frac{\lambda\alpha}{1 - (1-\lambda)\alpha} \quad (4)$$

where T_g is the glass transition temperature; $T_{g\infty}$ and T_{g0} are the glass transition temperatures of uncured and fully cured resin, respectively; and α is the degree-of-cure.

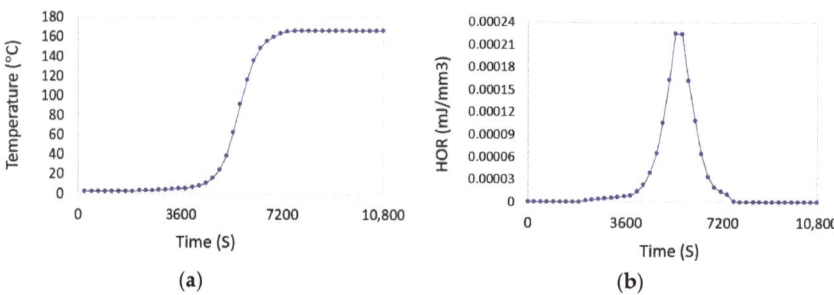

Figure 8. Resin transformation in the thermal analysis: (a) glass transition temperature and (b) heat of reaction (HOR).

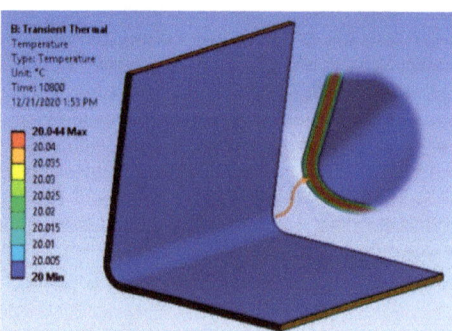

Figure 9. Temperature distribution in thermal analysis.

From the results of the thermal analysis, it was possible to calculate the instantaneous composite elastic constants. Similarly, the micromechanics approach could be used to calculate thermal and chemical strains for a given increment [37].

3.2. Static Structural Results

Static structural analysis was the final step implemented to calculate actual deformations and residual stresses. In the composite structure, the fibers created strength and stiffness, while the matrix provided bonding. Thus, the composite possessed good mechanical properties parallel to the fibers and was relatively weak in the perpendicular direction.

The structural analysis was performed on the layup with the $(0/45/90)_s$ stacking sequence. Figure 10 shows a graphical representation of the results based on the maximum shear stress and deformations. Figure 10a shows a graphical representation of the maximum amount of shear stress that could be generated in the structure. For the given diagram, the amount of shear stress generated after curing was 76.893 MPa. Figure 10b shows the total deformation that occurred in the same structure due to the generated residual stresses, and it can be seen in the diagram that the deformation occurred at up to 4.6191 mm for the layup.

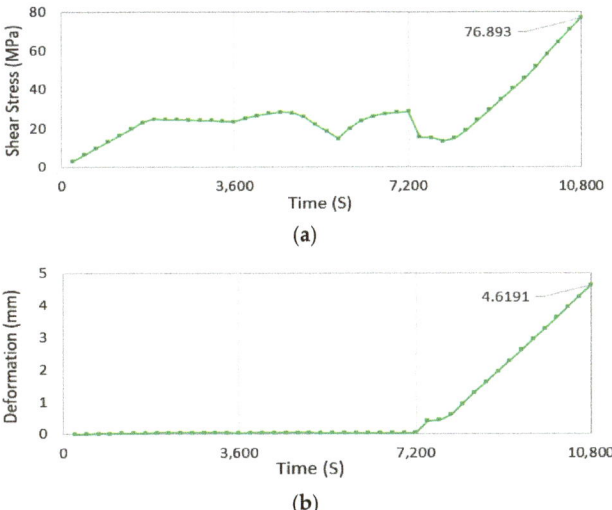

Figure 10. Graphical representation of results of stress analysis: (**a**) maximum shear stress and (**b**) total deformation.

Figure 11a,b shows the stress analysis results with spring-in, twist, and warpage. If a complex composite structure is investigated using the ACCS tool, bending, sagging, and/or twisting may occur (as Figure 11c shows this might behappened in similar studies), which could be deleterious for the final component's life.

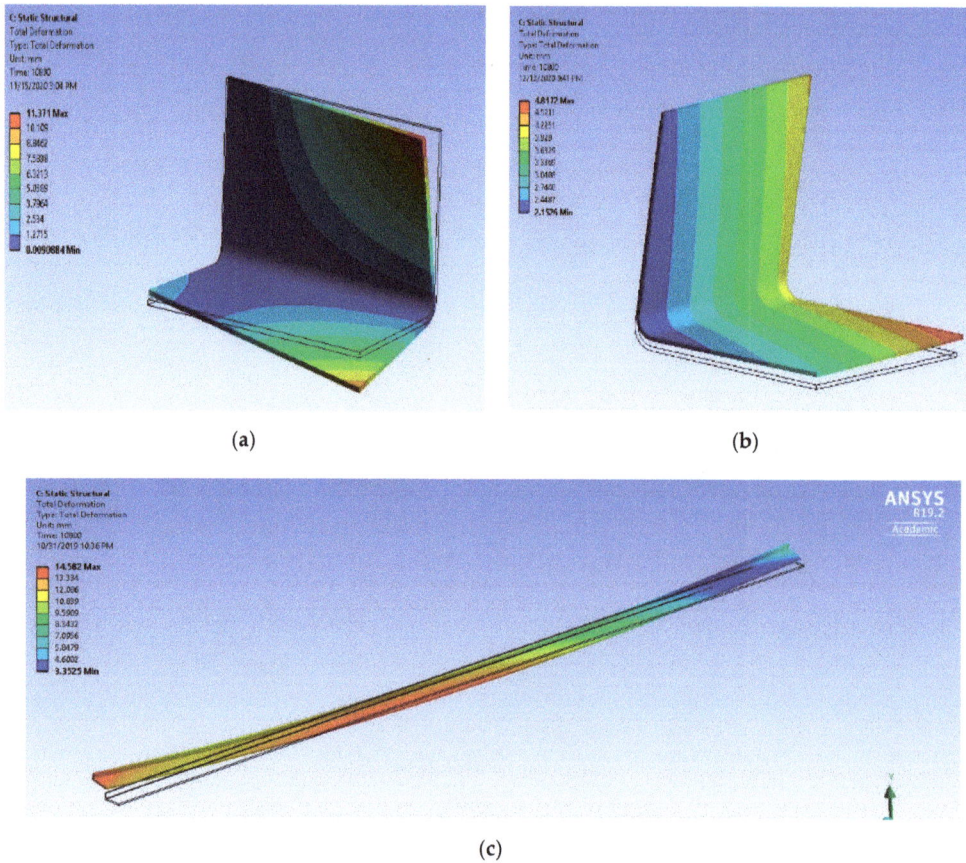

Figure 11. Possible deformations in the structure in the curing process: (**a**) warpage, (**b**) spring-in, and (**c**) sagging/bending.

3.3. Parameter Study Results

The workflow diagram shown in Figure 12 explains the network flow of the optimization process with the categories included in the process. It started with sample layup angles decided with the LHS plan. The selected α, i.e., the values of different layup angles, were further divided into several categories such as symmetry, cure cycle, and constraints. All the design parameters were then studied to analyze the output as minimum deformation and shear stress in the component. The sample data collected from the simulations were saved into the spreadsheet and are explained with a comparison to show the optimized structure. Table 4 shows the values of all the simulation results based on all the input design variables. The same data were further used in the scatter diagram, as shown in Figure 13, to represent the results [43–47].

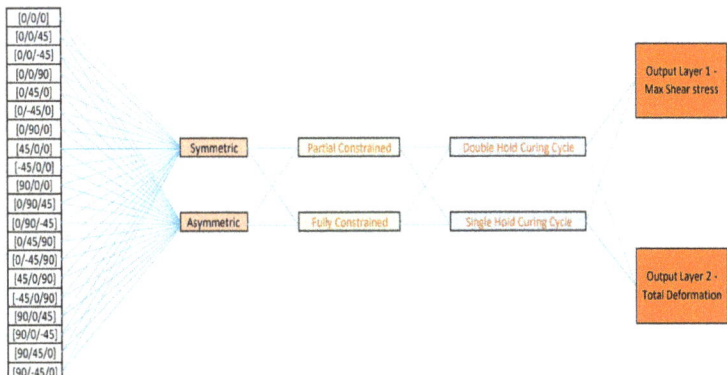

Figure 12. The details of categories that can be considered to build the neural network.

Table 4. Simulation results (Srs).

Sr No	Design Variable 1			Design Variable 2	Design Variable 3	Partially Constrained		Fully Constrained	
Design	a1	a2	a3	Symmetry	Cure Cycles	Deformations	Max Shear Stress	Deformations	Max Shear Stress
D1	0	45	90	Symmetric	Single	20.605	82.662	0.585062	87.257
D2	0	45	90	Symmetric	Double	4.6191	76.893	0.52853	96.252
D3	0	45	90	Asymmetric	Single	14.35	100.78	1.9054	161.43
D4	0	45	90	Asymmetric	Double	3.9376	93.543	1.7711	149.03
D5	0	−45	90	Symmetric	Single	6.7752	83.235	0.58509	87.833
D6	0	−45	90	Symmetric	Double	0.62761	75.102	0.54317	82.168
D7	0	−45	90	Asymmetric	Single	17.044	77.022	2.1651	190.16
D8	0	−45	90	Asymmetric	Double	3.5887	72.327	2.0126	176.04
D9	0	0	45	Symmetric	Single	22.091	460.74	0.64866	458.43
D10	0	0	45	Symmetric	Double	0.62892	428.89	0.61548	425.63
D11	0	0	45	Asymmetric	Single	16.951	388.01	4.7133	400.26
D12	0	0	45	Asymmetric	Double	9.1218	360.25	4.4283	371.67
D13	0	0	−45	Symmetric	Single	7.0188	458.4	0.6489	457.54
D14	0	0	−45	Symmetric	Double	2.2174	431.04	0.6297	425.68
D15	0	0	−45	Asymmetric	Single	51.262	387.19	4.7137	401.24
D16	0	0	−45	Asymmetric	Double	11.371	359.57	4.5054	372.67
D17	0	0	90	Symmetric	Single	2.5907	8.232	0.59803	68.097
D18	0	0	90	Symmetric	Double	1.5114	63.335	0.55558	63.197
D19	0	0	90	Asymmetric	Single	18.757	95.12	0.62251	95.06
D20	0	0	90	Asymmetric	Double	2.5089	88.294	0.57862	88.216
D21	0	45	0	Symmetric	Single	41.052	22.741	0.63468	552.69
D22	0	45	0	Symmetric	Double	17.676	558.9	3.3133	20.216
D23	0	45	0	Asymmetric	Single	12.784	575.29	5.4332	598.99
D24	0	45	0	Asymmetric	Double	15.849	536.4	5.1348	558.54
D25	0	−45	0	Symmetric	Single	15.58	599.15	0.6342	552.34
D26	0	−45	0	Symmetric	Double	16.522	557.7	0.60208	513.34
D27	0	−45	0	Asymmetric	Single	33.833	143.38	5.4338	596.4
D28	0	−45	0	Asymmetric	Double	11.343	573.84	29.024	31.582
D29	0	0	0	Symmetric	Single	111.51	562.62	0.63454	557.98
D30	0	0	0	Symmetric	Double	1.4481	522.28	0.59507	518.04

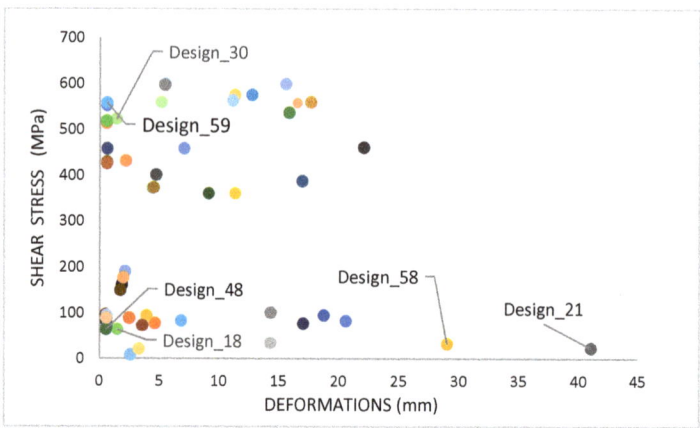

Figure 13. Scatter diagram showing a comparison of the ACCS results.

A graphical representation of all simulation results is shown in Figure 14. The relation between both objective functions is explained in the graph. The deformation values are on the X-axis, and shear stress is on the Y-axis. The values of all results are shown in Table 5. A feasible objective space defined in the graph illustrates the possible solutions for all design variables. It can be seen from the graph that few samples like Design_30 and Design_59 could meet the first objective, minimum deformation, but the value of shear stress-induced deformation was higher. Similarly, Design_21 and Design_58 had less shear stress, but they did not meet the objective of minimum deformation. Such designs could not be concluded as an optimum design. As mentioned by Ehsani, A., and H. Dalir [27] in their research, multi-objective optimization works on a trade-off analysis. Therefore, the result of these problems is typically presented by a curve with a set of optimal solutions known as a Pareto frontier. In this research work, all the design solutions had minimum deformations arranged with the given shear stress. As seen in the given figure, Design_48 and Design_18 showed the best results that could satisfy both the objectives of minimum deformation and shear stress.

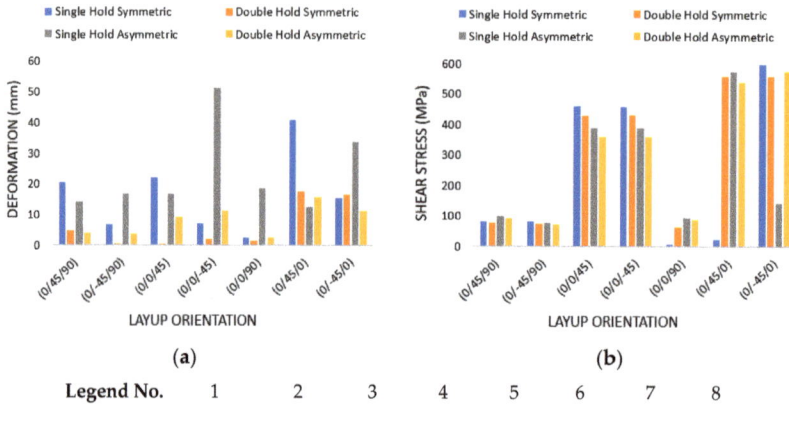

Figure 14. Comparison of residual stresses based on symmetric and asymmetric layup: (**a**) shear stress and (**b**) deformation.

Table 5. Deformation and shear stress of all designs.

Design_1	Design_2	Design_3	Design_4	Design_5	Design_6	Design_7	Design_8	Design_9	Design_10	Design_11	Design_12	Design_13	Design_14	Design_15
20.605	4.6191	14.35	3.9376	6.7752	0.62761	17.044	3.5887	22.091	0.62892	16.951	9.1218	7.0188	2.2174	51.262
82.662	76.893	100.78	93.543	83.235	75.102	77.022	72.327	460.74	428.89	388.01	360.25	458.4	431.04	387.19
Design_16	Design_17	Design_18	Design_19	Design_20	Design_21	Design_22	Design_23	Design_24	Design_25	Design_26	Design_27	Design_28	Design_29	Design_30
11.371	2.5907	1.5114	18.757	2.5089	41.052	17.676	12.784	15.849	15.58	16.522	33.833	11.343	111.51	1.4481
359.57	82.32	63.335	95.12	88.294	22.741	558.9	575.29	536.4	599.15	557.7	143.38	573.84	562.62	522.28
Design_31	Design_32	Design_33	Design_34	Design_35	Design_36	Design_37	Design_38	Design_39	Design_40	Design_41	Design_42	Design_43	Design_44	Design_45
0.585062	0.52853	1.9054	1.7711	0.58509	0.54317	2.1651	2.0126	0.64866	0.61548	4.7133	4.4283	0.6489	0.6297	4.7137
87.257	96.252	161.43	149.03	87.833	82.168	190.16	176.04	458.43	425.63	400.26	371.67	457.54	425.68	401.24
Design_46	Design_47	Design_48	Design_49	Design_50	Design_51	Design_52	Design_53	Design_54	Design_55	Design_56	Design_57	Design_58	Design_59	Design_60
4.5054	0.59803	0.55558	0.62251	0.57862	0.63468	3.3133	5.4332	5.1348	0.6342	0.60208	5.4338	29.024	0.63454	0.59507
372.67	68.097	63.197	95.06	88.216	552.69	20.216	598.99	558.54	552.34	513.34	596.4	31.582	557.98	518.04

Figure 15 shows the clustered column plot of results based on the curing cycle for different orientations. Figure 15a explains the difference in the deformation results that occurred with specific layup designs having single hold and double hold curing cycles. It can be seen from the graph that the single hold process could have more deformation in the structure than the double hold cure cycle. Similar plots are provided for the shear stress comparison. In the single hold cure cycle, the structure was held at 180 degrees for an hour, which resulted in more deformations than in the double hold cycle, where the structure could cure in two steps at 120 and 180 degrees for 30 min. Details of the deformation and shear stress data are shown in Table 6.

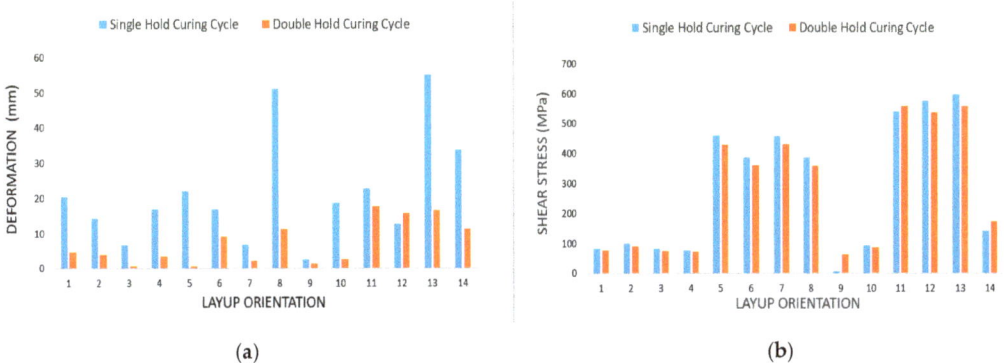

(a) (b)

Figure 15. Scatter diagram showing a comparison of results based on the curing cycle: (a) deformation and (b) shear stress.

The double cure cycle showed better results than the single cure cycle because of the resin-induced effects. Literature on cure cycling implied that if the curing reaction of the resin occurred too quickly, the resin flow time would be reduced and result in voids and deformations. It is essential to ensure the complete cure of final laminates, which guarantees a good laminate quality [19]. Dong [18] presented a study of laminate quality based on the initial cure temperature and cure cycle. They mentioned that the degree of cure increased with cure time until reaching a constant value. For our study in the double hold cure cycle, the maximum degree of cure at 120 °C was 0.78, which suggested an insufficient cross-linking network formation. When the temperature rose to 180 °C, the degree of cure (DOC) reached 0.95, which resulted in a better laminate quality with less deformation. Additionally, the porosity in the laminate was at a minimum for the double hold cure cycle with dwelling at 120 and 180 °C.

Table 6. ACCS results provided to show a curing cycle comparison.

Layup Orientation	Deformation		Shear Stress	
	Single Hold	Double Hold	Single Hold	Double Hold
$[0/45/90]_s$	20.605	4.6191	82.662	76.893
$[0/45/90]_{as}$	14.35	3.9376	100.78	93.543
$[0/-45/90]_s$	6.7752	0.62761	83.235	75.102
$[0/-45/90]_{as}$	17.044	3.5887	77.022	72.327
$[0/0/45]_s$	22.091	0.62892	460.74	428.89
$[0/0/45]_{as}$	16.951	9.1218	388.01	360.25
$[0/0/-45]_s$	7.0188	2.2174	458.4	431.04
$[0/0/-45]_{as}$	51.262	11.371	387.19	359.57
$[0/0/90]_s$	2.5907	1.5114	8.232	63.335
$[0/0/90]_{as}$	18.757	2.5089	95.12	88.294
$[0/45/0]_s$	22.741	17.676	541.052	558.9
$[0/45/0]_{as}$	12.784	15.849	575.29	536.4
$[0/-45/0]_s$	55.2	16.522	599.15	557.7
$[0/-45/0]_{as}$	33.833	11.343	143.38	573.84

All the design parameters selected in the process generated different deformations and shear stresses. A few of the common trends are shown in Figure 16, which shows results based on the symmetry and curing process for the given layup orientation. The graph demonstrates that the symmetric layup cured with a single hold cycle could generate more shear stress. The fifth layup with the $[0/0/90]_s$ orientation gave the best results for the shear stress. However, deformation results were diverse among the optimum cases. Moheimani et al. presented an important approach to study the failure of epoxy laminae using a cohesive multiscale model [38].

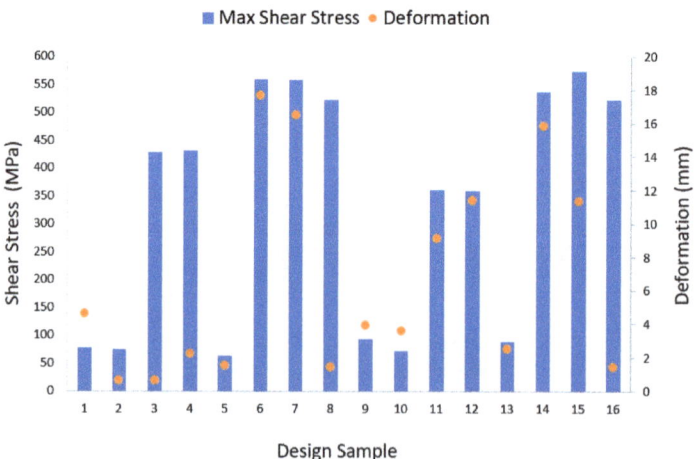

Figure 16. Layup angle comparison with reference to deformation and shear stress.

Figure 16 shows a combination bar and scatter graph, indicating the stress analysis results for the given layup orientation samples. The X-axis describes all the layup orientation samples considered in the process. The left Y-axis is used to measure the deformation results, while the right vertical axis is for the shear stresses. The bar graph combined with the scatter diagram shows the variations in the maximum shear stress (MPa) and deformation (mm).. The graph shows that the best results can be found in some layup orientation samples like design 5, 9, 10 and 13. These samples were considered in the feasible design space.

3.4. Validation of Simulation Results with Experiment

A comparison of the ACCS results with the experimental one was reviewed. It was found that few researchers have presented work on a comparison of deformation results. T. Garstka mentioned that they have done experimentation to validate spring-in results [39,40]. In the current work, a sample layup with the stacking sequence of $[0/45/90]_s$ was used to compare spring-in values based on analytical, numerical, and experimental procedures. The results showed that the analytical calculations gave a 0.90° spring-in. However, the ACCS simulation and experimental results showed spring-in deformations of 0.81° and 0.78°, respectively. It can be observed from the results that the simulation and experimental results did not have significant differences in their deformation values [11].

4. Conclusions

In the current research, the ACCS tool was utilized to generate data with different references such as the curing cycle, constraints, and layup symmetry, which were further investigated to find the best results after manufacturing. The fundamentals of the multi-objective optimization of a complex composite structure were described while considering two objective functions: the least amount of deformation and the shear stress generated during the thermal analysis. The process started with the DOE, in which the LHS sampling method was developed to generate the samples of fiber orientation angle. The design variables were categorized according to the used curing process, constraints, and stacking sequence. These results can be used to compensate for tool design to reduce errors in a prototype. The ACCS result data were analyzed to show a comparison based on the layup orientation.

The results signified that the curing process and the layup sequence used in the composites changed the deformation results and the amount of shear stress generated during cure. The data found with the ACCS tool led us to conclude that the single hold curing cycle caused more deformation than the double hold cycle. Furthermore, the scatter diagram of all studied cases showed that designs 33, 34, 38, and 46 had the best performance after manufacturing. These designs can be accommodated in a feasible design space. On the contrary, the samples made with designs 41 and 63 were found to be outside of the feasible design space; hence, they can be considered as non-feasible points. The comparison of layup orientations as shown in Figure 16 implied that the 5th $((0/0/45)_s)$ and 13th $((0/-45/0)_s)$ layup sequences had the most optimized results. A good agreement was obtained between with the empirical case study and the simulation modeling. We can claim that the foundation of this study like input and output parameters can be implemented in an ANN to find an optimal structure. In other words, the data executed with the FEA tool can be utilized to train parametric data in an artificial neural network.

Author Contributions: Methodology, software, formal analysis, and writing/draft preparation—N.K., methodology, review, supervision, and writing/editing—R.M., project supervision and administration—H.D. All authors have read and agreed to the published version of the manuscript.

Funding: The authors would like to express their gratitude to Indy Car (#073779) for the funding provided to do this research and for their assistance with the instrumentation and proofreading.

Acknowledgments: Authors would like to thank Simutech for their valuable support with the software tutorial and license. We would like to acknowledge partial funding from Indy Car Group for supporting this research work.

Conflicts of Interest: The authors declare no conflict of interest.

References

1. Liu, X.; Alizadeh, V.; Hansen, C.J. The compressive response of octet lattice structures with carbon fiber composite hollow struts. *Compos. Struct.* **2020**, *239*, 111999. [CrossRef]
2. Ehsani, A.; Rezaeepazhand, J. Stacking sequence optimization of laminated composite grid plates for maximum buckling load using genetic algorithm. *Int. J. Mech. Sci.* **2016**, *119*, 97–106. [CrossRef]

3. Ehsani, A.; Dalir, H. Multi-objective design optimization of variable ribs composite grid plates. *Struct. Multidiscip. Optim.* **2020**, *63*, 407–418. [CrossRef]
4. Kazemi, A.; Yang, S. Atomistic Study of the Effect of Magnesium Dopants on the Strength of Nanocrystalline Aluminum. *JOM* **2019**, *71*, 1209–1214. [CrossRef]
5. Patil, A.; Moheimani, R.; Shakhfeh, T.; Dalir, H. Analysis of Spring-in for Composite Plates Using ANSYS Composite Cure Simulation. In Proceedings of the American Society for Composites (ASC)—Thirty-Fourth Technical Conference on Composite Materials, Seattle, WA, USA, 23–25 September 2019. [CrossRef]
6. Pasharavesh, A.; Alizadeh Vaghasloo, Y.; Ahmadian, M.; Moheimani, R. Vibration of a Microbeam Under Ultra-Short-Pulsed Laser Excitation Considering Momentum and Heating Effect. In Proceedings of the ASME International Mechanical Engineering Congress and Exposition, Vancouver, BC, Canada, 12–18 November 2010; pp. 195–200.
7. Al-Dhaheri, M.; Khan, K.A.; Umer, R.; van Liempt, F.; Cantwell, W.J. Process-induced deformation in U-shaped honeycomb aerospace composite structures. *Compos. Struct.* **2020**, *248*. [CrossRef]
8. Onsorynezhad, S.; Abedini, A.; Wang, F. Parametric optimization of a frequency-up-conversion piezoelectric harvester via discontinuous analysis. *J. Vib. Control* **2020**, *26*, 1241–1252. [CrossRef]
9. Bellini, C.; Sorrentino, L.; Polini, W.; Corrado, A. Spring-in analysis of CFRP thin laminates: Numerical and experimental results. *Compos. Struct.* **2017**, *173*, 17–24. [CrossRef]
10. Bin Mohd Nasir, M.N.; Seman, M.A.; Mezeix, L.; Aminanda, Y.; Rivai, A.; Ali, K.M. Effect of the corner angle on spring-back deformation for unidirectional L-shaped laminate composites manufactured through autoclave processing. *ARPN J. Eng. Appl. Sci.* **2016**, *11*, 315–318.
11. Patil, A.S.; Moheimani, R.; Dalir, H. Thermomechanical analysis of composite plates curing process using ANSYS composite cure simulation. *Therm. Sci. Eng. Prog.* **2019**, *14*. [CrossRef]
12. Çiçek, K.F.; Erdal, M.; Kayran, A. Experimental and numerical study of process-induced total spring-in of corner-shaped composite parts. *J. Compos. Mater.* **2016**, *51*, 2347–2361. [CrossRef]
13. Shafiee, A.; Ahmadian, M.T.; Hoviattalab, M. Traumatic Brain Injury Caused by+ Gz Acceleration. In Proceedings of the ASME 2016 International Design Engineering Technical Conferences and Computers and Information in Engineering Conference, Charlotte, NC, USA, 21–24 August 2016.
14. Wisnom, M.R.; Potter, K.D.; Ersoy, N. Shear-lag analysis of the effect of thickness on spring-in of curved composites. *J. Compos. Mater.* **2007**, *41*, 1311–1324. [CrossRef]
15. Zhang, G.; Wang, J.; Ni, A. Process-Induced Stress and Deformation of Variable-Stiffness Composite Cylinders during Curing. *Materials* **2019**, *12*, 259. [CrossRef]
16. Mezeix, L.; Seman, A.; Nasir, M.; Aminanda, Y.; Rivai, A.; Castanié, B.; Olivier, P.; Ali, K. Spring-back simulation of unidirectional carbon/epoxy flat laminate composite manufactured through autoclave process. *Compos. Struct.* **2015**, *124*, 196–205. [CrossRef]
17. Fernlund, G.; Rahman, N.; Courdji, R.; Bresslauer, M.; Poursartip, A.; Willden, K.; Nelson, K. Experimental and numerical study of the effect of cure cycle, tool surface, geometry, and lay-up on the dimensional fidelity of autoclave-processed composite parts. *Compos. Part A Appl. Sci. Manuf.* **2002**, *33*, 341–351. [CrossRef]
18. Dong, A.; Zhao, Y.; Zhao, X.; Yu, Q. Cure Cycle Optimization of Rapidly Cured Out-Of-Autoclave Composites. *Materials* **2018**, *11*, 421. [CrossRef]
19. Liu, L.; Zhang, B.-M.; Wang, D.-F.; Wu, Z.-J. Effects of cure cycles on void content and mechanical properties of composite laminates. *Compos. Struct.* **2006**, *73*, 303–309. [CrossRef]
20. Ahmadi, A.; Sadeghi, F. A Novel Three-Dimensional Finite Element Model to Simulate Third Body Effects on Fretting Wear of Hertzian Point Contact in Partial Slip. *J. Tribol.* **2020**, *143*. [CrossRef]
21. Patham, B.; Huang, X. Multiscale modeling of residual stress development in continuous fiber-reinforced unidirectional thick thermoset composites. *J. Compos.* **2014**, *2014*, 172560. [CrossRef]
22. Kazemi, A.; Yang, S. Effects of magnesium dopants on grain boundary migration in aluminum-magnesium alloys. *Comput. Mater. Sci.* **2020**, *2020*, 110130. [CrossRef]
23. Khan, L.A.; Iqbal, Z.; Hussain, S.T.; Kausar, A.; Day, R.J. Determination of optimum cure parameters of 977-2A carbon/epoxy composites for quickstep processing. *J. Appl. Polym. Sci.* **2013**, *129*, 2638–2652. [CrossRef]
24. Kathiravan, R.; Ganguli, R. Strength design of composite beam using gradient and particle swarm optimization. *Compos. Struct.* **2007**, *81*, 471–479. [CrossRef]
25. Yang, J.; Zhan, Z.; Zheng, K.; Chen, C.; Hu, J.; Zheng, L. An uncertainty representation based sampling method for metamodeling in auto-motive design applications. *J. Mech. Sci. Technol.* **2016**, *30*, 4645–4655. [CrossRef]
26. Simpson, T.W.; Booker, A.J.; Ghosh, D.; Giunta, A.A.; Koch, P.N.; Yang, R.J. Approximation methods in multidisciplinary analysis and optimization: A panel discussion. *Struct. Multidiscip. Optim.* **2004**, *27*, 302–313. [CrossRef]
27. Storlie, C.B.; Reich, B.J.; Helton, J.C.; Swiler, L.P.; Sallaberry, C.J. Analysis of computationally demanding models with continuous and categorical inputs. *Reliab. Eng. Syst. Saf.* **2013**, *113*, 30–41. [CrossRef]
28. Ehsani, A.; Dalir, H. Multi-objective optimization of composite angle grid plates for maximum buckling load and minimum weight using genetic algorithms and neural networks. *Compos. Struct.* **2019**, *229*. [CrossRef]
29. Patil, H.; Jeyakarthikeyan, P.V. Mesh convergence study and estimation of discretization error of hub in clutch disc with integration of ANSYS. *IOP Conf. Ser. Mater. Sci. Eng.* **2018**, *402*. [CrossRef]

30. Khoun, L.; Centea, T.; Hubert, P. Characterization Methodology of Thermoset Resins for the Processing of Composite Materials—Case Study: CYCOM 890RTM Epoxy Resin. *J. Compos. Mater.* **2009**, *44*, 1397–1415. [CrossRef]
31. White, S.R.; Hahn, H.T. Cure Cycle Optimization for the Reduction of Processing-Induced Residual Stresses in Composite Materials. *J. Compos. Mater.* **1993**, *27*, 1352–1378. [CrossRef]
32. Hernández, S.; Sket, F.; González, C.; Llorca, J. Optimization of curing cycle in carbon fiber-reinforced laminates: Void distribution and mechanical properties. *Compos. Sci. Technol.* **2013**, *85*, 73–82. [CrossRef]
33. Shahverdi Moghaddam, H.; Keshavanarayana, S.; Yang, C.; Horner, A. Anisotropic hyperelastic constitutive modeling of in-plane finite deformation responses of commercial composite hexagonal honeycombs. *J. Sandw. Struct. Mater.* **2021**. [CrossRef]
34. Timoshin, A.; Kazemi, A.; Beni, M.H.; Jam, J.E.; Pham, B. Nonlinear strain gradient forced vibration analysis of shear deformable microplates via hermitian finite elements. *Thin Walled Struct.* **2021**, *161*, 107515. [CrossRef]
35. Yang, C.; Moghaddam, H.S.; Keshavanarayana, S.R.; Horner, A.L. An analytical approach to characterize uniaxial in-plane responses of commercial hexagonal honeycomb core under large deformations. *Compos. Struct.* **2019**, *211*, 100–111. [CrossRef]
36. Mian, H.H.; Wang, G.; Dar, U.A.; Zhang, W. Optimization of Composite Material System and Lay-up to Achieve Minimum Weight Pressure Vessel. *Appl. Compos. Mater.* **2013**, *20*, 873–889. [CrossRef]
37. Agatonovic-Kustrin, S.; Beresford, R. Basic concepts of artificial neural network (ANN) modeling and its application in pharmaceutical research. *J. Pharm. Biomed. Anal.* **2000**, *22*, 717–727. [CrossRef]
38. Bre, F.; Gimenez, J.M.; Fachinotti, V.D. Prediction of wind pressure coefficients on building surfaces using artificial neural networks. *Energy Build.* **2018**, *158*, 1429–1441. [CrossRef]
39. Myers, R.H.; Montgomery, D.C.; Anderson-Cook, C.M. *Response Surface Methodology: Process and Product Optimization Using Designed Experiments*; John Wiley & Sons: Hoboken, NJ, USA, 1995.
40. Lophaven, S.N.; Nielsen, H.B.; Søndergaard, J. *DACE: A Matlab Kriging Toolbox*; Omicron: Roskilde, Denmark, 2002; Volume 2.
41. McKay, M.D.; Beckman, R.J.; Conover, W.J. A comparison of three methods for selecting values of input variables in the analysis of output from a computer code. *Technometrics* **2000**, *42*, 55–61. [CrossRef]
42. Kamble, M.; Shakfeh, T.; Moheimani, R.; Dalir, H. Optimization of a composite monocoque chassis for structural performance: A comprehensive approach. *J. Fail. Anal. Prev.* **2019**, *19*, 1252–1263. [CrossRef]
43. Khezrloo, A.; Tayebi, M.; Shafiee, A.; Aghaie, A. Evaluation of compressive and split tensile strength of aluminosilicate geopolymer reinforced by waste polymeric materials using Taguchi method. *Mater. Res. Express* **2021**. [CrossRef]
44. Shahi, V.; Alizadeh, V.; Amirkhizi, A.V. Thermo-mechanical characterization of polyurea variants. *Mech. Time Depend. Mater.* **2020**, 1–25. [CrossRef]
45. Moheimani, R.; Sarayloo, R.; Dalir, H. Failure study of fiber/epoxy composite laminate interface using cohesive multiscale model. *Advanced Composites Letters* **2020**, *29*, 2633366X20910157. [CrossRef]
46. Garstka, T. *Numerical Tool Compensation and Composite Process Optimization*; Presentation; LMAT Lean Manufacturing Assembly Technologies: Bristol, UK, 2017.
47. Ersoy, N.; Garstka, T.; Potter, K.; Wisnom, M.R.; Porter, D.; Stringer, G. Modelling of the spring-in phenomenon in curved parts made of a thermosetting composite. *Compos. Part A Appl. Sci. Manuf.* **2010**, *41*, 410–418. [CrossRef]

Article

Strain Mapping and Damage Tracking in Carbon Fiber Reinforced Epoxy Composites during Dynamic Bending Until Fracture with Quantum Resistive Sensors in Array

Antoine Lemartinel [1,2], Mickaël Castro [2,*], Olivier Fouché [1], Julio-César De Luca [1] and Jean-François Feller [2]

1. Institut de Recherche Technologique Jules Verne, 44340 Bouguenais, France; antoine.lemartinel@gmail.com (A.L.); olivier.fouche@irt-jules-verne.fr (O.F.); julio-cesar.de-luca@irt-jules-verne.fr (J.-C.D.L.)
2. Smart Plastics Group, Université de Bretagne Sud, UMR CNRS 6027, IRDL, 56100 Lorient, France; jean-francois.feller@univ-ubs.fr
* Correspondence: mickael.castro@univ-ubs.fr

Abstract: The sustained development of wind energies requires a dramatic rising of turbine blade size especially for their off-shore implantation, which requires as well composite materials with higher performances. In this context, the monitoring of the health of these structures appears essential to decrease maintenance costs, and produce a cheaper kwh. Thus, the input of quantum resistive sensors (QRS) arrays, to monitor the strain gradient in area of interest and anticipate damage in the core of composite structures, without compromising their mechanical properties, sounds promising. QRS are nanostructured strain and damage sensors, transducing strain at the nanoscale into a macroscopic resistive signal for a consumption of only some µW. QRS can be positioned on the surface or in the core of the composite material between plies, and this homogeneously as they are made of the same resin as the composite. The embedded QRS had a gauge factor of 3, which was found more than enough to follow the strain from 0.01% to 1.4% at the final failure. The spatial deployment of four QRS in array made possible for the first time the experimental visualization of a strain field comparable to the numerical simulation. QRS proved also to be able to memorize damage accumulation within the sample and thus could be used to attest the mechanical history of composites.

Keywords: structural health monitoring (SHM); carbon nanotubes (CNT); quantum piezo-resistive sensor (QRS); in situ measurements; smart materials; embedded sensors

Citation: Lemartinel, A.; Castro, M.; Fouché, O.; De Luca, J.-C.; Feller, J.-F. Strain Mapping and Damage Tracking in Carbon Fiber Reinforced Epoxy Composites during Dynamic Bending Until Fracture with Quantum Resistive Sensors in Array. J. Compos. Sci. 2021, 5, 60. https://doi.org/10.3390/jcs5020060

Academic Editor: Jiadeng Zhu

Received: 18 January 2021
Accepted: 18 February 2021
Published: 20 February 2021

Publisher's Note: MDPI stays neutral with regard to jurisdictional claims in published maps and institutional affiliations.

Copyright: © 2021 by the authors. Licensee MDPI, Basel, Switzerland. This article is an open access article distributed under the terms and conditions of the Creative Commons Attribution (CC BY) license (https://creativecommons.org/licenses/by/4.0/).

1. Introduction

Recently, the need to reduce the carbon emissions resulting from the use of fossil fuels makes it necessary to strongly develop renewable energies worldwide [1]. In 2015, COP 21 agreed to achieve a balance between carbon emissions and removals by sinks of greenhouse gases [2]. In the case of wind energy, the size of wind turbine is continuously increased to provide more energy per unit. In the early 1980s, the length of the blade was 7.5 m; today it is 80 m, and could exceed 100 m in 2030 [3,4]. To ensure a continuous and powerful wind flow, the turbines are designed to withstand a harsher environment, and are therefore located further away from the coast. The lifetime of these structures should also be improved, with a view to less systematic maintenance. Nowadays, composites materials are essential for the manufacture of wind turbines blades, usually made of glass fibers assembled with an epoxy matrix. However, composites are anisotropic or orthotropic materials, and their failure is a complex combination of various mechanisms such as matrix cracking, delamination, fiber breakage, or interfacial debonding [5]. As a result, the initiation and propagation of damage remains difficult to predict. Detection of damage along with undergone strain could lead to better prediction of the ultimate failure of the structure, thereby reducing maintenance costs. For instance, of the current total cost of *onshore* wind turbines, which amounts to more than 1 M€/MW, about 10% is due to maintenance [6], and the shift to offshore plants could increase this cost to nearly 30% [7,8].

Within this frame, a Structural Health Monitoring system (SHM), which would provide clues on the state of health of the material, comes out like a very interesting tool [9–12]. SHM systems have multiple objectives, such as enabling optimal use of the structure, minimizing downtime and avoiding catastrophic failures. They should also help to replace scheduled and periodic maintenance with performance-based inspections; as well as reduce human intervention for less labor, fewer human errors, and thus greater reliability.

Over the past decades, various conventional techniques have been developed to detect and measure the deformation or initiation of cracks and their propagation through the structure being monitored. These techniques involve primarily the use of metal strain gauges, piezo-resistive sensors, optical fiber sensors, and acoustic sensors. In a complex structure such as a wind turbine blade, the maximum stresses are localized in defined areas, due to the shape of the structure [13]. Structural monitoring of these areas is critical to detect the degradation in the state of the structure. Therefore, the SHM device should provide details of the in situ state in order to prevent failure of the structure, since some parts of the structure, such as the spar cap in the case of windmills, may be several centimeters thick [14]. Ideally the location of strain and damage in the selected area would be done by volume mapping. To date, few of the existing monitoring techniques proved capable to perform internal and external measurements of strain and damage detection. A recent study by Holmes et al. [15] has addressed this issue based on fiber optic technology, specifically machined to reduce their thickness and providing through-thickness strain monitoring of advanced composites. However, the handling of such sensors remains tricky.

This challenging task has recently attracted the interest of the research community, especially since the discovery of new nanoparticles such as carbon nanotubes [16,17]. As a result, conductive nanocomposites obtained by dispersing these nanofillers in a polymer matrix have led to the development of promising self-sensing materials [18–27]. Thanks to their unique mechanical [25,28–30], thermal [31–33] and electrical properties [25,34,35], carbon nanotubes (CNT) are interesting candidates for the development of nanocomposites. Since the nanofillers can be structured into a 3D conductive network within the insulating polymer matrix, changes in the resistance of this network in response to mechanical solicitations can be used to detect and measure the deformation and possible damages to the composite [19,27,29,36–41]. Furthermore, the use of such smart materials leads to a perfect match between the durability of the host structure and the nanocomposite used to sense its state of health. Nevertheless, out of these different strategies to structure percolated networks into polymer matrices, one is particularly seducing, it consists to introduce locally a patch of matrix-CNT nanocomposite instead of dispersing CNT randomly in the whole sample matrix as frequently done. This local sensor could be deposited on the surface of the composite specimen by resin casting [42,43], spraying [18], or printing [44–47]. This smart microply can be inserted during processing, in the part's core. The sensor's sensitivity can be furthermore adjusted with the nanofiller content in the polymer according to the percolation law [19]. Using a CNT-filled epoxy nanocomposite, just above its percolation threshold around 0.5 wt%, allows to reach gauge factors (GF) as high as 78 [45], while at 0.7 wt% the GF decreases to 3.2 [43]. Such sensors typically have a thickness ranging from 1 µm [18] to 100 µm [43].

Michelis et al. [48] also proposed inkjet-printed CNT based strain sensors made directly on a polymer substrate with a GF of 0.98. A similar process has been used by Kaiyan et al. [49] with the addition of epoxy in the sensor, reaching a GF of 50. Dai et al. [50] chose a cost-effective process based on a non-woven aramid carrier fabric soaked in a CNT-filled solution to fabricate a sensor, achieving a linear response and an elastic gauge factor of 1.9.

Nevertheless, the sensors were located on the surface of samples, which avoided the in situ core measurements potentially affecting them by moisture and temperature. Another strategy developed by Feller et al. [19,29] is to embed in the composite during its processing a premade sensing patch obtained by spraying layer-by-layer (sLbL) CNT-epoxy solutions on a glass fiber textile.

In this case, the dimensions of the transducer as well as the electrode design can be fixed "on-demand", depending on the desired configuration in the host. These so-called quantum resistive sensors (QRS), can be made of the same resin as the composite and be cured to the same level of crosslinking degree to guaranty a complete homogeneity of thermal and mechanical properties. Conveniently, the resulting sensitivity to strain (GF) can be adjusted by changing the filler content or the number of sprayed layers [19]. Finally, QRS derive their great sensitivity to their environment, and in particular to nanodeformations from the percolated nature of their sensing architecture favoring tunneling conduction to the detriment of classical ohm conduction, which tend to produce an exponential variation of the resistive response upon an increase of the interparticular distance "d" as shown in Equation (1).

$$R_{tunnel} = \frac{V}{AJ} = \frac{h^2 d}{Ae^2\sqrt{2m\lambda}} \exp\left(\frac{4\pi d}{h}\sqrt{2m\lambda}\right) \quad (1)$$

where J is the tunneling current density; V is the electrical potential difference; A is the cross-sectional area of tunnel; e is the quantum of electricity; m is the mass of electron; h is Plank's constant; d is the distance between conductive particles; l is the height of energy barrier (for epoxy, 0.5–2.5 eV).

In this work, the use of CNT-epoxy Quantum Resistive Sensors (QRS), placed at different locations into the core of carbon-fiber reinforced polymer (CFRP) samples is reported, as illustrated in Figure 1. QRS are integrated during the laminate manufacturing process and are not limited by the shape or angle of the composite part. The evolution of the electrical behavior of several QRS during a mechanical deformation as a function of their location inside the structure is investigated (within a ply of the laminate, and between the plies, i. e. through thickness). The QRS network thus formed could be used from the elastic deformation to the final breakage of the composite and provide a location of the stress experienced in the structure.

Figure 1. Representation of three points bending experiments with Quantum Resistive Sensor (QRS) placed in the tensile zone.

2. Experimental Details

2.1. Materials

The Carbon fiber reinforced polymer (CFRP) prepreg was purchased from Hexcel. The carbon fiber is a taffetas fabric (0/90°, 300 g/m²) with 34% in weight in a M79 epoxy resin. Multiwall carbon nanotubes (NC-7000™) were kindly provided by Nanocyl® (Sambreville, Belgium). This grade corresponds to MWCNT with an average diameter of 10 nm and a mean length comprised between 100 and 1000 nm. Epolam 2020 epoxy resin & amine hardener were purchased from Axson Technologies (Saint-Ouen, France), glass fiber was purchased from Composites Distribution® (Orvault, France) and chloroform (99%) was purchased from Sigma-Aldrich (Saint-Quentin, France).

2.2. Manufacture of Samples

Calculated amounts of CNT and epoxy resin were homogenized in chloroform by ultra-sonication with a Branson 3510 sonicator for 90 min at 25 °C, and further degassed for 5 min. After dispersion of the CNTs in the epoxy resin, an amine hardener was added and the mixture was sonicated for 30 min. The obtained mixing contains 3% by weight of CNT in the epoxy. 7 mm × 3 mm thin-film sensors were obtained by the technique of spraying via layer-by-layer (sLbL) deposition and can be seen in Figure 2b. The electrical connections were made with silver-based ink. The solutions were sprayed with our homemade device onto the supporting glass fiber/epoxy film allowing a precise control of nozzle scanning speed (10 cm s^{-1}), solution flow rate (50 mm·sec^{-1}), stream pressure (0.20 MPa), and target to nozzle distance (8 cm). The QRS was then insulated with a layer of glass fiber/epoxy substrate. The QRS are embedded into the CFRP during the stacking sequence (16 plies). 4 locations are selected and their coordinates are related to the position of loading pin (X; Z), as illustrated in Figure 2a,c. A pair of QRSs aligned on the centreline of the loading pin (X0), under the first ply (X0; Z0) and in the mid-thickness of the CFRP (X0; Z-1/2), respectively. A second pair is located at the same depths, but offset by a quarter of the length of the sample, (X1/4; Z0) and (X1/4; Z-1/2) respectively. Curing is carried out by a 6 h heat treatment at 80 °C ensuring complete crosslinking and the resulting CFRP sample dimensions are 400 × 50 × 3.2 mm^3.

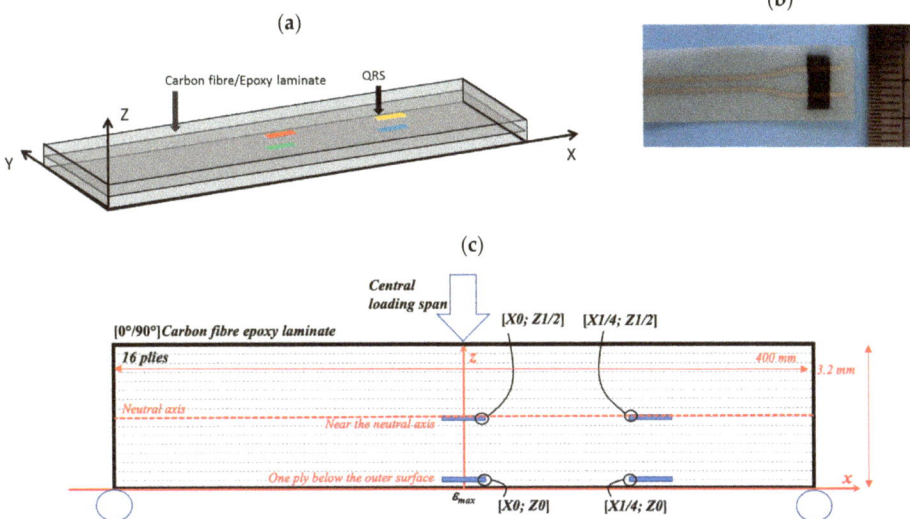

Figure 2. (a) Diagram of the CFRP specimen with embedded QRS below the surface and in the region of the neutral axis along with their name and position in the laminate. (b) Photography of a QRS, including the substrate (whitish area), the conductive silver-based tracks and the active CNT-filled QRS (dark area) (c) Representation of the CFRP specimen and the detailed location of the QRS during three-point bending experiments (QRS placed in the tension zone).

2.3. Characterization Techniques

An Instron 5566A servo-hydraulic was used to perform static three-point bending experiments, according to the standard NF EN ISO 14125. An Instron Electropuls E10000 was used to perform dynamic three-point bending experiments with the same apparatus as for the static tests.

The fracture profile was obtained by Scanning Electron Microscopy (SEM) using a Jeol series model (JSM-6031F) (JEOL Europe SAS, Croissy Sur Seine, France).

In situ electrical measurements were carried out by an HBM quantum 840 device with a direct current of 1 V applied to the QRS.

3. Results and Discussion

3.1. Mechanical Properties of the Sample with and without Embedded QRS

In order to study the effect of QRS on the mechanical performance of the CFRP sample, three-point static bending experiments were performed with and without QRS embedded in the sample core, as presented in Figure 3a. Both cases show similar behavior with first a linear increase in stress with strain up to 0.5% in the elastic range, followed by a slight reduction in the stress increase due to the initiation cracks at the 90° fibers interfaces and their propagation in the matrix until final failure around 1.4% [51]. The resulting mechanical properties, summarized in Table 1, are consistent with the literature data for the pristine 0/90° carbon fiber epoxy composite [52,53]. In addition, the introduction of QRS modifies the modulus, failure stress and failure strain by 1.2, 6.0 and 5.6% respectively, which are still within the measurement uncertainty.

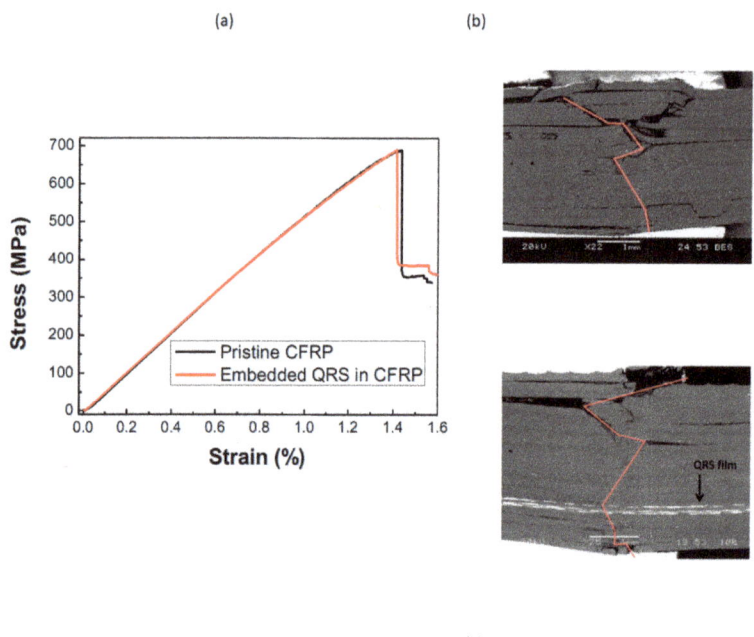

Figure 3. (a) Static three-point bending mechanical behavior of a CFRP composite with and without embedded QRS displaying equivalent behavior. Fracture profile under the central span (b) with and (c) without embedded QRS. The red line represents the fracture path.

Table 1. Mechanical properties of a CFRP composite (0/90°) with and without embedded QRS in static three-point bending, according to standard NF EN ISO 14125.

Type of Sample	Young Modulus (GPa)	Failure Stress (MPa)	Failure Strain (%)
Pristine CFRP	52.1 ± 1.8	669 ± 27	1.42 ± 0.06
CFRP with embedded QRS	51.5 ± 2.1	629 ± 61	1.34 ± 0.11

The regions where the rupture occurred were observed by the SEM as shown in Figure 3b,c. It was detected that the onset of failure occurred on both sides of the sample and propagated through the thickness while the delaminations are parallel to the plane of the fabric. This behavior is in line with that expected for the 0/90° composite under bending solicitation [53]. Samples with in-core QRS do not show additional delamination or failure compared to the blank samples, confirming that the mechanical properties are not affected by the introduction of QRS.

3.2. Electrical Response of QRS According to Their Location in the Specimen

QRS were incorporated at four different locations in the CFRP samples, as described earlier. Two were one ply below the surface, one below the central span and one offset 100 mm to the side of the sample. Two others QRS were placed at the same X-positions, but in depth, on the neutral axis of the sample. Because of these different locations, QRS are subject to different level of strain, as shown in the following equation:

$$\varepsilon = \frac{24\, f\, v\, x}{L^3} \qquad (2)$$

where ε is the strain, f is the deflection, v is the distance from the neutral axis, x is the position along the sample, and L is the length between the two external spans.

As already shown [19,32,37,54], QRS are displaying a change in electrical resistance with strain, which is characterized by the Gauge Factor (GF), defined as follows:

$$GF = \frac{Ar}{\varepsilon} = \frac{\Delta R/R_0}{\varepsilon} \qquad (3)$$

where ΔR is the change of electrical resistance compared to $\varepsilon = 0$, R_0 is the initial resistance at $\varepsilon = 0$, and Ar, the relative amplitude of the electrical resistance, i.e., the ratio $\Delta R/R_0$.

It has been previously estimated that for such embedded sensors with an initial resistance of about 20 kΩ, the GF is about 3 [19]. Therefore, in order to represent the effect of the sensor position, the electrical response has been normalized to the maximum value obtained for QRS (X0; Z0), which is assumed to be subject to the maximum strain and thus exhibit the highest response. Figure 3a represents the normalized amplitude of the electrical resistance of the four QRS, with the strain measured below the central span in a static three-points bending experiment, which is supposed to be the maximum strain value.

QRS placed a ply below the surface, e.g., QRS (X0, Z1/4), depicts a behavior similar to QRS (X0, Z0). An initial growth in resistance is visible up to 0.1%, followed by a steadier increase in resistance. The final value reached by the QRS (X0; Z0) is almost double that of the QRS (X1/4; Z0) one. This result is consistent with Equation (2) and Figure 4c,d, where the strain under the center span at the X1/2 position is about twice that of the position LX1/4. In addition, the stress undergone at the center QRS is higher, which may result in a noisier signal compared to others QRS. In the case of the QRS placed near the neutral axis, the resistance variation is almost equal to 0. This is consistent and corresponds to the fact that there is no deformation on the neutral axis. Focusing on the curves in Figure 4b. It can be seen that the QRS (X0; Z-1/2) is nevertheless subjected to an initial increase in resistance up to 0.3% followed by a slight decrease, while the QRS (X1/4; Z-1/2) displays a decrease in resistance throughout the bending test. This indicates their proximity to the neutral axis and the resulting nearly total absence of resulting strain. Thanks to the variation of the behavior as a function of the position, it demonstrates the ability of QRS to provide in situ measurements of the strain encountered, both along the sample and through the thickness. Although CFRP is an electrically conductive material, the absence of short circuits and the individual response of each QRS demonstrate the ability of the supporting film to isolate the QRS from the carbon fibers-reinforced composite.

In order to demonstrate the ability of the QRS to detect dynamic deformations, cycling tests were also performed in addition to the static test. Figure 5a shows the normalized change in QRS resistance with strain over 30 sinusoidal cycles. As expected and observed during the static test, although having the same gauge factor, one can see in Figure 5b that the amplitude of their electrical signal relates to their position in the composite. Indeed, the QRS positioned on the side (X1/4) displays a smaller change in resistance with the strain than the one placed under the center span (X0) and, near the neutral axis, almost no change in resistance is visible. The absence of hysteresis and the ability of QRS to follow the sinusoidal cycles confirm their possible use for dynamic in situ measurements. To further investigate the sensing capability of QRS in tension and compression, Figure 5c,d presents the dynamic three-points bending test of a CFRP specimen equipped with QRS located one ply below the surface, i.e., QRS (X0; Z0), QRS (X0; Z1/4) in the tensile side and QRS (X0; Z-1), QRS (X1/4; Z-1) in the compression side. The strain ranged from 0 to 0.1%. QRS displayed two symmetrically opposite electrical behaviors in tension and compression with a maximum normalized Ar of 0.089 and −0.085 for traction and compression. respectively. As observed earlier in static tension mode, when the sample is moved from the position X0 to the X1/4 position, the electrical response is two times less, which is consistent with Equation (2) and Figure 4c,d. As a result, QRS showed three distinct electrical behaviors, whether in traction, compression or on the neutral axis. They can therefore be used to measure the deformations experienced by the laminate composite, both on the surface and in the core of the part. The use of multiple QRS could be envisaged to map selected areas of the composite part in three dimensions.

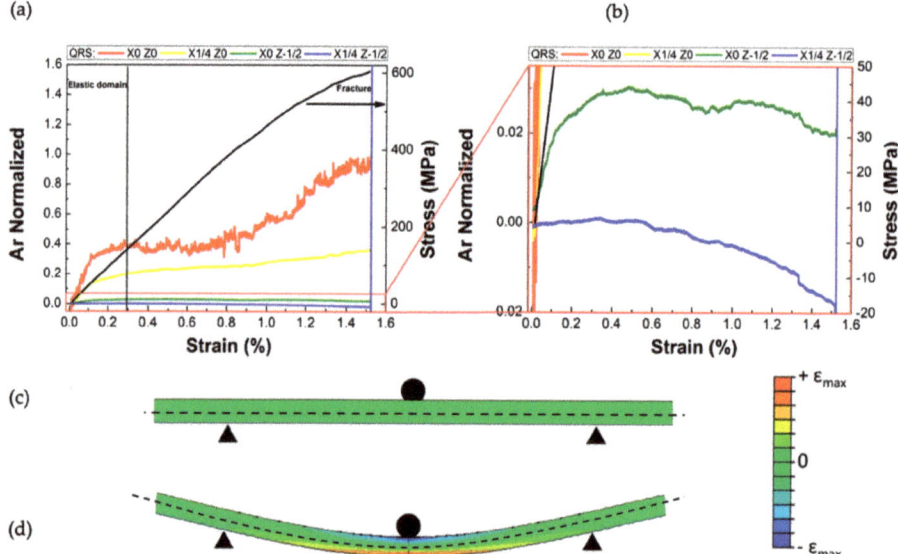

Figure 4. Electrical behavior of four QRS located at four different spots in the CFRP sample, below the central span (identification X0), or in the middle along the side of the sample (X1/4), and one ply below the surface (Z0) or close to the neutral axis (Z-1/2). (**a**) Normalized Ar of the four QRS as a function of the maximum strain below the central span during static three-points bending test. (**b**) Magnification of curves for (X0; Z-1/2) and (X1/4; Z-1/2). Modelling of the strain field during the three-points bending experiment at (**c**) initial state and (**d**) during bending.

To verify the reversibility of the QRS response, specimens were submitted to 300 consecutive cycles in the elastic domain and above the elastic limit. Figure 6a represents the strain imposed over time. Increasing successive stages are visible, with one below the elastic limit (average strain $\varepsilon_1 = 0.15\%$), one close to the elastic limit ($\varepsilon_2 = 0.30\%$), one above the elastic limit ($\varepsilon_3 = 0.80\%$) and on close to the final breakage ($\varepsilon_4 = 1.13\%$). These steps are interspersed with periods of solicitation referenced (labelled ε_1, ε_{1-2}, ε_{1-3}, ε_{1-4}) to a strain of $\varepsilon_1 = 0.15\%$. The range of deformation variation around the mean values during the step as was chosen at 0.06% to avoid any additional defects formation during the complete cycling test.

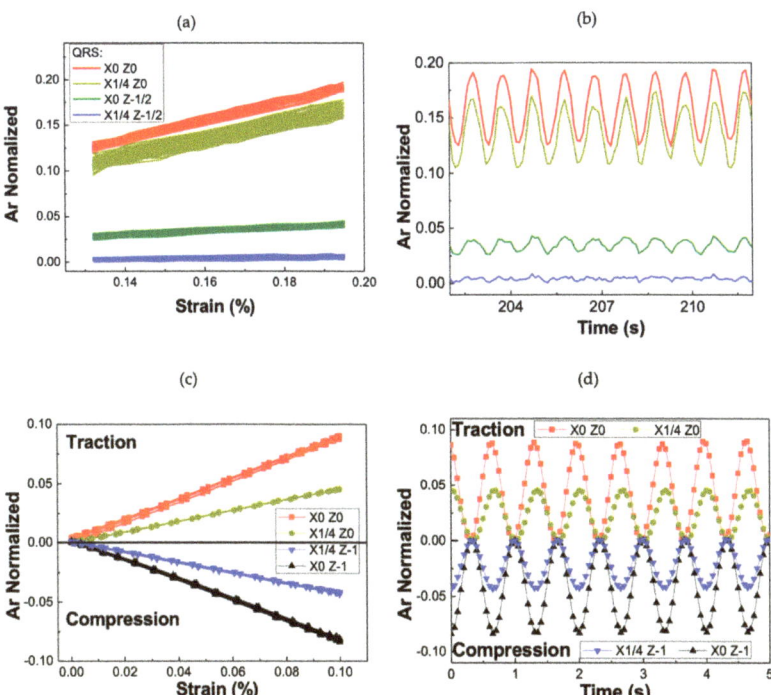

Figure 5. Electrical behavior of four QRS located at four different locations in the CFRP specimen, below the center span (identification X0), or in the middle along the side of the specimen (X1/4), and one ply below the surface (Z0) or near the neutral axis (Z-1.2). (**a**) Normalized electrical resistance (Ar) of the four QRS with maximum strain during dynamic elastic three-points bending test. (**b**) Representation of their electrical response with time according to the deflection of the beam. (**c**) Electrical behavior of the QRS located on the compression side, (X0; Z-1) and (X1/4; Z-1), or the tension side, (X0; Z0) and (X1/4; Z0), during the dynamic three-points bending test. (**d**) Representation of their electrical response associated with time for a given deflection.

Figure 6b shows the average value of the QRS response during the increasing stages of the cycle and the scale bars correspond to the total amplitude of the signal. QRS have a behavior equivalent to that of static deformation where the increase in Ar depends on the location of the QRS. The average increase in Ar and the increase in the amplitude of Ar moving on the one hand from the neutral axis to the tensile side, and on the other hand from the side to the underside of the central span. With the exception of QRS (X1/4; Z0) in the last step (ε_4), the amplitude of Ar is also constant for all steps, as shown in Table 2. This indicates that once a level of strain above the elastic limit is obtained, small mechanical solicitations do not deteriorate the specimen and the resulting QRS response. In addition, the modified behavior displayed by the QRS (X1/4; Z0) in the last step can be associated with the increase in nearby defects and in the QRS. The QRS can therefore attest to the local degradation of the material.

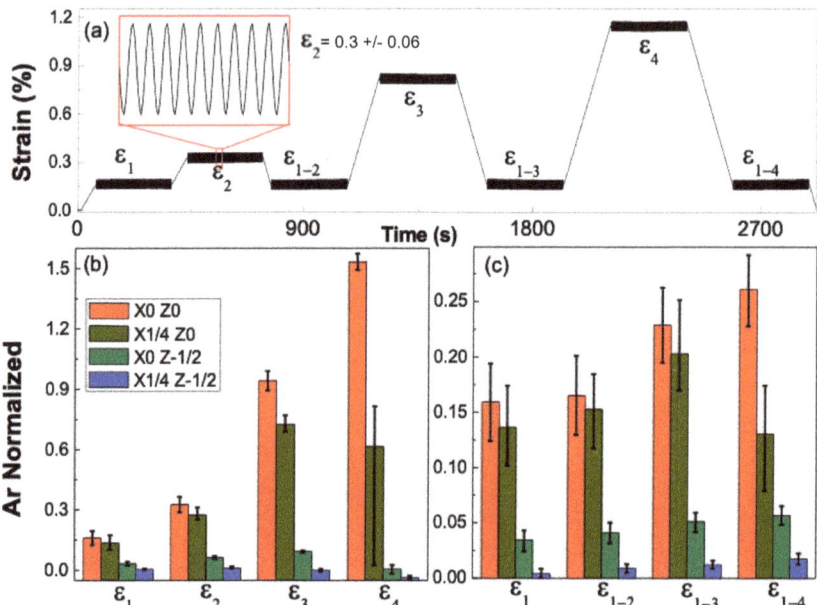

Figure 6. Dynamic deformation of a CFRP specimen with four QRS located below the center span (identification X0), or in the middle along the side of the specimen (X1/4), and one ply below the surface (Z0) or near the neutral axis (Z-1/2). (a) Four stages of increasing strain applied to the specimen and the reference strain periods/plateaux. The inset represents a magnification of the sinusoidal load applied to the sample during the test. In each sections, the sample is subjected to a dynamic sinusoidal bending load of a small amplitude (+/− 0.06) around a fixed value (ε_1, ε_2 etc.). (b) Average value of the relative electrical amplitude of QRS in the four increasing steps. The scale bars represent the amplitude of the dynamic response. (c) Average value of the relative amplitudes of the QRS during reference periods of low strain. The scale bars represent the amplitude of the dynamic response to a small variation of +/− 0.06% around a fixed ε_i value.

Table 2. QRS response amplitude during the cycling steps.

Step	(X0; Z0)	(X1/4; Z0)	(X0; Z-1/2)	(X1/4; Z-1/2)
ε_1	0.070	0.072	0.019	0.008
ε_2	0.075	0.072	0.015	0.011
ε_{1-2}	0.072	0.067	0.019	0.008
ε_3	0.096	0.082	0.012	0.014
ε_{1-3}	0.067	0.082	0.017	0.007
ε_4	0.080	0.79	0.044	0.015
ε_{1-4}	0.064	0.095	0.017	0.010

Figure 6c shows the QRS response during the steps of the reference cycle and the scale bars correspond to the total amplitude of the signal. For all QRSs, the first two reference steps (ε_1, ε_{1-2}) give equal responses. For consecutive reference steps (ε_{1-3}, ε_{1-4}), a residual Ar is visible for all QRS and is greater on the tensile side and under the center span. Chowdhury et al. [19] reported a fairly similar electrical behavior in the case of increasing static tensile cycles. Residual strain is due to the formation and propagation of cracks in the material. Once the elastic limit of the sample is exceeded, the defects and cracks created are mainly located in the region below the centre span. Thus, the residual Ar is higher for the QRS (X0; Z0) because it is located in the most damaged region. Therefore, the Ar drift is an indicator of the amount of core damage and it is conceivable that the use of a discrete number of QRS can be used to locate the main affected area in the specimen.

4. Conclusions

The integration of an array of four Quantum Resistive Sensors was carried out for the first time to map the deformation gradient inside epoxy-carbon fiber laminate composite samples, i.e., in plane and through thickness. It was also demonstrated that the presence of QRS did not affect the mechanical behavior and properties of the CFRP samples. The gauge factor of QRS was selected around 3, being a compromise between sensitivity and noise ratio. Although higher gauge factor can be found in the literature, this sensitivity was found sufficient to monitor very small deformation in the core of the composite (0.06%). The integration inside a carbon fiber reinforced composite was made possible thanks to insulation of the sensors which opens applications in the high-performance composites area previously reserved for optical fibers. QRS were tested in both static and dynamic experiments. The robustness of sensors to follow sinusoidal solicitation at different amplitude was demonstrated.

In addition, the electrical response of QRS, conditioned by their location in the samples, allowed to determine the local deformation. A positive response was observed on the side of the composite sample submitted to tension, whereas the sensors located near the neutral axis exhibited almost no response. For the two sensors located in the same ply, a lower response was found for the QRS located away from the loading span being consistent with the beam theory. The precise spatial location of sensors could thus be used to map the deformation gradient and eventually feed a simulation software.

A part from deformation measurements, the sensors proved to be able to follow damage up to failure in static mode. In dynamic mode, a transition in behavior was observed through the variation of sensitivity of sensors over a limit of 0.8%, assumed to be the elastic limit, suggesting that sensors could be used to track damage accumulation.

Further investigations will concern the implementation of a larger number of QRS in 3D in a laminate composite part in order to map the strain field in selected area. Other modes of solicitation such as compression, shear or impact, as well as long-term experimentation (fatigue) should be studied in order to broaden the fields of application and specially to get closer to the real environmental conditions of use

This could be of great interest to monitor the health of composite in the energy and transport fields, such as turbine blades or boat foils.

Author Contributions: Conceptualization, M.C., J.-F.F.; methodology, A.L., M.C., J.-F.F.; validation, A.L., M.C., J.-F.F., O.F.; formal analysis, A.L., M.C., J.-F.F.; investigation, A.L., M.C., J.-F.F.; resources M.C., J.-F.F.; writing—original draft preparation, A.L.; writing—review and editing, A.L., M.C., J.-F.F.; visualization, A.L., M.C., J.-F.F.; supervision, M.C.; project administration, O.F., J.-C.D.L., J.-F.F.; funding acquisition, J.-C.D.L., J.-F.F. All authors have read and agreed to the published version of the manuscript.

Funding: EVEREST Project (funded by IRT Jules Verne and G.E.).

Acknowledgments: The authors would like to specially thank Hervé Bellegou for developing the electrical acquisition system.

Conflicts of Interest: The authors declare no conflict of interest.

References

1. Burck, J.; Marten, F.; Bals, C.; Dertinger, A.; Uhlich, T. Climate Change Performance Index Results 2017. Available online: https://germanwatch.org/en/13042 (accessed on 19 February 2021).
2. UNFCCC. Adoption of the Paris Agreement: Proposal by the President to the United Nations Framework Convention on Climate Change. 2015. Available online: https://unfccc.int/documents/9064 (accessed on 19 February 2021).
3. Islam, M.; Mekhilef, S.; Saidur, R. Progress and recent trends of wind energy technology. *Renew. Sustain. Energy Rev.* **2013**, *21*, 456–468. [CrossRef]
4. Kumar, Y.; Ringenberg, J.; Depuru, S.S.; Devabhaktuni, V.K.; Lee, J.W.; Nikolaidis, E.; Andersen, B.; A Afjeh, A. Wind energy: Trends and enabling technologies. *Renew. Sustain. Energy Rev.* **2016**, *53*, 209–224. [CrossRef]
5. Jones, R.M. *Mechanics of Composite Materials*; Taylor & Francis: New York, NY, USA, 1998.

6. Syndicat des Energies Renouvelables, Etat des Coûts de Production de L'éolien Terrestre en France Analyse Economique de la Commission Eolienne du SER. 2014. Available online: https://www.syndicat-energies-renouvelables.fr/les-energies-renouvelables/eolien/eolien-terrestre/ (accessed on 19 February 2021).
7. The Tide Turns on Offshore Maintenance Costs. Windpower Offshore. Available online: https://www.windpowermonthly.com/article/1314299/tide-turns-offshore-maintenance-costs (accessed on 19 February 2021).
8. *Floating Offshore Wind Farms*; Springer International Publishing: Cham, Switzerland, 2016.
9. Schubel, P.; Crossley, R.; Boateng, E.; Hutchinson, J. Review of structural health and cure monitoring techniques for large wind turbine blades. *Renew. Energy* **2013**, *51*, 113–123. [CrossRef]
10. Structural Health Monitoring Person of the Year Award. *Struct. Heal. Monit.* **2008**, *7*, 89–90. [CrossRef]
11. Huston, D. *Structural Sensing, Health Monitoring and Performance Evaluation*; Series in Sensors; CRC Press: Boca Raton, FL, USA, 2010.
12. Farrar, C.R.; Worden, K. An introduction to structural health monitoring. *Philos. Trans. R. Soc. A Math. Phys. Eng. Sci.* **2006**, *365*, 303–315. [CrossRef]
13. Raman, V.; Drissi-Habti, M.; Guillaumat, L.; Khadhour, A. Numerical simulation analysis as a tool to identify areas of weakness in a turbine wind-blade and solutions for their reinforcement. *Compos. Part B Eng.* **2016**, *103*, 23–39. [CrossRef]
14. Resor, B.R. Definition of a 5 MW/61.5 m Wind Turbine Blade Reference Model. 2013. Available online: https://prod.sandia.gov/techlib-noauth/access-control.cgi/2013/132569.pdf (accessed on 19 February 2021).
15. Holmes, C.; Godfrey, M.; Bull, D.J.; Dulieu-Barton, J. Real-time through-thickness and in-plane strain measurement in carbon fibre reinforced polymer composites using planar optical Bragg gratings. *Opt. Lasers Eng.* **2020**, *133*, 106111. [CrossRef]
16. Iijima, S. Helical microtubules of graphitic carbon. *Nat. Cell Biol.* **1991**, *354*, 56–58. [CrossRef]
17. Radushkevich, L.V.; Lukyanovich, V.M. About the structure of carbon formed by thermal decomposition of carbon monoxide on iron substrate. *J. Phys. Chem.* **1952**, *26*, 88–95.
18. Robert, C.; Feller, J.F.; Castro, M. Sensing Skin for Strain Monitoring Made of PC–CNT Conductive Polymer Nanocomposite Sprayed Layer by Layer. *ACS Appl. Mater. Interfaces* **2012**, *4*, 3508–3516. [CrossRef]
19. Nag-Chowdhury, S.; Bellegou, H.; Pillin, I.; Castro, M.; Longrais, P.; Feller, J. Non-intrusive health monitoring of infused composites with embedded carbon quantum piezo-resistive sensors. *Compos. Sci. Technol.* **2016**, *123*, 286–294. [CrossRef]
20. Gao, L.; Thostenson, E.T.; Zhang, Z.; Chou, T.-W. Coupled carbon nanotube network and acoustic emission monitoring for sensing of damage development in composites. *Carbon* **2009**, *47*, 1381–1388. [CrossRef]
21. Gao, L.; Thostenson, E.T.; Zhang, Z.; Chou, T.-W. Sensing of Damage Mechanisms in Fiber-Reinforced Composites under Cyclic Loading using Carbon Nanotubes. *Adv. Funct. Mater.* **2009**, *19*, 123–130. [CrossRef]
22. Wichmann, M.H.G.; Buschhorn, S.T.; Gehrmann, J.; Schulte, K. Piezoresistive response of epoxy composites with carbon nanoparticles under tensile load. *Phys. Rev. B* **2009**, *80*, 245437. [CrossRef]
23. Böger, L.; Wichmann, M.H.; Meyer, L.O.; Schulte, K. Load and health monitoring in glass fibre reinforced composites with an electrically conductive nanocomposite epoxy matrix. *Compos. Sci. Technol.* **2008**, *68*, 1886–1894. [CrossRef]
24. Sebastian, J.; Schehl, N.; Bouchard, M.; Boehle, M.; Li, L.; Lagounov, A.; Lafdi, K. Health monitoring of structural composites with embedded carbon nanotube coated glass fiber sensors. *Carbon* **2014**, *66*, 191–200. [CrossRef]
25. Gojny, F.H.; Wichmann, M.H.; Fiedler, B.; Bauhofer, W.; Schulte, K. Influence of nano-modification on the mechanical and electrical properties of conventional fibre-reinforced composites. *Compos. Part A Appl. Sci. Manuf.* **2005**, *36*, 1525–1535. [CrossRef]
26. Tripathi, K.M.; Vincent, F.; Castro, M.; Feller, J.F. Flax fibers–epoxy with embedded nanocomposite sensors to design lightweight smart bio-composites. *Nanocomposites* **2016**, *2*, 125–134. [CrossRef]
27. Chowdhury, S.N.; Tung, T.T.; Ta, Q.T.H.; Kumar, G.; Castro, M.; Feller, J.-F.; Sonkar, S.K.; Tripathi, K.M. Upgrading of Diesel Engine Exhaust Waste into Onion-like Carbon Nanoparticles for Integrated Degradation Sensing in Nano-biocomposites. *New J. Chem.* **2021**. [CrossRef]
28. Treacy, M.M.J.; Ebbesen, T.W.; Gibson, J.M. Exceptionally high Young's modulus observed for individual carbon nanotubes. *Nat. Cell Biol.* **1996**, *381*, 678–680. [CrossRef]
29. Pillin, I.; Castro, M.; Chowdhury, S.N.; Feller, J.-F. Robustness of carbon nanotube-based sensor to probe composites' interfacial damage in situ. *J. Compos. Mater.* **2015**, *50*, 109–113. [CrossRef]
30. Robert, C.; Pillin, I.; Castro, M.; Feller, J.-F. Multifunctional Carbon Nanotubes Enhanced Structural Composites with Improved Toughness and Damage Monitoring. *J. Compos. Sci.* **2019**, *3*, 109. [CrossRef]
31. Fidelus, J.; Wiesel, E.; Gojny, F.; Schulte, K.; Wagner, H. Thermo-mechanical properties of randomly oriented carbon/epoxy nanocomposites. *Compos. Part A Appl. Sci. Manuf.* **2005**, *36*, 1555–1561. [CrossRef]
32. Gojny, F.H.; Wichmann, M.H.; Fiedler, B.; Kinloch, I.A.; Bauhofer, W.; Windle, A.H.; Schulte, K. Evaluation and identification of electrical and thermal conduction mechanisms in carbon nanotube/epoxy composites. *Polymer* **2006**, *47*, 2036–2045. [CrossRef]
33. Feller, J.-F. 6.10 Electrically Conductive Nanocomposites. In *Comprehensive Composite Materials II*; Elsevier BV: Amsterdam, The Netherlands, 2018; pp. 248–314.
34. Ebbesen, T.W.; Lezec, H.J.; Hiura, H.; Bennett, J.W.; Ghaemi, H.F.; Thio, T. Electrical conductivity of individual carbon nanotubes. *Nat. Cell Biol.* **1996**, *382*, 54–56. [CrossRef]
35. Feller, J.-F.; Kumar, B.; Castro, M. Conductive biopolymer nanocomposites for sensors. *Nanocompos. Biodegrad. Polym.* **2011**, 368–399. [CrossRef]

36. Levin, Z.S.; Robert, C.; Feller, J.F.; Castro, M.; Grunlan, J.C. Flexible latex—polyaniline segregated network composite coating capable of measuring large strain on epoxy. *Smart Mater. Struct.* **2012**, *22*, 1–9. [CrossRef]
37. Thostenson, E.T.; Chou, T.-W. Real-timein situsensing of damage evolution in advanced fiber composites using carbon nanotube networks. *Nanotechnology* **2008**, *19*, 215713. [CrossRef] [PubMed]
38. Tzounis, L.; Zappalorto, M.; Panozzo, F.; Tsirka, K.; Maragoni, L.; Paipetis, A.S.; Quaresimin, M. Highly conductive ultra-sensitive SWCNT-coated glass fiber reinforcements for laminate composites structural health monitoring. *Compos. Part B Eng.* **2019**, *169*, 37–44. [CrossRef]
39. Wang, Y.; Wang, Y.; Wan, B.; Han, B.; Cai, G.; Chang, R. Strain and damage self-sensing of basalt fiber reinforced polymer laminates fabricated with carbon nanofibers/epoxy composites under tension. *Compos. Part A Appl. Sci. Manuf.* **2018**, *113*, 40–52. [CrossRef]
40. Esmaeili, A.; Sbarufatti, C.; Ma, D.; Manes, A.; Jiménez-Suárez, A.; Ureña, A.; Dellasega, D.; Hamouda, A. Strain and crack growth sensing capability of SWCNT reinforced epoxy in tensile and mode I fracture tests. *Compos. Sci. Technol.* **2020**, *186*, 107918. [CrossRef]
41. Tung, T.T.; Karunagaran, R.; Tran, D.N.H.; Gao, B.; Nag-Chowdhury, S.; Pillin, I.; Castro, M.; Feller, J.-F.; Losic, D. Engineering of graphene/epoxy nanocomposites with improved distribution of graphene nanosheets for advanced piezo-resistive mechanical sensing. *J. Mater. Chem. C* **2016**, *4*, 3422–3430. [CrossRef]
42. Chiacchiarelli, L.M.; Rallini, M.; Monti, M.; Puglia, D.; Kenny, J.M.; Torre, L. The role of irreversible and reversible phenomena in the piezoresistive behavior of graphene epoxy nanocomposites applied to structural health monitoring. *Compos. Sci. Technol.* **2013**, *80*, 73–79. [CrossRef]
43. Meeuw, H.; Viets, C.; Liebig, W.; Schulte, K.; Fiedler, B. Morphological influence of carbon nanofillers on the piezoresistive response of carbon nanoparticle/epoxy composites under mechanical load. *Eur. Polym. J.* **2016**, *85*, 198–210. [CrossRef]
44. Kanoun, O.; Benchirouf, A.; Sanli, A.; Bouhamed, A.; Bu, L. Potential of Flexible Carbon Nanotube Films for High Performance Strain and Pressure Sensors, One Central Press 148–183. Available online: https://www.researchgate.net/publication/281438116_Potential_of_Flexible_Carbon_Nanotube_Films_for_High_Performance_Strain_and_Pressure_Sensors (accessed on 19 February 2021).
45. Sanli, A.; Benchirouf, A.; Müller, C.; Kanoun, O. Piezoresistive performance characterization of strain sensitive multi-walled carbon nanotube-epoxy nanocomposites. *Sens. Actuators A Phys.* **2017**, *254*, 61–68. [CrossRef]
46. Bouhamed, A.; Müller, C.; Choura, S.; Kanoun, O. Processing and characterization of MWCNTs/epoxy nanocomposites thin films for strain sensing applications. *Sens. Actuators A Phys.* **2017**, *257*, 65–72. [CrossRef]
47. Sanli, A.; Müller, C.; Kanoun, O.; Elibol, C.; Wagner, M.F.-X. Piezoresistive characterization of multi-walled carbon nanotube-epoxy based flexible strain sensitive films by impedance spectroscopy. *Compos. Sci. Technol.* **2016**, *122*, 18–26. [CrossRef]
48. Michelis, F.; Bodelot, L.; Bonnassieux, Y.; Lebental, B. Highly reproducible, hysteresis-free, flexible strain sensors by inkjet printing of carbon nanotubes. *Carbon* **2015**, *95*, 1020–1026. [CrossRef]
49. Kaiyan, H.; Weifeng, Y.; Shuying, T.; Haidong, L. A fabrication process to make CNT/EP composite strain sensors. *High Perform. Polym.* **2017**, *30*, 224–229. [CrossRef]
50. Dai, H.; Thostenson, E.T.; Schumacher, T. Processing and Characterization of a Novel Distributed Strain Sensor Using Carbon Nanotube-Based Nonwoven Composites. *Sensors* **2015**, *15*, 17728–17747. [CrossRef]
51. Chou, T.-W. *Microstructural Design of Fiber Composites*; Cambridge University Press: Cambridge, UK, 1992.
52. Corum, J. Basic Properties of Reference Crossply Carbon-Fiber Composite. 2001. Available online: https://www.osti.gov/biblio/777662-basic-properties-reference-crossply-carbon-fiber-composite (accessed on 19 February 2021).
53. Morioka, K.; Tomita, Y. Effect of lay-up sequences on mechanical properties and fracture behavior of CFRP laminate composites. *Mater. Charact.* **2000**, *45*, 125–136. [CrossRef]
54. Hu, N.; Karube, Y.; Arai, M.; Watanabe, T.; Yan, C.; Li, Y.; Liu, Y.; Fukunaga, H. Investigation on sensitivity of a polymer/carbon nanotube composite strain sensor. *Carbon* **2010**, *48*, 680–687. [CrossRef]

Article

Thermal Shock Behavior of Twill Woven Carbon Fiber Reinforced Polymer Composites

Farzin Azimpour-Shishevan [1], Hamit Akbulut [2] and M.A. Mohtadi-Bonab [3,*]

[1] Department of Mechanical Engineering, Faculty of Maragheh, Maragheh Branch, Technical and Vocational, Tehran, Iran; fazimpoor60@gmail.com
[2] Department of Mechanical Engineering, Ataturk University, Turkey University, 25030 Erzurum, Turkey; akbuluth@atauni.edu.tr
[3] Department of Mechanical Engineering, University of Bonab, Bonab, Iran
* Correspondence: m.mohtadi@ubonab.ac.ir

Abstract: In the current research, the effect of cyclic temperature variation on the mechanical and thermal properties of woven carbon-fiber-reinforced polymer (CFRP) composites was investigated. To this, carbon fiber textiles in twill 2/2 pattern were used as reinforced phase in epoxy, and CFRPs were fabricated by vacuum-assisted resin-infusion molding (VARIM) method. Thermal cycling process was carried out between −40 and +120 °C for 20, 40, 60 and 80 cycles, in order to evaluate the effect of thermal cycling on mechanical and thermal of CFRP specimens. In this regard, tensile, bending and short beam shear (SBS) experiments were carried out, to obtain modulus of elasticity, tensile strength, flexural modulus, flexural strength and inter-laminar shear strength (ILSS) at room temperature (RT), and then thermal treated composites were compared. A dynamic mechanical analysis (DMA) test was carried out to obtain thermal properties, and viscoelastic properties, such as storage modulus (E'), loss modulus (E") and loss factors (tan δ), were evaluated. It was observed that the characteristics of composites were affected by thermal cycling due to post-curing at a high temperature. This process worked to crosslink and improve the composite behavior or degrade it due to the different coefficients of thermal expansion (CTEs) of composite components. The response of composites to the thermal cycling process was determined by the interaction of these phenomena. Based on SEM observations, the delamination, fiber pull-out and bundle breakage were the dominant fracture modes in tensile-tested specimens.

Keywords: carbon fiber; epoxy matrix; thermal cycling; mechanical and thermal properties

1. Introduction

Carbon-fiber-reinforced composites are widely used in aeronautic applications, such as spacecraft structures, because of their superior mechanical properties, such as specific modulus of elasticity, high mechanical properties and low density of carbon fibers [1,2]. These components are exposed to various environmental conditions, such as cyclic thermal variation, and are degraded over time. Frequently, temperature variation between two degrees is called thermal shock cycling [3]. The specimens which are at an ambient temperature of 20 °C are placed in a heat oven whose temperature is close to 120 °C. In this process, the temperature of the specimen is gradually increased for 10 min. It rises and eventually reaches 120 °C. After 10 min, the specimens are taken out and placed in an industrial freezer with a temperature of −40 °C. In this process, the temperature of tested specimens gradually decreases and finally reaches −40 °C. A cycle of heat treatment of specimens consists of two processes for 10 min of gradual heating to 120 ° C and immediate cooling to −40 °C.

Thermal-shock cycling process produces a large thermal gradient in composite structures [4]. During the quenching process, the specimen is cooled from elevated to cryogenic temperatures, generally leading to tensile thermal stress at the surface of specimen and compressive stress at the interior sections [5]. The selection of appropriate materials in the structure of composite material is very important, since the nature of thermal stresses is from compressive type in the fibers and from tensile type in the matrix [6]. Different coefficients of thermal expansion (CTEs) of composite components may develop different strain and mismatch in the structure of composite [7–9]. The interface of fibers/matrix is considered to be the critical region, since the stress concentration is high [10]. When the temperature changes are repeated, frequently, mismatch of strain in the structure of composite is repeated too, and the stress concentration in the interface intensifies because of reducing the stress transition between matrix and reinforcement phase [11]. This phenomenon nucleates microcracks in the interface and initiation of delamination in this region. Delamination is one of the inherent weaknesses of laminated composites degrading their structure. On the other hand, thermal residual stress severely reduces the load-tolerating capacity in composites [12]. There are a number of research studies about the effect of thermal-shock cycling on mechanical and thermal properties of composite materials. For instance, Ray [4] studied the effect of thermal shock on interlaminar strength of glass-fiber-reinforced epoxy composites and concluded that the debonding effect of thermal cycling is not noticeable at low temperatures, and there is a possibility to improve the inter-laminar shear strength (ILSS) due to dominating post-curing phenomenon. Russell-Stevens et al. [13] evaluated the effect of thermal cycling on the properties of carbon-fiber-reinforced magnesium composites. The results of their studies showed that there was no change in the bending strength of these composites during thermal cycling. Azimpour et al. [14] studied the effects of thermal cycling on mechanical and thermal properties of basalt-fiber-reinforced polymer composites. In their studies, basalt fiber textiles have been used to fabricate BFRPs, and fabricated composites have been exposed to thermal cycling. Based on the findings of these authors, the mechanical properties of BFRPs were firstly improved; however, after 40 cycles, cracks nucleated and delamination occurred due to mismatch in CTE of fibers and formation of matrix microcracks. Wang et al. [15] investigated the effect of thermal cycling on the properties of paraffin/expanded graphite composites. They also added a suitable proportion of expanded graphite (EG) into the paraffin, to enhance the heat-transfer rate and improve the thermal efficiency of the whole heat-storage system. The effect of space environment temperatures on specific carbon fiber/bismaleimide composite laminates has been evaluated by Yang et al. [16]. They applied the simulated environment test method and performed experiments at −120 °C, room temperature, 150 °C, 170 °C and 200 °C. In this way, the material properties were characterized by using mechanical tests, including tensile, compressive, and in-plane shear. According to the results of research, matrix fracture and delamination was more likely to happen at high temperatures, while the interface strength was higher at low temperatures. Shin et al. [17] investigated the behavior of carbon-fiber-reinforced epoxy composites under thermal cycling between −70 and +100 °C. According to the results, thermal cycling decreased the strength and stiffness of the composites. Furthermore, the transverse flexural strength and stiffness of composites were affected more than other properties. In this research, the effect of thermal cycling on mechanical and thermal properties of woven carbon-fiber-reinforced epoxy composites was investigated. These types of composites are used in aeronautical applications in which the instruments are exposed to continuously changing of temperature. This condition may affect the mechanical properties of used composites, causing probable catastrophic events. Therefore, the evaluation of these processes is necessary. Few researches have been carried out on evaluation of these processes in the twill-carbon-fiber-reinforced epoxy composite materials. The current research tried to experimentally evaluate the effects of continuously changing of temperature in thermal testing laboratory according to in service conditions of the composites.

The main objective of this research was to investigate the effect of cyclic temperature variation on mechanical and thermal properties of woven carbon-fiber-reinforced polymer (CFRP). Moreover, modulus of elasticity, tensile strength, flexural modulus, flexural strength and inter-laminar shear strength (ILSS) were obtained by using tensile, bending and short beam shear (SBS) experiments at room temperature. Finally, thermal properties and viscoelastic properties were evaluated by using the dynamic mechanical analysis (DMA) test.

2. Experimental Procedure

2.1. Materials and Specimen Preparation

Used CFRP textiles were in a 2/2 twill pattern and were purchased from Spinteks Textile Co. Denizli, Turkey. Homogenous CFRP composites were fabricated by using VARIM method (Figure 1a). In this type of pattern, fibers were weaved in 0° and 90° angles (Figure 1b) and identified by its diagonal parallel lines known as wales. Designated 2/2 numerator indicated threads of fibers as warp and weft, one bundle as weft thread passing over two bundles as warp threads and then under two other threads. Areal mass of carbon fabrics was 212 gr/m^2 ± 5%with a thickness of 0.25 mm. Fabricated composite plates were 700 × 700 cm^2 and were cut with water jet, according to ASTM standards of reinforced plastic materials. Mechanical and physical properties of used materials were given in Table 1.

Figure 1. Carbon-fiber-reinforced polymer (CFRP) composites: (a) VARIM method and (b) fabricated CFRP plate.

Table 1. Mechanical and physical properties of used materials as components in the composite structure.

Material	Carbon Fiber	Epoxy
Density (gr/cm^3)	1.8	1.13
Modulus of elasticity (GPa)	537	3.3
Tensile strength (MPa)	5089	92
Thermal conductivity (W/mk)	498	0.6
Coefficient of thermal expansion (/°C) × 10^{-6}	−6	65

The used epoxy matrix was Huntsman Araldite 1564 mixed with Aradur 3487 as hardener and was prepared from Huntsman Advanced Materials in the USA. The modulus of elasticity and tensile strength of used epoxy were 3300 and 92 MPa, respectively. The CFRP homogenous composites were fabricated in Fibermak composites Ltd. CO., Izmir, Turkey. The curing process of the composites was carried out at −1 atm pressure and 80 °C, for 8 h. The schematic of the curing process is shown in Figure 2. The composite fabrication process (VARIM) and fabricated CFRP plate are shown in Figure 1.

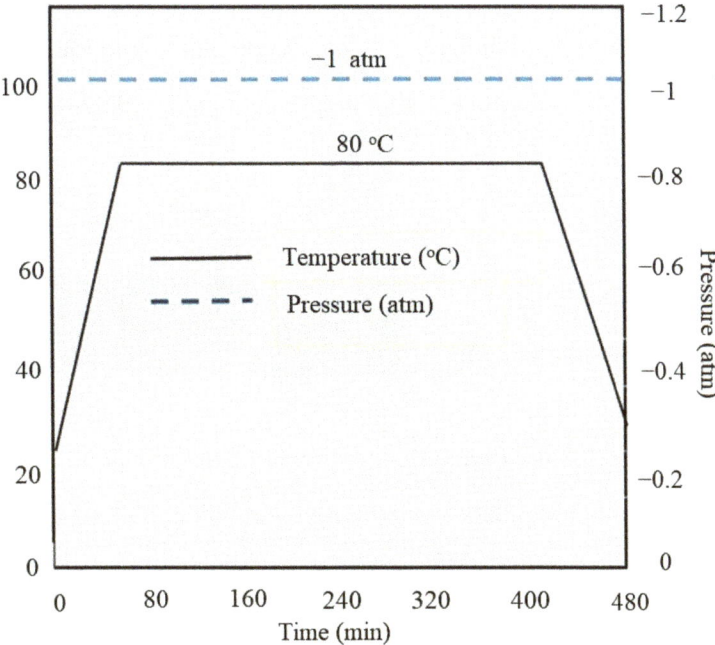

Figure 2. Schematic of curing process of fabricated CFRP composites.

2.2. Thermal-Cycling Process

In the thermal-cycling process, one cycle was determined as a sequence which was started from room temperature and increased to +120 °C, then decreased to −40 °C and finally returned to room temperature. As Figure 3 shows, the whole thermal-cycling process time was 20 min. This sequence was repeated for 20, 40, 60 and 80 cycles.

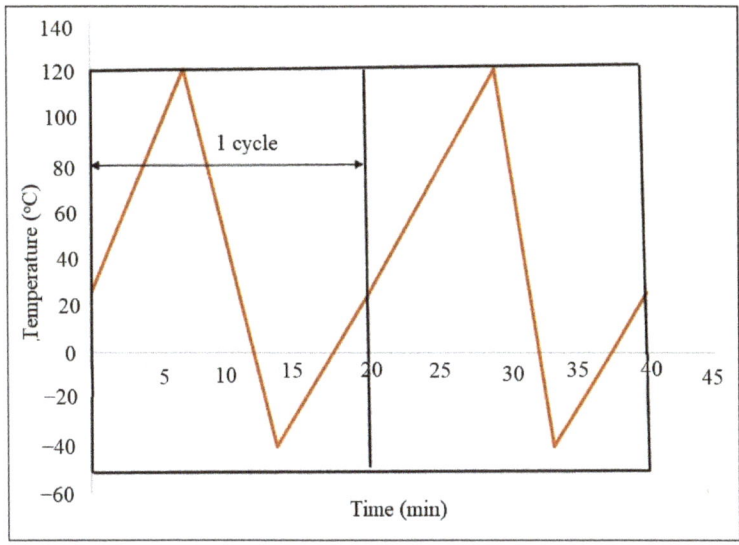

Figure 3. Temperature profile of thermal-cycling process for two cycles.

2.3. Mechanical Tests

Mechanical experiments, such as tensile, bending and SBS tests, were carried out to investigate the effect of thermal cycling on the mechanical properties of CFRP composites by using a rectangular specimen (see Figure 4 and Table 2). Therefore, tensile strength, modulus of elasticity, flexural modulus, flexural strength and ILSS were compared for different cycled specimens at room temperature.

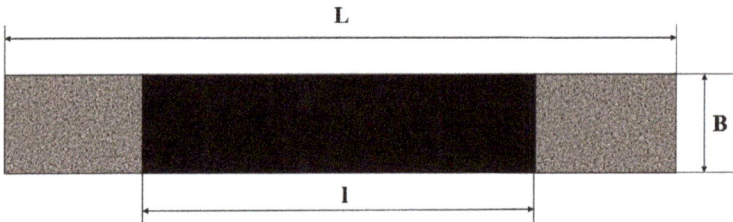

Figure 4. Geometry of mechanical-test specimens.

Table 2. Dimensions of test specimens for mechanical tests.

Test	L (mm)	l (mm)	B (mm)
Tensile	250	150	25
Bending	130	100	15
SBS	50	40	7

All the mechanical tests were carried out by using AGIS100 Shimadzu, Shimadzu Corporation/Japan instrument in mechanical engineering laboratory of Ataturk University, Turkey. The loading capacity of test instrument was 10 kN, and all mechanical experiments were accomplished at room temperature. According to ASTM standard, D3039, D790 and D2344 for tensile, bending and SBS tests, respectively, the cross-head speed of instrument during tensile and bending tests was 1 mm/min and for SBS test was 1.3 mm/min (see Figure 5a–c).

Figure 5. Mechanical tests of fabricated composites: (**a**) tensile test, (**b**) bending test and (**c**) inter-laminar shear strength (ILSS) test.

The tensile strength, flexural modulus and flexural strength were calculated from Equations (1)–(3), respectively. Moreover, Equation (4) was used to calculate the ILSS from the SBS test.

$$\sigma = \frac{P}{A} \qquad (1)$$

$$\sigma_f = \frac{3PL}{2bt^2} \tag{2}$$

$$E_B = \frac{mL^3}{4bt^3} \tag{3}$$

$$\tau = 0.75 \times \frac{P}{bt} \tag{4}$$

where P, b and t were the maximum applied load (N), width and thickness of samples. Moreover, according to ASTM D5379, m was the slope of force-deflection diagram in the bending test. Geometrical dimensions of the test specimen, such as L, b and t, were shown in Table 2, and the maximum force was extracted from experimental tests results.

2.4. Dynamic Mechanic Analyzes (DMA)

Perkin Elmer Pyris Diamond Dynamic Mechanic Analysis instrument was used to carry out DMA test for room temperature (RT) specimens and thermal-cycling-exposure-fabricated woven CFRP composites. The DMA test is an effective technique for characterization of viscoelastic behavior of composite materials. This technique assesses the viscoelastic properties of the materials, especially composites in response to the vibrational loads. Homogenous CFRP composites were tested under three-points bending mode at 1 Hz frequency, in the temperature range between 25 and 200 °C. The heat rate in this process is 2 °C/min, and the maximum mechanical load is 18 N. The test is carried out according to ASTM D7028 standard, where the dimensions of used square samples are 7×7 mm². The curves of viscoelasticity parameters, such as storage modulus (E'), loss modulus (E") and damping ratio (tan δ), are plotted versus temperature and evaluated by Pyris software. The viscoelastic properties, glass transition at fabricated RT and thermal exposure CFRP composites are determined by this test, as well as.

2.5. Hydrophobicity Test

The water-absorption characteristic of the fabricated CFRP composite before and after thermal cycling exposure was evaluated by using a hydrophobicity test. The water-absorption nature of the composite materials is a very important characteristic of composite materials, especially for those working in humid conditions or in direct contact with water, such as marine or aeronautic instruments. The hydrophobicity test was carried out by a KSV CAM 101 goniometer, and the contact angle for each specimen was determined at Chemical Laboratory of Kazim Karabekir Faculty of Ataturk University in Erzurum/Turkey. According to Figure 6, the material with contact angle (θ) higher than 90° is from the hydrophobic type, unless it is from the hydrophilic type.

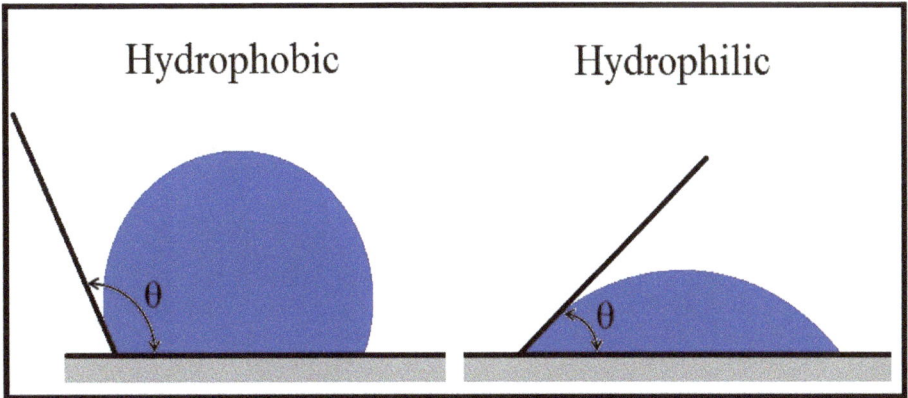

Figure 6. Contact angle in hydrophobicity test.

3. Results and Discussion

3.1. Mechanical Tests

As mentioned in the experimental tests section, the mechanical properties of fabricated CFRP composites were evaluated by using tensile, bending and SBS tests.

3.1.1. Tensile Test

The modulus of elasticity and tensile strength of CFRP composites before and after thermal exposure are shown and compared in Figure 7.

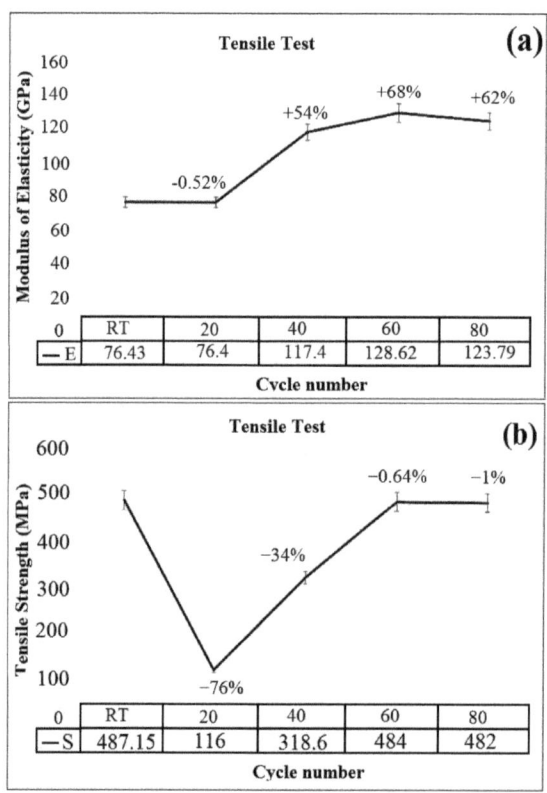

Figure 7. Tensile test results of room temperature (RT) and thermal exposure CFRP composites, (a) modulus of elasticity, (b) tensile strength.

As illustrated in Figure 7a, the modulus of elasticity of CFRP composites was slightly declined by the initiation of the thermal-cycling process from 76.43 to 76.4 GPa. The degradation effect of thermal cycling due to CTE difference between composite components reduced modulus of elasticity by 0.52% in CFRPs after 20 cycles of exposure between −40 and +120 °C. After that, the high-temperature effects of the thermal-cycling process made a dominant behavior of thermal-cycling process and increased crosslinking. This phenomenon improved properties of CFRPs, increasing by 68% in modulus of elasticity up to 128.616 GPa for 60 cycles. After this point, the effect of degradation was dominated due to CTE difference and modulus of elasticity reduced with the amount of 3.91 GPa after 80 cycles. About the tensile strength, as shown in Figure 7b, initiation of thermal-cycling process caused to a sharp decrease from 487.15 to 116 MPa. The stress–strain curves of fabricated CFRP composites before and after 20, 40, 60 and 80 cycles were plotted in Figure 8.

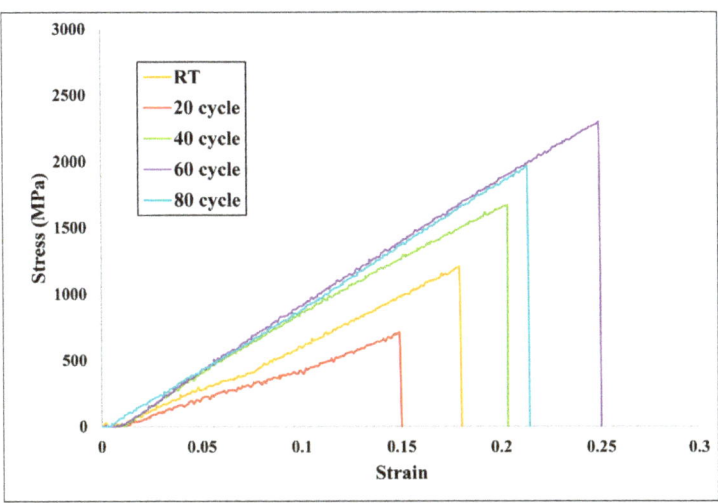

Figure 8. Stress–strain curves of fabricated CFRP composites before and after 20, 40, 60 and 80 cycles.

Initiation of microcracks in the structure of composite due to stress concentration in the interface of composite is an important reason for this reduction. By increasing the cycles number similar to modulus of elasticity, the dominant effect of post-curing improved the mechanical properties of CFRPs. Thus, the tensile strength of the composite was increased by 75.93% up to 60 cycles. Although, after this point, the post-curing effects were subsided, and microcracks started to grow and join together. This phenomenon caused delamination occurrence and decreased the tensile strength by about 41% in 80 cycles.

3.1.2. Bending Test

To study the effect of thermal cycling on the flexural modulus and flexural strength, the flexural modulus and strength variations were investigated in RT and thermal-exposure CFRPs (see Figure 9). As shown in Figure 9a, the bending modulus was decreased by 7.45% in the beginning of the thermal-cycling process. The reduction for the 20 cycles may be related to the debonding effect of thermal cycling. In this condition, the weakening effect of thermal cycling is dominant due to the lack of the conditioning time. Thereafter, when the conditioning time increased, the flexural modulus increased, as well, which may be due to greater post-curing effects of thermal conditioning. Moreover, the elasticity modulus increased by 8.27% up to 53.8 GPa after 60 cycles. The enhancement of flexural modulus did not continue up to 80 cycles of exposure. This could be related to higher thermal stresses that started dominating over the cryogenic compressive stresses for the longer conditioning of the thermal-cycling process. In the case of flexural strength, similar to the flexural modulus, initiation of thermal cycling decreased it from 489.6 MPa for RT condition to 426.94 MPa after 20 cycles of exposure. Thereafter, the post-curing effects started to dominate and increased the flexural strength by 16.48% after 60 cycles of exposure. At the longer conditioning of thermal cycling, the effect of post-curing, due to the high temperature in the cycling process, enhanced the crosslinking and led to increasing the flexural strength up to 495.21 MPa.

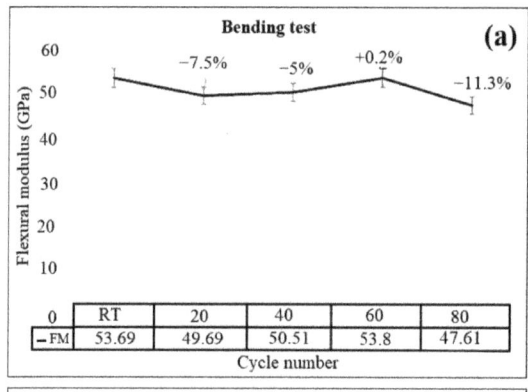
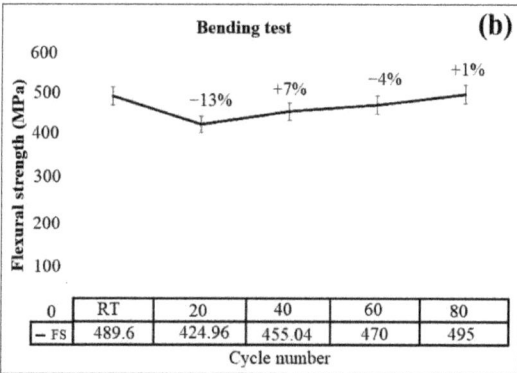

Figure 9. Bending test results of RT and thermal exposure CFRP composites, (a) flexural modulus, (b) flexural strength.

3.1.3. SBS Test

Voids or porosities are created if air is trapped inside the molded part. Such voids considerably deteriorate the ILSS magnitudes of fabricated composites. The ILSS values of CFRP composites at RT and after thermal cycling exposure are shown in Figure 10.

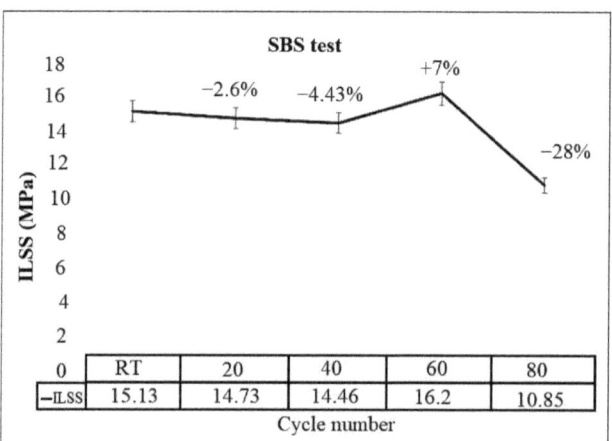

Figure 10. ILSS values of RT and thermal-cycling-exposure CFRP composites.

As shown in Figure 10, the shear strength was decreased slightly by the initiation of the thermal-cycling process up to 40 cycles. There was an initial reduction in ILSS for conditioning time up to 40 cycles and then an increment with increasing thermal-cycle numbers. In this condition, thermal cycling caused debonding, which resulted in the reduction of ILSS values. A low degree of post-curing strengthening was the result of the low conditioning time. When the thermal-cycles number increased to 60 cycles, the ILSS raised up to 16.2 MPa. Growing the ILSS value by rising cycle numbers may be due to the higher order of further polymerization. Development of stronger bonds at the fiber/matrix interface due to high-temperature conditioning during the thermal-cycling process, which increased the shear strength, may be the other reason for the ILSS enhancement [4]. After 60 cycles, development of compressive stresses due to quenching and also the debonding effect of thermal cycling caused the manifestation of residual stresses in the interface and reduced the stress transmissibility. This phenomenon increased thermal stress concentration in the interface region and decreased shear strength by 33.02%.

3.2. Thermal Test
3.2.1. DMA Test

The evaluation of the viscoelastic properties of RT and thermal-exposure CFRP composites was carried out by a DMA test. To this, the curves of viscoelastic properties, such as storage modulus, loss modulus and loss factor (tan δ), were plotted for fabricated RT and thermal-treated CFRP samples in Figure 11.

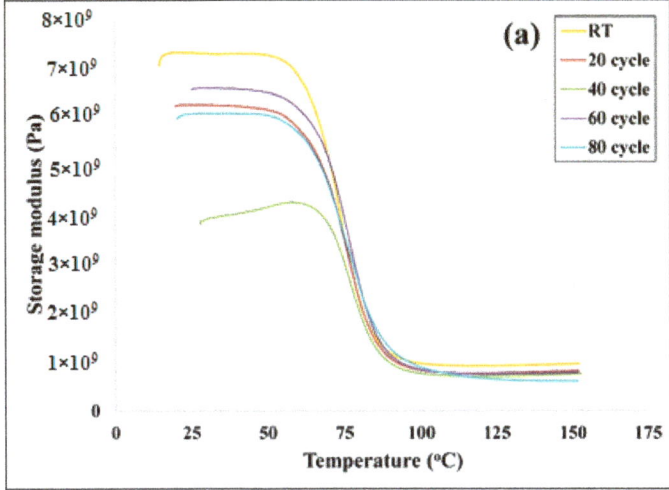

Figure 11. Cont.

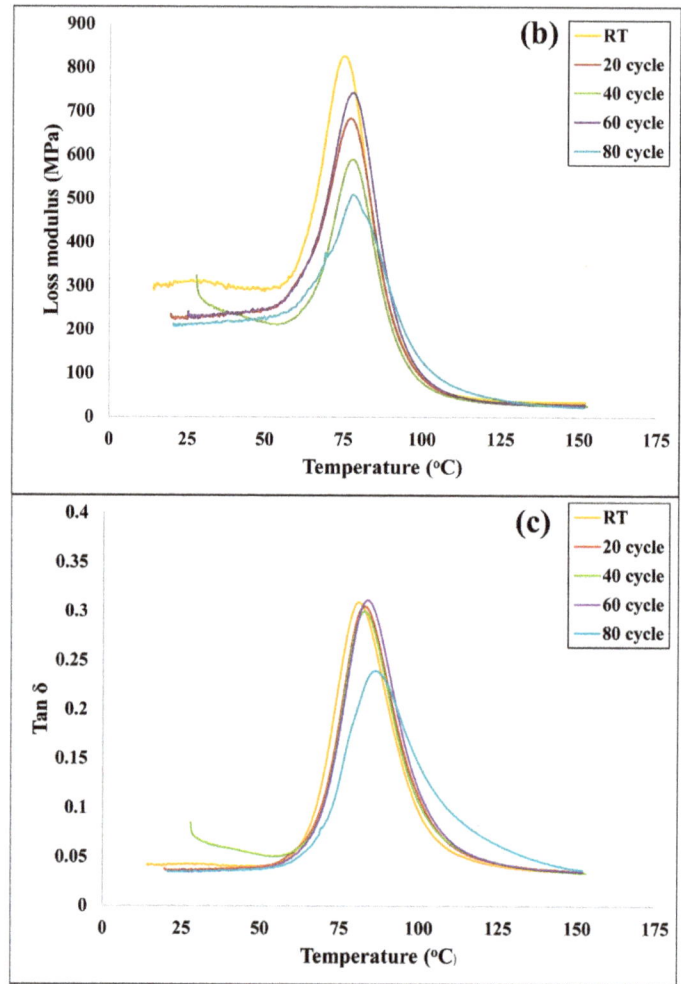

Figure 11. DMA results of RT and thermal-cycling-exposure CFRP samples: (**a**) storage modulus, (**b**) loss modulus and (**c**) loss factor.

The composite materials subjected to the DMA test are categorized in three states, namely glass, glass transition and rubbery states. In the glassy state, segmental mobility is limited in the structure of material, and the material has the highest storage modulus. In the glass transition state, the storage modulus is decreased steeply by partially changing in temperature. Glass transition temperature, which is known as one of the most important characteristics of polymeric composites, is determined by the analysis of the transition region. Moreover, the data attained from this state are used to characterize the viscoelastic properties of polymer composites. The final section of the DMA curve includes a rubbery state, where the storage modulus does not have considerable variation [18]. As shown in Figure 11a, the original sample of CFRPs gained the highest storage modulus in the glassy state. As it is clear, the loss modulus shows the viscous properties of a polymeric based material and is a criterion for energy loss as heat or dissipated during one cyclic load as well. It is worth mentioning that the loss modulus was almost steady at a low temperature and reached a peak value at the onset temperature, showing the maximum heat dissipated per unit deformation. The values of glass transition temperature and the maximum value of tanδ for CFRP composites in different thermal cycles extracted from the DMA traces are given in Table 3. It is observed that the Tg and tanδ for 40 cycles is the lowest. On the other hand, the sample exposed to 80 cycles gained the highest Tg. This phenomenon is similar to the results of flexural test, where the sample exposed to 80 cycles and gained the highest flexural strength because of the further effect of crosslinking [19].

Table 3. Glass transition temperature and loss factors of RT and thermal-exposure CFRP composites.

Samples	T_g	Tan δ_{max}
RT	80.79	0.31
20 cycle exposure	82.88	0.3051
40 cycle exposure	82.82	0.2998
60 cycle exposure	83.62	0.3114
80 cycle exposure	86.16	0.2395

3.2.2. Hydrophobicity Test

The hydrophobicity test results of original RT and thermal-cycle-exposure CFRP composites are shown in Figure 12 and Table 4. According to the results, the contact angle of original CFRP is 80.87°, indicating that the CFRP is from hydrophilic type and tends to be hydrophobic in nature due to the high hydrophobicity of carbon fibers. The contact angle of CFRP samples is slightly reduced by increasing the thermal cycles and the water-absorption capability. This phenomenon occurs due to evaporation of solvents of epoxy and increasing water-absorption nature of the matrix material. According to Figure 12, by increasing cycle numbers, especially from 60 to 80 cycles, the water-absorption desire of composite is reduced. The results reveal that the thermal-cycling process not only does not change the water-absorption nature of CFRP composites but also intensifies their hydrophilic nature.

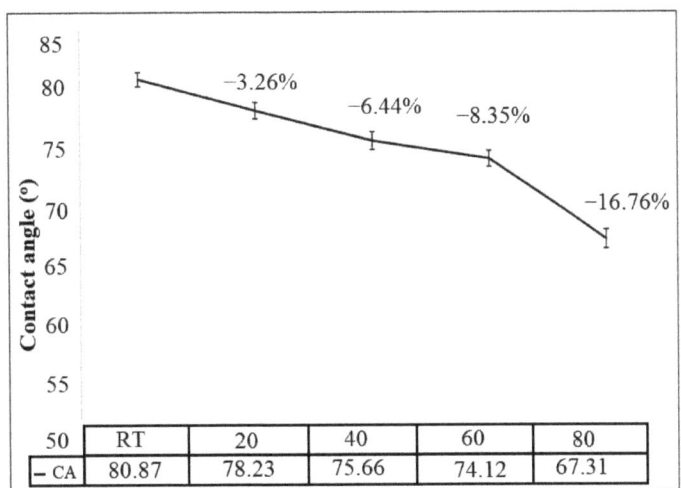

Figure 12. The contact angles and changing of them by increasing cycle numbers in thermal-cycling process.

Table 4. Contact angle and nature of RT and thermal-cycling-exposure CFRP composite materials.

Samples	Contact Angle (°)	Nature
RT	80.87	Hydrophil
20 cycle exposure	78.23	Hydrophil
40 cycle exposure	75.66	Hydrophil
60 cycle exposure	74.12	Hydrophil
80 cycle exposure	67.31	Hydrophil

3.3. Scanning Electron Microscopy (SEM)

The effect of thermal-cycling shock on microstructure of CFRP composite was evaluated by SEM micrographs (Figure 13).

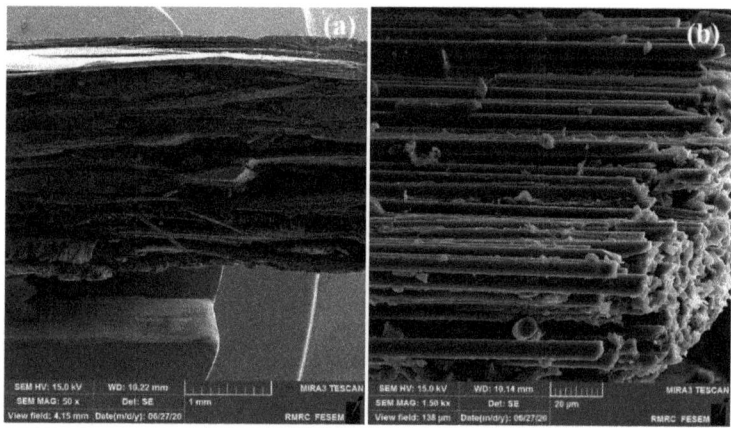

Figure 13. SEM micrographs of fracture surfaces of CFRP composites: (**a**) 50× and (**b**) 1.5k×.

SEM was also used to see the fracture surfaces. A small section of cross-section of composite laminates exposed to tensile test after thermal cycling was prepared for SEM observation. As shown in Figure 13, different failure modes were observed in tested specimens after 80 cycles between −40 and 120 °C. The fracture surfaces showed a macroscopic brittle character. According to Figure 13a, layer delamination is obvious, which occurred due to thermal cycling and difference of CTE for carbon fiber and epoxy matrix. This delamination nucleated and propagated some microcracks. In the other words, delamination was created during deformation due to the temperature variation and different behaviors of matrix and fiber. This phenomenon acted as a stress concentrator to initiate and propagate cracks under tensile loading. According to fracture surfaces, as shown in Figure 13, the delamination in composite layers, fiber pull-out due to weakness of bonds between fibers and polymer and bundle breakage were the dominant modes for the fracture in tensile-tested specimens.

4. Conclusions

The effect of thermal cycling on the mechanical and thermal properties of CFRP composites were studied. Mechanical tests such as tensile, bending and SBS were carried out, to characterize the mechanical properties, and a DMA test was also done, to evaluate the thermal effect of thermal cycling. It was concluded that the thermal cycling affected the polymer composites in two ways. First, it caused degradation because of CTE difference in composite components and manifestation of thermal-residual stress in the interface region that caused a debonding effect. Second, it increased crosslinking due to polymerization, which improved the interface behavior and mechanical characteristics of composite materials. As a result, the interaction of these mechanisms was very complex and indicated the material behavior at this condition. According to the results of the DMA test, the viscoelastic properties of CFRPs were not significantly affected by the increasing of thermal cycles. According to the hydrophobicity test results, the CFRP had hydrophilic nature, and its exposure to thermal cycling did not affect the nature of composite. In this regard, the contact angle of thermal-treated materials was slightly decreased by increasing cycle numbers. According to the SEM micrographs delamination, fiber pull-out and bundle breakage were the dominant modes for the fracture in the tensile test for CFRP composites.

Author Contributions: F.A.-S., conceptualization, investigation, methodology, data curation and writing—original draft; H.A., supervision; M.A.M.-B., investigation, data curation, formal analysis and writing—review and editing. All authors have read and agreed to the published version of the manuscript.

Funding: This research was financially supported by the Scientific and Technological Research Council of Turkey (TUBITAK), Project No. 213M600, and Ataturk University Scientific Research Grant, BAP 2012/448.

Data Availability Statement: The data presented in this study are available on request from the corresponding author.

Acknowledgments: The authors would like to thank Özgür Seydibeyoğlu and Volkan Acar for their scientific contributions.

Conflicts of Interest: The authors declare no conflict of interest.

References

1. Liu, Z.; Chen, P.; Han, D.; Lu, F.; Yu, Q.; Ding, Z. Atmospheric air plasma treated PBO fibers: Wettability, adhesion and aging behaviors. *Vacuum* **2013**, *92*, 13–19. [CrossRef]
2. Bora, M.O.; Çoban, O.; Sinmazcelik, T.; Gunay, V. Effect of Fiber Orientation on Scratch Resistance in Unidirectional Carbon-Fiber-Reinforced Polymer Matrix Composites. *J. Reinf. Plast. Compos.* **2009**, *29*, 1476–1490. [CrossRef]
3. Segerström, S.; Ruyte, E. Effect of thermal cycling on flexural properties of carbon-graphite fiber-reinforced polymers. *Dent. Mater.* **2009**, *25*, 845–851. [CrossRef] [PubMed]
4. Ray, B.C. Effect of thermal shock on interlaminar strength of thermally aged glass fiber-reinforced epoxy composites. *J. Polym. Sci.* **2006**, *100*, 2062–2066. [CrossRef]
5. Kot, P.; Baczmański, A.; Gadalińska, E.; Wroński, S.; Wroński, M.; Wróbel, M.; Bokuchava, G.; Scheffzük, C.; Wierzbanowski, K. Evolution of phase stresses in Al/SiCp composite during thermal cycling and compression test studied using diffraction and self-consistent models. *J. Mater. Sci. Technol.* **2020**, *36*, 176–189. [CrossRef]
6. Mikata, Y.; Taya, M. Stress Field in a Coated Continuous Fiber Composite Subjected to Thermo-Mechanical Loadings. *J. Compos. Mater.* **1985**, *19*, 554–578. [CrossRef]
7. PapanicolaouK, G.C.; Xepapadaki, A.G.; Pavlopoulou, S.; Zaoutsos, S.P. On the investigation of the stress threshold from linear to nonlinear viscoelastic behaviour of polymer-matrix particulate composites. *Mech. Time Depend. Mater.* **2009**, *13*, 261–274. [CrossRef]
8. Grzegorz, K.; Tomasz, N.; Robert, S. Modeling and Prediction of Thermal Cycle Induced Failure in Epoxy-Silica Composites. *Appl. Compos. Mater.* **2012**, *19*, 65–78.
9. Chu, I.; Lee, Y.; Amin, M.N.; Jang, B.-S.; Kim, J.-K. Application of a thermal stress device for the prediction of stresses due to hydration heat in mass concrete structure. *Constr. Build. Mater.* **2013**, *45*, 192–198. [CrossRef]
10. Chen, J.; Chen, J.J.; Jin, W.L. Experiment investigation of stress concentration factor of concrete-filled tubular T joints. *J. Steel Res.* **2010**. [CrossRef]
11. Ramanujam, N.; Vaddadi, P.; Nakamura, T.; Singh, R.P. Interlaminar fatigue crack growth of cross-ply composites under thermal cycles. *Compos. Struct.* **2008**, *85*, 175–187. [CrossRef]
12. Guerrero, J.; Mayugo, J.A.; Costa, J.; Turon, A. Failure of hybrid composites under longitudinal tension: Influence of dynamic effects and thermal residual stresses. *Compos. Struct.* **2020**, *233*, 111732. [CrossRef]
13. Russell-Stevens, M.; Todd, R.; Papakyriacou, M. Microstructural analysis of a carbon fibre reinforced AZ91D magnesium alloy composite. *Surf. Interface Anal.* **2005**, *37*, 336–342. [CrossRef]
14. Azimpour-Shishevan, F.; Akbulut, H.; Mohtadi-Bonab, M.A. Effect of thermal cycling on mechanical and thermal properties of basalt fibre-reinforced epoxy composites. *Bull. Mater. Sci.* **2020**, *43*, 1–10. [CrossRef]
15. Wang, Q.; Zhou, D.; Chen, Y.; Eames, P.; Wu, Z. Characterization and effects of thermal cycling on the properties of paraffin/expanded graphite composites. *Renew. Energy* **2020**, *147*, 1131–1138. [CrossRef]
16. Yang, B.; Yue, Z.; Wang, P.; Gan, J.; Liao, B. Effects of space environment temperature on the mechanical properties of carbon fiber/bismaleimide composites laminates. *Proc. Inst. Mech. Eng. Part G J. Aerosp. Eng.* **2017**, *232*, 3–16. [CrossRef]

17. Shin, Y.C.; Novin, E.; Kim, H. Electrical and thermal conductivities of carbon fiber composites with high concentrations of carbon nanotubes. *Int. J. Precis. Eng. Manuf.* **2015**, *16*, 465–470. [CrossRef]
18. Wang, C.; Smith, L.M.; Wang, G.; Shi, S.Q.; Cheng, H.; Zhang, S. Characterization of interfacial interactions in bamboo pulp fiber/high-density polyethylene composites treated by nano $CaCO_3$ impregnation modification using fractal theory and dynamic mechanical analysis. *Ind. Crop. Prod.* **2019**, *141*, 111712. [CrossRef]
19. Azimpour-Shishevan, F.; Akbulut, H.; Mohtadi-Bonab, M.A. Low Velocity Impact Behavior of Basalt Fiber-Reinforced Polymer Composites. *J. Mater. Eng. Perform.* **2017**, *32*, 337–2900. [CrossRef]

Journal of
Composites Science

Article

Droplet Spreading on Unidirectional Fiber Beds

Patricio Martinez [1,*], Bo Cheng Jin [1,2] and Steven Nutt [1]

[1] M.C. Gill Composites Center, Department of Chemical Engineering and Materials Science, University of Southern California, 3651 Watt Way, Los Angeles, CA 90089, USA; bochengj@usc.edu (B.C.J.); nutt@usc.edu (S.N.)
[2] Advanced Composites Simulation Lab, Department of Aerospace and Mechanical Engineering, University of Southern California, 3737 Watt Way, Los Angeles, CA 90089, USA
* Correspondence: mart136@usc.edu

Abstract: This study reports a method to analyze parametric effects on the spread flow kinetics of fluid droplets on unidirectional fiber beds. The investigation was undertaken in order to guide the design of droplet arrays for production of an out-of-autoclave (OoA) prepreg featuring discontinuous resin distribution, referred to here as semi-preg. Volume-controlled droplets of a resin facsimile fluid were deposited on carbon fiber beds and the flow behavior was recorded. The time to full sorption (after deposition) and the maximum droplet spread distance were measured. Experiments revealed that fluid viscosity dominated time to full sorption—doubling the viscosity resulted in an 8- to 20-fold increase in sorption time, whereas doubling fabric areal weight increased the time only by a factor of three. Droplet spread distance was nearly invariant with fiber bed architecture and fluid viscosity. A series of droplet arrays were designed, demonstrating how the results can be leveraged to achieve different resin distributions to produce semi-preg optimized for OoA cure.

Keywords: carbon fibers; prepreg processing; fluid flow; viscosity; wettability

Citation: Martinez, P.; Jin, B.C.; Nutt, S. Droplet Spreading on Unidirectional Fiber Beds. *J. Compos. Sci.* **2021**, *5*, 13. https://doi.org/10.3390/jcs5010013

Received: 20 December 2020
Accepted: 3 January 2021
Published: 6 January 2021

Publisher's Note: MDPI stays neutral with regard to jurisdictional claims in published maps and institutional affiliations.

Copyright: © 2021 by the authors. Licensee MDPI, Basel, Switzerland. This article is an open access article distributed under the terms and conditions of the Creative Commons Attribution (CC BY) license (https://creativecommons.org/licenses/by/4.0/).

1. Introduction

We investigate the effects of fiber bed architecture on the anisotropic flow behavior of fluid droplets on and into unidirectional (UD) fiber beds. In particular, we determine the effects of fiber bed areal weight and fluid viscosity on sorption time and spread distance. The work is motivated by a need to support the design of prepreg formats with discontinuous resin distributions (semi-pregs).

Conventional out-of-autoclave (OoA) prepregs typically feature continuous resin films partially impregnated into the fiber bed and thus rely on "edge breathing" for air removal [1–3]. Compared to autoclave prepregs, however, processing of OoA prepregs lacks robustness, particularly in challenging conditions, such as poor vacuum, ply ramps, embedded doublers and large parts [4,5]. In contrast, semi-pregs [6] feature discontinuous resin distributions that impart high through-thickness permeability and increase process robustness compared to conventional OoA prepregs [5,7–9]. Previous methods of fabricating semi-preg include hot-rolling resin onto the tow overlaps of woven fiber beds [10], using a release film mask to press a discontinuous film onto dry fibers [8] and dewetting a continuous resin film, then pressing onto dry fibers [7].

Recent work on semi-pregs has employed a polymer film dewetting approach to fabrication [7,9,11,12]. The inherent versatility of the method permits deposition of a variety of patterns on the fabric surface. However, the method has limitations. For example, the initial degree of impregnation (DoI) is negligible, leading to a higher bulk factor than conventional vacuum bag-only (VBO) prepreg [12]. In addition, the method as described requires filming of a continuous resin film prior to dewetting and pressing onto the fiber bed, adding cost to the manufacture. Finally, for thin resin films, difficulties can arise in generating uniform discontinuous resin distributions [12]. Thus, alternative methods of production are being evaluated, including gravure printing and droplet deposition onto fiber beds [13]. Out of these methods, droplet deposition allows for a resin distribution that is controlled based on the relationship between droplet position and fiber bed architecture and does not require a prior filming step as the dewetting method does. The present work constitutes a first step towards semi-preg production by droplet deposition. Experiments were undertaken to understand the flow of a single droplet placed on a fiber bed surface, with further scaling-up used to inform the future design of semi-preg with robust process characteristics.

Fluid flow on porous surfaces and through fiber beds has broad relevance for composites manufacturing [14,15] and has thus been studied extensively. However, most prior studies of surface flow have assumed material isotropy, treating pores as tubes oriented normal to the surface [16]. Fiber beds in composites, however, are anisotropic and permeability varies by orders of magnitude in directions normal and parallel to the fibers [17]. Thus, studies of fluid flow through fiber beds generally must consider flow through an anisotropic porous medium. Most often, such studies consider flow through the entire fiber bed, leading to full saturation (e.g., modeling the impregnation of an individual fiber tow [18]). These studies focus on infiltration, in which one fluid (air) is fully displaced by another (resin) in a porous medium [19], and as such ignore flow above and near the surface of the porous substrate.

Studies of fluid flow on a porous surface generally assume low-viscosity fluids (under 1Pa·s) [20]. Although, during cure, similar low viscosity values are achieved by typical prepreg resins, these studies have limited relevance with regard to surface flow during droplet deposition. The reason for this is that, to prevent advancing cure of the resin, droplet deposition is performed at lower temperatures than those used during cure, leading to higher viscosities. The present study considers higher viscosity fluids and focuses on local (near-surface) flow beneath individual droplets during initial wet-out. The droplets used in semi-preg production must only wet-out partially during the deposition stage in order to preserve connectivity of the dry spaces for air evacuation during the de-bulking stage of VBO processing. Full flow and subsequent saturation of the fiber bed only occurs after de-bulking, during cure of the laminate.

A combination of forces governs the fluid flow of a droplet on a solid surface, including gravitational forces, viscous forces and surface tension [20]. However, the effects of gravity can be ignored for small droplets, leaving only viscous forces, surface tension and capillary forces to govern the flow [21]. Similar forces control the flow of droplets dispersed on the surface of a porous medium saturated with the same fluid [22]. However, for dry fiber beds, capillary effects play a major role [14,15]. Specifically, standard wet-out phenomena lead to droplet spread on the surface, increasing the coverage, while capillary effects promote absorption into the substrate, reducing surface coverage and increasing impregnation [23].

In this study, we measured the surface flow of individual droplets on unidirectional fiber beds. Facsimile fluids with moderate viscosity values were selected (30–60 Pa·s) to resemble polymer resins used during hot-melt production of prepreg. Using the measured response of a single droplet, we also produced droplet arrays to maximize surface coverage, or to minimize interactions between neighboring droplets. The results relate to and can inform the design and production of semi-pregs, particularly the spacings and patterns of resin droplets on fiber beds. Note that despite the match in fluid viscosities, the facsimile fluid did not match the apparent contact angle nor the advancing droplet edge slope. Thus, while the results presented here are useful for determining droplet spread parameters and foreseeing how such droplets will behave in manufacturing conditions, further tests must be performed with actual resin to determine parameters for prepregging.

The experiments revealed parametric effects on droplet spread rates. The droplet absorption time depended strongly on viscosity: doubling the viscosity resulted in an 8- to 20-fold increase in absorption time. However, droplet spread along the surface showed little variation, at least within the fluid viscosity range tested. Similar relationships were noted with fiber bed areal weight, which had little effect on spread distances, but caused marked changes in sorption time. Finally, surface spread depended strongly on fiber bed architecture, particularly tow gaps caused by stitching. Using these results, we demonstrated how droplets can be positioned on fiber beds to ensure uniform impregnation, informing future methods for semi-preg production.

2. Materials and Methods

2.1. Resin Characterization and Fluid Selection

An epoxy resin designed for aerospace applications was selected (PMT-F4A, Patz Materials & Technologies, Benicia, CA, USA). The resin viscosity was typical of B-staged resins used in the production of conventional OoA prepreg. During prepreg production, the resin is pre-melted at 65–68 °C (150–155° F), filmed at 68–72 °C (155–162° F), then transferred to the fiber bed. The process from pre-melting to cooling takes less than 90 min. Thus, in the present work, rheological measurements were performed after each step in the thermal cycle, shown in Figure 1a. During the initial pre-melting stage, resin viscosity averaged 49.7 Pa·s, while during the filming stage, resin viscosity averaged 32.4 Pa·s, with a minimum of 27 Pa·s and a maximum of 52.5 Pa·s during the entire cycle.

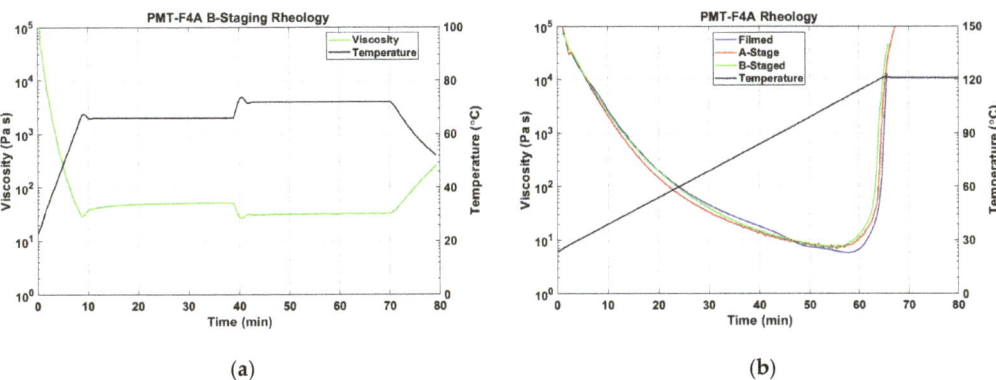

Figure 1. (a) Rheological profile for PMT-F4A following the filming process thermal cycle. (b) Rheological profiles for three separate samples of PMT-F4A.

To ensure that the process was consistent with the manufacturer's practices, the cure cycle rheology was compared with a sample provided by the manufacturer that had not undergone the melting process used for filming (A stage). The pre-melting process showed nearly identical cure cycle rheology (Figure 1b).

Facsimile fluids were chosen to match the viscosity values of the resin during the filming process (Figure 1a). Silicone viscosity standard fluids (General Purpose Silicone, Brookfield Ametek, Middleborough, MA, USA), with viscosities of 30 and 60 Pa·s (±1%), were selected. The use of facsimile fluids enables droplet flow testing at room temperature, minimizing difficulties in maintaining a uniform high temperature on the fiber bed while simultaneously recording data and ensuring the viscosity does not change due to advancing cure.

2.2. Contact Angle Comparisons

Surface tension can influence fluid flow along the surface of a substrate and facsimile fluids, despite matching resin viscosity, might exhibit different surface flow behavior. Consequently, measurements of the apparent contact angle were performed, comparing the facsimile fluids to the B-staged resin at the same viscosity. These measurements were performed using a goniometer (Ramé-Hart Model 500, Plymouth, MI, USA) from droplet deposition until apparent stabilization of the contact angle. By determining the difference in contact angle between the facsimile and the resin, a determination can be made of the validity of using the flow behavior observed for a facsimile fluid to predict that of an actual resin.

2.3. Fiber Beds

Four unidirectional non-crimp carbon fabrics (UD NCFs) with different areal weights were selected for use as dry fiber beds, including (A) 146gsm (4.3oz/sq.yard), (B) 136gsm (4.0oz/sq.yard), (C) 305gsm (9.0 oz/sq.yard) and (D) 756gsm (22.3oz/sq.yard) (FibreGlast Products #2596, #2585, #2583 and #2595, respectively). UD NCFs were chosen because of previous work with similar fabrics [6–8], their simplified geometry compared to woven fabrics, the similarity to tapes used in automated tape layups and their growing use in vacuum infusion processes both in aerospace and wind blades. Carbon fiber fabrics were selected due to their common use in prepreg for the aerospace industry. Using a different fiber material would result in different intra-tow capillary sizes based on fiber diameter, as well as differences in wettability based on fabric-fluid surface parameters, leading to differences in flow. The fabrics contained polyester binding to stitch layers together and impart ease of handling. The first three NCFs were bound using polyester stitching perpendicular to the fibers on one side, spaced ~10 mm apart. The heaviest weight fabric, however, featured binding in a diamond pattern and fiber tows with distinct edges, as shown in Figure 2.

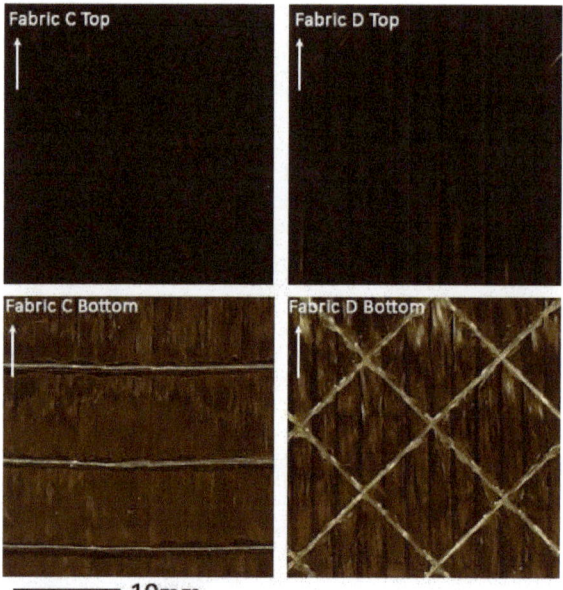

Figure 2. Top and bottom views of the two distinct types of fabrics. Fiber direction is indicated with arrows.

Prior to droplet deposition, a 19 mm strip of fabric was cut and the edges were secured with tape to prevent fraying, leaving a 19 × 19 mm square of exposed fabric. The tape was positioned perpendicular to the fiber direction, preventing free fiber edges from lifting and distorting the surface. The fiber bed squares were examined using a digital stereo microscope (Keyence VHX-5000, Osaka, Japan) to generate images and 3D contour maps of the surface. The fabric properties are summarized in Table 1.

Table 1. Comparison of properties for the different fiber beds. NCF, non-crimp carbon fabric.

Fabric	Areal Weight [g/m^2]	Tow Count	Fabric Style
A	136	12 K	Standard NCF
B	146	12 K	Standard NCF
C	305	24 K	Standard NCF
D	768	24 K	Quilted NCF

2.4. Droplet Deposition

Two devices were used to monitor the droplet deposition experiments: a goniometer and a camera (LUMIX GH4, Matsushita Electric Co., Osaka, Japan). The sample was positioned on the goniometer stage, with fibers aligned perpendicular to the goniometer light source and camera. Samples were positioned with stitching facing downward. The second camera was positioned directly above the sample (Figure 3). A syringe was used to deposit a single droplet of the silicone oil facsimile and both cameras started recording as the droplet contacted the fiber. After droplet deposition, the syringe was removed from the field of view.

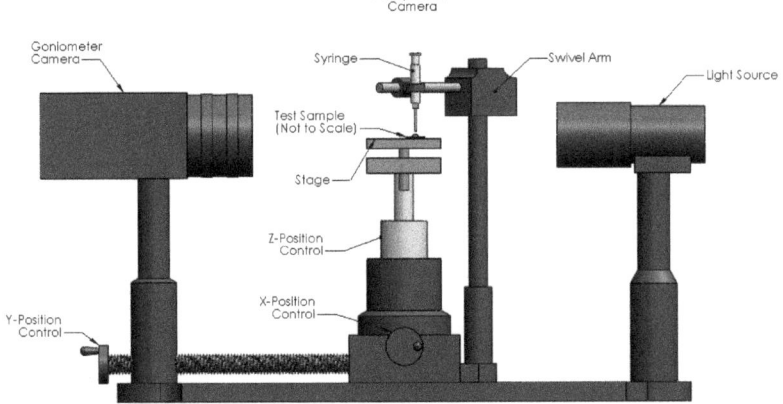

Figure 3. Diagram showing the droplet deposition test set up.

The goniometer camera was used to record images at two-second intervals, while the top-down camera recorded images at ten-second intervals. The images were compiled and analyzed separately. Software (MATLAB R2019a) was used to analyze the goniometer images, while the top-down images were analyzed manually using image editing software. Droplet perimeter was approximated using a brightness threshold, then further modified manually, as the contrast between wet and dry was not perfect. In both cases, the dimensions of the fluid spread were measured from images. Using the side view, droplet width and height were measured at each frame. Using the top view, the droplet spread was measured in directions parallel and transverse to the fibers. Figure 4 shows diagrams illustrating the measurements recorded, as well as an example of one such frame used for a single measurement. The goniometer camera detected fluid above the fiber substrate only, while the top-down camera allowed for the observation of fluid imbibed by the fiber bed.

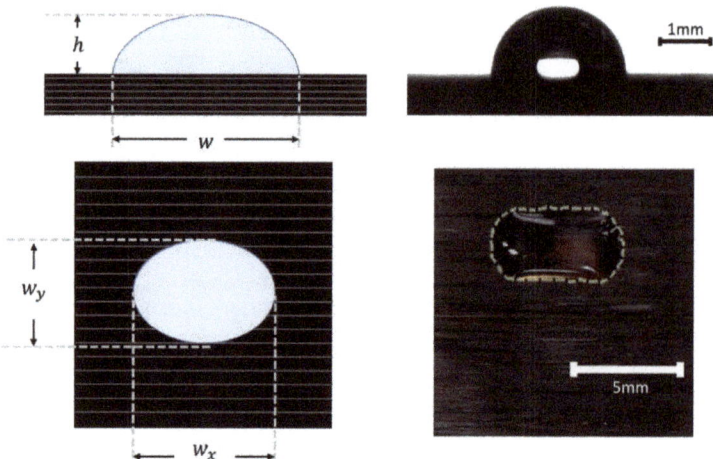

Figure 4. Diagram showing the measurements made from side and top-views of the droplet deposition test (**left**) and an examples of images used for such measurements (**right**), with the droplet edges outlined for clarity in the top-view.

2.5. Droplet Grid Tests

Droplet arrays were deposited to demonstrate how the results from previous sections can inform designs of resin patterns on semi-pregs. Droplet arrays of the facsimile fluid (30 Pa·s) were deposited on a 20 × 20 mm area of fabric A. The arrays were evaluated with respect to two parameters—(a) area covered by the fluid and (b) neighboring droplet interaction. Three distinct droplet arrays were deposited. The first pattern consisted of droplets uniformly distributed in a 3 × 3 square grid. The second pattern was based on data obtained from the single droplet tests, allowing the droplet positions to be arranged such that droplet overlap was minimized. The final grid pattern used the same data with minor changes to the positioning to ensure droplet-to-droplet interactions, showing how small deviations can result in differences in the final distribution.

Grid accuracy was maintained by producing a guide for the same syringe used for the single droplet deposition tests. Droplets were aligned using the grid guides and dispensed one at a time in raster fashion on the fiber bed. Droplet spreading was recorded using a top-down camera in the same manner as for the single droplet tests. Since the positioning guide shielded the camera view, data recording commenced 5 min after the first droplet was deposited. Images were captured at 10-sec intervals for up to 150 min. Using these images, the area covered by facsimile fluid was recorded over time. Furthermore, a time lapse was generated using the captured images for each droplet array.

3. Results
3.1. Contact Angle Comparisons

Using all three fluids—the resin and the two different viscosity silicone oils—differences in surface flow phenomena were observed and recorded. Experiments were conducted to measure the apparent contact angle as the droplet was deposited and absorbed into the fabric. As shown in Figure 5a, the facsimile fluid did not match the apparent contact angle of the resin. Given the non-static nature of the measured angle, it is more appropriate to refer to the metric not as the contact angle, but as the edge slope of the droplet. However, as shown in Figure 5b, the difference in edge slope between the lower and higher viscosity droplets was equivalent for both the epoxy resin and the facsimile fluid.

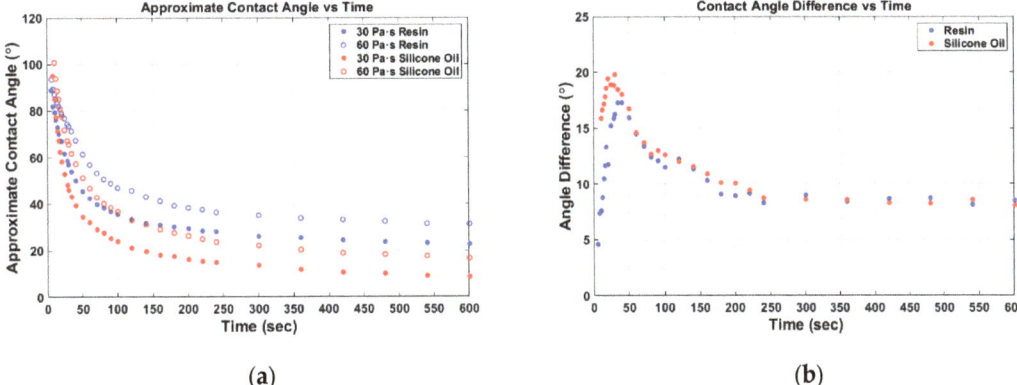

Figure 5. (a) Apparent contact angle of droplets of silicone oil and epoxy resin at 30 and 60 Pa·s viscosity. (b) Difference between the apparent contact angle of the 30 and 60 Pa·s droplets for both resin and facsimile fluid.

3.2. Single Droplet Test

Experiments were performed for each substrate–facsimile fluid combination, while recording droplet height (h), droplet width above the surface (w), fluid spread across fibers (w_y) and fluid spread along fibers (w_x). Note that w and h were recorded from the goniometer, while w_x and w_y were recorded from the top-down view. While both w and w_x represent droplet dimensions in the direction parallel to the fibers, w represents the width of the droplet seen above the surface, while w_x includes fluid flow visible at the surface from the top-down view. The time to full sorption (t_{h0}) was taken as the time required for the droplet height to reach a constant value. This time was used to normalize the remainder of the time for the figure, as follows:

$$\hat{t} = \frac{t}{t_{h0}}$$

Figure 6 shows (a) droplet height and width versus time and (b) test results normalized by time to full sorption. When plotted against the logarithm of normalized time, the height of the droplet decreased approximately linearly. Similarly, the spread distance along the fibers also increased linearly, which follows from Tanner's Law (1) for a two-dimensional droplet, where $R(t)$ is the radius of the droplet, γ is the surface tension, B is a constant, η is fluid viscosity and V is droplet volume [24,25]. In summary, the spread distance along the fibers, w_x in this case, follows a power law with time. In contrast, the spreading behavior across the fibers was distinctly nonlinear (albeit noisy), exhibiting spread at the start followed by a quasi-stable plateau. Droplet spreading generally followed patterns similar to those depicted in Figure 6, with variations in the rate of growth and decay.

$$R(t) \approx \left[\frac{10\gamma}{9B\eta}\left(\frac{4V}{\pi}\right)^3\right]^{\frac{1}{10}} \propto t^n, \tag{1}$$

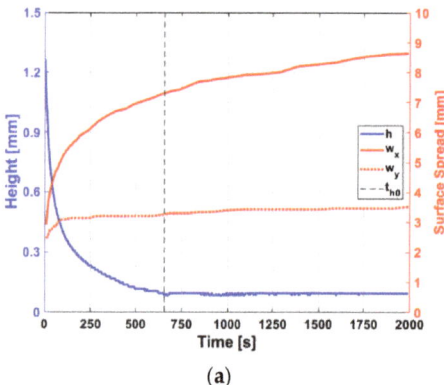
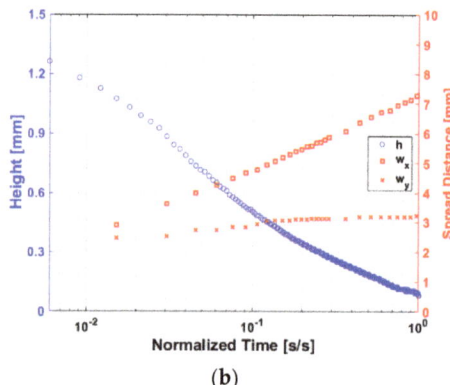

(a)   (b)

Figure 6. (a) Results from depositing 30 Pa·s viscosity fluid on fabric C. The left y-axis shows results derived from the side view, while the right y-axis shows results from top-down view. (b) Same results as (a), with time scale normalized to t_{h0}.

3.3. Aggregate Tests Results

To visualize the evolution of droplet height with time, select images were assembled in an array, as shown in Figure 7. For these images, the height of the droplet was mapped against fractions of time to full sorption, t_{h0}. These images show that as fabric areal weight increased, in-plane spreading of droplets generally occurred more rapidly, the only exception being the fabric with the heaviest areal weight. In the case of the standard NCFs with the highest viscosity droplets, most of the spreading occurred early; that is, the fluid spread out rapidly before it started being absorbed into the fabric. The effects of gravity in assisting flow were strongest in early stages, when droplet mass was most centralized.

Figure 7. Time lapse of side view for all combinations of fluid viscosity and fabric types. Each snapshot is at a fraction of the time to full sorption for that test. Red line corresponds to the baseline from which height is measured, equal to the height of the droplet at $t = t_{h0}$ ($h = 0$).

Figure 8 was generated using the time to full sorption, t_{h0}, for each test. For the first three fabric samples, lower areal weight correlated with increased time to full sorption for the low-viscosity oil. When fluid viscosity increased (from 30 to 60 Pa·s), t_{h0} markedly increased (between 8- and 20-fold). In contrast, no similar correlation appeared for the heaviest fabric: fluid viscosity did not affect time to full sorption for the 756 gsm fabric.

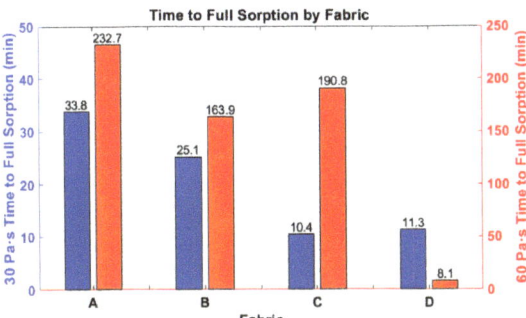

Figure 8. Relationship between time to full sorption and fabric type. Note that the 60 Pa·s samples have been cut off, but the relative height between each of the 60 Pa·s tests (excluding fabric D) remains.

Figure 9 compares the maximum fluid flow distance in both directions of interest for each test. As expected, the fluids spread longer distances along the fibers than across the fibers by a factor of 2–3×. For the first three fabrics, spread distances along the fibers fell within a 1.9 mm range and spread distances across the fibers were within a narrower 1.1 mm range. Spread distances along the fibers did not fall in this range for fabric D, at least for the lower viscosity fluid.

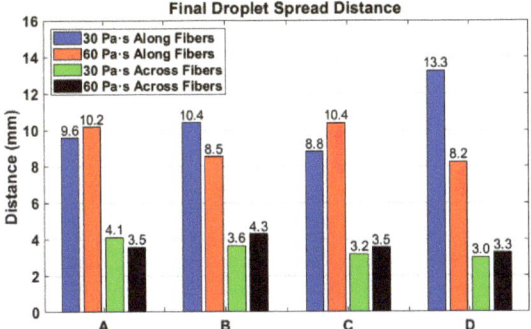

Figure 9. Relationship between fiber bed type and spread, in directions across and along the fibers.

3.4. Quilted Fiber Bed

Fabric D behaved differently from other fabrics due to its distinct architecture. This fabric featured individual tows secured by a grid pattern of stitches, as opposed to simple unidirectional stitches (shown previously in Figure 2). The gaps between individual tows afforded pathways for fluid flow into the fiber bed, acting effectively as macrochannels between fiber bundles. Top-down images of fabric D revealed that most of the fluid flowed into the gaps (shown overlaid on a topographical image of the fiber bed in Figure 10). The results in the previous section showed that fabric D yielded the largest fluid flow distance along the fibers, accompanied by minimal flow across the fibers. These observations support the assertions that irregularities in flow along the fibers were caused by greater fluid penetration into the surface, and that penetration occurred through and along the larger inter-tow gaps created by stitching. Comparing this result to a similar test on one of the non-quilted fabrics, the bottom image in Figure 10 shows that fluid also flowed into gaps/irregularities within the fiber bed, but that the flow was spread more uniformly along the fibers.

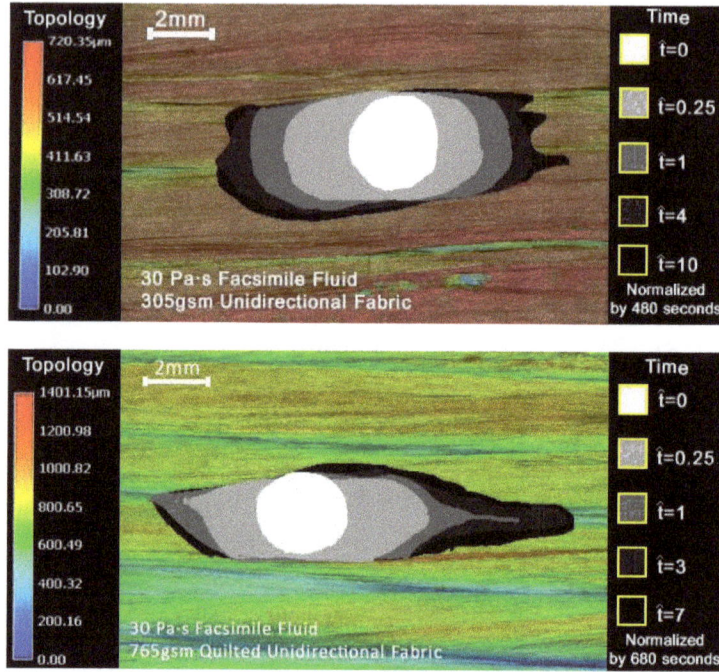

Figure 10. Images showing the contours of the area infiltrated by fluid over time, overlaid onto a topographical representation of the fiber bed for the 30 Pa·s viscosity fluid tests on fabric C (**above**) and fabric D (**below**).

The observations in Figure 10 highlight challenges when attempting to correlate the results reported here to other fabric types. Fine-scale variations in the fiber bed, such as inter-tow gaps, pinholes and other such irregularities common in woven fabrics, are intrinsic and affect fluid flow. However, the impact of such variations on the flow of an individual droplet is likely to be more pronounced. Due to the similarity in length scales of droplet size and surface features, droplet flow is dominated by the positioning of the droplet on the surface. For example, in the case of the quilted unidirectional fabric above, a droplet at the center of a tow bundle is likely to behave differently from one deposited directly atop a tow gap, as shown in the top image of Figure 10. Since the non-quilted UD NCFs, fabrics A–C, do not exhibit large-scale surface irregularities, the results from these fabrics can be analyzed with fiber areal weight as the only variable.

3.5. Droplet Grid Optimization

Using the results from Figure 9, three droplet arrays were designed, including a control pattern, with droplets arranged in a square grid, a staggered array to maximize coverage and minimize droplet overlap and a tight-staggered array that ensured droplet interaction while maximizing spread distance (grids 1–3, Figure 11). Coverage refers to the area fraction of the surface that was covered by the facsimile resin upon full sorption and overlap refers to neighboring droplets impinging on each other and coalescing into a single fluid pool.

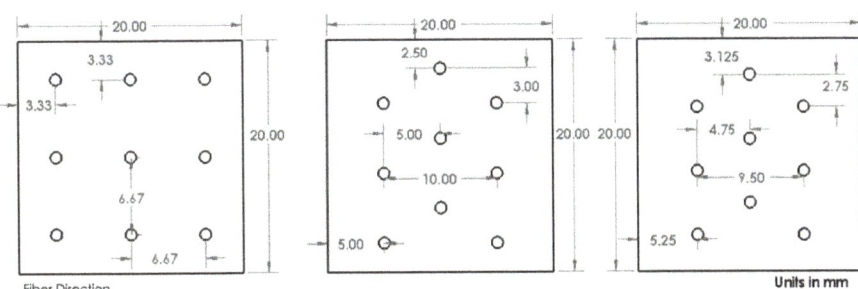

Figure 11. Diagrams of droplet grid guides used. From left to right: 1. square grid, 2. staggered grid, 3. tight-staggered grid.

The dimensions of the grid arrays can be used to estimate the fiber loading and resin content for a fully impregnated prepreg. Using the controlled volume of the droplet, the areal weight of fabric A and a fiber density of ~1.7×10^6 g/m^3, eighteen droplets on a 20×20 mm area correspond to a fiber volume fraction of 60–70%, assuming full impregnation and no voids. This range of fiber loadings lies within the range used for commercial unidirectional prepregs. Since commercial prepregs are produced by applying resin to both sides of a fiber bed, half the droplets needed to achieve a proper volume fraction were used and nine droplets were placed within a 20×20 mm area for the droplet grids.

Diagrams of the three grids are shown in Figure 11. Grid 1 featured droplets separated by 6.67 mm in a square array that extended across and along the fibers. The design for grid 2 was informed by the observations that droplets spread 9.6 mm along the fibers and 4.1 mm across them. The droplets were spaced 10 mm apart along the fibers and 3 mm apart across the fibers, with an offset of 5 mm along the fibers between rows. Grid 3 featured a modification of grid 2, with droplets spaced 9.5 mm apart along the fibers and 2.75 mm apart across the fibers.

Figure 12 shows the final images captured, as well as contours for the area covered by the fluid for the first and last images. The images show that all three arrangements resulted in droplet overlap. However, by design, grid 2 exhibited the least overlap, while grid 3 yielded the most. The square grid, grid 1, showed overlap along the fibers and dry gaps between droplets (across the fibers). Green and red outlines in Figure 12 show initial and final perimeters of the nine initial droplets, respectively, as seen from the surface. At the end of each test, the red outlines show that grid 1 resulted in three distinct fluid-covered regions, grid 2 with six and grid 3 with two. Furthermore, upon completion of each test, the portion of the fluid area that extended beyond the prescribed region was 5.3%, 0.6% and 3.2% for grids 1, 2 and 3, respectively. These fluid regions were determined only above the surface and do not include subsurface flow.

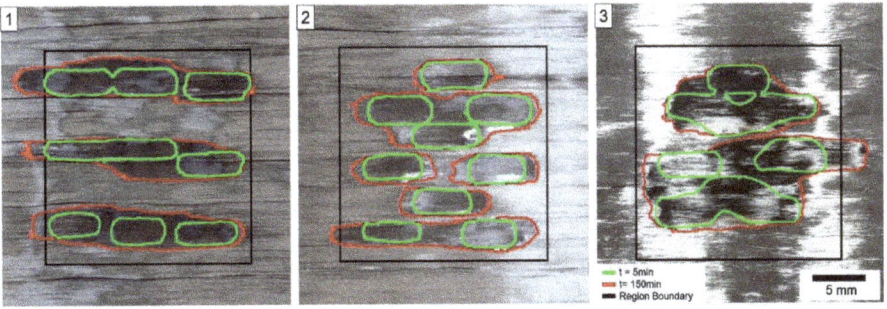

Figure 12. Images showing the extent of fluid coverage for all three grids, including 5 min and 150 min after deposition.

The area covered by the droplets was recorded and plotted versus time (Figure 13). This graph shows that grids 2 and 3 covered greater areas than grid 1, but the difference was small (3–4%). Upon completion of the tests, grids 1–3 covered 192.44, 198.05 and 200.48 mm², respectively.

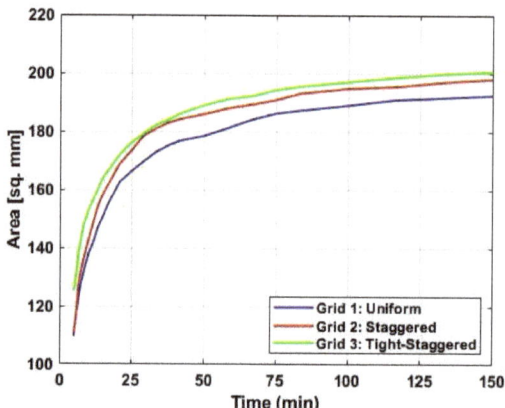

Figure 13. Graph showing the area covered by the droplets using all three grids.

4. Discussion

4.1. Absorption Kinetics

The time to full sorption, t_{h0}, was inversely proportional to fiber bed areal weight. As shown in Figure 8, for 30 Pa·s droplets, fabrics A, B and C exhibited t_{h0} values of 33.8, 25.3 and 10.5 min, respectively. Furthermore, only fabric A exhibited full through-thickness penetration of the fluid, as evidenced by the presence of facsimile fluid residue on the underside of the fabric. From these results, we assert that heavier fabrics allowed for deeper through-thickness flow of the droplet, reducing the time-to-saturation for droplets of the same size.

In contrast, increasing fluid viscosity generally increased sorption time, although the relationship between viscosity and t_{h0} was not proportional (based on the two silicone oils tested). Figure 8 shows that, except for fabric D, increasing fluid viscosity by a factor of two led to an 8- to 20-fold increase in sorption time. In fabric D, high-viscosity resin readily penetrated the macrolevel features (features that were absent in fabrics A through C).

The mechanism by which droplet height changed depended on fluid viscosity. For example, the height change of the 60 Pa·s droplets was driven primarily by spreading on the fabric surface, while the height change of 30 Pa·s droplets was dominated by more rapid penetration into the fabric, shown previously in Figure 7. Higher viscosity fluid droplets spread longer distances along the fiber bed surface, in some cases extending beyond the field of view. In contrast, droplets of the lower viscosity fluid remained within the field of view until full sorption was reached. The 60 Pa·s droplets showed reduced absorption and spread to a large area, resulting in extended absorption times.

The most important metric for production of semi-pregs via droplet deposition is the time to full sorption, referring to the time until the droplet is fully within the fiber bed. As such, by ensuring that a deposited droplet can achieve maximum spread within the time to full sorption, it should be possible to produce prepreg with a degree of impregnation high enough to achieve bulk factors similar to standard OoA prepregs (~10%). As shown in Figure 6, droplet flow continued after t_{h0}, allowing the droplet to flow further within the fiber bed and ensuring full saturation of the fiber bed during cure. However, in practice, a balance must be achieved between the DoI, bulk factor and droplet separation. In particular, the spreading time must be close enough to t_{h0} to achieve both a high DoI and low bulk factor, while still maintaining droplet separation and allowing for full ply saturation during cure. Furthermore, given the differences in t_{h0} achieved by the minor increase in viscosity used in this study, droplet spreading can be virtually arrested by simply cooling the prepreg.

4.2. Spread Kinetics

Fabrics A–C were all uniform, unidirectional fabrics with similar smooth surfaces, as shown in Figure 2 (left), and as such similar droplet spread behavior was expected for these fabrics. Figure 9 shows that droplet spread distances fell within a narrow range, regardless of fiber areal weight. Furthermore, there was no correlation between spread distance and areal weight. Droplet spread distance was governed solely by surface topography, while areal weight had a negligible effect.

The droplet spread was characterized by two stages as well as two different directions. The two spreading stages were (a) rapid spreading while the droplet remained atop the fabric surface, followed by (b) slower spreading once the droplet was fully imbibed. In the test shown in Figure 6, 80% of spreading along the fibers occurred before full sorption, and 90% of spreading across the fibers occurred before full sorption. Spreading above the surface was driven primarily by gravity and surface tension, while spreading within the fiber bed was driven by capillary effects. The spreading was also different along and across the fiber direction and droplets spread much longer distances in the fiber direction than across fibers. Furthermore, most of the spreading occurred before sorption in the through-thickness direction. The early stage spreading was aided by capillarity between aligned fibers, facilitating fluid flow along fibers and impairing flow across the fibers after full absorption.

For fabrics A–C, the capillary radius was expected to remain the same, and when using the same viscosity fluid, the remaining parameters were similarly unchanged. Therefore, the distance traveled within the capillary, and as such the final distance spread, remained the same, as expected. The horizontal capillary flow of a droplet is described by Washburn's Equation (2), where L is the distance traveled within the capillary, γ is the surface tension, r is the capillary pore radius, t is time, θ refers to the contact angle and η is the viscosity [26].

$$L = \sqrt{\frac{\gamma r t \cos(\theta)}{2\eta}}, \qquad (2)$$

An increase in viscosity was shown in Figure 5 to result in an increase in contact angle, meaning the change in silicone oil resulted in two terms changing within Washburn's equation. Furthermore, referring to Equation (2), it can be seen that the increase in viscosity and increase in angle effectively counteracted one another, resulting in only minor deviations in spread distance.

4.3. Effects of Macrofeatures

Macroscopic features of fabrics, particularly inter-tow gaps, strongly influenced droplet spread. For example, Figure 10 showed that fluid flowed quickly into inter-tow regions of fabric D. These regions served as channels for macroflow, as opposed to microflow within tows. Similar results can be expected for woven fabrics, where the placement of the droplet, particularly with regard to the proximity to pinholes in the weave, affects droplet spread more strongly than fabric areal weight or intra-tow capillary sizes.

With regard to developing a method for predicting droplet spread, an analytical solution or a numerical simulation would be useful for unidirectional fabrics. However, for woven fabrics, which feature complex surface topography, the parameters of the fabric may have only minor effects on surface spread compared to the position of the droplet on the fabric terrain. Generating an analytical solution or simulation for such fabrics would require an added level of complexity and rely on the specific surface features of the given weave.

4.4. Droplet Arrays

Droplet positioning and array designs can be guided by the results presented here, as shown by the observations of spread behavior in the three different droplet grids. The findings support the hypothesis that fluid flow occurs more rapidly along the fibers than across and provide approximate indications of how a droplet will spread along the surface. The spread distances can be used to arrange droplets in arrays that prevent/minimize fluid overlap and that ensure maximum, controlled flow distances. However, these same results can also be used to ensure full surface coverage, maximizing overlap and flow distances, as shown in Figures 12 and 13. For example, grid 3 exhibits overlapping fluid covered regions while maintaining similar coverage as the other arrays. Due to the strong degree of anisotropy of unidirectional fiber beds, droplet positioning must be staggered to ensure maximum surface coverage.

The droplet grid tests also revealed effects of droplet impingement. Neighboring droplets influenced the direction of spread and in some cases altered the spreading behavior that would be expected for a single droplet. As shown in Figure 12, once contact between droplets occurred, droplets quickly spread to fill regions between neighboring droplets. Flow occurred even across the fiber bed when driven by the surface tension of the fluid. Transverse flow was responsible for the increase in surface area covered by grids 1 and 2, which included droplet interaction across the fiber bed.

Using these insights into single-droplet behavior, droplets can be positioned in arrays that minimize droplet interaction and maximize through-thickness air evacuation pathways. Alternatively, droplets can be positioned to intentionally create a degree of interaction, achieving a higher surface coverage. These capabilities can be leveraged for use in semi-preg design, which requires discontinuous resin distribution, and more careful control of the resin distribution is achievable with this method than with currently implemented methods. Furthermore, should other novel composite designs emerge that rely on discreet resin droplets, this study may serve as an initial guide.

4.5. Facsimile Fluid

Given the impact of droplet–substrate surface parameters on the driving forces behind droplet spreading, matching viscosity is not sufficient to ensure accurate simulation of resin droplets. Experiments showed that the apparent contact angles, or the evolving edge slopes, did not coincide between the facsimile fluid and the resin, and as such the observed results of spreading distance and absorption times for the facsimile fluid cannot be assigned one-to-one to the resin. However, as Figure 5b shows, the relationship between the 30 Pa·s droplet and the 60 Pa·s droplet was maintained both for resin and the silicone oil facsimile. Based on this result, we assert that the relative effects of increasing the viscosity are equivalent when using actual resin droplets.

5. Conclusions

We have demonstrated the effects of unidirectional fiber beds on droplet flow on the surface. The fabric feature that most strongly influenced the surface flow was the fiber bed architecture. Macrolevel features of the quilted UD fabric (fabric D) dominated the surface flow of droplets. Decreasing fiber areal weight and increasing viscosity both led to an increase in time to full absorption. However, neither of these factors appreciably influenced the droplet surface coverage.

The findings presented entail multiple implications. First, predicting droplet surface flow is most reliable for unidirectional fiber beds and the anisotropy of UD can be exploited to design patterns that prevent or minimize droplet impingement. By heating or cooling the fluid, viscosity values can be adjusted to allow longer working times for droplet deposition. Within the narrow viscosity range used in this study, the actual droplet spread distances would see little variation. These findings are potentially useful, yet challenges remain. For droplet deposition onto woven fabrics, the specific interactions between droplets and macrolevel details of the fabric must be tracked. In addition, an actual resin must eventually be used, as opposed to a facsimile fluid. Finally, single droplet size was monitored throughout the study. However, the volume of resin needed to achieve a proper volume fraction for prepreg differs with the fabric weight and smaller droplets are likely more reliable for thinner fabrics. Attending to these challenges will inevitably fall to those designing a droplet deposition system.

The findings highlight the groundwork required to develop practical methods to produce semi-pregs and indicate a pathway to achieving more efficient and robust OoA production. The spread of resin after deposition can be manipulated by determining the time allowed for flow into the fiber bed, as well as droplet position in relation to other droplets and to the fiber bed, allowing for intelligent design of resin distribution in semi-preg. Such designs will facilitate the scaling-up for manufacture of semi-pregs via droplet deposition, as opposed to current methods that rely on prepreg manufacture one ply at a time. Semi-preg is a robust intermediate material for OoA processing and can potentially restore robustness to levels comparable to conventional autoclave manufacturing. In addition, OoA processes reduce costs and increase part throughput, allowing for increased part complexity and accessibility.

Author Contributions: Conceptualization, P.M., B.C.J. and S.N.; methodology, P.M.; software, P.M.; validation, P.M. and B.C.J.; formal analysis, P.M.; investigation, P.M.; resources, S.N.; data curation, P.M.; writing—original draft preparation, P.M.; writing—review and editing, P.M., B.C.J. and S.N.; visualization, P.M.; supervision, B.C.J. and S.N.; project administration, P.M. and S.N.; funding acquisition, S.N. All authors have read and agreed to the published version of the manuscript.

Funding: This research was funded by the National Science Foundation Partnerships for Innovation, grant number 53-4504-7788.

Acknowledgments: The authors thank Mark Anders at the M.C. Gill Composites Center for guidance and support, as well as the M.C. Gill Composites Center for material support.

Conflicts of Interest: The authors declare no conflict of interest.

References

1. Steele, M.; Corden, T.; Gibbs, A. The development of out-of-autoclave composite prepreg technology for aerospace applications. In Proceedings of the SAMPE, Long Beach, CA, USA, 23–26 May 2011.
2. Thorfinnson, B.; Biermann, T. Degree of Impregnation of Prepregs—Effects on Porosity. *Int. SAMPE Symp. Exhib.* **1987**, *32*, 1500–1509.
3. Repecka, L.; Boyd, J. Vacuum-bag-only-curable prepregs that produce void-free parts. In Proceedings of the 47th International SAMPE Symposium and Exhibition, Long Beach, CA, USA, 12–16 May 2002.
4. Centea, T.; Hubert, P. Out-of-autoclave prepreg consolidation under deficient pressure conditions. *J. Compos. Mater.* **2014**, *48*, 2033–2045. [CrossRef]
5. Cender, T.A.; Simacek, P.; Davis, S.; Advani, S.G. Gas Evacuation from Partially Saturated Woven Fiber Laminates. *Transp. Porous Media* **2016**, *115*, 541–562. [CrossRef]
6. Tavares, S.S.; Michaud, V.; Månson, J.A.E. Through thickness air permeability of prepregs during cure. *Compos. Part A Appl. Sci. Manuf.* **2009**, *40*, 1587–1596. [CrossRef]
7. Schechter, S.G.K.; Centea, T.; Nutt, S.R. Polymer film dewetting for fabrication of out-of-autoclave prepreg with high through-thickness permeability. *Compos. Part A Appl. Sci. Manuf.* **2018**, *114*, 86–96. [CrossRef]
8. Edwards, W.T.; Martinez, P.; Nutt, S.R. Process robustness and defect formation mechanisms in unidirectional semipreg. *Adv. Manuf. Polym Compos. Sci.* **2020**. [CrossRef]
9. Martinez, P.; Nutt, S.R. Robust manufacturing of complex-shaped parts using out-of-autoclave prepregs with discontinuous formats. CAMX 2018—Compos. In Proceedings of the Composites and Advanced Materials Expo, Dallas, TX, USA, 15–18 October 2018.
10. Grunenfelder, L.K.; Fisher, C.; Cabble, C.; Thomas, S.; Nutt, S.R. Defect Control in Out-of-Autoclave Manufacturing of Structural Elements. In Proceedings of the SAMPE, Charleston, SC, USA, 22–25 October 2012.
11. Schechter, S.G.K.; Centea, T.; Nutt, S. Effects of resin distribution patterns on through-thickness air removal in vacuum-bag-only prepregs. *Compos. Part A* **2020**, *130*, 105723. [CrossRef]
12. Schechter, S.G.K.; Grunenfelder, L.K.; Nutt, S.R. Design and application of discontinuous resin distribution patterns for semi-pregs. *Adv. Manuf. Polym Compos. Sci.* **2020**, *6*, 72–85. [CrossRef]
13. Nutt, S.; Grunenfelder, L.; Centea, T. High-Permeability Composite Prepreg Constructions and Methods for Making the Same. U.S. Patent Application WO 2018/129378 A1, 28 November 2019.

14. Davis, S.H.; Hocking, L.M. Spreading and imbibition of viscous liquid on a porous base. *Phys. Fluids* **1999**, *11*, 48–57. [CrossRef]
15. Davis, S.H.; Hocking, L.M. Spreading and imbibition of viscous liquid on a porous base. II. *Phys. Fluids* **2000**, *12*, 1646–1655. [CrossRef]
16. Alleborn, N.; Raszillier, H. Spreading and sorption of a droplet on a porous substrate. *Chem. Eng. Sci.* **2004**, *59*, 2071–2088. [CrossRef]
17. Comas-Cardona, S.; Binetruy, C.; Krawczak, P. Unidirectional compression of fibre reinforcements. Part 2: A continuous permeability tensor measurement. *Compos. Sci. Technol.* **2007**, *67*, 638–645. [CrossRef]
18. Centea, T.; Hubert, P. Modelling the effect of material properties and process parameters on tow impregnation in out-of-autoclave prepregs. *Compos. Part. A Appl. Sci. Manuf.* **2012**, *43*, 1505–1513. [CrossRef]
19. Michaud, V.; Mortensen, A. Infiltration processing of fibre reinforced composites governing: Phenomena. *Compos. Part. A Appl. Sci. Manuf.* **2001**, *32*, 981–996. [CrossRef]
20. Neogi, P.; Miller, C.A. Spreading kinetics of a drop on a rough solid surface. *J. Colloid Interface Sci.* **1983**, *92*, 338–349. [CrossRef]
21. Hocking, L.M. The wetting of a plane surface by a fluid. *Phys. Fluids* **1995**, *7*, 1214–1220. [CrossRef]
22. Starov, V.M.; Kosvintsev, S.R.; Sobolev, V.D.; Velarde, M.G.; Zhdanov, S.A. Spreading of liquid drops over saturated porous layers. *J. Colloid Interface Sci.* **2002**, *246*, 372–379. [CrossRef]
23. Starov, V.M.; Zhdanov, S.A.; Velarde, M.G. Spreading of liquid drops over dry porous layers: Complete wetting case. *Langmuir* **2002**, *18*, 9744–9750. [CrossRef]
24. Tanner, L.H. The spreading of silicone oil drops on horizontal surfaces. *J. Phys. D Appl. Phys.* **1979**, *12*, 1473–1484. [CrossRef]
25. Bonn, D.; Eggers, J.; Indekeu, J.; Meunier, J. Wetting and spreading. *Rev. Mod. Phys.* **2009**, *81*, 739–805. [CrossRef]
26. Washburn, E.W. The Dynamics of Capillary Flow. *Phys. Rev.* **1921**, *17*, 273–283. [CrossRef]

Article

Preparation and Evaluation of the Tensile Characteristics of Carbon Fiber Rod Reinforced 3D Printed Thermoplastic Composites

Arivazhagan Selvam [1,*], Suresh Mayilswamy [2], Ruban Whenish [3], Rajkumar Velu [4,*] and Bharath Subramanian [3]

1. Department of Mechanical Engineering, SCAD Institute of Technology, Coimbatore 641664, India
2. Department of Robotics and Automation Engineering, PSG College of Technology, Coimbatore 641004, India; kvm.suresh@gmail.com
3. Department of Mechanical Engineering, Sri Krishna College of Technology, Coimbatore 641042, India; wruban1990@gmail.com (R.W.); bharathsubramaniyan1497@gmail.com (B.S.)
4. Centre for Laser Aided Intelligent Manufacturing, University of Michigan, Ann Arbor, MI 48109, USA
* Correspondence: arivuazhagan001@gmail.com (A.S.); rajkumar7.v@gmail.com (R.V.)

Citation: Selvam, A.; Mayilswamy, S.; Whenish, R.; Velu, R.; Subramanian, B. Preparation and Evaluation of the Tensile Characteristics of Carbon Fiber Rod Reinforced 3D Printed Thermoplastic Composites. *J. Compos. Sci.* **2021**, *5*, 8. https://doi.org/10.3390/jcs5010008

Received: 20 October 2020
Accepted: 24 December 2020
Published: 31 December 2020

Publisher's Note: MDPI stays neutral with regard to jurisdictional claims in published maps and institutional affiliations.

Copyright: © 2020 by the authors. Licensee MDPI, Basel, Switzerland. This article is an open access article distributed under the terms and conditions of the Creative Commons Attribution (CC BY) license (https://creativecommons.org/licenses/by/4.0/).

Abstract: The most common method to fabricate both simple and complex structures in the additive manufacturing process is fused deposition modeling (FDM). Many researchers have studied the strengthening of FDM components by adding short carbon fibers (CF) or by reinforcing solid carbon fiber rods. In the current research, we sought to enhance the mechanical properties of FDM components by adding bioinspired solid CF rods during the fabrication process. An effective bonding interface of bioinspired CF rods and polylactic acid (PLA) was achieved by triangular interlocking sutures and by employing synthetic glue as the binding agent. In particular, the tensile strength of solid CF rod reinforced PLA samples was studied. Critical parameters such as layer thickness, extruder temperature, extruder speed, and shell thickness were considered for optimization. Significant process parameters were identified through leverage plots using the response surface methodology (RSM). The optimum parameters were found to be layer thickness of 0.04 mm, extruder temperature of 215 °C, extruder speed of 60 mm/s, and shell thickness of 1.2 mm. The results revealed that the bioinspired solid CF rod reinforced PLA (CFRPLA) composite exhibited a tensile strength of 82.06 MPa, which was approximately three times higher than the pure PLA (28 MPa, 66% lower than CFRPLA), acrylonitrile butadiene styrene (ABS) (28 MPa, 66% lower than CFRPLA), polyethylene terephthalate glycol (PETG) (34 MPa, 60% lower than CFRPLA), and nylon (34 MPa, 60% lower than CFRPLA) samples.

Keywords: CF rod reinforced PLA (CFRPLA); solid carbon fiber rod; fused deposition modeling (FDM); process parameters; tensile strength

1. Introduction

Additive manufacturing (AM) has brought the next industrial revolution in many key areas, such as aerospace, medicine, and automotive. The extensive usage of AM has widened to technologies applied in biomedical implants, architecture, and full-body organs using materials such as polymers, metals, composites, ceramics, wood, etc. [1,2]. Fused deposition modeling (FDM) is one of the simplest and most commonly used technologies in AM. Different types of polymers are used for building 3D objects as per computer-aided design (CAD) by adding one layer over another with specified process parameters. The quality of the printed polymer objects depends on various process parameters, such as the printing material and part orientation during printing. These parameters influence the mechanical characteristics, building time, and volume, which affect the final cost of the part [3,4]. FDM-based fabricated components play a key role in medical, automotive, and aerospace applications due to their simplicity and cost-effectiveness. Polylactic acid

(PLA) is widely used as a processing material in FDM technology due to its desirable characteristics and biocompatibility. Most of the research work carried out with pure PLA and PLA composites have become the hotspot in 3D printing [5]. In their research, Cicala et al. selected three PLA-based commercial filaments to study the influence of the mechanical properties of FDM specimens. A comparative approach was taken to differentiate pure PLA vs. PLA with added mineral fillers in order to identify better mechanical properties and printing quality. The mineral fillers dispersed in the filament improved the tensile properties of the printed specimens compared to pure PLA [6]. However, the strength of the pure thermoplastic components fabricated by the FDM process is comparatively low, which has significant limitations and restricts the application of 3D printed parts as functional components [7]. To overcome these limitations, researchers have recently shifted their focus toward dispersing additives in neat polymers to enhance their properties. Plain polymers produced via 3D printing have limited physicochemical properties. These properties can be improved by introducing fibers or particles along with the polymer matrix. In his study, Blanco showed how these reinforced polymers possessed better properties and could be associated with many fields of application [8]. Similarly, the strength of 3D printing samples was found to increase with the addition of short fibers when compared to samples without fiber reinforcement [9,10]. The mechanical properties and strength of thermoplastics such as acrylonitrile butadiene styrene (ABS) and polyamide-6 produced through FDM was improved by reinforcing different modifiers like plasticizers and short glass fiber [11,12].

Hao et al. analyzed the bending interlaminar strength and lifetime of carbon/epoxy laminated curved beams using the digital speckle correlation method (DSCM) with a four-point bending test configuration by varying the thickness and radius/thickness ratio. The bending interlaminar strength of carbon/epoxy laminated curved beams were analyzed by ultrasonic C-scan images. The failure strength and maximum interlaminar radial stresses of carbon/epoxy laminated curved beams were found when deformation occurred by applying delamination mechanisms [13,14]. Ruban et al. carried out an experiment on the effect of process parameters such as layer thickness, hatch spacing, laser power, and part bed temperature for Laserform stainless steel (ST-100) by adopting the selective laser sintering (SLS) technology. It was observed that better mechanical properties were obtained when the layer thickness was kept as 0.08 mm [15]. Zeng et al. fabricated wood–plastic composite (WPC) parts using SLS. They analyzed the relationship between laser intensity and WPC parts. When the laser intensity increased from 226 to 311 W/mm^2, the tensile strength improved by 191% and the bending strength improved by 17%. However, when the laser intensity exceeded 340 W/mm^2, both the tensile and bending strength decreased [16]. Gray studied polypropylene (PP) thermoplastic material reinforced with thermotropic liquid crystalline polymer (TLCP) fibrils produced by the dual extruder process in FDM technology. The melting point of PP was 165 °C, whereas it was 283 °C for TLCP fibrils. TLCP has long and short fibers, and both were reinforced with PP and produced separately. With respect to mechanical properties, long TLCP fibers were found to have better tensile properties than short TLCP fibers [17]. Fischer et al. designed and developed a nozzle-type 3D fiber print head to produce a polymer matrix composite layer by layer [18]. Melenka et al. fabricated 3D printed structures reinforced with continuous carbon, fiberglass, or Kevlar fibers and evaluated the elastic properties using the volume average stiffness (VAS) method. The fibers were added to the matrix in different volume fractions, and the difference between experimental and predicted elastic properties was identified. This method helped researchers design the functional components to choose the right combination of matrix and reinforcement [19]. Christ et al. analyzed the green strength of large and fragile printed structures while performing the depowdering procedure in a powder bed. By reinforcing different short fibers with the polymer matrix of cellulose-modified gypsum powder, the binding strength increased by about 180%, which added much strength and avoided collapse of the printed green structures, while the work of fracture values was 10 times higher [20]. Considering the bonding interface between carbon fiber (CF) and PLA resin produced by the FDM process, Li et al. designed a nozzle to mix uniformly continuous CF with PLA. The printed part had less bonding interface between CF and PLA, and continuous CF was preprocessed and added to the PLA resin, which improved the tensile and flexural strength of composites by 13.8% and 164%, respectively [21].

Ćwikła et al. investigated the influence of printing parameters on the mechanical properties of ABS produced through FDM technology. Infill pattern, infill density, shell thickness, printing temperature, and the type of material were selected as parameters, and the samples were tested for a honeycomb pattern. Infill density of about 40–50% and shell thickness of 2–3 layers/line were found to be better parameters considering the strength of the printed pattern [22]. In their experiment, Ning et al. evaluated the effect of process parameters such as raster angle, layer thickness, infill speed, and nozzle temperature for producing carbon fiber reinforced plastics (CFRP) using FDM technology. Raster angle and infill speed had a significant influence on the tensile characteristics of the printed parts, while a layer thickness of 0.15 mm produced good tensile properties [23,24]. Another study by Tian et al. analyzed carbon fiber reinforced PLA (CFRPLA) produced by the FDM process considering critical process parameters such as temperature and pressure. The optimized bonding strength between layers was achieved using optimized process parameters such as a layer thickness of 0.4–0.6 mm, hatch spacing of 0.6 mm, and temperature of 200–230 °C. The maximum flexural strength of 335 MPa and flexural modulus of 30 GPa was achieved when the fiber content of the composite specimens was 27% [25]. Lin et al. analyzed process parameters such as tip angle and bonded tip region presence/absence as well as geometries such as trapezoidal, rectangular, triangle, and antitrapezoidal for 3D printed polymer sutures. Triangular sutures exhibited uniform stress distribution throughout the structure, with the teeth areas especially possessing better mechanical properties. The experimental results and analytical models of 3D printed sutures showed excellent agreement [26]. Padzi et al. did a comparative study on the suitability of ABS produced by FDM for industrial applications. ABS with a dog bone shape was produced by FDM or a molding process as per the ASTM D638 standard. Fatigue tests were carried out at 40, 60, and 80% tensile strength. The results showed that ABS produced through 3D printing had lower fatigue life than through the molding process, indicating its suitability for low-strength applications [27]. Changes in process parameters and addition of composites might increase the strength of ABS prototypes produced by FDM [28,29]. Wu optimized the process parameters of FDM by designing a cylindrical model. Parameters such as slice height, printing time, consumables, and dimensional accuracy were optimized, and the results showed that a slice height of 0.14 mm reduced the printing time without sacrificing the quality of printing [30]. Srinivasan et al. examined the influence of infill density on the tensile characteristics of polyethylene terephthalate glycol (PETG) and ABS by FDM technology. Specimens were printed by maintaining the other parameters as constant but varying the infill density levels. It was observed that the infill density had a considerable influence on the tensile strength of the PETG and ABS parts [31–33].

Selvam et al. discussed ways to improve the strength of 3D printed bioinspired interlock sutures made up of PLA reinforced CF by particle swarm optimization (PSO). The strength of 3D printed sutures was highly influenced by building parameters such as printing speed, layer thickness, and temperature. It was observed that a printing speed of 60 mm/s, layer thickness of 0.1 mm, and temperature of 190 °C yielded maximum strength outcomes [34]. Dey et al. studied the tensile properties of ABS and PLA material reinforced with fibers produced by FDM technology and found that they were highly influenced by layer thickness, extruder speed, raster angle, extruder temperature, and shell thickness [35].

In the aforementioned literature, researchers reinforced short fibers or continuous fibers during the fabrication process to improve the strength of 3D printed components. Instead of reinforcing fibers during the fabrication process, this research focused on inserting solid bioinspired carbon fiber rods into the sample after the fabrication process. In addition, triangular-shaped interlock suture from a stickleback fish pelvic girdle was biomimicked and applied to the CF rod and PLA sample. The maximum tensile strength was obtained with optimum process parameters of bioinspired 3D printed PLA using the response surface methodology (RSM) technique.

2. Materials and Methods

2.1. Bioinspired Suture Selection

Various biological structures of animals and plants can be mimicked to acquire unique properties with effective energy absorption capacity. The ancient sea ammonite shells are one of the best examples of suture interfaces that show various types of geometric profile starting from the basic curve to complicated fractal-like designs. Martin et al. referred to a bioinspired reinforced architecture created with 3D magnetic printing by combining additive technologies with discontinuous fibers, which delivered enhanced mechanical performance compared to the monolithic structure [36]. The suture interface geometry of the anatomy of different creatures was investigated and selected to design a solid carbon fiber rod with interlock sutures.

The triangular interlock sutures were inspired from the pelvic girdle of marine sticklebacks (fish). The three-spine marine sticklebacks protect themselves from predators with their biological armor structured body. The armor body of sticklebacks consists of the basal plate, spines, dorsal spines, lateral plates, and pelvic girdle. Many studies have been done on the structure of stickleback armor to develop bioinspired protective shields, vehicle body, building structures, protective coating, abrasion/rock penetration resistance pipelines, safety helmets, etc. The pelvic girdle of sticklebacks is made up of two ventral pelvic plates joined together by a triangular interlocking suture. Many researchers have used different bioinspired types of sutures like antitrapezoidal, rectangular, sawtooth, jigsaw, and triangular to enhance the mechanical properties of FDM components [37]. These studies reveal that, compared to other interlock sutures, triangular geometry sutures have a simple design and are more effective as they have superior tensile strength, stiffness, flexural strength, and toughness. From the above observation, a triangular-shaped suture was made manually on a plain carbon fiber rod using a specialized triangular file, as shown in Figure 1. The triangular-shaped file tool had a three-sided design that could sharpen and make a saw teeth profile (triangular suture).

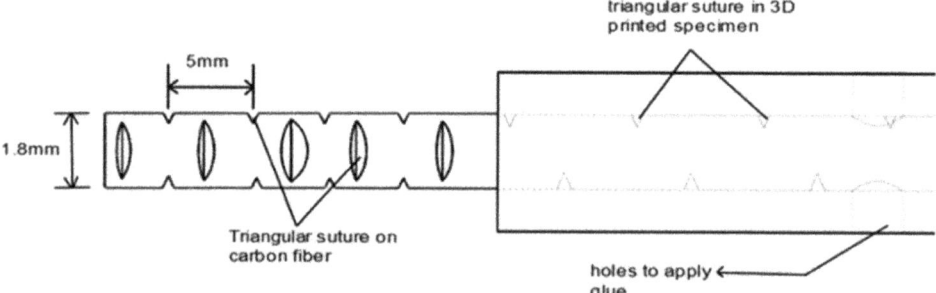

Figure 1. Solid carbon fiber (CF) rod and 3D printed specimen with triangular interlock sutures.

2.2. Critical Process Parameters

In the additive manufacturing process of solid CF rod reinforced FDM samples, there are various significant process parameters that influence and improve the mechanical properties. Based on the literature study, layer thickness, extruder temperature, extruder speed, and shell thickness were chosen as the parameters [33].

2.3. Material Selection and Mechanical Property Investigation

PLA was selected as a printing material for the interlocking suture due to its biodegradable nature, availability, and diverse properties. A short carbon fiber rod was employed with pure PLA to enhance the mechanical properties. The addition of CF as a composite material is not conventional, so a new approach was applied by employing CF as a solid rod. The properties of PLA and the CF material are mentioned in Table 1 below.

Table 1. Mechanical properties of polylactic acid (PLA) and carbon fiber rod.

Mechanical Properties of PLA	Mechanical Properties of Carbon Fiber Rod
Tensile strength: 59 MPa	Tensile strength: 3800 MPa
Flexural strength: 106 MPa	Density: 1.81 g/cm^3
Elongation at break: 7.0%	Carbon content: 95%
Rockwell hardness: 88HR	Elastic modulus: 242 GPa

2.4. Specimen Fabrication and Use of Solid Carbon Rod

A tensile specimen of pure PLA was designed with 3D modeling software and converted into STL format. The fabrication was initiated with reference process parameters that were chosen from various literatures. For the printing of pure PLA specimen, process parameters such as raster angle of 45° and horizontal build orientation [33] were kept constant throughout the experiments as per ASTM D638-10 tensile test specimen, as shown in Figure 2. To increase the tensile strength of the fabricated reference sample, bioinspired solid CF rods of diameter 1.8 mm was reinforced manually into the three holes provided in the specimen. The inner surface of the holes was projected with a triangular suture. The interface bonding between the PLA and the CF rods were improved by applying adhesive (cyanoacrylate) through the holes exactly above the path of the solid CF rods. The process parameters mentioned in Table 2 were used to fabricate the pure PLA tensile specimen.

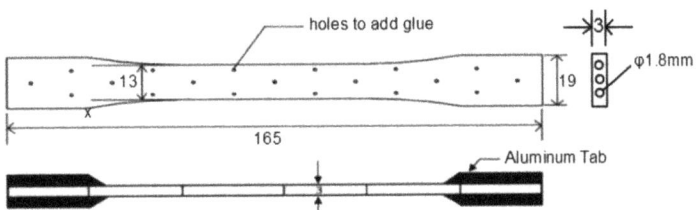

Figure 2. ASTM D638-10 tensile test specimen reinforced with solid carbon fiber rods.

Table 2. Reference sample parameters [29].

Layer Thickness (mm)	Extruder Temperature(°C)	Extruder Speed (mm/s)	Shell Thickness (mm)	Tensile Strength (MPa)
0.08	205	80	0.8	71.6

If the thickness of the tensile specimen is 3 mm, the diameter of the solid CF rod should be less than 3 mm. At the same time, it should occupy more than 50% of the thickness of a tensile specimen in order to enhance the tensile strength and triangular suture characteristics. Solid CF rods are available in different diameters, but based on the above criteria, a rod diameter of 1.8 mm, or 60% of the tensile specimen value, was selected. The plain solid CF rod was preprocessed by a triangular file with 60° angle to generate triangular interlock sutures over the surface with a regular interval of 5 mm. Next, the solid CF rod was inserted into the tensile specimen hole with the help of a plastic mallet (interference fit). The linear and angular movement of the solid carbon rod was arrested by pouring the glue into the holes on the tensile specimen surface as well as in the 1.8 mm hole. The bonding strength between the CF rod and the pure PLA tensile specimen was ensured by a bonding agent (cyanoacrylate). The flow ability of the bonding agent was higher and covered the whole inner surface of the tensile specimen. Hence, no other external force was needed to increase the bonding strength. The layer view of solid CFRPLA is presented in Figure 3. Figure 4 represents the raster angle of 45°, and Figure 5 illustrates the types of built orientation for fabricating tensile specimens.

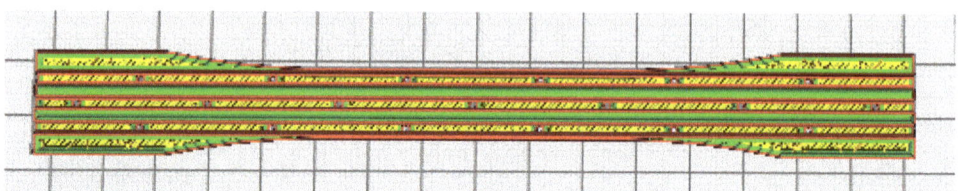

Figure 3. Layer view of solid CF rod reinforced with polylactic acid(PLA) specimen.

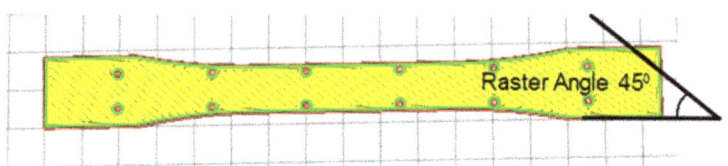

Figure 4. Raster angle.

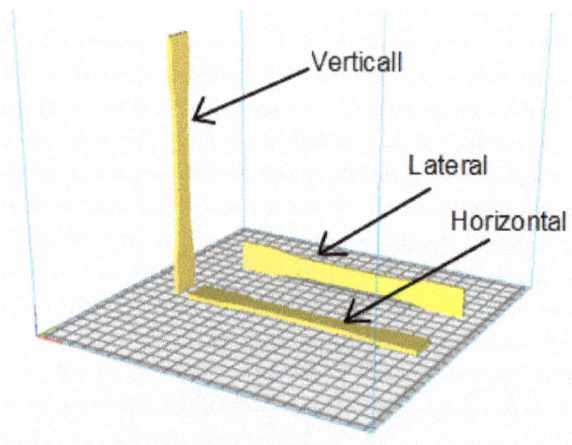

Figure 5. Built orientation.

2.5. Tensile Testing of the CFRPLA Rod

After the fabrication of pure PLA, it was employed with the CF rod using the binding agent. The tensile strength was measured to understand the mechanical behavior of PLA employed with the CF rod. The Autograph AG-Is 50 KN machine was used for the tensile test. The tensile test for the specimen was carried out gradually at a jaw speed of 0.4 mm/s until the specimen broke. An extensometer was employed to measure the strain. The tensile test was performed to measure the tensile strength of the reference sample and compare it with the pure PLA sample, as shown in Figure 6. The results showed that the tensile strength of bioinspired solid CF rod reinforced FDM samples was 2.5 times superior to 3D printed pure PLA samples. The Young's modulus (tensile modulus) of the CFRPLA was 2.5 to 4 times higher than the plain PLA based on the stress–strain curve. Figure 7 shows the CFRPLA tensile specimen after the tensile test. To maximize the tensile strength of 3D printed components and determine the detailed relationship between each parameter and the tensile characteristics of specimens, a trial and error method was followed.

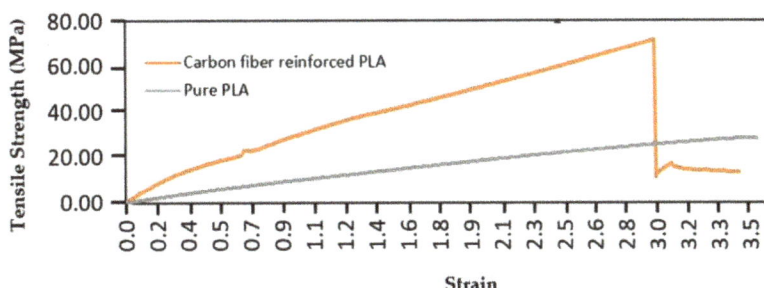

Figure 6. Pure PLA vs. carbon fiber reinforced specimen.

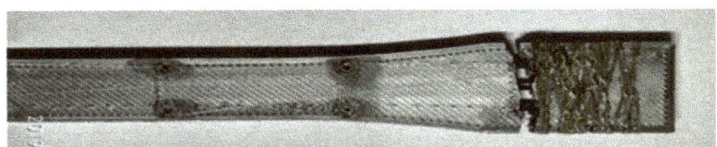

Figure 7. Tensile specimen of the CF rod reinforced PLA (CFRPLA) rod.

2.5.1. Tensile Strength vs. Layer Thickness

The effect of layer thickness was evaluated by fabricating five different samples of varying layer thickness ranging from 0.04 to 0.12 mm, as specified in Table 3. In the reference sample, the layer thickness was kept as 0.08 mm referring to the literature. In addition, two upper and two lower values with intervals of 0.02 mm were used. While fabricating the samples, the other four parameters were kept constant as mentioned in the reference sample. As shown in Figure 8, the sample printed with a layer thickness of 0.06 mm had the maximum tensile strength. From the above result, it can be concluded that the optimum layer thickness 0.06 mm would give the highest tensile strength for CF rod reinforced 3D printed samples.

Table 3. Tensile strength vs. layer thickness.

Layer thickness (mm)	0.04	0.06	0.08	0.1	0.12
Tensile strength (MPa)	69	72	71.6	68	67

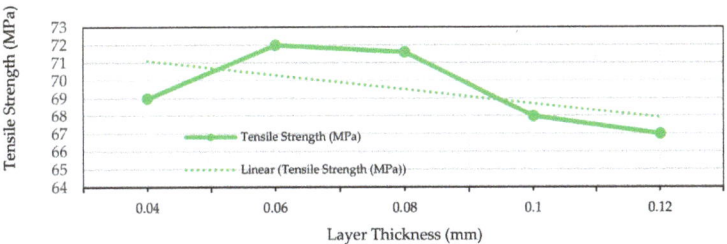

Figure 8. Tensile strength vs. layer thickness.

2.5.2. Tensile Strength vs. Extruder Temperature

The effect of extruder temperature was investigated by fabricating five different samples with extruder temperature ranging from 195 to 215 °C, as specified in Table 4. In addition to the sample specimen extruder temperature 205 °C, two upper and two lower values with intervals of 5 °C were used. While printing the samples, the previously obtained optimum layer thickness of 0.06 mm was used, and the remaining three parameter values mentioned in the reference sample were kept constant. As shown in Figure 9, the sample printed with the extruder temperature of 210 °C resulted in the highest tensile strength. From the above result, it was concluded that the optimum extruder temperature of 210 °C would give the highest tensile strength compared to the other extruder temperatures.

Table 4. Tensile strength vs. extruder temperature.

Extruder temperature (°C)	195	200	205	210	215
Tensile strength (MPa)	65	70	72	74.6	74.1

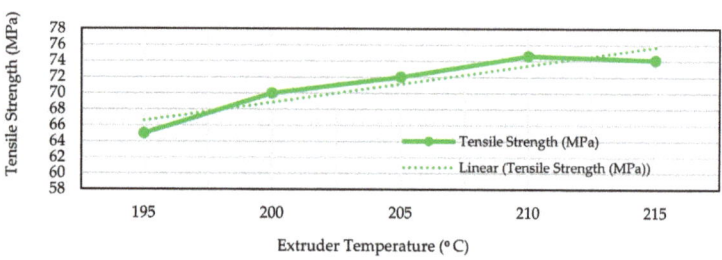

Figure 9. Tensile strength vs. extruder temperature.

2.5.3. Tensile Strength vs. Extruder Speed

The effect of extruder speed was investigated by fabricating five different samples with extruder speeds ranging from 60 to 100 mm/s, as specified in Table 5. In addition to the sample specimen extruder speed of 80 mm/s, two upper and two lower values with intervals of 10 mm/s were used. While printing the samples, the previously obtained optimum layer thickness of 0.06 mm and extruder temperature of 210 °C were used, and the remaining two parameter values mentioned in the reference sample were kept constant. As shown in Figure 10, the tensile strength of the solid CF rod reinforced 3D printed sample was highest when the extruder speed was 60 mm/s. From the above result, it was concluded that the optimum extruder speed of 60 mm/s would give the highest tensile strength of solid CF rod reinforced 3D printed samples.

Table 5. Tensile strength vs. extruder speed.

Extruder speed (mm/s)	60	70	80	90	100
Tensile strength (MPa)	76.1	75.6	74.5	72	70

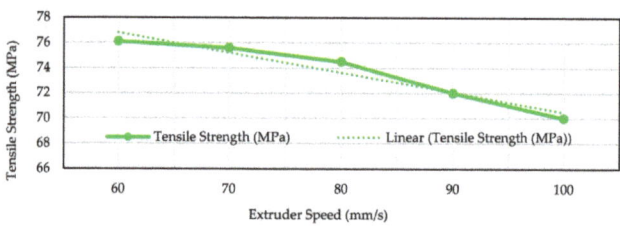

Figure 10. Tensile strength vs. extruder speed.

2.5.4. Tensile Strength vs. Shell Thickness

The effect of shell thickness was investigated by fabricating five different samples with shell thickness ranging from 0.4 to 1.2 mm, as specified in Table 6. In addition to the sample specimen shell thickness of 0.8 mm. two upper and two lower values with intervals of 0.2 mm were used. While printing the samples, the previously obtained optimum parameter values, i.e., layer thickness of 0.06 mm, extruder temperature of 210 °C, and extruder speed of 60 mm/s, were used, and the remaining one parameter mentioned in the reference sample was kept constant. As shown in Figure 11, the tensile strength was enhanced when the sample was printed with a shell thickness of 1.2 mm. From the above result, it was concluded that the optimum shell thickness of 1.2 mm would give the highest tensile strength.

Table 6. Tensile strength vs. shell thickness.

Shell thickness (mm)	0.4	0.6	0.8	1	1.2
Tensile strength (MPa)	71.4	72.8	74.6	76.1	78.8

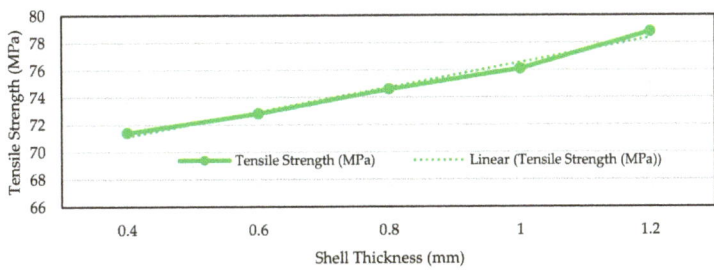

Figure 11. Tensile strength vs. shell thickness.

Based on the experimental results, the process parameters and tensile strength values of each specimen was validated analytically with RSM. The optimum parameters and their influence could be identified and evaluated based on the experimental results, as shown in Table 7.

Table 7. Tensile measurement.

Experimental Runs	Layer Thickness (mm)	Extruder Temperature (°C)	Extruder Speed (mm/s)	Shell Thickness (mm)	Tensile Strength (MPa)
1	0.08	205	80	0.8	71.6
2	0.04	205	80	0.8	69
3	0.06	205	80	0.8	72
4	0.08	205	80	0.8	71.6
5	0.1	205	80	0.8	68
6	0.12	205	80	0.8	67
7	0.06	195	80	0.8	65
8	0.06	200	80	0.8	70
9	0.06	205	80	0.8	72
10	0.06	210	80	0.8	74.6
11	0.06	215	80	0.8	74.1
12	0.06	210	60	0.8	76.1
13	0.06	210	70	0.8	75.6
14	0.06	210	80	0.8	74.5
15	0.06	210	90	0.8	72
16	0.06	210	100	0.8	70
17	0.06	210	60	0.4	71.4
18	0.06	210	60	0.6	72.8
19	0.06	210	60	0.8	74.6
20	0.06	210	60	1	76.1
21	0.06	210	60	1.2	78.8

2.5.5. Mathematical Model

From the experimental data depicted in Tables 8 and 9, a regression model was developed using JMP (John's Macintosh Project) software version 15, 2019. JMP was used in this research for the design of experiments (DOE) and reliability modeling of CFRPLA based on the tensile characteristics. The tensile strength of CFRPLA with respect to the significant process parameters were expressed with the general regression model, as shown in Equation (1).

$$\hat{y} = y - \varepsilon = b0 \times 0 + b1 \times 1 + b2 \times 2 + b3 \times 3 + b4 \times 4 \tag{1}$$

Table 8. Parameter estimates.

Term	Estimate	Std Error	t Ratio	Prob > |t|
Intercept	−17.684	17.071	−1.04	0.316
Layer thickness (mm)	−44.529	19.480	−2.29	0.036
Extruder temperature (°C)	0.451	0.077	5.83	<0.0001
Extruder speed (mm/s)	−0.105	0.030	−3.53	0.003
Shell thickness (mm)	9.05	2.272	3.98	0.001

Table 9. Analysis of variance.

Source	DF	Sum of Squares	Mean Square	F Ratio
Model	4	191.195	47.799	23.155
Error	16	33.028	2.064	Prod > F
C. Total	20	224.223		<0.0001

From the above table, the coefficient of each parameter was substituted in the general regression Equation (1) to develop the mathematical model to predict the optimum process parameter values to maximize the tensile strength.

$$-17.684 - 44.5 \times 1 + 0.450 \times 2 - 0.105 \times 3 + 9.05 \times 4 = \text{Tensile Strength} \tag{2}$$

From the above relationship between the tensile strength of predicted value vs. actual value, the R^2 value of tensile strength was found to be 0.85. A p-value < 0.0001 was used to indicate the significance and influence of process parameters on the tensile characteristic of CFRPLA. The plot between actual tensile strength vs. predicted value is shown in Figure 12.

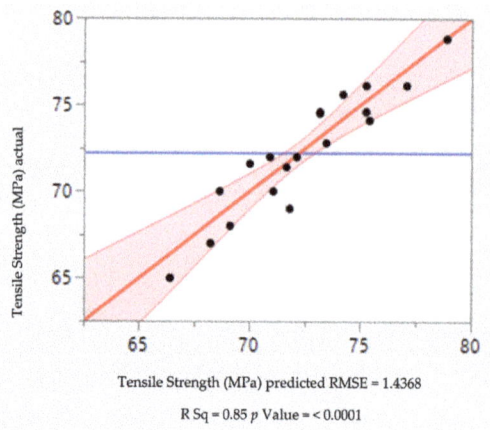

Figure 12. Actual tensile strength vs. predicted value.

3. Result and Discussion

3.1. Influence of Process Parameters on Tensile Strength

The influence of each parameter on the tensile strength of CFRPLA was evaluated by a leverage plot. The *p*-value of each parameter revealed how the parameter influenced the tensile strength.

3.2. Leverage Plot of Layer Thickness vs. Tensile Strength

The plot in Figure 13 shows that $p = 0.036$, which is greater than 0.005, indicating that layer thickness was not a highly significant process parameter.

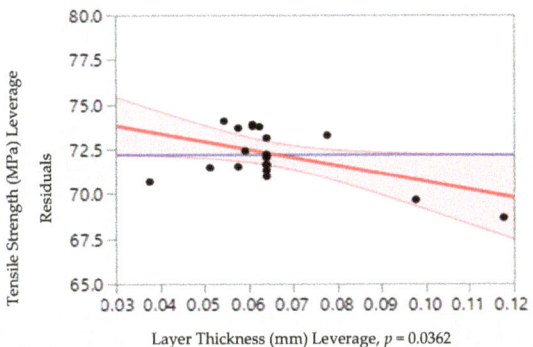

Figure 13. Tensile strength vs. layer thickness.

3.3. Leverage Plot of Extruder Temperature vs. Tensile Strength

The plot in Figure 14 shows $p < 0.0001$, which is less than 0.005, indicating that extruder temperature was a highly significant process parameter.

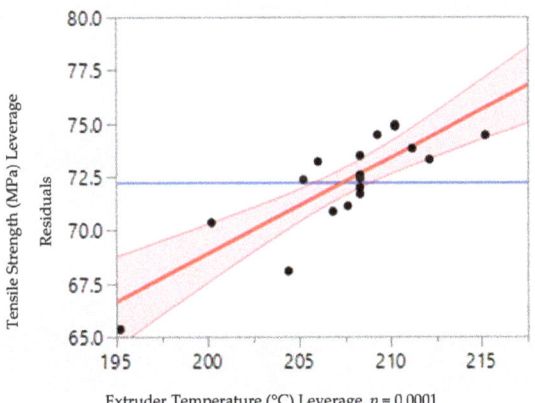

Figure 14. Tensile strength vs. extruder temperature.

3.4. Leverage Plot of Extruder Speed vs. Tensile Strength

The plot in Figure 15 shows that $p = 0.003$, which is less than 0.005, indicating that extruder speed was a highly significant process parameter.

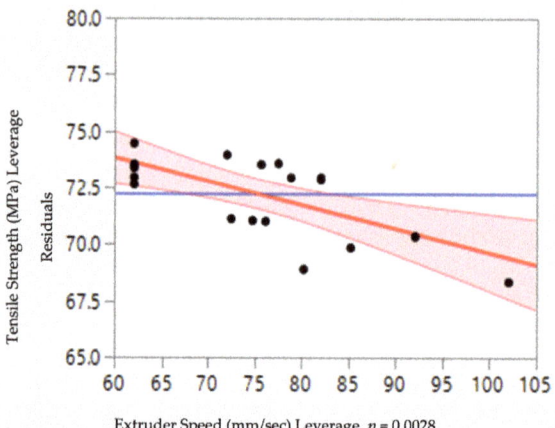

Figure 15. Tensile strength vs. extruder speed.

3.5. *Leverage Plot of Shell Thickness vs. Tensile Strength*

The plot in Figure 16 shows $p = 0.001$, which is less than 0.005, indicating that shell thickness was a highly significant process parameter.

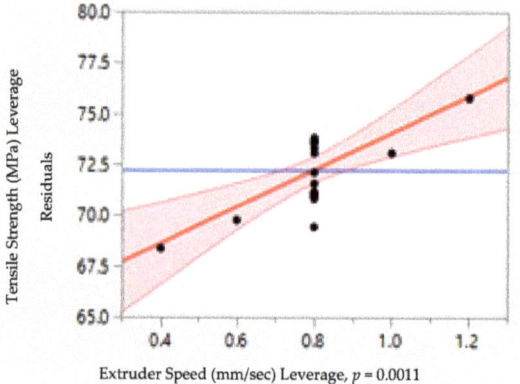

Figure 16. Tensile Strength vs. Shell Thickness.

From the leverage plots, the p-value for each process parameters was obtained to determine their significance. The extruder temperature, extruder speed, and shell thickness highly influenced the tensile strength of the CFRPLA specimen while layer thickness had less impact on the tensile strength.

- The interrelationship between the process parameters influencing the tensile strength of the CFRPLA specimen is exhibited in the interaction plot in Figure 17.
- The layer thickness and extruder temperature were directly proportional to each other in terms of tensile strength.
- The layer thickness and extruder speed were inversely proportional to each other in terms of tensile strength.
- The layer thickness and shell thickness were directly proportional to each other in terms of tensile strength.

- The extruder temperature and extruder speed were inversely proportional to each other in terms of tensile strength.
- The extruder temperature and shell thickness were directly proportional to each other in terms of tensile strength.
- The extruder speed and shell thickness were directly proportional to each other in terms of tensile strength.

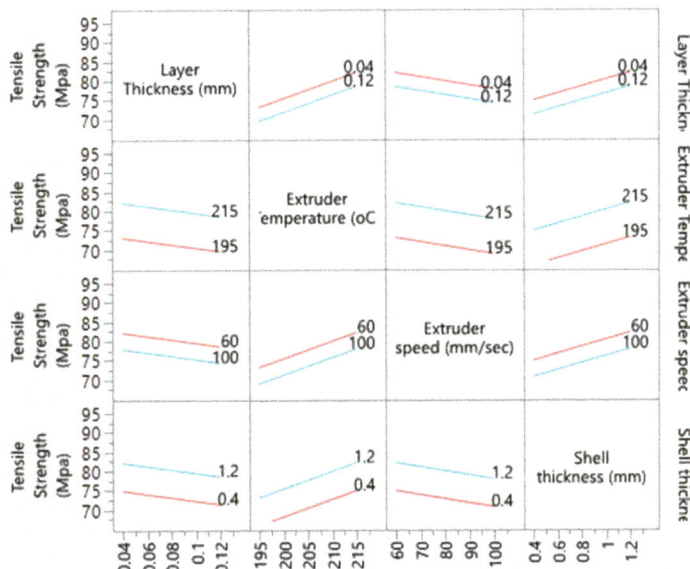

Figure 17. Interaction plot.

From the experimental results, the maximum tensile strength of 78.8 MPa was obtained by maintaining the optimum process parameters values, i.e., layer thickness of 0.06 mm, extruder temperature of 210 °C, extruder speed of 60 mm/s, and shell thickness of 1.2 mm. From the desirability plot shown in Figure 18, the maximum tensile strength that could be obtained was 82.03 MPa using the following process parameters values: layer thickness of 0.04 mm, extruder temperature of 215 °C, extruder temperature of 60 °C, and shell thickness of 1.2 mm. It was observed that maintaining a minimum layer thickness and increasing the number of layers increased the density and also enhanced the bonding strength between the layers of PLA. The maximum extruder temperature of 215 °C and the minimum extruder speed of 60 mm/s increased the curing time and stabilized the printed layer in order to receive further material deposition over it. The maximum shell thickness of 1.2 mm provided protection against external loads over the printed layers, which maximized the tensile strength of CFRPLA.

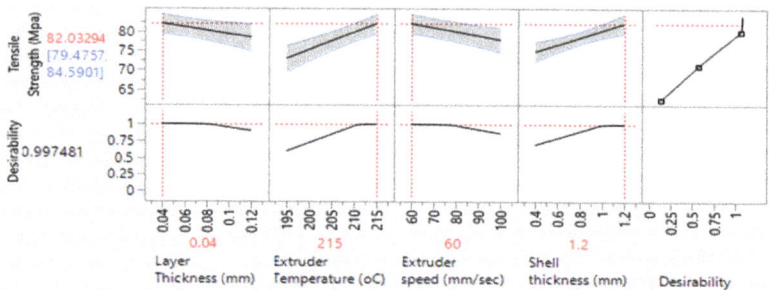

Figure 18. Desirability plot.

4. Conclusions

The ultimate aim of this work was to maximize the tensile strength of FDM-processed PLA with a solid CF rod. The following process parameters were identified to obtain the maximum tensile strength of 82.06 MPa: layer thickness of 0.04 mm, extruder temperature of 215 °C, extruder speed of 60 mm/s, and shell thickness of 1.2 mm. The results clearly showed that the tensile strength of bioinspired solid CFRPLA FDM components were high compared to pure PLA 3D printed components. The results revealed that the tensile strength of pure PLA was 28 MPa, whereas the tensile strength of 3D printed CFRPLA specimen with optimized parameters was 82.06 MPa. This was approximately three times higher compared to pure PLA, indicating a drastic improvement. Future studies will be carried out to determine other mechanical characteristics, such as flexural strength, impact strength, and wear behavior of CFRPLA.

Author Contributions: Conceptualization, A.S.; Formal analysis, A.S.; Investigation, A.S. and R.W.; Resources, R.W.; Supervision, S.M. and R.V.; Writing—original draft, A.S. and R.W.; Writing—review & editing, S.M., R.V and B.S. All authors have read and agreed to the published version of the manuscript.

Funding: This research received no external funding.

Conflicts of Interest: The authors declare no conflict of interest.

References

1. Additive Manufacturing: Opportunities and Constraints. Available online: https://www.raeng.org.uk/publications/reports/additive-manufacturing (accessed on 23 May 2013).
2. Rajkumar, V.; Felix, R.; Sarat, S. 3D printing technologies and composite materials for structural applications. In *Green Composites for Automotive Applications*, 1st ed.; Georgios, K., Arlindo, S., Eds.; Woodhead Publishing: Cambridge, UK, 2019; pp. 171–196.
3. Khaled, G.M.; Carlo, M.; Ahmed Jawad, Q. Strength to cost ratio analysis of FDM Nylon 12 3D Printed Parts. *Procedia Manuf.* **2018**, *26*, 753–762.
4. Rajkumar, V.; Nahaad Mohammed, V.; Chadurvedi, V.; Felix, R.; Murali, K. Experimental Investigation on Fabrication of Thermoset Prepreg Composites Using Automated Fibre Placement Process and 3D Printed Substrate; Procedia CIRP: Amsterdam, The Netherlands, 2019; Volume 85, pp. 296–301.
5. Liu, Z.; Wang, Y.; Wu, B. A critical review of fused deposition modeling 3D printing technology in manufacturing PLA parts. *Int. J. Adv. Manuf. Technol.* **2019**, *102*, 2877–2889. [CrossRef]
6. Cicala, G.; Giordano, D.; Tosto, C.; Filippone, G.; Recca, A.; Blanco, I. Polylactide (PLA) Filaments a Biobased Solution for Additive Manufacturing: Correlating Rheology and Thermomechanical Properties with Printing Quality. *Materials* **2018**, *11*, 1191. [CrossRef] [PubMed]
7. Karsli, N.G.; Aytac, A. Tensile and thermo mechanical properties of short carbon fiber reinforced polyamide 6 composites. *Compos. Part B Eng.* **2013**, *51*, 270–275. [CrossRef]
8. Blanco, I. The Use of Composite Materials in 3D Printing. *J. Compos. Sci.* **2020**, *4*, 42. [CrossRef]
9. Zhong, W.; Li, F.; Zhang, Z.; Song, L.; Li, Z. Short fiber reinforced composites for fused deposition modelling. *Mater. Sci. Eng. A-Struct.* **2001**, *301*, 125–130. [CrossRef]

10. Hao, W.; Liu, Y.; Zhou, H.; Chen, H.; Fang, D. Preparation and characterization of 3D printed continuous carbon fiber reinforced thermosetting composites. *Polym. Test.* **2018**, *65*, 29–34. [CrossRef]
11. Shofner, M.L.; Lozano, K.; Rodríguez-Macías, F.J.; Barrera, E.V. Nanofiber-reinforced polymers prepared by fused deposition modeling. *J. Appl. Polym. Sci.* **2003**, *89*, 3081–3090. [CrossRef]
12. Standard Terminology for Additive Manufacturing Technologies. West Conshohocken: ASTM International. 2012. Available online: https://www.astm.org/Standards/F2792.htm (accessed on 31 December 2020).
13. Hao, W.; Ge, D.; Ma, Y.; Yao, X.; Shi, Y. Experimental investigation on deformation and strength of carbon/epoxy laminated curved beams. *Polym. Test.* **2012**, *31*, 520–526. [CrossRef]
14. Hao, W.; Yuan, Y.; Zhu, J.; Chen, L. Effect of impact damage on the curved beam interlaminar strength of carbon/epoxy laminates. *J. Adhes. Sci. Technol.* **2016**, *30*, 1189–1200. [CrossRef]
15. Ruban, W.; Vijayakumar, V.; Dhanabal, P.; Pridhar, T. Effective process parameters in selective laser sintering. *Int. J. Rapid Manuf.* **2014**, *4*, 148–164. [CrossRef]
16. Zeng, W.L.; Guo, Y.L.; Jiang, K.Y.; Yu, Z.X.; Liu, Y.; Shen, Y.D. Laser intensity effect on mechanical properties of Wood-Plastic composite parts fabricated by selective laser sintering. *J. Thermoplast. Compos.* **2012**, *26*, 125–136. [CrossRef]
17. Gray, W.; Baird, D.G.; Bøhn, J.H. Thermoplastic composites reinforced with long fiber thermotropic liquid crystalline polymers for fused deposition modelling. *Polym. Compos.* **1998**, *19*, 383–394. [CrossRef]
18. Fischer, A.; Rommel, S.; Bauernhansl, T.; Kovács, G.; Kochan, D. New Fiber Matrix Process with 3D Fiber Printer—A Strategic In-process Integration of Endless Fibers Using Fused Deposition Modeling (FDM). *IFIP Adv. Inf. Commun. Technol.* **2013**, *411*, 167–175.
19. Melenka, G.W.; Cheung, B.K.O.; Schofield, J.S.; Dawson, M.R.; Carey, J.P. Evaluation and prediction of the tensile properties of continuous fiber-reinforced 3D printed structures. *Compos. Struct.* **2016**, *153*, 866–875. [CrossRef]
20. Susanne, C.; Martin, S.; Elke, V.; Jurgen, G.; Uwe, G. Fiber reinforcement during 3D printing. *Mater. Lett.* **2015**, *139*, 165–168.
21. Nanya, L.; Li, Y.; Liu, S. Rapid prototyping of continuous carbon fiber reinforced polylacticacid composites by 3D printing. *J. Mater. Process. Technol.* **2016**, *238*, 218–225.
22. Cwikla, G.; Grabowik, C.; Kalinowski, K.; Paprocka, I.; Ociepka, P. The influence of printing parameters on selected mechanical properties of FDM/FFF 3D-printed parts. *Mater. Sci. Eng.* **2017**, *227*, 012033.
23. Ning, F.; Cong, W.; Hu, Y.; Wang, H. Additive manufacturing of carbon Fiber-reinforced plastic composites using fused deposition modeling: Effects of process parameters on tensile properties. *J. Compos. Mater.* **2017**, *51*, 451–462. [CrossRef]
24. Farzad, R.; Godfrey, C.O. Fused deposition modelling (FDM) process parameter prediction and optimization using group method for data handling (GMDH) and differential evolution (DE). *Int. J. Adv. Manuf. Technol.* **2014**, *73*, 509–519.
25. Tian, X.; Liu, T.; Yang, C.; Wang, Q.; Li, D. Interface and performance of 3D printed continuous carbon fiber reinforced PLA composites. *Compos. Part A* **2016**, *88*, 198–205. [CrossRef]
26. Lin, E.; Li, Y.; Ortiz, C.; Boyce, M.C. 3D printed, bio-inspired prototypes and analytical models for structured suture interfaces with geometrically-tuned deformation and failure behaviour. *J. Mech. Phys. Solids* **2014**, *73*, 166–182. [CrossRef]
27. Huang, B.; Meng, S.; He, H.; Jia, Y.; Xu, Y.; Huang, H. Study of Processing Parameters in Fused Deposition Modeling Based on Mechanical Properties of Acrylonitrile-Butadiene-Styrene Filament. *Polym. Sci. Eng.* **2019**, *59*, 120–128. [CrossRef]
28. Padzi, M.M.; Bazin, M.M.; Muhamad, W.M.W. Fatigue Characteristics of 3D Printed Acrylonitrile Butadiene Styrene (ABS). *Mater. Sci. Eng.* **2017**, *269*, 012060. [CrossRef]
29. Zhao, H.; Liu, X.; Zhao, W.; Wang, G.; Liu, B. An Overview of Research on FDM 3D Printing Process of Continuous Fiber Reinforced Composites. *J. Phys. Conf. Ser.* **2019**, *1213*, 052037. [CrossRef]
30. Wu, J. Study on optimization of 3D printing parameters. *Mater. Sci. Eng.* **2018**, *392*, 062050. [CrossRef]
31. Srinivasan, R.; Ruban, W.; Deepanraj, A.; Bhuvanesh, R.; Bhuvanesh, T. Effect on infill density on mechanical properties of PETG part fabricated by fused deposition modelling. *Mater. Today Proc.* **2020**, 1–5. [CrossRef]
32. Srinivasan, R.; Pridhar, T.; Ramprasath, L.S.; Sree Charan, N.; Ruban, W. Prediction of tensile strength in FDM printed ABS parts using response surface methodology (RSM). *Mater. Today Proc.* **2020**, 1–6. [CrossRef]
33. Srinivasan, R.; Nirmal Kumar, K.; Jenish Ibrahim, A.; Anandu, K.V.; Gurudhevan, R. Impact of fused deposition process parameter (infill pattern) on the strength of PETG part. *Mater. Today Proc.* **2020**, 1–5. [CrossRef]
34. Selvam, A.; Mayilswamy, S.; Whenish, R. Strength improvement of additive manufacturing components by reinforcing carbon fiber and by employing bioinspired interlock sutures. *J. Vinyl Addit. Technol.* **2020**, 1–13. [CrossRef]
35. Dey, A.; Yodo, N. A Systematic Survey of FDM Process Parameter Optimization and Their Influence on Part Characteristics. *J. Manuf. Mater. Process.* **2019**, *3*, 64. [CrossRef]
36. Martin, J.; Fiore, B.; Erb, R. Designing bioinspired composite reinforcement architectures via 3D magnetic printing. *Nat. Commun.* **2015**, *6*, 8641. [CrossRef] [PubMed]
37. Zhang, Y.; Yao, H.; Christine, O.; Xu, J.; Dao, M. Bio-inspired interfacial strengthening strategy through geometrically interlocking designs. *J. Mech. Behav. Biomed. Mater.* **2012**, *15*, 70–77. [CrossRef] [PubMed]

Article

Mechanical Analysis of Flexible Riser with Carbon Fiber Composite Tension Armor

Haichen Zhang, Lili Tong * and Michael Anim Addo

College of Aerospace and Civil Engineering, Harbin Engineering University, Harbin 150001, China; zhc@hrbeu.edu.cn (H.Z.); adoreen64@yahoo.com (M.A.A.)
* Correspondence: tonglili@hrbeu.edu.cn

Abstract: As oil and gas exploration moves to deeper areas of the ocean, the weight of flexible risers becomes an important factor in design. To reduce the weight of flexible risers and ease the load on the offshore platform, this paper present a cylindrical tensile armor layer made of composite materials that can replace the helical tensile armor layer made of carbon steel. The ACP (pre) of the workbench is used to model the composite tension armor. Firstly, the composite lamination of the tensile armor is discussed. Then, considering the progressive damage theory of composite material, the whole flexible riser is analyzed mechanically and compared with the original flexible riser. The weight of the flexible riser decreases by 9.73 kg/m, and the axial tensile stiffness decreases by 17.1%, while the axial tensile strength increases by 130%. At the same time, the flexible riser can meet the design strength requirements of torsion and bending.

Keywords: flexible riser; tensile armor; carbon fiber; composite tensile armor

Citation: Zhang, H.; Tong, L.; Addo, M.A. Mechanical Analysis of Flexible Riser with Carbon Fiber Composite Tension Armor. *J. Compos. Sci.* **2021**, *5*, 3. https://dx.doi.org/10.3390/jcs5010003

Received: 1 December 2020
Accepted: 22 December 2020
Published: 25 December 2020

Publisher's Note: MDPI stays neutral with regard to jurisdictional claims in published maps and institutional affiliations.

Copyright: © 2020 by the authors. Licensee MDPI, Basel, Switzerland. This article is an open access article distributed under the terms and conditions of the Creative Commons Attribution (CC BY) license (https://creativecommons.org/licenses/by/4.0/).

1. Introduction

Nowadays, the demand for resources is increasing rapidly in the world, and there are many oil and gas resources in the ocean. Therefore, more and more researchers pay attention to the exploitation of marine resources [1]. A flexible riser is a lifeline for transporting offshore oil and for natural gas resources. The force acting on it during installation and operation is very complex, including bearing tension and internal and external pressure loads. With increases in depth, the suspension weight and fatigue performance of risers have gradually become important. Flexible risers consist of 8 or 9 layers generally, each of which is of a different material, geometry, and function. The carcass and the pressure armor are both self-locking. The carcass is made of AISI304, and the function is to prevent the pipeline from crushing due to hydrostatic pressure or gas accumulation in the annulus of the pipeline. When the axial deformation is too large, the axial stiffness of the riser will be greatly increased after the formation of self-locking [2]. The pressure armor is made of carbon steel strips, and its function is to withstand the stress caused by the fluid pressure in the pipe. Tensile armor is usually made of carbon steel strips that are cross-wound. The cross angle is 30°–55°. It mainly bears the tension, torsional load, and bending moment of flexible pipe. The inner and outer plastic and anti-wear tape are all made of polymers, which provide leak-proof qualities for flexible risers while avoiding direct contact between the armor [3,4].

As oil and gas exploration moves to deeper areas of the ocean, the weight of flexible risers becomes an important factor in the design. To reduce the weight of flexible risers, ease the load on the offshore platform, and alleviate the influence of gravity on risers, the float bowl can be installed outside the risers to provide a certain buoyancy. A common float bowl system currently is the catenary anchor leg mooring (CALM) system. The density of buoy material is as small as possible so that it can provide as large a buoyancy as possible per unit volume. At the same time, it is required to withstand hydrostatic pressure and have lower absorption of water and elasticity modulus so that it can provide stable buoyancy.

However, the buoy increases the volume of the riser. Under waves and current loads, the riser will generate greater motion response and vortex-induced vibration. Ryu [5] used numerical and experimental methods to analyze the time-domain coupling of the CALM system. The free-floating buoy and the buoy with the mooring system were studied, respectively. The frequency–domain analysis model considering the buoy damping force was established, which proved that the damping force plays an important role in the motion response of the buoy. Kang [6] and Hovde [7] studied a deep-water CALM system in West Africa. Kang [6] analyzed the coupled response of the CALM system. It can be seen that the installation of a buoy device reduces the suspension gravity of the riser to a certain extent, but it causes vortex-induced vibration, resulting in greater fatigue damage in the area with the buoy.

Fiber material has high strength weight, which makes the pipeline structure with equivalent structural bearing capacity lighter, and it has higher fatigue and corrosion resistance. Flexible composite pipe (FCP) is made up of reinforced thermoplastic liner coated protective layers [3,4]. Toh et al. [8] analyzed flexible composite pipes and explained that the flexible composite pipes could reduce their weight to alleviate the load-bearing pressure of offshore platforms under the premise of maintaining the mechanical properties by appropriate design. Amaechi et al. [9] discussed the effects of carbon fiber/epoxy (T700), glass fiber/epoxy (S-2), and composite lamination on the safety factor of reinforced thermoplastic flexible composite pipes with an ANSYS ACP (ANSYS Composite PrepPost) module and presented an optimum design method by analyzing the stress of different functional layers. It is more and more common to use FRP (Fiber Reinforced Polymer) material or composite armor instead of metal compression armor and metal tension armor of non-bonded flexible pipe. Technip has developed flexible tubes with four carbon fiber armor layers (CFA). The flexible tube integrated with four carbon fiber armor layers has higher performance and is suitable for light flexible risers, which can reduce or eliminate the need for buoyancy cylinders in ultra-deep-water. Liu [10] developed a model of composite tensile armor layers, introduced it into the whole model, and analyzed the tensile properties of the flexible risers under tensile action. Moreover, for composite cylindrical shells, Wang [11] introduced a reliability-based optimization framework and used it to design filament-wound cylindrical shells with variable angle tows. Almeida et al. [12] evaluated the damage and failure in a carbon fiber reinforced filament wound composite tube and developed a non-linear finite element model, and they found that the model was in good agreement with the experimental structure under external pressure. For progressive damage of composite tubes, Almeida et al. [13], in 2017, used a genetic algorithm to analyze the damage degradation of composite pipes.

There are three main methods to analyze flexible risers: experimental methods, theoretical methods, and numerical methods. De Sousa [14] analyzed the stress of flexible risers under tension, torsion, and internal pressure by an experiment. Because of the high cost of flexible riser experiments, theoretical analysis and numerical simulation analysis have become the prevalent analysis methods. Martindale [15] proposed a theoretical model and used it to investigate the effect of end fittings on the sliding size of steel strips in the tension armor layer of flexible risers. Zhou [16] used a quasi-linear method to establish a theoretical model of risers and analyzed the stress of helical layers under bending. Yoo [2] used Workbench to analyze the ultimate strength of flexible risers. Tang [17] used ABAQUS to analyze the force of flexible risers under combined loads.

There are two kinds of traditional unbonded flexible risers: one is a tensile armor layer made of carbon steel, and the other is a composite flexible riser that is more portable. However, due to its own excessive weight, the former leads to overburden of offshore working platforms, while the latter has poor compressive performance. Therefore, a cylindrical composite tensile armor is proposed in this paper. By taking the structure and function characteristics of flexible risers into account, a numerical model is developed using the finite element model. Considering the progressive damage, firstly, in order to discuss the composite lamination of the tensile armor, the third, fourth and fifth layer models are used. The whole model is used to predict the mechanical properties of the flexible riser with composite tensile armor. The stiffness and stress of flexible risers under different loads are also calculated. These results are compared with the normal flexible riser with carbon steel helical tensile armor.

2. Theory and Model
2.1. Theory
2.1.1. Elastic Modulus of Composite Material

The composite is composed of matrix material and reinforced material. Assuming that the component materials are uniformly distributed, the macroscopic mechanical behavior of the composite can be regarded as the combined performance of the constituent materials.

In Jones [18], based on the equivalent stress assumption, the elastic parameters of the composite were simplified.

The Young's modulus in the axial (x-axis), circumferential (y-axis) and radial (y-axis) directions as follows:

$$E_x = E_m V_m + E_F V_f$$

$$E_y = E_m(1 - V_1) + \frac{E_m E_f V_1}{\frac{E_m V_f}{V_1} + E_f\left(1 - \frac{V_f}{V_1}\right)}$$

$$E_z = E_m\left(1 - \frac{V_f}{V_1}\right) + \frac{E_m E_f \frac{V_f}{V_1}}{E_m V_f + E_f(1 - V_1)}$$

$$E_z = E_m\left(1 - \frac{V_f}{V_1}\right) + \frac{E_m E_f \frac{V_f}{V_1}}{E_m V_f + E_f(1 - V_1)}$$

The Poisson's ratio is as follows:

$$\mu_{xy} = \mu_{xz} = \mu_f V_f + \mu_m V_m$$

$$\mu_{yz} = \frac{\mu_f V_f + \mu_m \frac{E_f}{E_m}(V_1 - V_f)}{\frac{V_f}{V_1} + \frac{E_f}{E_m}\left(1 - \frac{V_f}{V_1}\right)} + \mu_m V_2$$

The shear modulus is as follows:

$$G_{xy} = \frac{G_f G_m \frac{V_f}{V_1}}{G_m V_1 + G_f V_2} + G_m\left(1 - \frac{V_f}{V_1}\right)$$

$$G_{yz} = \frac{G_f G_m V_1}{G_m \frac{V_f}{V_1} + G_f\left(1 - \frac{V_f}{V_1}\right)} + G_m V_2$$

$$G_{xz} = \frac{G_f G_m \frac{V_f}{V_1}}{G_m V_1 + G_f V_2} + G_m\left(1 - \frac{V_f}{V_1}\right)$$

2.1.2. Analytical Model of Unbonded Flexible Riser

When establishing the analysis model, the flexible riser is usually divided into a spiral layer and cylindrical layer. The cylindrical layer consists of an anti-wear layer and internal and external plastic layers. The helical layer consists of a tensile armor layer, a compressive armor layer, and a carcass. In this paper, the carbon fiber tensile armor is considered as a cylindrical layer. Limited by the length of this paper, only the establishment of the cylindrical layer model is introduced.

The polymer material is considered to be linearly elastic and isotropic. The thick-walled cylinder theories are used to analyze the cylindrical layer [2,18,19].

Stress and deformation are shown in Figure 1, where

$$\varepsilon_1 = \frac{\mu_z}{L};\ \varepsilon_2 = \frac{\partial \mu_R}{\partial R};\ \varepsilon_3 = \frac{\mu_R}{R};\ \gamma_{12} = R\frac{\mu_\theta}{L}$$

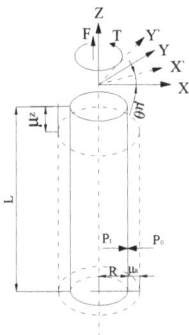

Figure 1. Loads and deformations of cylindrical layer.

This external potential energy [20] is calculated as follows:

$$U_e = P_I \Delta V_I - P_o \Delta V_o + F\mu_z + T\mu_\theta \qquad (1)$$

The strain energy of the cylindrical layer is

$$U_c = \frac{1}{2}\int_v (\sigma 1 \varepsilon 1 + \sigma 2 \varepsilon 2 + \sigma 3 \varepsilon 3 + \tau 12 \gamma 12) dv \qquad (2)$$

In order to simplify the calculation process, it is necessary to simplify the parameters.

$$X_Z = \frac{\mu_z}{L};\ X_\theta = R\frac{\mu_\theta}{L};\ X_{\varphi I} = \frac{\mu_{R_I}}{R_I};\ X_{\varphi O} = \frac{\mu_{R_O}}{R_O}$$

The total potential energy of the cylindrical layer is as follows, and the strain energy U_c is obtained from Equation (2); the external potential energy U is obtained from Equation (1):

$$\Pi_c = U - U_e$$

Taking the variation of the total potential energy and making it equal to 0, the equation is as follows:

$$\begin{bmatrix} K_{c11} & K_{c12} & K_{c13} & K_{c14} \\ K_{c21} & K_{c22} & K_{c23} & K_{c24} \\ K_{c31} & K_{c32} & K_{c33} & K_{c34} \\ K_{c41} & K_{c42} & K_{c43} & K_{c44} \end{bmatrix} \begin{bmatrix} X_Z \\ X_{\varphi I} \\ X_{\varphi O} \\ X_\theta \end{bmatrix} = \begin{bmatrix} F + \pi P_I R_I^2 - \pi P_O R_O^2 \\ 2\pi P_I R_I^2 \\ -2\pi P_O R_O^2 \\ T \end{bmatrix}$$

2.1.3. Failure Criteria for Composites

When the matrix of the composites is damaged, and the fibers are not broken, the composites can still bear the load. Therefore, the strength degradation caused by matrix failure should be considered in the mechanical analysis.

The failure process of composites is complex, and the damage forms are various. There are many failure criteria for composites. The Hashin failure criterion is based on the material parameter degradation criterion, and the cumulative damage of a single-layer plate is considered. The Hashin criterion is used to judge the composite tensile armor [21,22]. The results of Hashin criteria are composed of four separate failure modes [23]:

The tensile fibers mode $\sigma_{11} > 0$ is

$$\left(\frac{\sigma_{11}}{T_{11}^2}\right)^2 + \left(\frac{\tau_{12}^2 + \tau_{13}^2}{S_{12}^2}\right) = 1$$

Compression fibers mode $\sigma_{11} < 0$ is

$$-\frac{\sigma_{11}}{C_{12}} = 1$$

Tensile matrix mode $\sigma_{22} + \sigma_{33} > 0$ is

$$\left(\frac{\sigma_{22} + \sigma_{33}}{T_{22}}\right)^2 + \frac{(\tau_{23}^2 - \sigma_{22}\sigma_{33})}{S_{23}^2} + \frac{(\tau_{12}^2 + \tau_{13}^2)}{S_{12}^2} = 1$$

Compression matrix mode $\sigma_{22} + \sigma_{33} < 0$ is

$$\frac{1}{C_{22}}\left[\left(\frac{C_{22}}{2S_{23}}\right)^2\right](\sigma_{22} + \sigma_{33}) + \frac{(\sigma_{22} + \sigma_{33})^2}{4S_{23}^2} + \frac{(\tau_{23}^2 - \sigma_{22}\sigma_{33})}{S_{23}^2} + \frac{(\tau_{12}^2 + \tau_{13}^2)}{S_{12}^2} = 1$$

The matrix stiffness degradation criterion used in this paper is as follows [24]:
Tensile cracking of matrix:

$$E_{22} = 0.2E_{22}, \ G_{12} = 0.2G_{12}, \ G_{23} = 0.2G_{23}$$

Compression cracking matrix:

$$E_{22} = 0.4E_{22}, \ G_{12} = 0.4G_{12}, \ G_{23} = 0.4G_{23}$$

Fiber shear matrix:

$$G_{12} = v_{12} = 0$$

2.1.4. Contact Formulation

Coulomb type friction between bodies is

$$F = \mu \times R$$

For the contact of faces, pure penalty or augmented Lagrange formulations can be used. The main difference between pure penalty and augmented Lagrange methods is that augmented Lagrange includes the contact force (pressure) calculations:

Pure Penalty:

$$F = K_N \times x_P$$

Augmented Lagrange:

$$F = K_N \times x_P + \lambda$$

where K_N is the contact stiffness, x_P is the penetration, and λ is the Lagrange multiplier.

2.2. Finite Element Model

Three cases of flexible risers are used in this paper, as shown in Table 1. The first case (case 1) is flexible risers with a carbon steel structure, as seen in Figure 2, and the second case (case 2) is flexible tubes with four carbon fiber armor layers (CFA) developed by Technip. The third case (case 3) is flexible risers with a carbon fiber composite tensile armor layer.

Table 1. Three cases in this paper.

Case	Amount of Layer	Tensile Armor	Figure
case 1	8	2 carbon steel tensile armor	As shown in Figure 2
case 2	12	4 spiral carbon fiber composite tensile armor	Technip and reference
case 3	8	2 cylindrical carbon fiber composite tensile armor	As shown in Figure 3

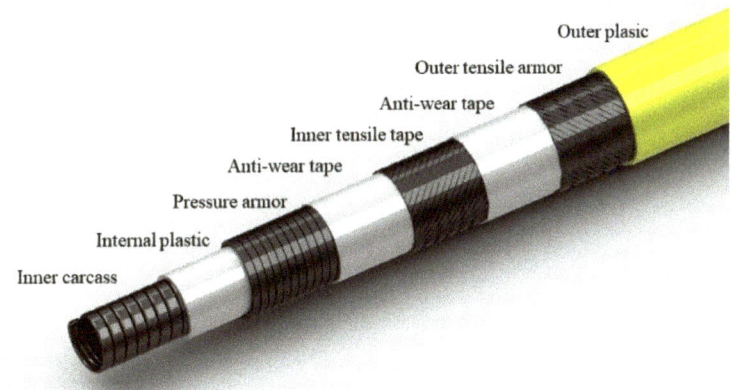

Figure 2. Flexible riser with carbon fiber composite tension armor.

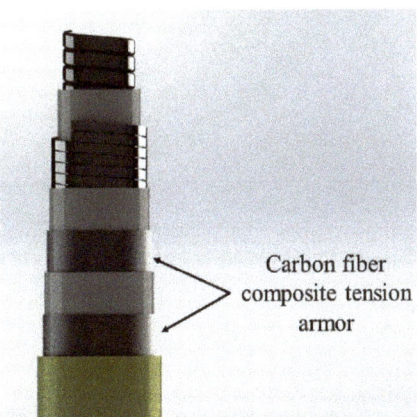

Figure 3. Flexible riser with eight layers.

2.2.1. Model Parameters

In this paper, the geometrical and material parameters of the flexible risers are given in the experimental papers on non-bonded flexible risers published by de Sousa [14], as shown in Tables 2 and 3.

Table 2. Geometric parameters of flexible risers tested.

No.	Type	Material	Inner Radius (mm)	External Radius (mm)	Angle	Thickness (mm)	Moment of Inertia (mm^4)	Width (mm)
1	Carcass	AISI 304	31.75	35.25	+87.6° (1)	3.5	23.1	5.6
2	Internal plastic	Polyamide 11	35.25	40.25	-	5.0	-	-
3	Pressure armor	Carbon steel	40.25	46.85	85.6°	6.6	173.4	8.73
4	Anti-wear tape	Polyamide 11	46.85	48.85	-	2	-	-
5	Internal tensile armor	Carbon steel	48.85	51.35	+30° (32)	2.5		8
6	Anti-wear tape	Polyamide 11	51.35	52.85	-	1.5	-	-
7	Outer tensile armor	Carbon steel	52.85	55.35	−30° (34)	2.5		8
8	Fabric tape	Polyamide	55.35	55.85	-	0.5	-	-
9	Out plastic	Polyamide	55.85	60.85	-	5.0	-	-

Table 3. Material parameters.

Material	Density (kg/m^2)	E (MPa)	ν
AISI 304	7930	205×10^3	0.3
Polyamide 11	803	345	0.3
Carbon steel	7820	205×10^3	0.3
Polyamide (eighth)	803	345	0.3
Polyamide (ninth)	803	215	0.3

The ultimate strength of carbon steel is 750–800 MPa [3,4].

Details of the material properties used in this investigation are presented in Table 4.

Table 4. Mechanical properties of carbon fiber reinforced composites.

Carbon fiber/Epoxy(T700)	Density (kg/m^2)	1580	σ_1^C (MPa)	900
	E1 (GPa)	120	σ_2^T (MPa)	20
	E2 =E3 (GPa)	10	σ_2^C (MPa)	240
	G12 = G13 (GPa)	5	τ_{12} (MPa)	18
	G23 (GPa)	5	$\nu_{12} = \nu_{13}$	0.2
	σ_1^T (MPa)	1800	ν_{23}	0.27

Flexible pipes usually have high axial stiffness, and their allowable elongation is about 0.5–1.5% [3,4]. The design must satisfy the requirements of the required axial stiffness and torsional properties and control the clearance between metal wires.

The composite lamination of the tensile armor could be designed by the stress characteristics of the tensile armor and the mechanical properties of the composite material. The direction of the pipe axis is taken as the reference direction, and the composite tension armor includes 0° angle fibers, 30° angle fibers, and 90° angle fibers. The axial load is mainly supported by the fibers in the angle of 0° and 30°. The torsion load is mainly supported by the fibers in the angle of 30° in order to deal with the phenomenon of "birdcage" caused by the axial pressure mentioned in references [2,17,25,26], and to maintain the integrity of the composite tensile armor of 90 angle directional fibers. The sequence and orientation of composite tensile armor are shown in Table 5.

Table 5. The sequence and orientation of composite tensile armor.

Name	Thickness (mm)	Sequence and Orientation
Internal tensile armor	2.4	$[90_2, \alpha_2, 0_6, 0_6, -\alpha_2, 90_2]$
Outer tensile armor	2.4	$[90_2, -\alpha_2, 0_6, 0_6, \alpha_2, 90_2]$

2.2.2. Flexible Riser Model

The structure of a flexible riser, which is very complex, can be divided into the spiral layer (carcass, compressive armor, tensile armor) and cylindrical layer (inner and outer sheath layer, anti-wear layer). In order to enhance the convergence, improve the computational efficiency, and save the computational cost, the shell element is used to simulate the cylindrical structure in this paper. The self-locking carcass and the pressure armor are simplified as an anisotropic cylindrical shell [27].

According to the structure and material parameters in Tables 4 and 5, a three-dimensional model of flexible pipe with a length of 1 m was established firstly. Then, the model was introduced to the ANSYS workbench for mesh generation. The mesh of tensile armor size was 10 mm, and other layers were 15 mm. The tensile armor layer and the overall model are shown in Figure 4.

(a)   (b)

Figure 4. Mesh of flexible riser model. (a) Tensile armor layer, (b) Overall model

2.2.3. Composite Tension Armor

The setup flows as follows: setting the material properties → establishment of reference coordinate system → setting the angle of the fiber. The direction of the reference axis in this paper is the direction of the tube axis, and the normal direction is the thickness direction of the tensile armor, as shown in Figure 5a. The yellow arrow indicates the direction of the fiber reference, and the purple arrow represents the normal direction. The direction of the fiber layer in 0° direction is shown in Figure 5b.

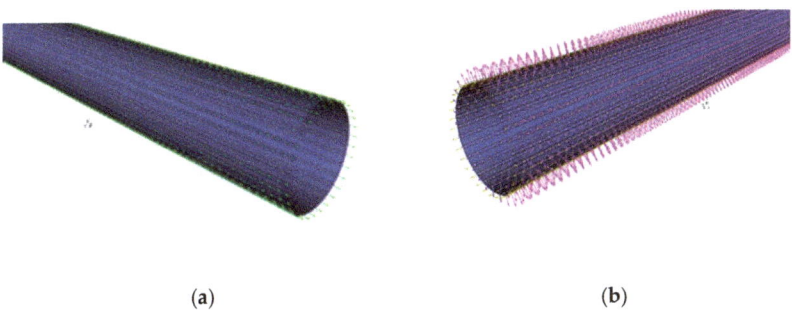

(a)   (b)

Figure 5. Definition of composite tensile armor. (a) Reference axis, (b) 0° fiber.

2.2.4. Contact

Contact mechanics is a considerable challenge when creating numerical models. The distribution of normal and tangential force, penetration and local constraint conditions, and variation in material and mechanical properties of each layer present difficulties in obtaining solution convergence and successful modeling outcomes [28,29].

To converge the calculation and ensure the accuracy of the results, the augmented Lagrange method is chosen in this paper. The difference between simulation using solid element and shell element is discussed in [2]. The stiffness calculated by the shell element is considered to be slightly larger, however, when the intrusion tolerance of the contact surface is 5%; the calculated stiffness of the shell element is slightly less than that of the solid element, but the difference can be neglected. With respect to the tangential contact, the friction coefficient used in Yoo [2] and Asousa [14] is 0.1, and Tang [17] is 0.2. The friction coefficient used in this paper is 0.1.

2.2.5. Boundary Conditions and Load

According to [3], the design of the tensile armor of flexible risers must meet the requirements of the required axial strength and any torsional properties. In this paper, we set fixed boundary conditions at one end of the pipe. At the other end, axial loads and torsional loads were applied. External pressure was applied to the outer surface of the plastic. For bending cases, the loads and constraints are shown in Figure 6.

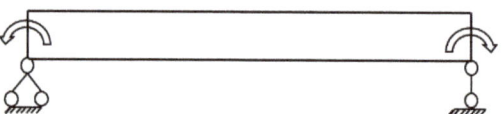

Figure 6. Boundary conditions of bending condition.

3. Results

3.1. The Composite Lamination of the Tensile Armor

It was proved in [15,27] that under the action of axial tensile load, the carcass, inner plastic, and compressive armor do not affect the stress analysis. The fourth, fifth, and sixth layers structures were only analyzed. To simulate the action of carcass and compressive armor against internal and external pressure, radial restraint was applied to the inner surface of the fourth layer. The stress distribution diagram of $0°$ angle fiber in composite tensile armor is shown in Figure 7.

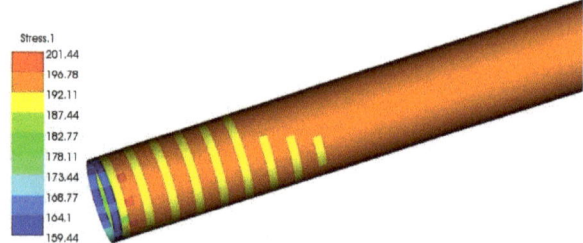

Figure 7. The stress of $0°$ fiber.

The relationship between the axial deformation and the axial tension is shown in Figure 8a. The stiffness of the whole riser was obtained by dividing the deformation by the load. Moreover, when the axial tension was small, the relationship between tension and deformation was linear, and with increases in tension, the axial stiffness decreased. The reason is that the elastic modulus of composites was reduced, according to the failure criteria and strength degradation criteria explained in Section 2.1.2. Figure 8b shows the relationship between the torsion angle and torque. As can be seen from the figure, the relationship was linear. When α was $30°, 45°$, or $60°$, the torsional stiffness of the cylindrical tension armor was the largest. Therefore, the α in this paper was $30°$.

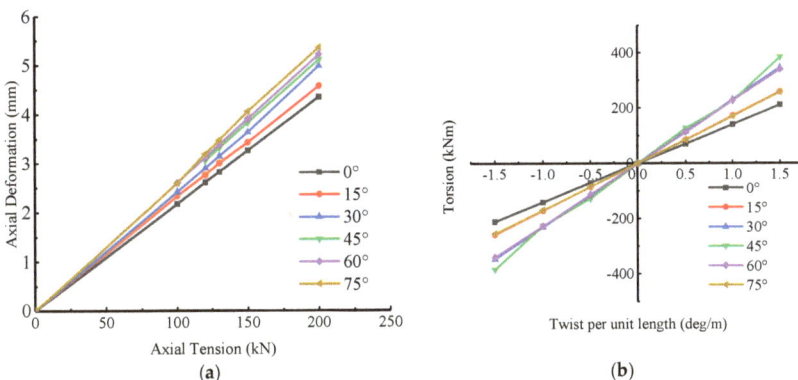

Figure 8. Displacement of flexible riser under tension and torsion. (**a**) Axial deformation versus tension per unit length, (**b**) twist versus torsion per unit length.

3.2. Pure Axial Tensile

The axial deformation was applied at the free end, and the axial tension was measured by the reaction force at the fixed end. The stress of the tensile armor was observed until the strength limit of the material. The relationship between axial tension and deformation of two cases of flexible risers is shown in Figure 9a, and the stress of tensile armor versus axial tension is shown in Figure 9b.

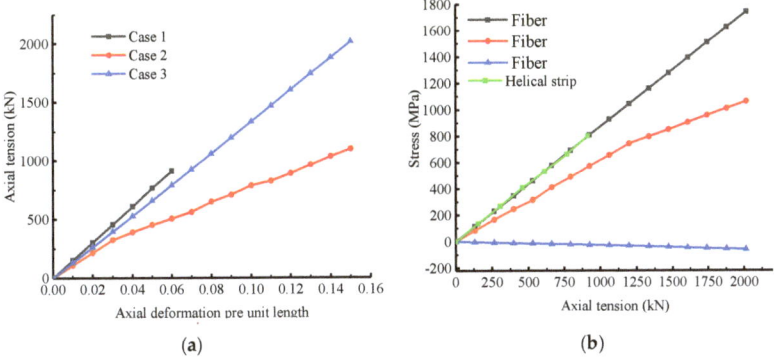

Figure 9. Axial deformation versus tension per unit length and stress of tensile armor. (**a**) Axial deformation versus tension per unit length; (**b**) the stress of tensile armor versus tension per unit length.

As shown in Figure 9a, the results of case 1 were compared with the results of an experiment in [14]. The simulation error of the model for the flexible riser was less than 5%, which proves the availability of the model. It can be seen from the figure that the relationship between the axial tension and the axial deformation of flexible risers was linear. However, when the load increased, case 2 and case 3 had a nonlinear relationship. The reason is that the matrix was cracked by the tension when the *displacement* was 10 mm. The axial tension of flexible risers in case 2 was smaller than those in case 3 with the same axial deformation. This indicates that the overall tensile strength of the flexible risers in case 2 was lower than those in case 3. This indicates that the overall stiffness of the flexible riser with the composite tensile layer was lower than that with the carbon steel tensile layer, and the overall stiffness of the flexible riser with cylindrical carbon fiber tensile armor layers in case 3 was higher than the flexible riser with four carbon fiber armor layers.

Figure 9b shows the stress of helical strips in case 1 and fiber in case 3. As seen in this figure, the tensile load in case 2 was mainly borne by fibers in the direction of 0 and 30. The fiber stress in the angle of 0° increased linearly with the increase of load. When the tension was greater than 1197 kN, the rate of increase of the fiber stress in the angle of 30° decreased. When the axial deformation was 6 mm, the helical strip stress of case 1 was 800 MPa, which is the ultimate strength of carbon steel. When the load turned to 2000 kN, the axial *displacement* was 15 mm, and the stress of the fiber at 90° was 1744 MPa which is the ultimate strength of carbon fiber. Comparing case 1 with case 3, the ultimate tensile strength of flexible risers with cylindrical composite tensile armor layers improved greatly, about one time.

3.3. Tension and External Pressure

The flexible riser was mainly subjected to the combined load (internal and external pressure and tension) during its operation and installation. Therefore, this section discusses the effect of external pressure on axial stiffness. Firstly, the different external pressure was applied on the outer surface of the flexible riser, and then the axial displacement was applied at the free end. The reaction force at the fixed end was the axial tension.

Figure 10a shows the variation of axial tension of the flexible riser with external pressure. It can be seen from the figure that the axial stiffness of the flexible riser decreased under external pressure. Compared with case 1, case 2, and case 3, the effect of external pressure on the overall axial tensile stiffness of the flexible riser with cylinder carbon fiber tensile armor (in case 3) was smaller than the risers with helical carbon steel tensile armor (in case 1), and the effect on flexible risers with four carbon fiber armors (in case 2) was the biggest. This suggests that the effect is enhanced when the form of the tensile armor layer is helical.

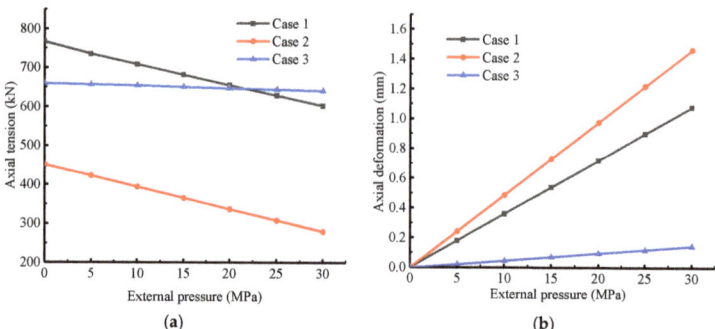

Figure 10. Influence of external pressure on axial deformation. (**a**) Axial tension versus external pressure per unit length, (**b**) axial deformation versus external pressure per unit length.

With the increase of external pressure, the reduction of axial load is linear [14]. The flexible riser itself caused axial deformation under external pressure. Figure 10b shows the axial deformation under pure external pressure. As seen in Figure 10, the flexible riser in case 1 produced 1.08 mm axial deformation under 30 MPa external pressure, and the riser in case 2 produced 1.436 mm; the riser in case 2 only produced 0.144 mm axial deformation under the same external pressure. This is the reason why the reduction rate of axial stiffness is different under external pressure.

3.4. Torque

The allowable torsion deformation of flexible pipe is 0.5–1.5 (deg/m) [4]. The flexible riser was fixed at one end, and torsional displacement was applied at the other end. Clockwise was positive (the same winding direction as the outer tension armor layer), and counterclockwise was negative. The results are shown in Figure 11a.

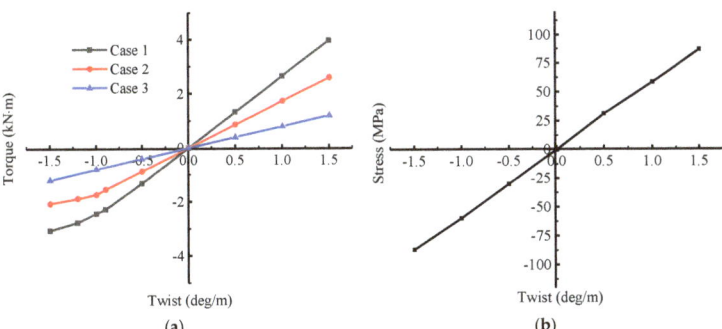

Figure 11. Torque and stress of 0° fiber versus twist per unit length. (**a**) Torque versus twist per unit length; (**b**) stress of 30° fiber versus twist per unit length.

As can be seen from this figure, when the torsion angle was positive, the torsion and the torsion moment of the two flexible risers showed a linear relationship. When the torsion was negative, the torsional stiffness of the flexible riser in case 1 decreased with the increase of the torsion, which was caused by the phenomenon of separation between layers of the pipeline. In case 2, the torsional stiffness of flexible riser did not change due to the different directions of the torsional angle. Overall, the torsional stiffness of flexible risers with composite tension armor was reduced by three times. Within the allowable torsion angle of the flexible riser (0.5–1.5 deg/m), the composite tensile armor did not show any damage.

The stress of each layer's fiber in the composite tensile armor can be obtained from the post module. It can be seen that when the flexible riser was twisted at the free end, the fibers in the composite tensile armor layer were the major fibers, and the stress was high, but the fibers in the direction of 0 and 90 were very low. Figure 12 shows the stress of fiber at the angle of 30° under the twist of 1.5 deg/m at the free end. Figure 11b is the stress of 30° fibers varying with the torsion angle of the free end. It can be seen that the relationship between stress of 30° fibers and twist of the free end is linear.

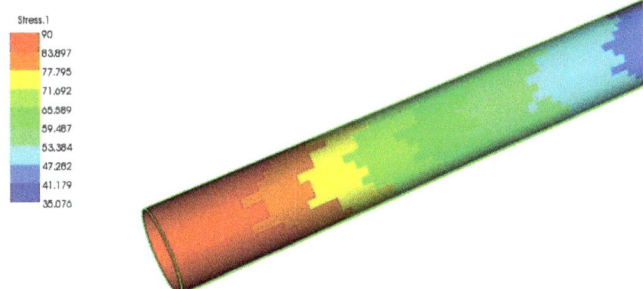

Figure 12. The stress of 30° fiber under 1.5 deg/m twist.

3.5. Bending

The flexible riser bends in the process of transportation, installation, and operation; as a result, the composite tension armor layer needs to meet the requirements of the flexible riser's bending deformation.

As shown in Figure 13a,b, both the flexible risers in the two cases could be divided into three phases during bending: non-slip phase, partial slip phase, and complete slip phase. Furthermore, the bending stiffness of the flexible riser in case 2 was higher than that in case 1. From the non-slip phase to the slip phase, the change of bending stiffness in case 2 had a smaller change than that in case 1.

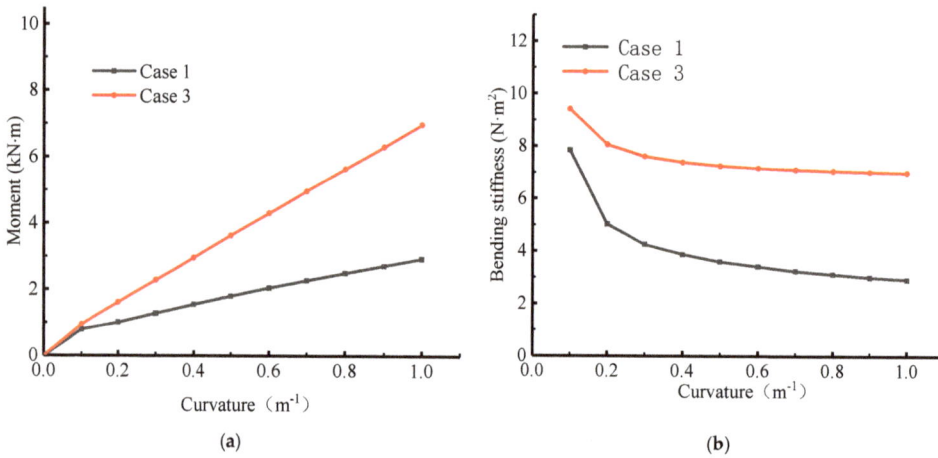

Figure 13. Bending moment and stiffness. (**a**) Moment–curvature; (**b**) bending stiffness.

Figure 14 shows the fiber stress of the composite tensile armor layer when the riser was bending. Figure 14a shows the fiber stress of the composite tensile armor layer during bending deformation of the riser. For the most part, 0 and 30-degree fibers were working. With the increase of curvature, the stress increased linearly. The lamination of the composite tensile armor layer was [90_2, 30_2, 0_6, 0_6, -30_2, 90_2]. Figure 14b shows the stress of all fibers in the angle of 0° in the composite tensile armor when the curvature was 1 rad/m. The layer numbers from 1 to 12 indicate from inside to the outside of the diameter. It can be found from the figure that the stress of the 0° fibers in the two composite tensile armors increased with the layer number. According to the failure criterion in Section 2.1.2, the failure of composite tension armor is matrix compression and shear failure. Figure 15 shows the stresses of fibers in the angle of 0° and 30°. The maximum stress value of 0° fiber of composite tensile armor was 889 MPa, which did not reach the tensile strength of the carbon fiber, and there was no tensile fracture phenomenon of the fiber. It shows that the flexible pipe with composite tensile armor can withstand bending deformation without failure and meet the design requirements.

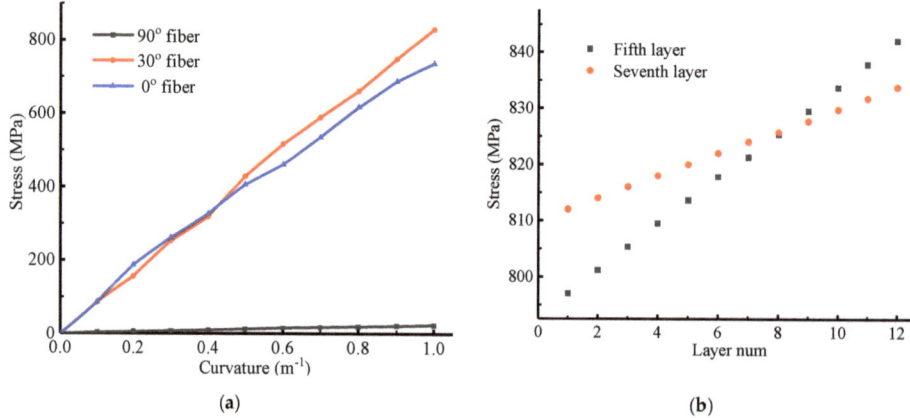

Figure 14. Fiber stress of composite tension armor. (**a**) Fiber stress versus curvature per unit length; (**b**) the stress of fiber at 0° versus curvature per unit length.

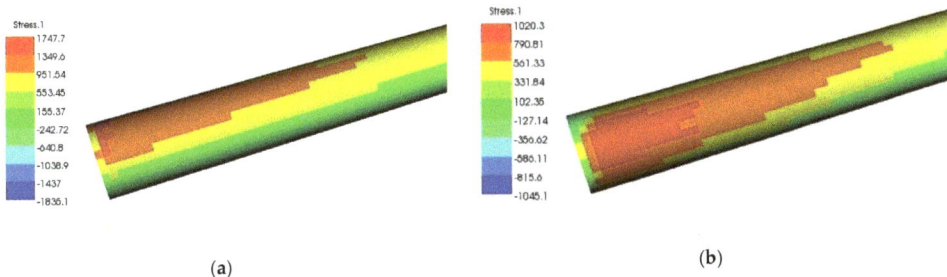

Figure 15. Stress state of fiber. (**a**) Stress fiber in 0° angle; (**b**) stress fiber in 30° angle.

4. Conclusions

To reduce the self-weight of a flexible riser, a cylinder composite tensile armor is proposed in this paper. From the analysis and calculation, the conclusions are as follows:

The suitable laminate for the composite tensile armor is [90_2, 30_2, 0_6, 0_6, -30_2, 90_2]. The weight of the flexible riser with composite armor decreased by 9.73%. It can effectively reduce the burden of offshore platforms.

The composite matrix will be damaged in the process of use, but the fibers can continue to function. Therefore, the flexible riser with composite tensile armor sometimes experiences the phenomenon of stiffness reduction in the process of loading. The stiffness and strength of flexible risers with cylinder composite tensile armor changed; the tensile stiffness was reduced by 17%, the torsional stiffness was reduced by 60%, and the flexural stiffness was increased by 130%. Moreover, the effect of external pressure on the overall tensile stiffness was significantly reduced. In terms of strength, the tensile strength of flexible risers increased by 150%. At the same time, flexible risers could also withstand torsional and bending deformations within a specified range.

Author Contributions: Conceptualization, H.Z. and L.T.; methodology, H.Z. and L.T.; software, H.Z.; validation, H.Z., L.T., and M.A.A.; formal analysis, H.Z. and L.T.; investigation, H.Z.; resources, H.Z. and L.T.; data curation, H.Z. and L.T.; writing—original draft preparation, H.Z. and M.A.A.; writing—review and editing, H.Z., L.T., and M.A.A.; visualization, H.Z.; supervision, L.T.; All authors have read and agreed to the published version of the manuscript.

Funding: This research received no external funding.

Informed Consent Statement: Informed consent was obtained from all subjects involved in the study.

Data Availability Statement: The data presented in this study are available on request from the corresponding author. The data are not publicly available due to privacy or ethical restrictions.

Acknowledgments: The authors are grateful to Lili Tong for her help in the analysis and writing of this paper and Michael Anim Addo for his help in grammar and writing.

Conflicts of Interest: We declare that we have no financial and personal relationships with other people or organizations that can inappropriately influence our work, and that there is no professional or other personal interest of any nature or kind in any product, service, and/or company that could be construed as influencing the position presented in, or the review of, the manuscript entitled.

References

1. Dudley, B. *BP Energy Outlook*; BP Amoco: London, UK, 2018.
2. Yoo, D.-H.; Jang, B.-S.; Yim, K.-H. Nonlinear finite element analysis of failure modes and ultimate strength of flexible pipes. *Mar. Struct.* **2017**, *54*, 50–72. [CrossRef]
3. API. *Recommended Practice for Flexible Pipe*, 5th ed.; API RP 17B; American Petroleum Institute: Washington, DC, USA, 2014.
4. API. *Specification for Unbounded Flexible Pipe*, *APIA Specification 17*, 4th ed.; American Petroleum Institute: Washington, DC, USA, 2014.
5. Ryu, S.; Duggal, A.S.; Heyl, C.N.; Liu, Y. Prediction of deepwater oil offloading buoy response and experimental validation. *Int. J. Offshore Polar Eng.* **2006**, *16*, 290–296.

6. Kang, Y.; Sun, L.; Kang, Z.; Chai, S. Coupled analysis of FPSO and CALM buoy offloading system in West Africa. In Proceedings of the ASME 2014 33rd International Conference on Ocean, Offshore and Arctic Engineering, San Francisco, CA, USA, 8–13 June 2014.
7. Hovde, G.O.; Kaalstad, J.P.; Skjaastad, O. Offloading in Deep and Ultradeep Water—Main Drivers and Need for Improved Systems. In Proceedings of the Offshore Technology Conference, Houston, TX, USA, 2 May 2005.
8. Toh, W.; Bin Tan, L.; Jaiman, R.K.; Tay, T.E.; Tan, V. A comprehensive study on composite risers: Material solution, local end fitting design and global response. *Mar. Struct.* **2018**, *61*, 155–169. [CrossRef]
9. Amaechi, C.V.; Gillett, N.; Odijie, A.C.; Hou, X.; Ye, J. Composite risers for deep waters using a numerical modelling approach. *Compos. Struct.* **2019**, *210*, 486–499. [CrossRef]
10. Liu, Q.; Xue, H.; Tang, W.; Yuan, Y. Theoretical and numerical methods to predict the behaviour of unbonded flexible riser with composite armour layers subjected to axial tension. *Ocean Eng.* **2020**, *199*, 107038. [CrossRef]
11. Wang, Z.; Almeida, J.H.S., Jr.; St-Pierre, L.; Wang, Z.; Castro, S.G. Reliability-based buckling optimization with an accelerated Kriging metamodel for filament-wound variable angle tow composite cylinders. *Compos. Struct.* **2020**, *254*, 112821. [CrossRef]
12. Almeida, J.H.S.; Ribeiro, M.L.; Tita, V.; Amico, S.C. Damage and failure in carbon/epoxy filament wound composite tubes under external pressure: Experimental and numerical approaches. *Mater. Des.* **2016**, *96*, 431–438. [CrossRef]
13. Almeida, J.H.S.; Ribeiro, M.L.; Almeida, J.H.S.; Amico, S.C. Stacking sequence optimization in composite tubes under internal pressure based on genetic algorithm accounting for progressive damage. *Compos. Struct.* **2017**, *178*, 20–26. [CrossRef]
14. De Sousa, J.R.; Magluta, C.; Roitman, N.; Vargas-Londono, T.; Campello, G. A Study on the Response of a Flexible Pipe to Combined Axisymmetric Loads. In Proceedings of the ASME 2013 32nd International Conference on Ocean: Offshore and Arctic Engineering, Nantes, France, 9–14 June 2013.
15. Out, J.; Von Morgen, B. Slippage of helical reinforcing on a bent cylinder. *Eng. Struct.* **1997**, *19*, 507–515. [CrossRef]
16. Zhou, Y.; Vaz, M.A. A quasi-linear method for frictional model in helical layers of bent flexible risers. *Mar. Struct.* **2017**, *51*, 152–173. [CrossRef]
17. Tang, L.; He, W.; Zhu, X.; Zhou, Y. Mechanical analysis of un-bonded flexible pipe tensile armor under combined loads. *Int. J. Press. Vessel. Pip.* **2019**, *171*, 217–223. [CrossRef]
18. Bai, Y.; Lu, Y.; Cheng, P. Analytical prediction of umbilical behavior under combined tension and internal pressure. *Ocean Eng.* **2015**, *109*, 135–144. [CrossRef]
19. Merino, H.E.M.; de Sousa, J.R.M.; Magluta, C.; Roitman, N. Numerical and Experimental Study of a Flexible Pipe under Torsion. In Proceedings of the International Conference on Offshore Mechanics and Arctic Engineering, Shanghai, China, 6–11 June 2010; pp. 911–922.
20. Bussetta, P.; Marceau, D.; Ponthot, J.-P. The adapted augmented Lagrangian method: A new method for the resolution of the mechanical frictional contact problem. *Comput. Mech.* **2011**, *49*, 259–275. [CrossRef]
21. Gu, J.; Chen, P. Some modifications of Hashin's failure criteria for unidirectional composite materials. *Compos. Struct.* **2017**, *182*, 143–152. [CrossRef]
22. Tserpes, K.I.; Papanikos, P.; Kermanidis, T. A three-dimensional progressive damage model for bolted joints in composite laminates subjected to tensile loading. *Fatigue Fract. Eng. Mater. Struct.* **2001**, *24*, 663–675. [CrossRef]
23. Zhang, X.; Gou, R.; Yang, W.; Chang, X. Vortex-induced vibration dynamics of a flexible fluid-conveying marine riser subjected to axial harmonic tension. *J. Braz. Soc. Mech. Sci. Eng.* **2018**, *40*, 365. [CrossRef]
24. Liu, P.; Gu, Z.; Yang, Y.; Peng, X. A nonlocal finite element model for progressive failure analysis of composite laminates. *Compos. Part B Eng.* **2016**, *86*, 178–196. [CrossRef]
25. Novitsky, A.; Gray, F. Flexible and Rigid Pipe Solutions in the Development of Ultra-Deepwater Fields. *Int. Conf. Offshore Mech. Arct. Eng.* **2003**, *36827*, 755–770.
26. Vaz, M.; Rizzo, N. A finite element model for flexible pipe armor wire instability. *Mar. Struct.* **2011**, *24*, 275–291. [CrossRef]
27. Li, X.; Jiang, X.; Hopman, H. A strain energy-based equivalent layer method for the prediction of critical collapse pressure of flexible risers. *Ocean Eng.* **2018**, *164*, 248–255. [CrossRef]
28. Ebrahimi, A.; Kenny, S.; Hussein, A. Finite Element Investigation on the Tensile Armor Wire Response of Flexible Pipe for Axisymmetric Loading Conditions Using an Implicit Solver. *J. Offshore Mech. Arct. Eng.* **2018**, *140*, 041402. [CrossRef]
29. De Sousa, J.R.; Magluta, C.; Roitman, N.; Ellwanger, G.B.; Lima, E.C.; Papaleo, A. On the response of flexible risers to loads imposed by hydraulic collars. *Appl. Ocean Res.* **2009**, *31*, 157–170. [CrossRef]

Article

Boosting Inter-ply Fracture Toughness Data on Carbon Nanotube-Engineered Carbon Composites for Prognostics

Sunil C. Joshi

School of Mechanical and Aerospace Engineering, Nanyang Technological University, Nanyang Avenue, Singapore 639798, Singapore; mscjoshi@ntu.edu.sg; Tel.: +65-6790-5170

Received: 13 September 2020; Accepted: 19 November 2020; Published: 20 November 2020

Abstract: In order to build predictive analytic for engineering materials, large data is required for machine learning (ML). Gathering such a data can be demanding due to the challenges involved in producing specialty specimen and conducting ample experiments. Additionally, numerical simulations require efforts. Smaller datasets are still viable, however, they need to be boosted systematically for ML. A newly developed, knowledge-based data boosting (KBDB) process, named COMPOSITES, helps in logically enhancing the dataset size without further experimentation or detailed simulation. This process and its successful usage are discussed in this paper, using a combination of mode-I and mode-II inter-ply fracture toughness (IPFT) data on carbon nanotube (CNT) engineered carbon fiber reinforced polymer (CFRP) composites. The amount of CNT added to strengthen the mid-ply interface of CFRP vs the improvement in IPFT is studied. A simpler way of combining mode-I and mode-II values of IPFT to predict delamination resistance is presented. Every step of the 10-step KBDB process, its significance and implementation are explained and the results presented. The KBDB helped in not only adding a number of data points reliably, but also in finding boundaries and limitations of the augmented dataset. Such an authentically boosted dataset is vital for successful ML.

Keywords: fracture toughness; carbon composites; knowledge-based data boosting; predictive analytic; machine learning

1. Introduction

Advanced polymeric composites are being used in many weight-sensitive, load-carrying applications. Layered configurations, or laminates, are preferred because of flexibility in tailoring properties, optimizing weight and achieving high specific properties [1]. However, exploiting laminate's full potential is always marred by delamination issues [2].

Researchers have studied inter-ply fracture modes and have suggested a failure criterion based on inter-ply fracture toughness (IPFT), which quantitatively is the strain energy released per unit area (i.e., G kJ/m^2) during crack propagation. Loading applied in an opening fracture mode or mode-I is the most damaging. Mode-II is a sliding mode and less detrimental. The tearing mode, commonly known as mode-III, is less of a concern, as the laminates are seen to offer high resistance in this mode. It is likely that most composite laminates experience mixed mode I/II loading in service. Benzeggagh and Kenane [3] suggested a criterion to assess delamination growth under mixed mode I/II as below.

$$G_I + G_{II} = G_{IC} + (G_{IIc} - G_{IC}) * [G_{II}/(G_I + G_{II})]^m \quad (1)$$

where G_{IC} and G_{IIc} are the critical IPFT in modes I and II, respectively. G_I and G_{II} are the energy release rates in respective modes under mixed mode loading condition. Constant m is an empirical constant.

Ali, Amin and Sepideh [4] adopted this formula further and revised it as a failure criterion against delamination. According to them, the total IPFT (G_{Tc}) is

$$G_{TC} = G_{IC} + (G_{IIc} - G_{IC}) * [G_{II}/(G_I + G_{II})]^m \qquad (2)$$

In Equation (2), 'm' is the Benzeggah and Kenane [3] parameter, which is empirical.

Irrespective of the value of 'm', it is clear that $[G_{II}/(G_I + G_{II})] < 1.0$ when $G_I \neq 0.0$ and $G_{IIc} \neq 0.0$. Thus, in all cases G_{TC} shall lie between G_{IC} and G_{IIc}. This also means that one will be able to build a prognostic for possible delamination resistance based on G_{IC} and G_{IIc} values in absence of mixed mode data.

The use of machine learning (ML) in prognostic is widespread in engineering, including composites and materials [5–8]. Success of ML depends on accurate algorithms and data; in particular, a machine cannot train or learn without data. Such data shall be accurate, and large enough for reliable ML-based predictive analytic for engineering materials like composites. Gathering such data can be demanding, due to the challenges involved in producing special specimen and conducting extensive experiments. Numerical simulation also requires efforts. Smaller datasets, however, may be boosted systematically for ML. Data augmentation works well with images form of data. Statistical methods are helpful, but the accuracy of the generated data cannot be guaranteed, which involves understanding of the engineering materials, parameters, phenomena and even the processes involved. All these warranted a systematic approach to ensure the quality of the new data points (NP).

In this paper, a knowledge-based data boosting (KBDB) process, proposed [9] and discussed [10] by the same author in earlier publications, is adopted. This procedure is meant to address data sparsity systematically and reliably without compromising the data quality. The KBDB process, acrostically named 'COMPOSITES' consists of 10 steps that are essential, starting from identifying data, understanding physics, sanitizing available data, to final scrutiny of the boosted dataset. This provides a pragmatic novel way of creating near-real engineering data for ML. KBDB can be adopted for any case that comprises countable variables and quantities, and requires the available lean data to be augmented to the size useful for ML.

A study on IPFT of carbon nanotubes (CNT) engineered carbon fiber reinforced polymer (CFRP) laminates is used to explain and discuss the entire KBDB process. An experimental data highlighting the effect of CNT content on the critical IPFT values is taken from [11] to develop the case study. This original research involved nano-scale strengthening of inter-ply interfaces in CFRP using multi-wall carbon nanotubes (MWCNT) for studying improvement in its IPFT under mode-I and mode-II type of loading. This is still an area under research, offering limited case studies and datasets. For composites, experiments are generally possible only in small numbers, due to the specialty materials, processing and testing methods involved.

The KBDB study presented here facilitated understanding of the data values, operational boundaries, and limitations of the augmented dataset. This demonstrated the utility of the COMPOSITES process for engineering data augmentation.

2. Materials and Methods

A study is conducted for a CFRP composites, a schematic of which is shown in Figure 1. Such composites typically consist of many layers or plies stacked one above the other and cured using heat, vacuum, and pressure to bond the plies together and consolidate them into a laminate.

The inter-ply region between the individual plies is matrix dominated lacking other reinforcement that can resist any form of initiation, formation, and propagation of cracks. The addition of nano-fillers, such as carbon nanotubes (CNT), is seen to improve the resistance to cracks at the inter-ply interfaces [12]. One of the already published [13–16] research works on such nano-engineered woven CFRP composites from the author's research group is used in this KBDB exercise.

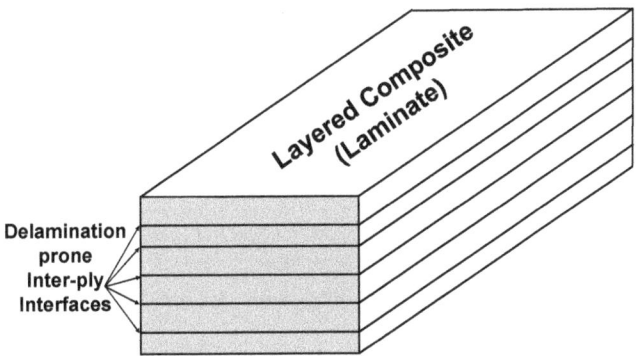

Figure 1. Schematic of layered composites.

The specimens were fabricated using a bi-directional, 12 k, plain weave, carbon fabric prepreg that consisted of Epoxy L-930HT matrix. Various type of MWCNT used to nano-reinforce the inter-ply interface were found to bond well with the prepreg [11].

Due to their nano-size, CNT do agglomerate making it difficult to disperse and penetrate into the adjacent plies forming a good reinforcement. In order to achieve uniform dispersion, "dry CNT on transfer media" process was used, as reported in [11,13]. CNT were mixed separately in Ethanol in 1:500 proportion. The mixture was sonicated for 2 hours, and then was spread on a Teflon sheet. After the Ethanol evaporated out, a layer of carbon prepreg was laid on the top such that the CNT stuck onto the prepreg surface in contact. The Teflon sheet was removed after flipping the prepreg upside down. The prepreg layer thus prepared was then stacked in a pre-decided lay-up sequence to form the test laminates. Thin impervious Teflon film used as crack initiator was also laid appropriately. These laminates were cured in an autoclave as recommended by the prepreg supplier [11,13] under 85 psi pressure and 120 deg C temperature for 2 hours. A vacuum of 15–25 mmHg was also applied for debulking and removing volatiles.

Even though CNT agglomeration was eliminated, their spread over the prepreg surface was random, and also, the cutline into the prepreg and the fibers might not have been exactly aligned in all plies. All these might have resulted in the scatter seen in the test results, which was already accounted in for by selecting the minima, mean and maxima values of IPFT. Note that the amount of CNT used varied from 0.0 g/sqm to less than 1.0 g/sqm of the prepreg area.

The DCB and ENF test samples cut from these laminates were used to conduct the tests according to ASTM D 5528 and JIS K7086 respectively [11].

Generation of any engineering data requires understanding of the phenomena and the various parameters involved. Adding new data points to any such available dataset requires the same attention and rigor, especially when the points are to be generated artificially, as before, to uphold the same accuracy. The new knowledge-based data boosting (KBDB) process, reported in [9,10] and used in this paper, is schematically shown in Figure 2.

Below described are these 10 generic steps in the proposed KBDB process. As seen, these also form an acrostic 'COMPOSITES' [9,10].

Collect   Gather authentic engineering, semi-processed/raw, data.
Organize   Choose/select data points based on certain parameter or criterion.
Mathematics   Tabulate and plot data points to check scatter, trend and deviation.
Physics   Examine underlying reasons for the data point values.
Oddities   Identify and remove outliers and extremity points.
Space   Mark the space or domain within which the data may be boosted.
Infer   Form guidelines to be observed while boosting the data.
Translate   Form mathematical propositions based on inferred guidelines.
Employ   Apply those propositions to augment the data and visualize it.
Scrutinize   Examine the new dataset considering all details and build suggestions.

The implementation of the process for the CFRP composites case is demonstrated in the next section.

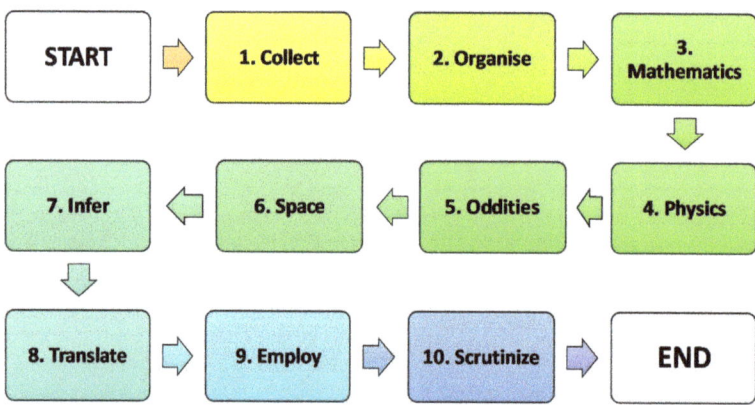

Figure 2. Schematic of the knowledge-based data boosting (KBDB) process.

3. Results

In this study, only mode-I and mode-II data for critical IPFT are considered, which are then systematically augmented using the COMPOSITES process.

3.1. Collect:

Double cantilever beam (DCB) and end notched flexure (ENF) test data characterizing mode-I and mode-II IPFT (i.e., G_{IC} and G_{IIc} kJ/m^2), respectively, is collected from [11], representing delamination resistance of a bi-directional CFRP composites without nano-fillers and with a variety of CNT added in to strengthen the inter-ply interfaces. The fabrication and testing of the DCB and ENF coupons, for which the data was gathered, were conducted in a controlled environment. A total of five different cases for each mode were identified, where the processing parameters, such as the autoclave curing cycle, applied vacuum as well as sonication period for CNT were consistent. All cases have sufficient scatter. In order to take care of the scatter, all lower-bound, mean and upper-bound values of IPFT are captured as part of the dataset. Out of the five cases for each mode, one is on zero CNT, one with non-functionalized multi-wall (MW) CNT (or MWCNT), and three with functionalized MWCNT with –COOH, –OH1 or –OH2 groups.

3.2. Organize:

The data is arranged such way that the amount of CNT (g/sqm surface of prepreg) used as nano-reinforcement for the mid-plane is directly related to the respective IPFT values. Additionally, the G ratio, G_{IIc}/G_{IC}, is calculated to understand whether the composites offered more resistance in mode-I or mode-II. This also helps in understanding in which mode the CNT are more effective.

In all total, 30 data points for IPFT and 15 for the G ratio were gathered and plotted in a X-Y domain. The plots for lower, mean, and upper limit values were drawn separately (refer to Figure 3) to get an overall picture of the variations in G_{IC} and G_{IIc} kJ/m² values as a function of weight of CNT per unit area of the inter-ply interface.

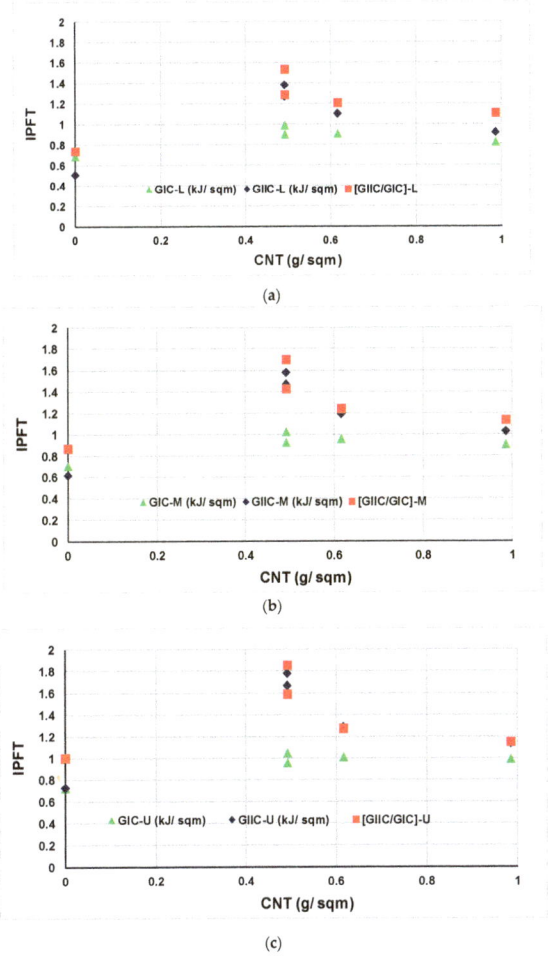

Figure 3. Chosen set of data points, (a) minima or lower-bound, (b) mean or average, (c) maxima or upper-bound, for Mode-I inter-ply fracture toughness.

This amount of data (15 for each mode) was certainly very lean and not sufficient for any predictive modeling. This certainly required boosting of the entire data set systematically, for which the remaining steps of the COMPOSITES process are implemented.

3.3. Mathematics:

It is clear from the data points that the relationship between IPFT and the CNT content is non-linear. Subsequently trend line functions, such as, moving average and polynomials were tried to get the best-fit lines. The third order polynomial functions fit the data in a reasonable manner. The plots are shown in Figure 4. Separate trend lines were plotted for the minima (L), mean (M) and the maxima (U) for clarity and understanding the variance. The trends seemed to best fit and the R^2 (i.e., Coefficient of Determination, or, in short, CoD) values were high enough in all three cases.

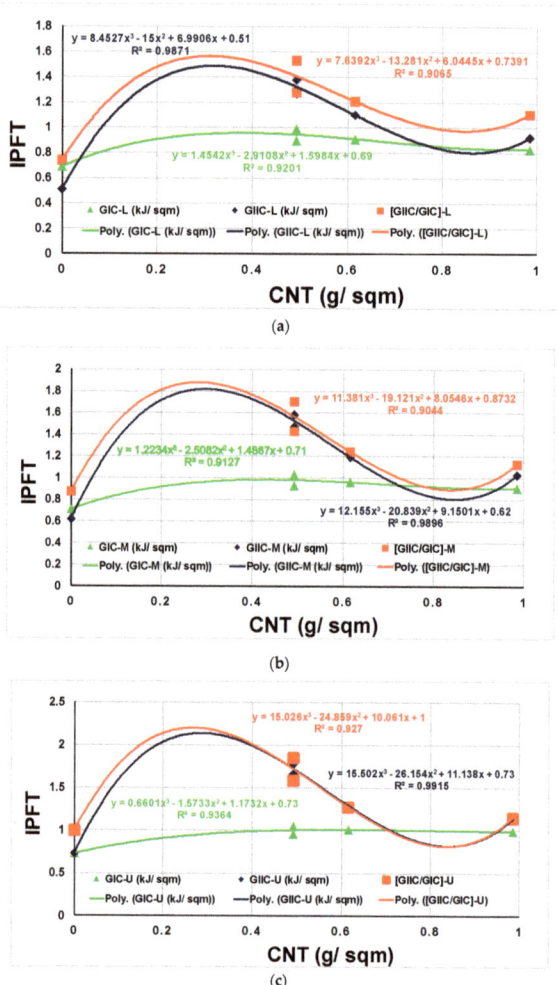

Figure 4. Trends and coefficient of determination (CoD) for, (**a**) minima or lower-bound, (**b**) mean or average, (**c**) maxima or upper-bound, Mode-I inter-ply fracture toughness data for carbon fiber reinforced polymer (CFRP).

However, the dataset is very sparse and there are intermittent regions of no data, especially where peaks and deeps for the polynomial fits lie. These certainly warrant logical augmentation of the dataset.

3.4. Physics:

It is clear from Figure 4 that, without the addition of CNT, the IPFT in mode-I is slightly higher than in mode-II. However, the CNT changed this trend in that IPFT in mode-II seemed to have improved significantly as compared to IPFT in mode-I. It is certain that with a right proportion of CNT at the inter-ply interface, significant amount of fiber bridging between fibers from the adjacent plies occurs. This fiber bridging helps in sustaining loads for a much longer time, resulting in higher

IPFT. In conclusion, the mode-I and mode-II data gathered are representative enough for studying delamination resistance.

It is clear from Equation (2) that, under a mixed mode loading, composites would offer a better resistance than in a pure mode-I condition with a certain amount of CNT added in at the inter-ply interfaces. With this, equation (2) can be simplified as below to derive G_{TC}. Equation (3) also satisfies the $G_{IC} \leq G_{TC} \leq G_{IIC}$ condition.

$$G_{TC} = (G_{IC} * z) + G_{IIc} * (1-z) \text{ where } 0 > z > 1 \qquad (3)$$

In this study, $z = 0.75$ is adopted. Figure 5 shows the plots and the trend line equations for the G_{TC} thus calculated; the CoD, in all cases, is excellent.

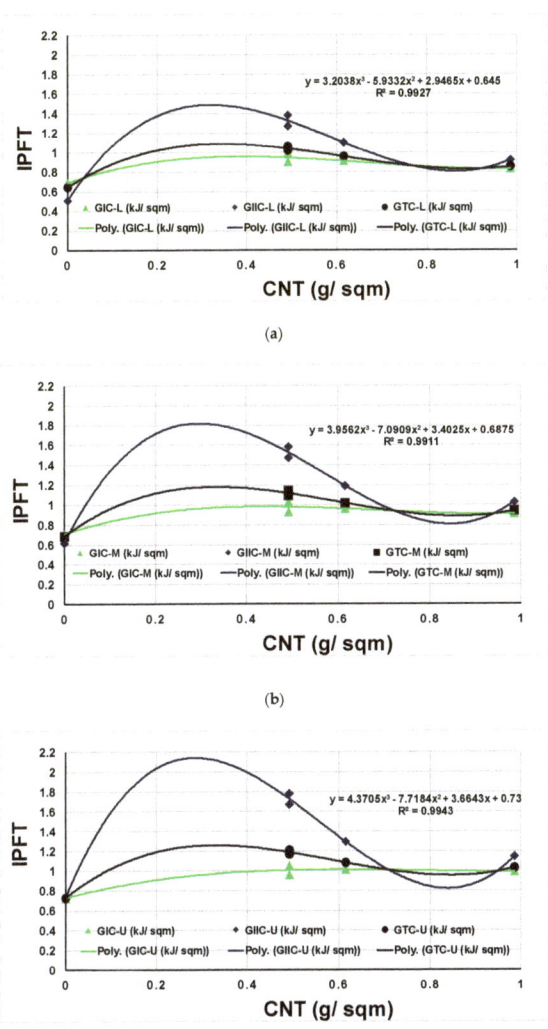

Figure 5. Trends and CoD for total inter-ply fracture toughness data for CFRP (a) minima or lower-bound, (b) mean or average, (c) maxima or upper-bound.

Although, the way that G_{TC} was calculated might be conservative, still it is more practical than using only single mode values. This also means that for prognostic related to delamination resistance, the ML shall be based on the G_{TC} dataset than G_{IC} and G_{IIc} individually and independently. Notwithstanding, the data augmentation has to be conducted still at the level of individual fracture modes in order to maintain the reliability of the G_{TC} data.

The choice of factor 'z' and its impact on the final dataset is discussed later in Section 4.

3.5. Oddities:

As presented in Figures 3–5, out of the 15 cases for each mode, three points are the result of zero CNT, three with non-functionalized multi-wall or MWCNT, and nine with functionalized MWCNT with –COOH, –OH$_1$ or –OH$_2$. There are six points that belong to the same percentage of MWCNT, but with two different functionlized groups, viz; -COOH and –OH$_2$. However, the resultant G_{TC} are very close. Therefore, both the data points were retained.

One oddity to observe is the best fit for G_{IIc}. Its apex exceeds the maxima of the original dataset (i.e., 1.78 kJ/m^2), which does not seem to be realistic. This oddity needs to be and will be scrutinized after the data boosting exercise.

3.6. Space:

It was noticed that beyond 0.6–0.7 g/sqm of CNT content, the gain in IPFT results is minimal. It is mentioned in [11] that CNT content higher than 1 g/sqm thickens the interface, and thereby reduces the IPFT. In the current study, the data values chosen are only up to 0.98 g/sqm of CNT content, hence, that automatically forms the valid limit. In the same way, the minima and maxima of G_{IC} and G_{IIC} data points together (0.51 kJ/m^2 and 1.78 kJ/m^2) form the other limits for the G_{TC}. Thus, it will be good enough to observe these boundaries while data boosting.

3.7. Infer:

Understanding of the data and the physics behind the chosen data points helped in guiding data boosting. First, new $G_{IC}, G_{IIC},$ and G_{TC} data points are to be added in without changing the CNT. In this 'vertical KBDB', new points (NP) are to be created in between the minima and mean, as well as the mean and maxima values. This will be linear interpolation. Subsequently, NP to be generated between two close-by CNT content points. It is fair to assume that between any two close-enough points, the variations in G_{IC} and G_{IIC} are linear. This may be termed as 'horizontal KBDB'. The same may be applied to the diagonally opposite, adjacent data points. These are also the proximity points, one from the L data range and the other from the U data range. This operation is named as 'cross KBDB'. In summary, the vertical, horizontal and cross KBDB shall logically add NP within the defined space.

These NP creations are justifiable given the fact that the CNT content used in nano-engineering the composites interfaces is small (less than 1 g/sqm), and some variation is expected and bound to happen during actual manufacturing.

3.8. Translate:

The above inferences were then translated into the mathematical simple formulae based on the schematics shown in Figure 6. The effect of these operations can be studied upon generating and adding NP to the original datasets. The trend-lines, their CoD, as well as space boundaries may be examined then.

Typically –

$$LM1(x) = \frac{L1(x) + M1(x)}{2}$$

$$LM1(y) = \frac{L1(y) + M1(y)}{2}$$

Typically –

$$L12(x) = \frac{L1(x) + L2(x)}{2}$$

$$L12(y) = \frac{L1(y) + L2(y)}{2}$$

Typically –

$$LM12(x) = \frac{L1(x) + M2(x)}{2}$$

$$LM12(y) = \frac{L1(y) + M2(y)}{2}$$

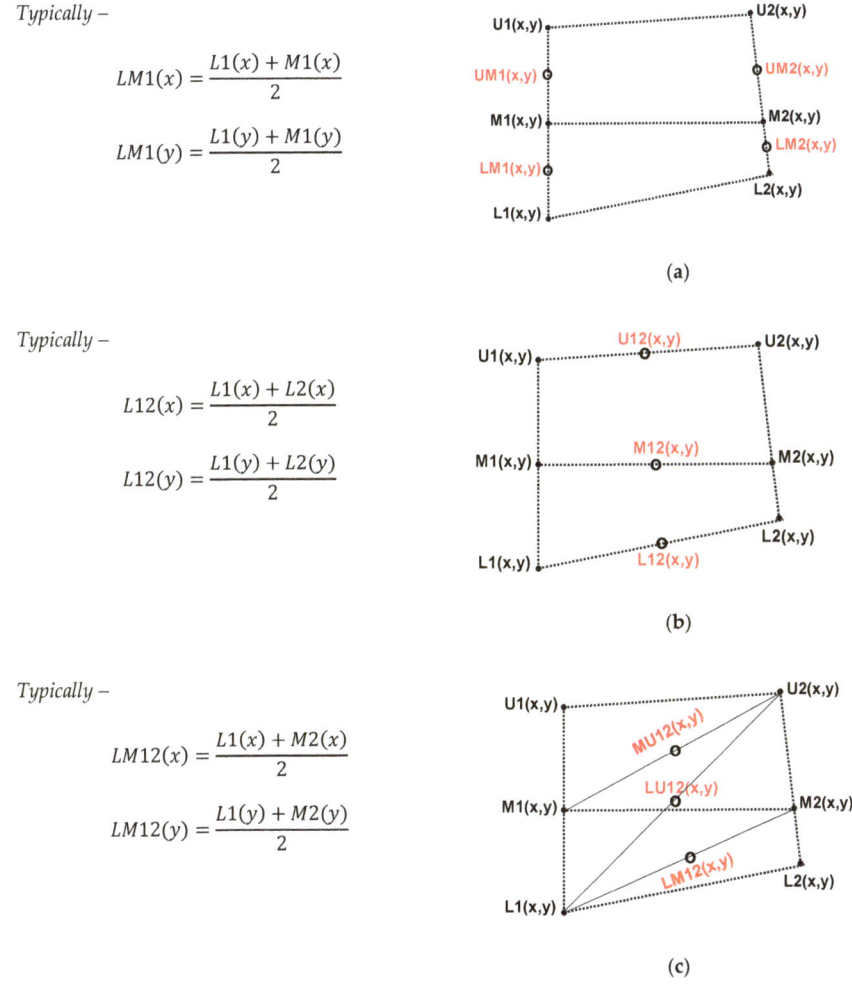

Figure 6. KBDB strategy used for generating (**a**) vertical, (**b**) horizontal, and (**c**) cross new points (NP).

The cross KBDB can be either forward (as opted in the current case), backward or cross diagonal, which may lead to some variation in data points. This aspect is deliberated in Section 4.

3.9. Employ:

As planned, the NP were created systematically and added to the original dataset. The results for G_{IC}, G_{IIC} and G ratio are shown in Figure 7, while the results for G_{TC} are shown in Figure 8.

It is clear from Figure 8 that the KBDB implemented was reasonable, and did not distort the data. It, in fact, helped to fill in empty space rationally closing the gaps in the original data.

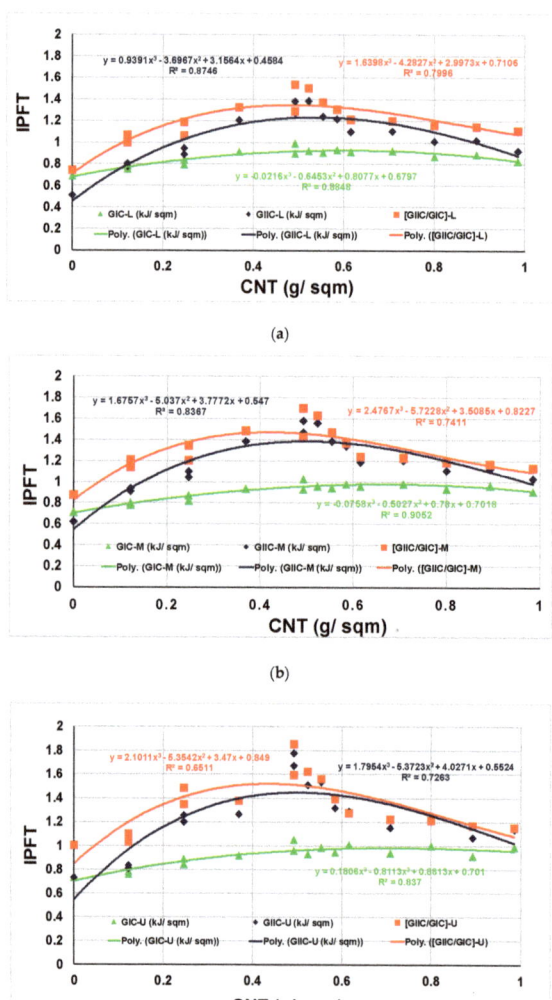

Figure 7. G_{IC}, G_{IIC} and G ratio values, trends and CoD after the KBDB for (**a**) minima or lower-bound, (**b**) mean or average, (**c**) maxima or upper-bound data.

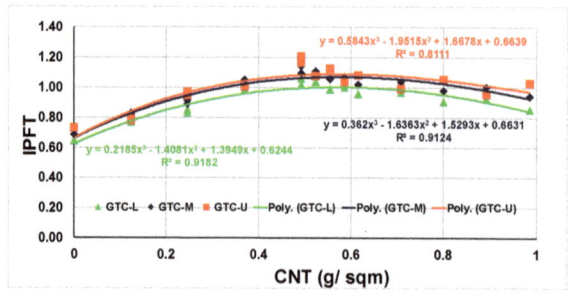

Figure 8. G_{TC} (minimum, mean and maximum) values, trends and CoD after the KBDB.

3.10. Scrutinize:

Although the initially collected data was organized in minima, mean and maxima for convenience of analysis, all three points still belonged to only one CNT content. It essentially means that, during predictive modeling, one need not differentiate between L, M and U values. Rather, one can merge all L, M and U values, and use their collective trend for a prognostic. As seen in Figure 9, the entire dataset aligns well irrespective of second or third order trend fitting. In addition, all points fall within the space marked earlier.

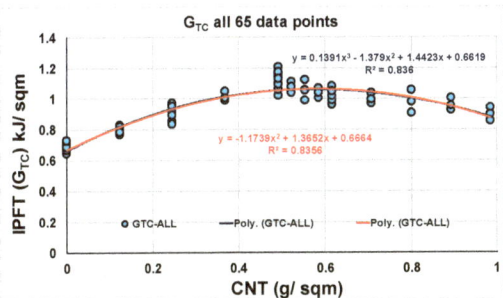

Figure 9. All G_{TC} values, with second and third order polynomial best fit trends and CoD.

This corroborates that all NP are valid and justifiable. The data augmentation factor achieved at this stage through the COMPOSITES approach is 4.33 (i.e., the ratio of augmented dataset of 65 points to the original dataset of 15 values). It was interesting to see how the augmented dataset compared with the original dataset. Referring to Figure 10 for the comparison, it is clear that the dataset augmentation has removed the spuriousness observed in the third order best fit line, due to the sparsity of data.

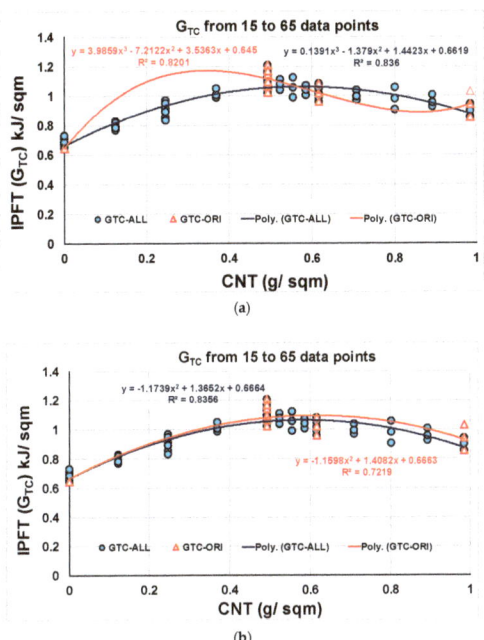

Figure 10. Comparison between original and augmented G_{TC} datasets, their trends and CoD values (a) third order polynomial trends (b) second order polynomial trends.

The second order trend lines have no such issue; however, the CoD for the augmented dataset is better than the original. This will also mean that the boosted dataset follows the original dataset well and will help better facilitate the ML for predictive modeling.

It may also be noted that the second and third order polynomial trend lines show similar trends and CoD values for the new dataset. This is a sort of bench marking and the testimony that the data boosting carried out is reliable. Further error analysis was not conducted since no ML exercise was planned to be part of this study.

4. Discussion

As mentioned earlier, the value of parameter 'z' was preselected as 0.75 for this study. This, however, can be changed to fit the loading conditions. For the loading dominated by opening mode, one shall use higher 'z' values whereas when loading is closer to sliding mode, 'z' shall be low. For mixed mode loading, users may use their preference. Notwithstanding, 'z' that gives slightly conservative outcome is always safer.

The final G_{TC} datasets for three different values of 'z' are shown in Figure 11. It shows that the shift in trend lines is systematic and logical. This means that the prognostic will remain valid at all such values, as long as the mode-I and mode-II dataset remain the same. The user may even use the same database for either mode-I ($z = 1$) or mode-II ($z = 0$) prognostic. This is an added advantage.

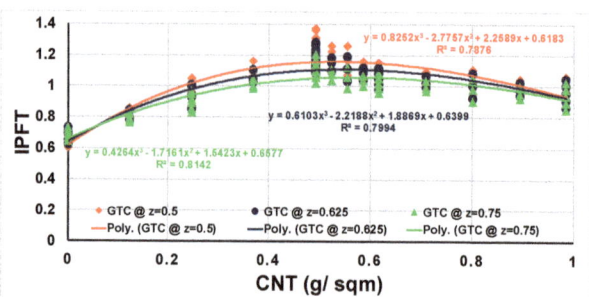

Figure 11. Comparison of all G_{TC} values, with different values of 'z' parameter.

As shown in Figure 6, cross KBDB uses forward interpolation. One may alternatively use reversed or X interpolation as shown in Figure 12 to generate NP.

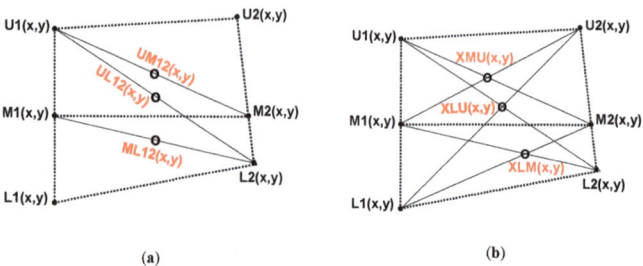

Figure 12. Alternative (a) reversed, (b) X interpolation KBDB strategies for generating cross NP.

As a result of the small ranges of x and y values, the magnitudes of $G_{IC}, G_{IIC},$ and G_{TC} will, however, not change significantly. In addition, the user may be able to boost the dataset further with additional interpolations within the extent of the original experimental data scatter. However, all these operations fall within the same COMPOSITES approach, and hence are not attempted in this paper.

It may be noted from the polynomial equations in Figure 11 that the fourth or the last term typically indicates the G_{TC} value at zero CNT content (e.g., 0.6577 kJ/m^2 in Equation 4). The rest of the terms in the same equation, associated with CNT content, give the additional delamination resistance attainable by adding the corresponding amount of CNT at the inter-ply interface.

$$G_{TC} = 0.4264(CNT)^3 - 1.7161(CNT)^2 + 1.5423(CNT) + 0.6577 \text{ for } z = 0.75 \quad (4)$$

The type of functionalized CNT does have impact on the percentage improvement achievable in G_{TC}. However, in this dataset, the variation is within the experimental scatter of the entire dataset. Hence, the user may not need to pay specific attention to the type of functionalization (–COOH, –OH$_1$ or –OH$_2$) for the MWCNT to be used.

5. Conclusions

The IPFT data on composites in the study is boosted by a factor 4.33 successfully using the 10-step, COMPOSITES process. This proves the usefulness and versatility of the KBDB process. The new dataset, in fact, became more representative, and both second and third order polynomials seemed to best fit without much variation in CoD. It is demonstrated that this KBDB process not only helped in adding NP to the originally lean dataset, but also helped in identifying and defining the dataset boundaries within the experienced scatter.

Besides this, a simpler way of estimating the total IPFT using the mode-I and mode-II data is presented. This avoided the use of an empirical parameter as used by other researchers. This provided additional flexibility in working with different ratio of mixing between mode-I and mode-II. As such, the same modeling exercise can therefore be used for cases from pure mode I, mixed mode and pure mode II.

Funding: This research received no external funding.

Conflicts of Interest: The author declares no conflict of interest.

References

1. Gürdal, Z.; Haftka, R.T.; Hajela, P. *Design and Optimization of Laminated Composite Materials*; John Wiley & Sons: Hoboken, NJ, USA, 1999.
2. Sridharan, S. *Delamination Behaviour of Composites*; Woodhead Publishing and Maney Publishing: Cambridge, UK, 2008.
3. Benzeggagh, M.L.; Kenane, M. Measurement of mixed-mode delamination fracture toughness of unidirectional glass/epoxy composites with mixed-mode bending apparatus. *Compos. Sci. Technol.* **1996**, *56*, 439–449. [CrossRef]
4. Ali, D.-N.; Amin, F.; Sepideh, R.J. An energy based approach for reliability analysis of delamination growth under mode I, mode II and mixed mode I/II loading in composite laminates. *Int. J. Mech. Sci.* **2018**, *145*, 287–298. [CrossRef]
5. Chen, C.-T.; Gu, G.X. Machine learning for composite materials. *MRS Commun.* **2019**, *9*, 556–566. [CrossRef]
6. Navid, Z.; Johannes, R.; Reza, V. Theory-guided machine learning for damage characterization of composites. *Compos. Struct.* **2020**, *246*, 112407. [CrossRef]
7. Daghigh, V.; Lacy, T.E., Jr.; Daghigh, H.; Gu, G.; Baghaei, K.T.; Mark, F.; Horstemeyer, M.F.; Pittman, C.U., Jr. Machine learning predictions on fracture toughness of multiscale bio-nano-composites. *J. Reinf. Plast. Comp.* **2020**, *39*, 587–598. [CrossRef]
8. Gossett, E.; Toher, C.; Oses, C.; Isayev, O.; Legrain, F.; Rose, F.; Zurek, E.; Carrete, J.; Mingo, N.; Tropsa, A.; et al. AFLOW-ML: A RESTful API for machine-learning predictions of materials properties. *Comp. Mater. Sci.* **2018**, *152*, 134–145. [CrossRef]
9. Joshi, S.C. COMPOSITES: A pragmatic knowledge-based engineering data boosting process. *J. Eng. Sci.* **2020**, *1*, 1–2.

10. Joshi, S.C. Knowledge based data boosting exposition on CNT-engineered carbon composites for machine learning. *Adv. Comp. Hybrid. Mat.* **2020**, *3*, 354–364. [CrossRef]
11. Dikshit, V. Manufacturing and Performance Studies of Laminated Composites with Nano-reinforced Inter-ply Interfaces. Ph.D. Thesis, Nanyang Technological University Singapore, Singapore, 2014.
12. Fereidoon, A.; Memarian, F.; Ehsani, Z. Effect of CNT on the delamination resistance of composites. *Fuller. Nanotub. Carb. Nanostruct.* **2013**, *21*, 712–724. [CrossRef]
13. Joshi, S.C.; Dikshit, V. Enhancing interlaminar fracture characteristics of woven CFRP prepreg composites through CNT dispersion. *J. Comp. Mat.* **2011**, *46*, 665–675. [CrossRef]
14. Dikshit, V.; Bhudolia, S.K.; Joshi, S.C. Multiscale polymer composites: A review of the interlaminar fracture toughness improvement. *Fibers* **2017**, *5*, 38. [CrossRef]
15. Boon, Y.D.; Joshi, S.C. A review of methods for improving interlaminar interfaces and fracture toughness of laminated composites. *Mat. Today Com.* **2020**, *22*, 100830. [CrossRef]
16. Dikshit, V.; Joshi, S.C. Manufacturing of multiscale interlaminar interface composites and quantitative analysis of interlaminar fracture toughness. In *Fiber-Reinforced Nanocomposites: Fundamentals and Applications*; Elsevier: New York, NY, USA, 2020; pp. 261–278. [CrossRef]

Publisher's Note: MDPI stays neutral with regard to jurisdictional claims in published maps and institutional affiliations.

© 2020 by the author. Licensee MDPI, Basel, Switzerland. This article is an open access article distributed under the terms and conditions of the Creative Commons Attribution (CC BY) license (http://creativecommons.org/licenses/by/4.0/).

Article

Effect of Graphene Additive on Flexural and Interlaminar Shear Strength Properties of Carbon Fiber-Reinforced Polymer Composite

Mohamed Ali Charfi [1], Ronan Mathieu [1], Jean-François Chatelain [1,*], Claudiane Ouellet-Plamondon [2] and Gilbert Lebrun [3]

[1] École de Technologie Supérieure, Mechanical Engineering Department, Montréal, QC H3C1K3, Canada; mohamed-ali.charfi.1@ens.etsmtl.ca (M.A.C.); ronan.mathieu.1@ens.etsmtl.ca (R.M.)
[2] École de Technologie Supérieure, Construction Engineering Department, Montréal, QC H3C1K3, Canada; claudiane.ouellet-plamondon@etsmtl.ca
[3] Mechanical Engineering Department, Université du Québec à Trois-Rivières (UQTR), Trois-Rivières, QC G8Z 4M3, Canada; gilbert.lebrun@uqtr.ca
* Correspondence: jean-francois.chatelain@etsmtl.ca; Tel.: +1-514-396-8512

Received: 28 September 2020; Accepted: 26 October 2020; Published: 30 October 2020

Abstract: Composite materials are widely used in various manufacturing fields from aeronautic and aerospace industries to the automotive industry. This is due to their outstanding mechanical properties with respect to their light weight. However, some studies showed that the major flaws of these materials are located at the fiber/matrix interface. Therefore, enhancing matrix adhesion properties could significantly improve the overall material characteristics. This study aims to analyze the effect of graphene particles on the adhesion properties of carbon fiber-reinforced polymer (CFRP) through interlaminar shear strength (ILSS) and flexural testing. Seven modified epoxy resins were prepared with different graphene contents. The CFRP laminates were next manufactured using a method that guarantees a repeatable and consistent fiber volume fraction with a low porosity level. Short beam shear and flexural tests were performed to compare the effect of graphene on the mechanical properties of the different laminates. It was found that 0.25 wt.% of graphene filler enhanced the flexural strength by 5%, whilst the higher concentrations (2 and 3 wt.%) decreased the flexural strength by about 7%. Regarding the ILSS, samples with low concentrations (0.25 and 0.5 wt.%) demonstrated a decent increase. Meanwhile, 3 wt.% slightly decreases the ILSS.

Keywords: CFRP; composite; graphene; contact molding; mechanical properties

1. Introduction

Carbon fiber-reinforced polymer composites (CFRP) are increasingly being used in a wide range of domestic and industrial applications, such as aerospace, automobile, wind energy, sport, and goods industries, to name a few [1,2]. Owing to their advantageous properties like corrosion resistance, temperature resistance, light weight, and high mechanical properties, more than 50% of new aircrafts (Airbus A350 and Boeing 787) are composed of CFRP [3,4]. Their strength/weight and stiffness/weight ratios can be five to eight times greater than ordinary metals [5]. However, the matrix/fiber interface is considered as the weakest link of composite materials [6], with typical flaws such as voids and uncovered fibers. This would eventually initiate the failure of composite parts [7].

As a solution for interfacial weaknesses, researchers have sought to incorporate fillers in the matrix. These fillers have great potential to ameliorate the mechanical, chemical, and physical properties of the polymer. Therefore, enhancing the matrix bonding properties should improve the overall composite quality [8,9]. Graphene is one of the most promising fillers in polymers. Since its first discovery in

2004 [10], this material has gained enormous attention [11,12]. It possesses exceptional characteristics like high thermal and electrical conductivity [13,14] and being lightweight, as well as astonishing mechanical properties [15,16]. Graphene is usually obtained through the exfoliation of graphite. However, its manufacturing is cost-prohibitive and presents safety risks in large-scale production [17]. Graphene materials are good fillers for polymer matrices, with nano-clay and carbon nanotubes being the most relevant competitors. Both graphene and nano-clay are platelet-type materials, characterized by a layered structure with high aspect ratios (>1000) [18,19]. Composite polymers based on platelet fillers demonstrated outstanding mechanical properties. However, graphene outperforms nano-clay by its excellent thermal and electrical properties. In contrast, carbon nanotubes possess similar thermal and electrical properties to graphene. However, they are not considered as suitable fillers because of their relatively expensive, high mixture viscosity causing the entangling of nanotubes, and immense anisotropic properties [20]. Another comparative study [21] was performed between graphene nano-platelets (GnPs), single-walled carbon nanotubes (SWCNT), and multi-walled carbon nanotubes (MWCNT). The Young's modulus of the graphene nanocomposite is 31% higher than that of the pristine epoxy as opposed to only 3% enhancement for the SWCNT. Moreover, the tensile strength of the graphene nanocomposite is 26% higher than that based on MWCNT [21]. This renders graphene nanoplatelets the ideal filler for our experiments. Homogeneous dispersion, graphene exfoliation, and load percentage play a vital role in the composite quality. For instance, while flexural stress and short beam shear stress tend to peak at lower graphene percentages, electrical and thermal conductivity significantly increase at higher filler percentages. Moreover, dispersion and exfoliation processes have an undeniable impact on the composite quality [22].

Han et al. [23] studied the impact of graphene oxide (GO) concentration on the interlaminar shear strength (ILSS) of CFRP laminates. From 0 to 0.1 wt.%, the ILSS increased but beyond 0.2 wt.%, the ILSS was inversely proportional to the GO content. Indeed, a maximum improvement of more than 8% was recorded at 0.1 wt.% compared to the same composite without graphene. The same tendency was observed by Kamar et al. [24], who used glass fabric composites with GnPs particles. The optimum graphene concentration was 0.25 wt.%, which induced 29% and 25% of improvement in flexural strength and mode I fracture toughness, respectively. Increasing the concentration to more than 0.5 wt.% considerably reduces the fracture toughness and decreases the interlaminar adhesion which in turn leads to delamination and micro-buckling.

Tang et al. [25] investigated the influence of graphene dispersion in epoxy. They prepared two graphene/epoxy mixtures with 0.2 wt.% by means of the ball mill mixing technique. The graphene particles were better dispersed in one mixture than the other. As a result, highly dispersed graphene induces higher strength and fracture toughness. The fracture improvement was 52% for the highly-dispersed mixture as opposed to only 24% for the poorly dispersed mixture. A similar study conducted by Raza et al. [26] concluded that graphene dispersed by mechanical mixing produces better thermal and electrical conductivity than when prepared by a dual asymmetric centrifuge speed mixer. This trend is attributed to the intensive shearing of mechanical mixing. Chandrasekaran et al. [27] compared the electrical properties of two different processing methods: the high-shear mixing plus three-roll milling method (3RM) against the high-shear mixing plus sonication method, emphasizing that samples prepared by 3RM have a higher electrical conductivity than those prepared by sonication.

The fiber volume fraction variation (V_f) and porosity contents are crucial factors to drive a clear conclusion of this study. The composite properties are dominated by the fiber properties and not by the matrix (i.e., Young's modulus, flexural strength, etc.) [28]. Thereby, any trivial variation of the fiber volume fraction would conceal the impact of graphene. Similarly, the void ratio negatively affects the properties. For example, interlaminar shear strength decreases by about 7% per 1% of void content [29]. Therefore, all samples must have the same V_f with a minimum of variation as well as low and consistent porosity.

To the best of the authors' knowledge, most of the manufactured laminates presented by previous studies possess either a relatively large V_f variation or a non-uniform filler dispersion which might affect the repeatability of their results. For example, in [24,30,31], the authors incorporated different types of particles into the laminate by means of vacuum resin transfer molding (VARTM). Even though this technique might lead to a consistent fiber volume fraction, it may lead to non-uniform dispersion of the filler particles because of a "filtering" mechanism of the filler by the fibers all along the resin transfer. Mclaughlin [28] used hand-layup and vacuum bagging with weight control of the fibers and the epoxy. Nevertheless, the fiber volume fraction was not found repeatable from one laminate to another. For these reasons, the first objective of this study is to propose an innovative and reliable manufacturing method, at a reasonable cost for experiment purposes, that leads to clear conclusions about the effect of graphene on the mechanical properties. It is important to note that this study is a part of a larger scientific study, with the main goal of improving the machinability of CFRP. This material is inherently rough and abrasive which makes its machinability more difficult than other materials. Tool wear and poor surface finish frequently occur in this matter [31]. So, enhancing the machinability of CFRP is a necessity. However, we still have to improve or, at least maintain, the mechanical properties of the modified material as compared to the unfilled resin composite. We also investigate the optimum filler percentage that induces the best short beam shear (SBS) and flexural strengths. The results found of this study are the starting point to find the best filler percentage that improves the machinability of CFRP.

2. Materials and Methods

In this section, we present the samples preparation in four principle paragraphs: the graphene/epoxy mixing process, the CFRP laminate manufacturing method, composite quality verification, and the mechanical tests. Concerning the quality verification, this was carried out by analyzing the fiber volume fraction and the porosity content. The mechanical properties were evaluated through SBS and flexural tests. Both SBS and flexural coupons possess a rectangular shape. However, SBS is smaller than the flexural coupon, with a length and a width of only 18 and 8 mm, respectively.

2.1. Graphene-Epoxy Mixing Process

Graphene particles (0XB) were provided by Nano-Xplore Inc. (Saint-Laurent, QC, Canada). These particles were mixed and exfoliated within a Marine 820 epoxy resin from Axson Technologies (Madison Heights, MI, USA). The filler percentages varied from 0 to 3 wt.%, as presented in Table 1.

Table 1. Graphene's percentages.

Test Number	Graphene (wt. %)
1	0
2	0.25
3	0.50
4	0.75
5	1.00
6	2.00
7	3.00

Filler percentages were calculated with respect to the total weight of the mixture which includes the hardener weight, epoxy weight, and filler weight. According to the epoxy supplier, the weight ratio of hardener/epoxy should be 18%. The following equations present the weight of each part of the mixture.

$$Wh = 0.18 \times Wep \tag{1}$$

$$Wg = wt.\% \times Wt \tag{2}$$

$$Wt = Wep + Wh + Wg \tag{3}$$

$$Wg = (wt.\% \times 1.18 Wep)/(1 - wt.\%) \tag{4}$$

Here, Wh is the hardener weight (g); Wg is the graphene weight percentage (g); Wep is the epoxy weight (g), and Wt is the total weight (g).

Based on previous studies [23,24], we used three-roll milling (3RM) and high-shear mixing to incorporate and homogenize graphene particles in the epoxy. Next, a Silverson L5M-A (Silverson Machines, Inc., East Longmeadow, MA, USA) high-shear mixer was used to mechanically blend the filled epoxy. The intensive shear force deeply exfoliates the graphene particles which considerably reduces the flakes' thickness. As shown in Table 2, the mechanical mixing was divided into seven segments, during which the temperature was almost maintained with the help of an ice bath. However, the ice bath was removed in the last step to allow the temperature to rise to a maximum of 50 °C. The elevated temperature decreased the mixture viscosity, facilitating air bubble extraction. [31].

Table 2. Shear mixing sequence.

Time	Speed (RPM)
2 min	1000
2 min	3500
2 min	6000
2 min	8000
2 × 3 min	10,000

Final degassing was done in a vacuum oven for one hour, and the loaded epoxy was finally mixed with the hardener according to the supplier's instructions.

2.2. Laminate Manufacturing

The following constraints are imposed on the manufacturing method in this study: homogeneous distribution of graphene, consistent laminate thickness, minimum void, and minimum fluctuation of the fiber orientation. Even though there are multiple processing methods, most of them are either expensive or difficult to operate and unadjustable to meet the above-mentioned requirements, especially the one requiring a homogeneous distribution of filler particles. In this context, contact molding was the only alternative to liquid injection molding processes for this scientific application. Since the operator impregnates the fiber layer by layer with the filled epoxy, the graphene particles are evenly dispersed on the surface of each layer as well as across the laminate thickness. Nonetheless, the basic contact molding process usually produces important void contents, stochastic fiber angle fluctuation, and uncontrollable thickness variations. The void ratio can be reduced by applying a vacuum during the cure of the plate to remove most of the trapped air from the laminate [32]. In terms of angle fluctuations of the fiber, three wooden sticks with several grooves were fixed on top of the mold (Figure 1). These sticks form a frame that prevents the fiber layers from slipping during the hand-layup and the curing processes. These grooves allow the excess of epoxy to be expelled without displacing the wooden sticks.

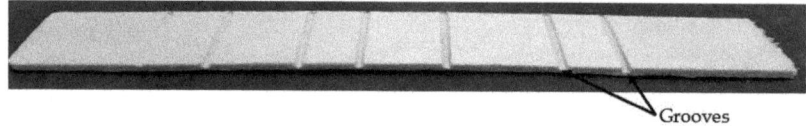

Figure 1. Guide frame wooden stick.

A uniform laminate thickness was obtained using four spacers having the same thickness. These spacers were placed on the mold, distanced from the wooden frame by approximately 10 mm on each side. This technique improved the hand-layup process and made it adequate for this study. Figure 2 presents a 3D sketch of the developed method.

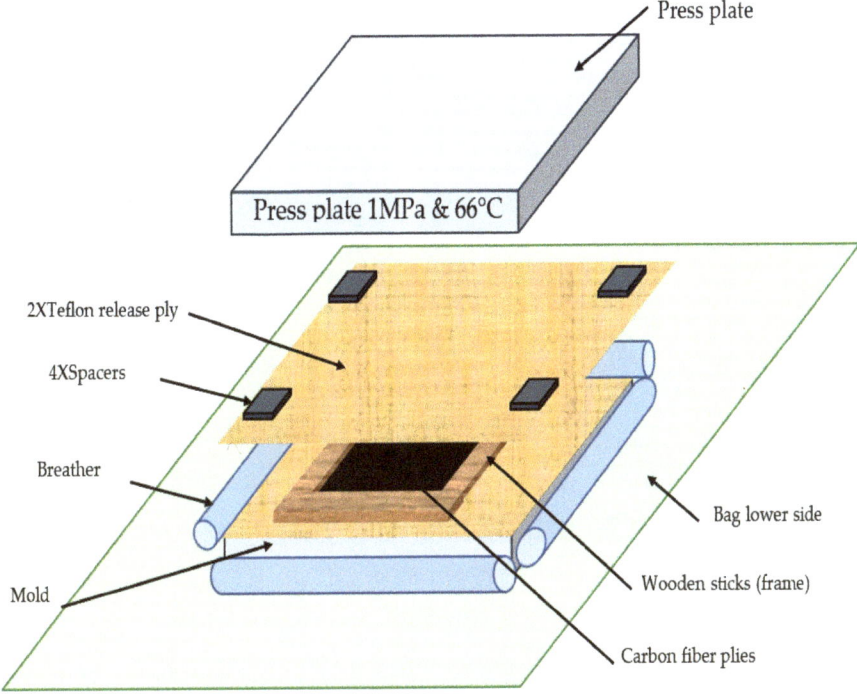

Figure 2. Exploded view of the layup assembly.

Fourteen plies of TC-09-U unidirectional high-modulus carbon fiber from Texonic Inc. (Saint-Jean-sur-Richelieu, QC, Canada) were used to form $[0]_{14}$ laminates with a thickness of around 4 mm. The single-layer reinforcement surface density is 320 g/m². In terms of curing parameters, the vacuum pressure was set at a maximum of 29 inches of mercury, and the hydraulic press was equipped with heater plates which allow for curing the laminate according to the supplier recommendation, at 66 °C for 3 h and under a pressure of 1 MPa to have a consistent thickness. Figure 3 shows a typical photograph of the developed method. As for the demolding part, we used a Teflon sheet instead of a release agent and peel ply. Indeed, the release agent was not effective under such curing conditions. In addition, the peel ply randomly absorbs a certain quantity of epoxy which affects the fiber volume fraction. Conversely, the Teflon sheet is quite efficient during the demolding and its intrinsic sealant nature guarantees a smooth surface on both sides of the plaque.

A post-curing was required to uniformize curing through the thickness of the laminates. Therefore, all the plates were put inside a Despatch oven (Ontario Ovens Inc., Brampton, ON, Canada) which was programed to gradually increase the temperature with a segment ramp of 10 °C per hour until it reaches a maximum of 66 °C. The temperature was held constant for 24 h. Eventually, it gradually decreases with a negative slope of 10 °C per hour until reaching room temperature. Verification of the laminate quality was performed through measurement of the fiber volume fraction (V_f) and void ratio. To avoid oxidation of carbon fibers using a high temperature of pyrolysis when measuring V_f [29,30], it was instead calculated by measuring the laminate thickness and implementing it into Equation (5).

$$V_f = (M_f \times N)/(\varphi \times h) \tag{5}$$

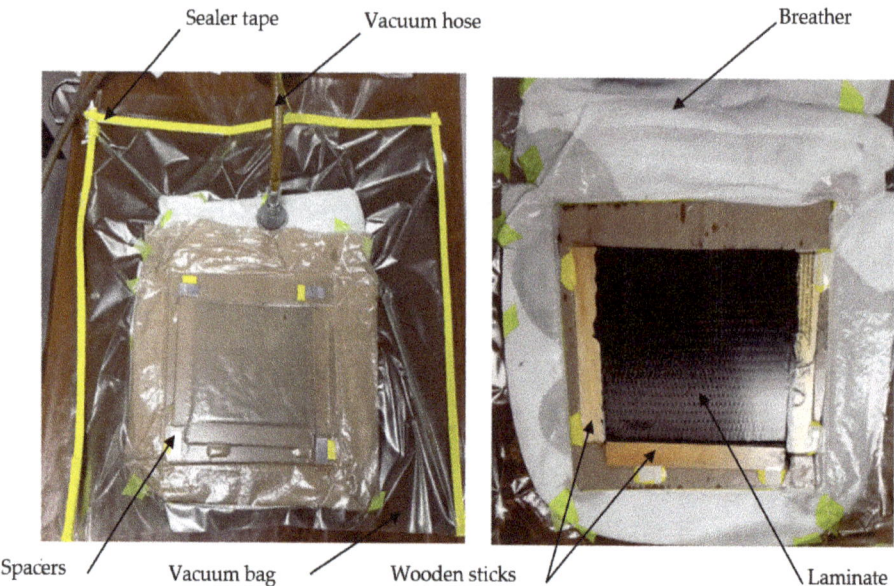

Figure 3. Contact molding assisted with vacuum bag and hot press.

Here, Mf is the surface density of the dry reinforcement (g/m^2); N is the number of plies; φ is the fiber density (g/m^3), and h is the laminate thickness (m).

Unlike the resin burn-off technique, this method does not reveal the emptiness ratio. Therefore, a porosity test was required. Micrographic image processing is deemed to be a good approach to evaluate the porosity level in laminates. Kite et al. [33] emphasized that the outcomes of this method correlate well with matrix digestion. Thus, 6 samples of each plaque were cut and prepared to be polished by a motopol 2000 automatic polisher. Next, an optical microscope was used to take a sufficient number of pictures of the whole sample's area with a magnification of 50X (Figure 4). Finally, an open-source software named ImageJ was used to segment this image into three zones: the background, the composite, and voids. This step was performed through machine learning, where the operator manually introduces all of these different sections to the software. Doing so, the software can determine the percentages of zones of the entire image.

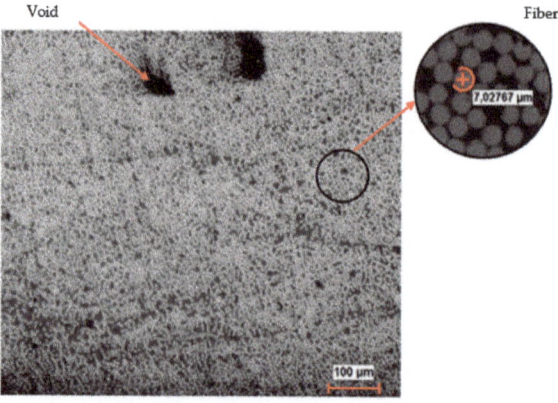

Figure 4. Microscopic picture X50.

2.3. Mechanical Test

The purpose of the short beam shear test is to investigate the matrix adhesion quality. This test is similar to the flexural test. However, the length of the short beam's coupon is short compared to the flexural coupon. Thus, the vertical force induces a shear stress in the plane of specimens. The ASTM D2344 standard requires precise geometrical and dimensional tolerances of the coupon. In order to respect these constraints, a high-precision cutting machine (Struers Secotom 50, Struers, Mississauga, ON, Canada) with a diamond saw was used. To guarantee representative results, ten coupons from each laminate with 0° fiber orientation were tested. The interlaminar shear strength (ILSS) tests were performed on a universal testing machine (MTS alliance RF/200, MTS systems corporation, Eden prairies, MN, USA) which was equipped with a 10 kN load cell. The crosshead speed was set at 1 mm/min and the interlaminar shear strength was calculated according to this equation.

$$ILSS = 0.75 \times (Fmax/A) \qquad (6)$$

where ILSS is the interlaminar shear strength (N/mm^2); Fmax is the maximum force (N), and A is the surface area of the coupon (mm^2).

Ten flexural samples were prepared according to ASTM D7264. Like the short beam shear sample, the fiber orientation was kept at 0° and the crosshead speed of 1 mm/min was maintained until the first drop load occurred which indicates the breakage point. Based on the standard recommendations, the strain points of 0.001 and 0.003 were used to calculate the chord modulus of elasticity. Moreover, the ultimate flexural strength and chord modulus of elasticity were determined by implementing the maximum applied forces in Equations (7) and (8):

$$\sigma f = (3 \times Fmax \times L)/(2 \times b \times h^2) \qquad (7)$$

$$E_F = \Delta\sigma/\Delta\varepsilon \qquad (8)$$

where Fmax is the maximum force (N); L is the specimen length (mm); h is the specimen thickness (mm); b is the specimen width (mm), and ε is the strain (mm/mm).

3. Results and Discussion

3.1. Fiber Volume Fraction

The developed manufacturing method induces repeatable results. Consistent fiber volume fraction is a must to distinguish between the effect of thickness variations and the effect of graphene particles. A digital micrometer was used to take four measurements of each laminate, and these values were then averaged and used in equation (5) to calculate the fiber volume fraction. The V_f mean value, between laminates, equals 64% ± 0.41%. This trivial variation can be ascribed to variation in the Teflon sheet thickness, the spacers thickness, and the mold surface flatness. Nonetheless, this variation is still acceptable, and its impact can be averted through a statistical normalization around the average value of V_f.

In the porosity analysis, the software ImageJ results in a "classifier" file, and this file is a template that contains data through which we can classify similar microscopic pictures. Subsequently, these pictures are converted into a binary image with black and white colors (Figure 5). White pixels with color under a certain threshold will be counted as voids; others with color above the threshold will be counted as composites. The software will then calculate the void percentage, which is the white area divided by the total area.

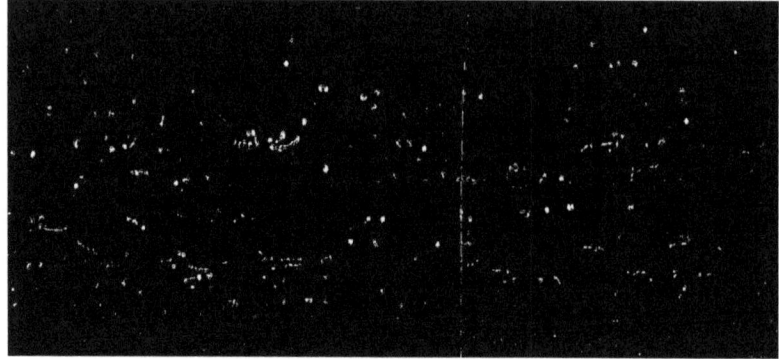

Figure 5. Binary microscopic image.

Figure 6 presents the void percentage of the composites versus the percentage of filler. As can be seen, the results are almost constant with a slight variation. The average value of porosity is equal to 0.61% with the smallest value of 0.4% and a maximum of 0.86%. The results are relatively good compared to the common standards. For instance, in aerospace applications, aircraft parts with a porosity level between 2.5% and 5% are usually accepted [34]. Nevertheless, Costa et al. [35] emphasized in their review that the interlaminar shear strength can be significantly affected if the porosity content goes beyond 0.9%. In addition, Hakim et al. [36] highlighted that higher porosity levels make the composite part sensitive to water penetration and environmental factors which detrimentally impact the static and fatigue strength. The minimum void percentage that we found can be attributed to the high vacuum pressure applied during the curing process. Hakim et al. [36] evaluated the impact of vacuum pressure on the porosity level. They examined three levels of vacuum pressure: poor (0 mmHg), medium (330 mmHg), and high (686 mmHg) and concluded that poor vacuum pressure induces 3.43% of porosity versus only 1.43% with high vacuum pressure. Furthermore, larger pores were more discernable with low vacuum pressure than with high vacuum pressure. According to the bar chart (Figure 6), the void percentage decreases as the filler percentages increase. For instance, laminates with 0 wt.% have 0.86% of void, whereas laminates with 3 wt.% have only 0.4% of void. Typically, increasing the load percentage would increase the matrix viscosity which hinders the extraction of the air bubbles and thus increases the porosity level. Consequently, the depicted pattern in the bar chart does not corroborate with the mentioned hypothesis. Indeed, this trend can be assigned to the detection accuracy of the software Image J. After a sheer number of iterations, the software gained the aptitude to identify various parts of the micrographic image with higher precision. Therefore, the depicted values of the void with 3 wt.% are calculated with relatively more accuracy than the ones of 0 wt.%. Nonetheless, all the presented values are less than 1% which is considered as the threshold for a good composite quality.

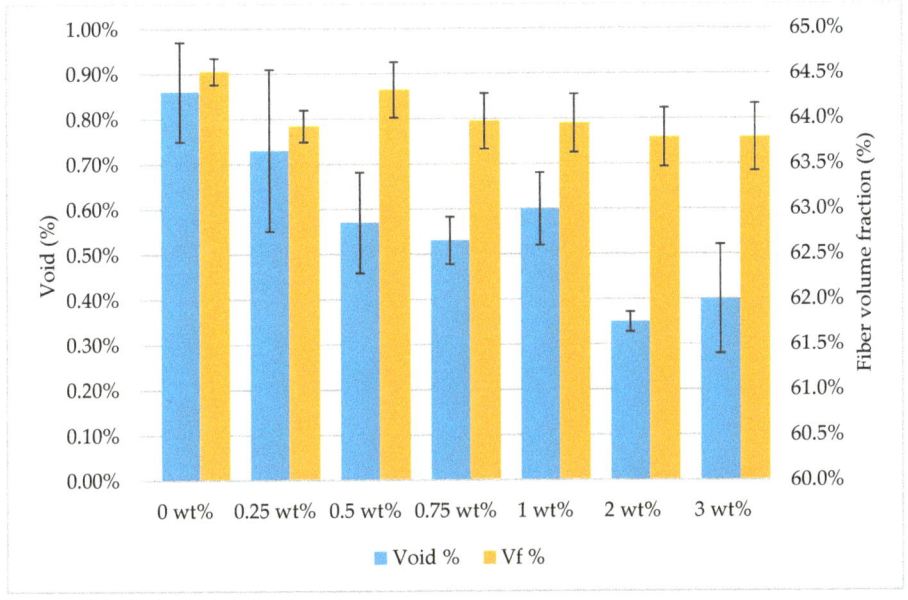

Figure 6. Fiber volume fraction and void of the composite.

3.2. Mechanical Test

A statistical study was required to detect and eliminate the outliers from the list of coupon results. These filtered results were then normalized using equation (9) to a $V_f = 64\%$ [37].

$$\text{Normalized value} = (\text{test value} \times 0.64)/(\text{sample's } V_f) \quad (9)$$

A Pearson correlation test was performed on the normalized values versus the fiber volume fraction (V_f) variation. This test was done under a significance level of 5%. The null hypothesis states that there is no significant correlation between the mechanical properties outcomes versus the V_f variation. Most of the results did not reveal any significant correlation, therefore we can conclude that these results solely present the impact of the filler percentages. All flexural tests behaved linearly until the first drop of the load. This point was used to calculate the flexural strength for each sample. Figure 7 shows that the ultimate flexural strength of specimens with 0.25 wt.% improved by 5% compared to the pristine composite. It is important to mention that the fiber volume fraction of coupons with 0.25 wt.% is only 63.9% as opposed to 64.5% for plain coupons, which indicates that the 5% enhancement of flexural strength is purely assigned to the effect of the filler. On the other hand, the other laminates with a filler concentration higher than 0.25 wt.% demonstrated a slight decrease. The worst cases were with 2 and 3 wt.% with a drop of approximately 7% of the strength. A similar pattern was found by Kamar et al. [24], where they specified 1 wt.% as a threshold of the graphene percentage. Beyond this point, the filler particles start to agglomerate into relatively big bundles which weaken the interlaminar adhesion, thus leading to delamination and micro-buckling defects.

The stiffness of samples was calculated through Equation (8). Figure 7 presents the recorded results. No improvement was found by increasing the filler percentage. In contrast, the stiffness of the loaded specimens mildly decreased to a minimum of 125.7 GPa for the coupons with 0.5 wt.% as compared to a maximum of 131.4 GPa for the neat coupons. Both the stiffness and the ultimate shear strength of samples with 0.5 wt.% and higher filler contents are observed to decrease. This cannot be explained by the fiber volume fraction variation but rather by the effect of the graphene content, as can be seen by comparing the stiffness of samples with 0.25 wt.% and samples with 0.5 wt.% of graphene content. The latter has a slightly higher fiber volume fraction and yet exhibits lower stiffness. In the case of the 0.25 wt.% concentration, the ultimate strength of this specimen showed a certain enhancement, but the stiffness is reduced. This does not meet the anticipated outcomes. For example, Hung et al. [38] investigated the impact of graphene oxide on the mechanical properties of CFRP, reporting an enhancement of around 18% and 5% in the flexural strength and the stiffness, respectively. Nonetheless, in the case of 0.25 wt.%, the Pearson correlation test demonstrated a vivid relation between the stiffness and the V_f variations (Figure 8). This was corroborated with previous studies [9,25], highlighting that flexural and tensile specimens with 0° fiber orientation are not sensitive to the matrix adhesion quality but rather to the fiber volume fraction and the fiber mechanical properties. Therefore, this trend can be assigned to the impact of V_f variation from one sample to another and not to the graphene percentage.

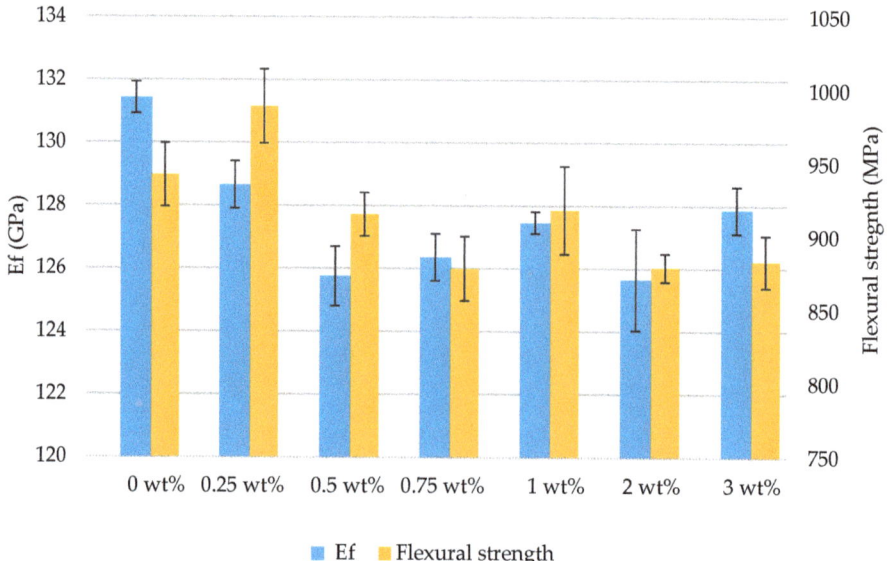

Figure 7. Stiffness and flexural strength of the composite.

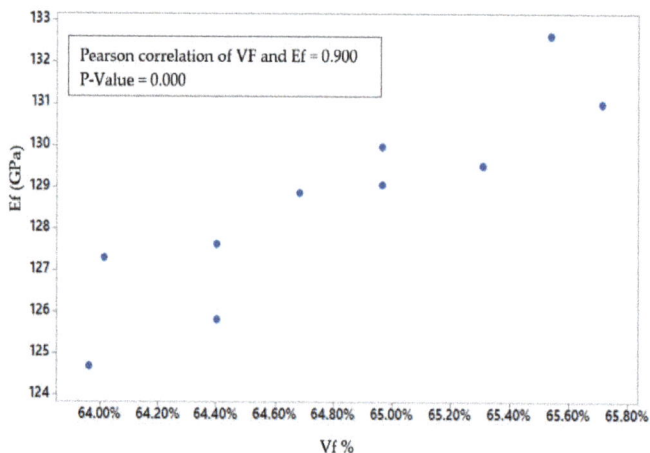

Figure 8. Scatter plot stiffness versus V_f variations of samples with 0.25 wt.%.

Figure 9 displays the mean values of the interlaminar shear strength (ILSS) of each filler percentage. Samples with 0.25 and 0.5 wt.% showed a decent increase ranging from 56 to approximately 59 MPa. The fiber content of the 0.25 wt.% coupons is less than the fiber content of plain coupons. Nonetheless, the recorded improvement counts as 5% in comparison to the neat composite. Meanwhile, 0.75 and 2 wt.% of graphene concentrations almost showed the same improvement of the ILSS by about 2%. However, the ILSS of all the specimens decreased with a minimum of 52 MPa in the case of 3 wt.%. The behavior of the ILSS largely depends on the graphene content, and in the case of low concentrations (0 to 0.5 wt.%), the ILSS behaves linearly with a positive slope. Beyond 0.75 wt.%, this linear relation appears to be inversely proportional, except for 2 wt.% which presents a certain enhancement. A similar trend was revealed by [23], though their tested graphene range was smaller, going from 0.1 to 0.4 wt.%. More experiments must be carried out to better understand this trend.

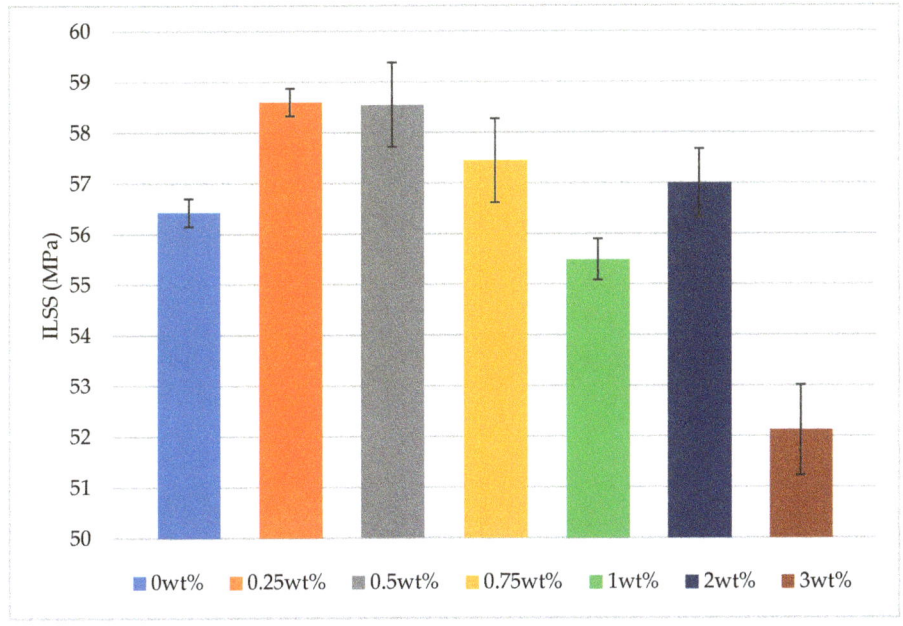

Figure 9. Interlaminar shear strength of the composites.

For further clarification, we have presented the relative details and figures of the previous diagrams in Table 3.

Table 3. Figures recapitulation.

GnPs wt.% %	V_f %	Void%	Flexural Strength (MPa)	Stiffness (GPa)	ILSS (MPa)
0	64.5	0.86	942.3	131.4	56.4
0.25	63.9	0.73	989.1	128.6	58.6
0.50	64.3	0.57	915.6	125.7	58.5
0.75	64.0	0.53	878.7	126.3	57.4
1.00	63.9	0.60	918.5	127.4	55.5
2.00	63.8	0.35	879.3	125.6	57
3.00	63.8	0.40	883.6	127.8	52.1

4. Conclusions

In this study, seven laminates of CFRP were manufactured by hand-layup assisted with vacuum bagging and hydraulic press, and these laminates contain different GnPs percentages ranging from 0 to 3 wt.%. The methodology process was described and has been approved as repeatable with low porosity and V_f variation. The manufactured laminates possess a consistent fiber volume fraction with an average value of 64%. The micrographic porosity test revealed an acceptable void content with a mean of 0.61%. Hence, the fiber volume fraction and the porosity level have little influence on the mechanical properties. Ten flexural samples and ten short beam samples were cut from each laminate, and this was performed according to the ASTM standards. The outcomes presented in this study lead to the following conclusion:

The best graphene percentage was 0.25 wt.%; samples with this filler content induced 5% improvement of the ILSS.

It appears that 0.5 wt.% presents a threshold of the graphene filler, so more than this value, the particles will coalesce which, in turn, results in a poor graphene dispersion and therefore the mechanical properties will be negatively affected.

The same trend was observed regarding the flexural test. Samples with 0.25 wt.% filler showed a maximum improvement of 5%. It was expected that the chord modulus of elasticity would reveal the same trend as the flexural strength. However, it seems that the stiffness is more prone to the fiber volume fraction content. Pearson correlation test showed that the stiffness and the V_f variation using this filler concentration are linearly proportional. Hence, it is difficult to conclude regarding this aspect.

For future studies, we should test other graphene percentages between 0 and 0.25 wt.%. Additionally, further tests should be carried on with 90° fiber orientation which might be more effective to evaluate the adhesion quality of the filled epoxy.

Author Contributions: J.-F.C., G.L. and C.O.-P. designed and directed the study. M.A.C. and R.M. developed the CFRP manufacturing method. M.A.C. performed all the experiments, measurements, and analysis of the results. The writing and revision of the manuscript were performed according to the order of the authors. All authors have read and agreed to the published version of the manuscript.

Funding: This work was co-funded by Nano-Xplore company and Natural Sciences and Engineering Research Council of Canada.

Acknowledgments: We would like to express our special gratitude to Giovanna Gutierrez, R&D Manager and Nima Moghimian, Director of R&D at Nano-Xplore for their technical support and the graphene particles supplied during this research. We also express our gratitude to Nabil Mazeghrane, and Éric Marcoux, from ETS, who provided technical assistance during the experiments.

Conflicts of Interest: The authors declare no conflict of interest.

References

1. Othman, R.; Ismail, N.I.; Basri, H.M.; Sharudin, H.; Hemdi, A.R. Application of carbon fiber reinforced plastics in automotive industry: A review. *J. Mech. Manuf.* **2018**, *1*, 12.
2. Zhang, J.Z. Study on Carbon Fiber Composite Materials in Sports Equipment. *Appl. Mech. Mater.* **2013**, *329*, 105–108. [CrossRef]
3. Breuer, U.P. *Commercial Aircraft Composite Technology*; Springer International Publishing: Cham, Switzerland, 2016.
4. Bouvet, C. *Mechanics of Aeronautical Composite Materials*; ISTE Ltd./John Wiley and Sons Inc.: Hoboken, NJ, USA, 2017; ISBN 978-1-78630-114-7.
5. Budynas, R.G.; Nisbett, J.K. *Shigley's Mechanical Engineering Design*, 9th ed.; McGraw-Hill: New York, NY, USA, 2011; ISBN 978-0-07-352928-8.
6. Shesan, O.J.; Stephen, A.C.; Chioma, A.G.; Neerish, R.; Rotimi, S.E. Fiber-Matrix Relationship for Composites Preparation. In *Renewable and Sustainable Composites*; Pereira, A.B., Fernandes, F.A.O., Eds.; IntechOpen: Rijeka, Croatia, 2019.

7. Liu, X.; Chen, F. A Review of Void Formation and its Effects on the Mechanical Performance of Carbon Fiber Reinforced Plastic. *Eng. Trans.* **2016**, *64*, 33–51.
8. Pathak, A.K.; Borah, M.; Gupta, A.; Yokozeki, T.; Dhakate, S. Improved mechanical properties of carbon fiber/graphene oxide-epoxy hybrid composites. *Compos. Sci. Technol.* **2016**, *135*, 28–38. [CrossRef]
9. Qin, W.; Vautard, F.; Drzal, L.T.; Yu, J. Mechanical and electrical properties of carbon fiber composites with incorporation of graphene nanoplatelets at the fiber–matrix interphase. *Compos. Part B Eng.* **2015**, *69*, 335–341. [CrossRef]
10. Novoselov, K.S.; Geim, A.K.; Morozov, S.V.; Jiang, D.; Zhang, Y.; Dubonos, S.V.; Grigorieva, I.V.; Firsov, A.A. Electric Field Effect in Atomically Thin Carbon Films. *Science* **2004**, *306*, 666–669. [CrossRef] [PubMed]
11. Kim, H.; Abdala, A.A.; Macosko, C.W. Graphene/Polymer Nanocomposites. *Macromolecules* **2010**, *43*, 6515–6530. [CrossRef]
12. Stankovich, S.; Dikin, D.A.; Dommett, G.H.B.; Kohlhaas, K.M.; Zimney, E.J.; Stach, E.A.; Piner, R.D.; Nguyen, S.T.; Ruoff, R.S. Graphene-based composite materials. *Nat. Cell Biol.* **2006**, *442*, 282–286. [CrossRef]
13. Balandin, A.A.; Ghosh, S.; Bao, W.; Calizo, I.; Teweldebrhan, D.; Miao, F.; Lau, C.N. Superior Thermal Conductivity of Single-Layer Graphene. *Nano Lett.* **2008**, *8*, 902–907. [CrossRef]
14. Lee, C.; Wei, X.; Kysar, J.W.; Hone, J. Measurement of the Elastic Properties and Intrinsic Strength of Monolayer Graphene. *Science* **2008**, *321*, 385–388. [CrossRef]
15. Yang, W.; Zhao, Q.; Xin, L.; Qiao, J.; Zou, J.; Shao, P.; Yu, Z.; Zhang, Q.; Wu, G. Microstructure and mechanical properties of graphene nanoplates reinforced pure Al matrix composites prepared by pressure infiltration method. *J. Alloys Compd.* **2018**, *732*, 748–758. [CrossRef]
16. Stoller, M.D.; Park, S.; Zhu, Y.; An, J.; Ruoff, R.S. Graphene-Based Ultracapacitors. *Nano Lett.* **2008**, *8*, 3498–3502. [CrossRef] [PubMed]
17. Nazarpour, S.; Waite, S.R. (Eds.) *Graphene Technology: From Laboratory to Fabrication*; Wiley-VCH Verlag GmbH & Co. KGaA: Weinheim, Germany, 2016.
18. Chen, G.-H.; Wu, D.-J.; Weng, W.-G.; Yan, W.-L. Preparation of polymer/graphite conducting nanocomposite by intercalation polymerization. *J. Appl. Polym. Sci.* **2001**, *82*, 2506–2513. [CrossRef]
19. Giannelis, E.P. Polymer Layered Silicate Nanocomposites. *Adv. Mater.* **1996**, *8*, 29–35. [CrossRef]
20. Sandler, J.; Pegel, S.; Cadek, M.; Gojny, F.; Van Es, M.; Lohmar, J.; Blau, W.; Schulte, K.; Windle, A.; Shaffer, M.S.P. A comparative study of melt spun polyamide-12 fibres reinforced with carbon nanotubes and nanofibres. *Polymer* **2004**, *45*, 2001–2015. [CrossRef]
21. Rafiee, M.A.; Rafiee, J.; Wang, Z.; Song, H.; Yu, Z.-Z.; Koratkar, N. Enhanced Mechanical Properties of Nanocomposites at Low Graphene Content. *ACS Nano* **2009**, *3*, 3884–3890. [CrossRef]
22. Al Imran, K. Enhancement of Electrical Conductivity of Carbon/Epoxy Composites by Graphene and Assessment of Thermal and Mechanical Properties. Ph.D. Thesis, North Carolina A&T State University, Greensboro, NC, USA, 2016.
23. Han, X.; Zhao, Y.; Sun, J.; Li, Y.; Zhang, J.; Hao, Y. Effect of graphene oxide addition on the interlaminar shear property of carbon fiber-reinforced epoxy composites. *New Carbon Mater.* **2017**, *32*, 48–55. [CrossRef]
24. Kamar, N.T.; Hossain, M.M.; Khomenko, A.; Haq, M.; Drzal, L.T.; Loos, A. Interlaminar reinforcement of glass fiber/epoxy composites with graphene nanoplatelets. *Compos. Part A Appl. Sci. Manuf.* **2015**, *70*, 82–92. [CrossRef]
25. Tang, L.-C.; Wan, Y.-J.; Yan, D.; Pei, Y.-B.; Zhao, L.; Li, Y.-B.; Wu, L.-B.; Jiang, J.-X.; Lai, G.-Q. The effect of graphene dispersion on the mechanical properties of graphene/epoxy composites. *Carbon* **2013**, *60*, 16–27. [CrossRef]
26. Raza, M.; Westwood, A.; Stirling, C. Effect of processing technique on the transport and mechanical properties of graphite nanoplatelet/rubbery epoxy composites for thermal interface applications. *Mater. Chem. Phys.* **2012**, *132*, 63–73. [CrossRef]
27. Chandrasekaran, S.; Seidel, C.; Schulte, K. Preparation and characterization of graphite nano-platelet (GNP)/epoxy nano-composite: Mechanical, electrical and thermal properties. *Eur. Polym. J.* **2013**, *49*, 3878–3888. [CrossRef]
28. Mclaughlin, A.M. The Effect of Exfpmoated Graphite on Carbon Fiber Reinforced Composites for Cryogenic Applications. Ph.D. Thesis, University of Massachusetts Lowell, Ann Arbor, MA, USA, 2013.
29. Mayr, G.; Plank, B.; Sekelja, J.; Hendorfer, G. Active thermography as a quantitative method for non-destructive evaluation of porous carbon fiber reinforced polymers. *NDT E Int.* **2011**, *44*, 537–543. [CrossRef]

30. Chowdhury, F.; Hosur, M.; Jeelani, S. Studies on the flexural and thermomechanical properties of woven carbon/nanoclay-epoxy laminates. *Mater. Sci. Eng. A* **2006**, *421*, 298–306. [CrossRef]
31. El-Ghaoui, K.; Chatelain, J.F.; Ouellet-Plamondon, C.; Mathieu, R. Effects of Nano Organoclay and Wax on the Machining Temperature and Mechanical Properties of Carbon Fiber Reinforced Plastics (CFRP). *J. Compos. Sci.* **2019**, *3*, 85. [CrossRef]
32. Mehdikhani, M.; Gorbatikh, L.; Verpoest, I.; Lomov, S.V. Voids in fiber-reinforced polymer composites: A review on their formation, characteristics, and effects on mechanical performance. *J. Compos. Mater.* **2018**, *53*, 1579–1669. [CrossRef]
33. Kite, A.H.; Hsu, D.K.; Barnard, D.J.; Thompson, D.O.; Chimenti, D.E. Determination of Porosity Content in Composites by Micrograph Image Processing. In *AIP Conference Proceedings*; American Institute of Physics: Golden, CO, USA, 2008; Volume 975, pp. 942–949. [CrossRef]
34. Kastner, J.; Plank, B.; Salaberger, D.; Sekelja, J. Defect and Porosity Determination of Fibre Reinforced Polymers by X-ray Computed Tomography. In Proceedings of the 2nd International Symposium on NDT in Aerospace, Hamburg, Germany, 22–24 November 2010; pp. 1–12.
35. Costa, M.L.; Almeida, S.F.M.; Rezende, M.C. The influence of porosity on the interlaminar shear strength of carbon/epoxy and carbon/bismaleimide fabric laminates. *Compos. Sci. Technol.* **2001**, *61*, 2101–2108. [CrossRef]
36. Hakim, I.; Donaldson, S.L.; Meyendorf, N.; Browning, C. Porosity Effects on Interlaminar Fracture Behavior in Carbon Fiber-Reinforced Polymer Composites. *Mater. Sci. Appl.* **2017**, *8*, 170–187. [CrossRef]
37. Technomic Publishing Company (Ed.) *The Composite Materials Handbook: MIL 17, Volume 3: Materials, Usage, Design, and Analysis*; Technomic Publ.: Lancaster, UK, 1999.
38. Hung, P.-Y.; Lau, K.-T.; Qiao, K.; Fox, B.; Hameed, N. Property enhancement of CFRP composites with different graphene oxide employment methods at a cryogenic temperature. *Compos. Part A Appl. Sci. Manuf.* **2019**, *120*, 56–63. [CrossRef]

Publisher's Note: MDPI stays neutral with regard to jurisdictional claims in published maps and institutional affiliations.

© 2020 by the authors. Licensee MDPI, Basel, Switzerland. This article is an open access article distributed under the terms and conditions of the Creative Commons Attribution (CC BY) license (http://creativecommons.org/licenses/by/4.0/).

Article

Experimental and Computational Analysis of Low-Velocity Impact on Carbon-, Glass- and Mixed-Fiber Composite Plates

Ahmed S. AlOmari [1,*], Khaled S. Al-Athel [1], Abul Fazal M. Arif [2] and Faleh. A. Al-Sulaiman [1]

[1] Mechanical Engineering Department, King Fahd University of Petroleum and Minerals, Dhahran 31261, Saudi Arabia; kathel@kfupm.edu.sa (K.S.A.-A.); falehas@kfupm.edu.sa (F.A.A.-S.)
[2] McMaster Manufacturing Research Institute, McMaster University, Hamilton, ON L8S 4L8, Canada; afmarif@mcmaster.ca
* Correspondence: alomari@kfshrc.edu.sa; Tel.: +966-11-5577267; Fax: +966-11-5572598

Received: 22 June 2020; Accepted: 10 July 2020; Published: 13 October 2020

Abstract: One of the problems with composites is their weak impact damage resistance and post-impact mechanical properties. Composites are prone to delamination damage when impacted by low-speed projectiles because of the weak through-thickness strength. To combat the problem of delamination damage, composite parts are often over-designed with extra layers. However, this increases the cost, weight, and volume of the composite and, in some cases, may only provide moderate improvements to impact damage resistance. The selection of the optimal parameters for composite plates that give high impact resistance under low-velocity impact loads should consider several factors related to the properties of the materials as well as to how the composite product is manufactured. To obtain the desired impact resistance, it is essential to know the interrelationships between these parameters and the energy absorbed by the composite. Knowing which parameters affect the improvement of the composite impact resistance and which parameters give the most significant effect are the main issues in the composite industry. In this work, the impact response of composite laminates with various stacking sequences and resins was studied with the Instron 9250G drop-tower to determine the energy absorption. Three types of composites were used: carbon-fiber, glass-fiber, and mixed-fiber composite laminates. Also, these composites were characterized by different stacking sequences and resin types. The effect of several composite structural parameters on the absorbed energy of composite plates is studied. A finite element model was then used to find an optimized design with improved impact resistance based on the best attributes found from the experimental testing.

Keywords: composite plate; carbon fiber; glass fiber; impact; finite elements

1. Introduction

Fiber-reinforced composite polymers are used in almost all types of advanced engineering structure. They combine glass or carbon reinforcing fibers with a matrix material such as epoxy, phenolic, or polyester. Composite materials are complex, mainly due to the degree of anisotropy induced by the reinforcing fibers. Thermosetting polymers consist of chain molecules that chemically bond or cross-link with the others when heated together. Composite materials have a high resistance to internal and external corrosion. They are light with a very smooth inside for higher throughputs. Manufacturing a composite structure starts by incorporating a large number of fibers in a thin layer of the matrix to form a ply. The required load in a fiber-reinforced composite structure is obtained by stacking a plurality of layers in a specified order and then grouping them to form a ply. Various layers in a ply can contain fibers in different directions. It is also possible to combine different types of fibers (for example, glass and carbon) to form a hybrid laminate.

A low-velocity impact by foreign objects is a significant concern for composite laminate as this can cause damage to the interior of materials, which significantly reduces the strength of the composite component and may not be easily detected. The complexity of such an impact on composite laminate is due to the different failure modes that occur in composites compared with metals. The selection of optimal parameters for composite plates that give a high resistance to low-velocity impact loads should consider various factors related to the material properties, as well as the manner of manufacturing the composite product. To obtain the desired impact resistance, it is essential to know the correlation between these parameters and the energy absorbed by the composite plate. The development process of this correlation is not an easy task because unknown process parameters are non-linear. Knowing the parameters affecting the impact resistance and the degree to which these parameters affect are the most significant problems of the composite laminate industry.

Drop-weight impact testing is the standard test procedure used to study the impact of resistance and the behavior of composite laminates. Drop-weight testing also tends to be the preferred method when performing low-velocity impact testing. American Society for Testing and Materials (ASTM) Test Method D7136/D7136M [1] is the governing international standard used to study the impact testing on a rectangular plate. This test technique determines the damage resistance of multidirectional polymer matrix composite laminated plates exposed to a drop-weight impact event. The standard test utilizes constant impact energy normalized by specimen thickness. The properties obtained using this test method can guide researchers concerning the anticipated damage resistance of composite structures of similar material, thickness, stacking sequence, and so forth. To compare samples quantitatively, several equations may be used, which can be found in ASTM D7136.

The total amount of energy introduced to a composite specimen and the energy absorbed by the composite specimen through the impact event are essential parameters to assess the impact response of the composite structures. The introduction of new fiber materials is a promising method for strengthening interfacial bonding between the matrix and fibers in hybrid composite laminates. This alteration of the material has been used to enhance the impact resistance of polymer composite materials. A considerable improvement in the impact resistance was achieved by using hybrid composites. The formation of delamination generally relates to matrix cracking. Generally, in any impact situation, matrix cracking occurs first, followed by delamination [2]. Many useful techniques have been successfully devised to improve delamination resistance in the past three decades, namely three-dimensional (3D)-weaving, stitching, braiding, embroidery, Z-pin anchoring, fiber hybridization, toughening the matrix resin, and interleaving with tough polymer, short fibers or micro-scale particles. These methods enhanced the interlaminar properties but at the cost of in-plane mechanical properties [3]. Multiwall carbon nanotube-reinforced carbon-fiber laminates have better energy absorption capacity as compared to neat carbon-reinforced fiber laminate [4]. On the other hand, glass-fiber composites exhibited evident delamination between the plies, matrix transverse cracks within plies, and significant fiber damage at relatively low impact energies [5].

Several low-velocity impact tests considering target size, projectile diameter, and test temperature were carried out by many authors to determine the response of four different combinations of hybrid laminates to low-velocity impact loading using an instrumented impact testing machine [6–9]. Impact resistance is proportional to the thickness of the composite panel, and it was not affected by the geometry of the plate.

New types of fiber materials and different staking were considered [10]. Fiber metal laminate exhibits outstanding impact absorption capacity under various energy levels, where its energy resistance is lower than standard woven fabrics [11]. There is a recent increase in the use of ecofriendly, natural fibers as reinforcement for the fabrication of lightweight with an increasing trend in research publications and activity in the area of basalt fibers. Natural fiber composite has the potential to be widely applied in the alternative to fiberglass composites in sustainable energy impact-absorption structures [12,13]. Deposition of micro- and nano-fillers, such as aluminum powder, colloidal silica, and silicon carbide powder, in glass fiber-reinforced epoxy composites can enhance the impact resistance and impact energy absorption of the hybrid composite laminates [14,15]. Microencapsulated epoxy and healing agents can be incorporated into a glass fiber-reinforced epoxy matrix to produce a polymer composite capable of self-healing with excellent mechanical strength [16–19].

Sensitivity analysis is one of the approaches that can be used to ascertain the degree of influence of various mechanical and material parameters on the impact performance of the composite laminated plates. Many researchers have tried optimization of the impact performance of the composite plates using the design of experiment (DOE) and artificial neutral network (ANN) model with Finite Element Modelling (FEM) techniques. Based on several studies using ANN models to find the optimal laminate combination, the low-velocity impact resistance of fiber-reinforced polymer composite plates depends more significantly on the thickness and the stacking sequence and the effect of the elastic moduli of the fibers. At the same time, the matrix has less effect than the strength of the fiber and matrix materials of the composite [20–23].

Analysis of absorbed energy and velocities during impact testing of composites may not be all that is needed for characterization. For damage mechanism characterization and type of failure identification, post-impact analysis is required to be carried out for the damaged sample [24]. Several techniques were tried and tested and have been proven to provide useful results to characterize the damaged areas in a polymer composite resulting from a low-velocity impact. Visual inspection can be used to analyze the impact tested samples for specific damage types that include dent/depression, cracking/splitting, fiber failure, and delamination. Correlation between detected volumes and absorbed energy using optical measurement is an excellent tool to estimate impact effects [25]. Nowadays, plenty of Non-Destructible Testing (NDT) techniques are investigated for composite inspection [26–31]. Infrared thermography has shown great potential and advantages, which has greater inspection speed, higher resolution and sensitivity, and detectability of inner defect due to heat conduction.

Finite element modeling and simulations are commonly used as well to evaluate the impact resistance of single- and multi-layer fibers [32]. The finite element (FE) model, in conjunction with the material model, is capable of capturing the behavior of composite for multi-layer and staking configurations under low-velocity impact [33,34].

Others have comprehensively studied the behavior of low-velocity impact on the composites in the literature. However, the current work aims to investigate the impact response of the angle-ply laminated plates using different fibers (carbon and glass). A combination of two types of fibers was also examined. Several types of stacking sequence and resin were considered. Absorbed energy-time curves were presented to understand the behavior of the low impact velocity loading. A flowchart of the entire procedure, experimental and modeling, is given in Figure 1.

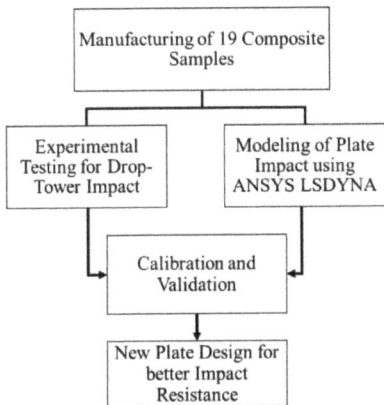

Figure 1. Flowchart for the entire experimental and modeling procedure in this study.

2. Experimental Work

2.1. Materials and Specimen Preparation

The materials in this study consist of woven carbon, glass, and mixed fiber-reinforced laminates, which were manufactured with an asymmetric, quasi-isotropic layup of plies. The glue ratio was 1.1 for the carbon and 0.6 for the glass fiber. Different ply thicknesses and resin types were also considered. Table 1 summarizes the detailed information for the 19-symmetric laminated composite plates that were impact tested. Different stacking sequences characterized these composites. The amount of fiber and resin for each specimen was also considered with different ratios. Note that C stands for carbon-fiber, G stands for glass-fiber, and M stands for mixed-fiber. For matrix material, three types of resin were used, which are epoxy, phenolic (PH), and polyester (PL). The final size of each test specimen is 129 mm by 129 mm.

Table 1. Properties of composite laminates.

Sample No.	Fiber Type (Matrix)	Glue Type (Resin)	Percent Age of Fiber	Percent Age of Epoxy	No. of Layers	Stacking Sequence	Measured Thickness (mm)
C1	Carbon	Epoxy	42.48	57.52	16	[90/−60/−30/0/90/−60/−30/0]s	3.3
C2	Carbon	Epoxy	43.59	56.41	16	[90/0/45/−45/90/0/45/−45]s	3.8
C3	Carbon	Epoxy	45.49	54.51	20	[45/−45/90/0/45/−45/90/0/45/−45]s	4.7
C4	Carbon	Epoxy	45.74	54.26	24	[90/0/45/−45/90/0/45/−45/90/0/45/−45]s	5.5
C4 PH	Carbon	Phenolic	45.74	54.26	24	[90/0/45/−45/90/0/45/−45/90/0/45/−45]s	5.15
C4 PL	Carbon	Polyester	45.74	54.26	24	[90/0/45/−45/90/0/45/−45/90/0/45/−45]s	5.5
C5	Carbon	Epoxy	47.68	52.32	28	[45/−45/90/0/45/−45/90/0/45/−45/90/0/45/−45]s	6.4
C6	Carbon	Epoxy	47.58	52.42	32	[90/0/45/−45/90/0/45/−45/90/0/45/−45/90/0/45/−45]s	7.4
C7	Carbon	Epoxy	45.52	54.48	28	[−45/−60/60/45/−45/−60/60/45/−45/−60/60/45/−45/−60]s	6.6
C8	Carbon	Epoxy	50.00	50.00	32	[60/45/−45/−60/60/45/−45/−60/60/45/−45/−60/60/45/−45/−60]s	7.3
G1	Glass	Epoxy	50.04	49.96	24	[90/0/45/−45/90/0/45/−45/90/0/45/−45]s	4.9
G1 PH	Glass	Phenolic	50.04	49.96	24	[90/0/45/−45/90/0/45/−45/90/0/45/−45]s	5
G1 PL	Glass	Polyester	50.04	49.96	24	[90/0/45/−45/90/0/45/−45/90/0/45/−45]s	4.7
G4	Glass	Epoxy	52.51	47.49	24	[90/−60/−30/0/90/−60/−30/0/90/−60/−30/0]s	5.1
G5	Glass	Epoxy	58.79	41.21	36	[−30/−45/45/30/−30/−45/45/30/−30/−45/45/30/−30/−45/45/30/−30/−45]s	7.1
G6	Glass	Epoxy	66.60	33.40	36	[45/−45/90/0/45/−45/90/0/45/−45/90/0/45/−45/90/0/45/−45]s	7.4
M1	Mixed Carbon–Glass (2 C in the Middle)	Epoxy	62.50	37.50	32	[60/45/−45/−60/60/45/−45/−60/60/45/−45/−60/60/45/−45/−60]s	6.5
M2	Mixed Carbon–Glass (2 C in the Bottom)	Epoxy	62.53	37.47	32	[60/45/−45/−60/60/45/−45/−60/60/45/−45/−60/60/45/−45/−60]s	6.7
M3	Mixed Carbon–Glass (2 C in the Top)	Epoxy	64.51	35.49	32	[60/45/−45/−60/60/45/−45/−60/60/45/−45/−60/60/45/−45/−60]s	6.7

2.2. Low-Velocity Impact Testing

The low-velocity impact tests were performed by an INSTRON Dynatup 9250G (Norwood, MA, USA) drop tower. The impact machine was equipped with a hemispherical impactor head with a diameter of 1 inch. During this test, two types of damages can occur. The first is visible impact damage (CVID), which can easily be seen by the naked eye. The second is barely visible impact damage (BVID), which requires equipment or techniques to capture it. Weights are added to alter the energy of the impact. For all impact tests in this study, the mass of the impactor was 9.2 kg with a constant impact energy level of 20 J, which corresponds to an impact velocity of 2.06 m/s. The impact velocity was measured by a photocell device that is placed in the path of the striker before the impactor strikes the composite plates. The strains measured during impact were loaded into the acquisition software.

The force-time history was measured from the point of initial contact with the plate until the impactor leaves the plate. The energy was calculated from the integration of the force-time signal. The data acquisition system recorded the force-time and energy-time histories. Two rebound arrestors were located on both sides of the composite plate to avoid multiple impacts after the striker rebounds on the plate.

The arrestors were pneumatically actuated, and spring up to separate the impactor from the composite plate after the first impact. Figure 2a shows a schematic picture of the drop weight test. The composite plates to be impacted were positioned under the drop tower using an in-house manufactured specimen fixture where the exposed composite area within the fixture is 110 mm × 110 mm. The composite plates were clamped along all edges. Clamping force was provided by steel plates on the top and bottom edges, as shown in Figure 2b. The clamping force was applied by tightening two bolts at the edge of the fixture.

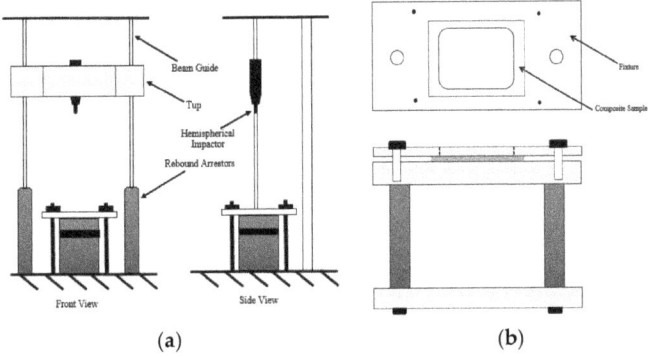

Figure 2. Schematic of (a) drop-weight test, and (b) fixture and sample location.

For the FE model, the plate impact theory is based on the developed theory by Jang et al. [35]. Newton's second law is used, and the solution for acceleration $a(t)$ is given by:

$$a(t) = g - \frac{P(t)}{M} \quad (1)$$

where g is the gravitational acceleration constant, $P(t)$ is the load as a function of time, and M is the mass of the impactor. At the time when the impact testing starts, $t = 0$ which is known as the initial conditions,

$$v(t) = V \text{ at } t = 0$$

$$x(t) = 0 \text{ at } t = 0$$

where V is the velocity just before impact. The acceleration equation can be integrated to obtain an expression for $v(t)$, and then the velocity equation can also be integrated to obtain a solution for $x(t)$:

$$v(t) = V + \int_0^t g - \frac{P(t)}{M} dt \quad (2)$$

$$x(t) = 0 + \int_0^t v(t) dt \quad (3)$$

It is important to remember that the $x(t)$ equation works as long as the composite plate is not punctured. Once we get $x(t)$, it is easy to solve for the absorbed energy "$E_{absorbed}$" as a function of time.

$$E(t) = \int_0^t P(t)v(t)dt \tag{4}$$

The integration of the energy equation between zero and t, where the impactor is no longer in contact with the composite plate, yields the total energy (E_{total}). The initial velocity v_0 is given as a function of gravity and freefall height H as:

$$v_0 = \sqrt{2gH} \tag{5}$$

Impactor velocity v and displacement x are calculated by integrating the impact force:

$$v(t) = v_0 - \left(\frac{1}{m}\right)\int_0^t P(t)dt \tag{6}$$

$$x(t) = \int_0^t \left[v_0 - \left(\frac{1}{m}\right)\int_0^t P(t)dt\right]dt \tag{7}$$

The kinetic energy of the impactor and the absorbed energy are given by:

$$E_{imp} = \frac{1}{2}mv^2 \tag{8}$$

$$E_{ab}(t) = \frac{1}{2}mv_0^2 - \frac{1}{2}m\left(v_0 - \left(\frac{1}{m}\right)\int_0^t P(t)dt\right)^2 \tag{9}$$

Different results can be obtained from the low-velocity impact test. Typical time versus impact energy and peak loads plots are illustrated in Figure 3 [36]. For the impact energy-time history curve, the highest peak of the curve shows the maximum impact energy, and the end of the curve shows the absorbed energy. The maximum impact is the first impactor kinetic energy, and as the impactor contacts the plate, part of the energy is transferred to the plate. At the end of the impact, not all of the initial kinetic energy is returned to the impactor as part of it is absorbed by processes like plastic deformation and failure of the composite plate. The impact force can be defined by the reaction force between the composite plate and the impactor.

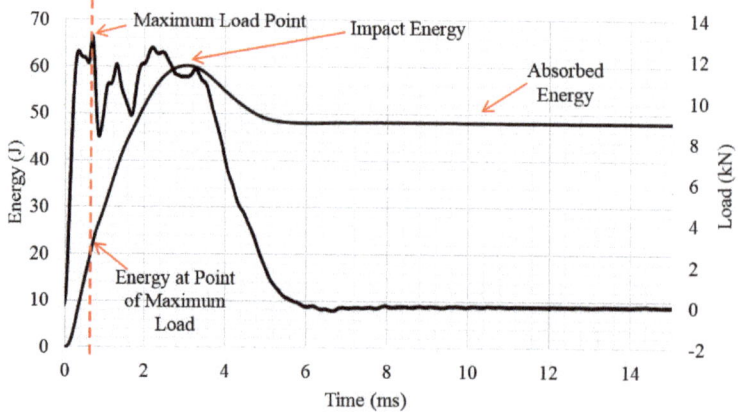

Figure 3. Typical load and energy versus time curve and characteristic points for post-impact analysis.

3. Experimental Results and Discussion

3.1. Low-Velocity Impact Testing

The impact tests were performed using an INSTRON Dynatup 9250G drop tower to determine the amount of impact energy lost in damage during the impact process for each of the defined cases listed in Table 2. Figures 4 and 5 summarize the measured absorbed energies and peak loads, respectively, for all composite samples.

Table 2. Low-velocity impact properties of composite samples.

Sample No.	Peak Load (kN)	Deflection at Peak Load (mm)	Absorbed Energy (J)
C1	4.48	4.97	15.68
C2	4.10	4.69	16.27
C3	5.83	4.67	14.25
C4	7.66	3.42	12.75
C4 PH	5.62	6.63	13.78
C4 PL	8.34	4.16	11.18
C5	9.57	3.03	12.89
C6	11.25	2.68	10.69
C7	9.45	3.18	12.62
C8	11.44	2.73	14.16
G1	7.59	4.90	9.01
G1 PH	5.53	7.47	13.20
G1 PL	7.57	4.97	9.49
G4	7.45	4.76	10.22
G5	10.09	3.24	9.46
G6	9.66	3.19	11.08
M1	9.66	3.19	11.08
M2	8.74	3.53	11.51
M3	8.91	3.47	11.28

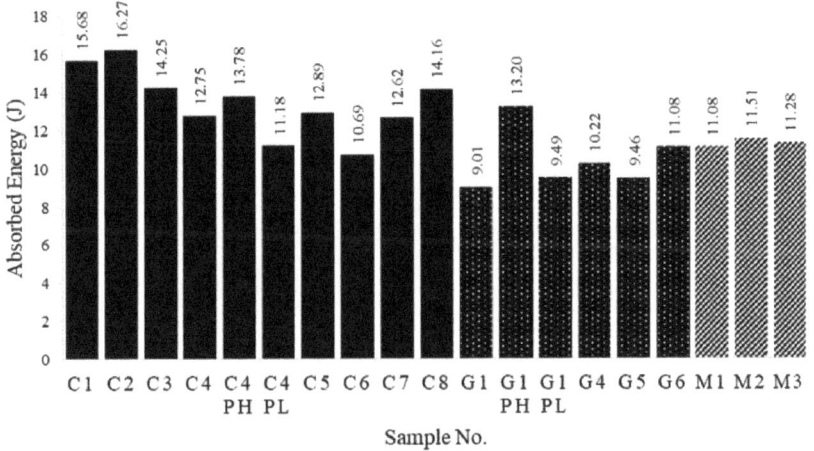

Figure 4. Summary of the measured absorbed energy for all composite plates.

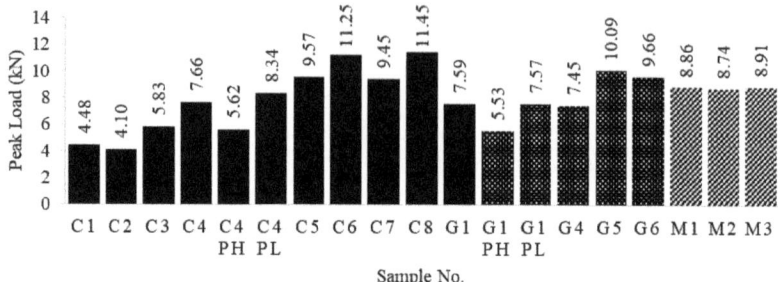

Figure 5. Summary of the measured peak loads for all composite plates.

3.2. Carbon Fiber-Reinforced Polymer (CFRP) Plates

Table 3 and Figure 6 summarize the measured absorbed energies for the tested carbon-fiber plates. The tests show that the impact resistance was affected by the thickness of the plate and the number of plies. It is clear that carbon-fiber composite samples C1, C2, and C3 exhibit the highest absorbed energy. Carbon-fiber plate C2 has the highest measured absorbed energy, then C1, C3, C4 PH, C4, C5, C4 PL, and C7, while C6 has the lowest absorbed energy. The maximum impact force is found to be approximately 11.45 KN for carbon-fiber composite sample C8, while the minimum impact force was measured for carbon-fiber sample C2. The higher impact force is attributed to the energy release due to the composite damage (delamination and matrix cracking). It worth mentioning that there are almost negligible differences in the measured absorbed energies for C5, C4 PL, and C7.

Table 3. Low-velocity impact test conditions and results for carbon-fiber plates.

Sample No.	Absorbed Energy (J)
C1	15.68
C2	16.27
C3	14.25
C4	12.75
C5	12.89
C6	10.69
C7	12.62
C8	10.76

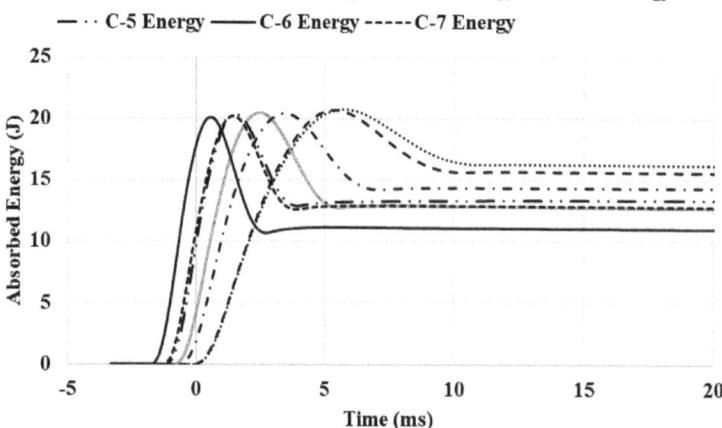

Figure 6. Absorbed energy history of low-velocity impact for different carbon-fiber epoxy plates.

In general, a further increase in the overall thickness of the carbon-fiber plate, either through an increase in layer thickness or by increasing the number of layers, resulting in a decrease in performance. Physically, when the thickness of the composite plate is small, the plate behaves more like a membrane and stretches during impact until all the kinetic energy is transferred to the plate. Then, it pushes back the impactor giving some of the energy back to the impactor, where the rest is dissipated in the form of damage within the plate. The more the thickness of the plate is increased, the stiffer it becomes, and the ability to bend under impact loads is reduced, which increases the bending stress, and hence, the plate suffers more damage. At very high thickness, the plate becomes very strong, which results in deficient amounts of energy absorbed.

Moreover, the energy absorbed in the carbon composites changes with the change in stacking sequence due to the stress redistribution within the laminate. The results show that the minimum amount of energy absorbed is for the case where the laminate configuration is such that the laminate behaves as quasi-isotropic material. The number of layers does not show a linear relation like the stacking sequence. Increasing or decreasing the number of layers by 5%, while keeping the total laminate thickness constant, results in an increased impact energy absorption. Hence, it can be deduced that there must be an optimal number of layers for a fixed thickness, which gives the best impact performance. Thus, there must be an optimal condition for which the amount of absorbed energy and the resulting damage is minimized.

3.3. Glass Fiber-Reinforced Polymer (GFRP) Plates

Table 4 lists the absorbed energies for the glass-fiber plates, while Figure 7 shows the time history of these measured absorbed energies. It was found that G1 PH has the highest measured absorbed energy than G1 PL, G4, G1, G6, while G5 has the lowest absorbed energy. It was expected that the composite plates with glass fiber as the reinforcement material would behave in a similar way to the carbon fiber-based plates. However, it was noticed that the increase in thickness of individual layers increases the absorbed energy. Also, higher energy absorption is seen in [45/−45/90/0]s glass-fiber plates than other stacking sequences, which agrees with the high measured forces of [45/−45/90/0]s glass-fiber composite plate.

Table 4. Measured absorbed energies for glass-fiber plates.

Sample No.	Absorbed Energy (J)
G1	9.00
G4	10.22
G5	9.45
G6	11.08

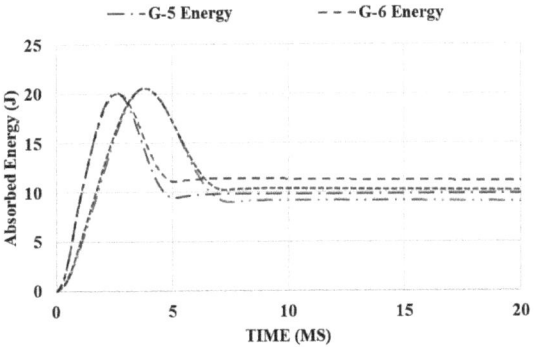

Figure 7. Absorbed energy for different glass-fiber epoxy.

Delamination, crack, or indentation are usually the observed damage forms. If the energy absorbed by the specimen is not too high, the impactor is pushed back, and a rebound occurs. In the case of a rebound, the first drop of force indicates damage to the first material, and the second force drop indicates the failure of the first laminate. Figure 8 shows the extent of the damage for the glass-fiber composite samples (G1, G4, G5, and G6) from the front and back sides. The extent of the damage with the measured absorbed energy can be correlated where the glass-fiber plates with higher absorbed energy show less damage, as in the case of G4.

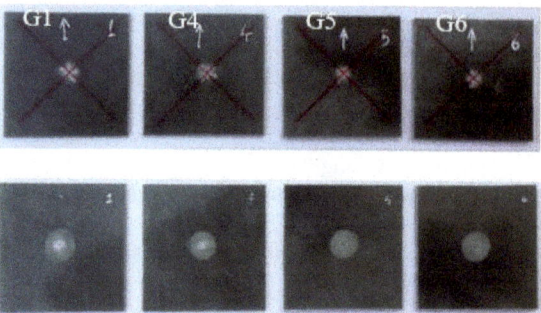

Figure 8. Glass-fiber epoxy plates, (Top: front side, bottom: backside).

3.4. Mixed Fiber-Reinforced Polymer Plates

To understand the relation between the placement of the inclusions and the impact performance, different placements were tried for the carbon layers in the composite plate that mainly consisted of glass fibers. The inclusion of other materials alone cannot guarantee a better performance of the structure, and the placement of the fibers is equally critical. Since the material to be included is based on superior strength and better performance, it should be placed where the damage initiates. The following different combinations were tested with the position of woven carbon lamina as the middle two layers, bottom two layers, and top two-layer. For the aforementioned mixed composite combinations, the glass-fiber composite with two carbon-fiber plies on the bottom has the highest value of the absorbed energy in compression with the same plate with carbon-fiber plies in the middle or at the top. However, the difference is not that pronounced, as can been seen from Table 5 and Figure 9.

Table 5. Low-velocity impact test conditions and results for mixed-fiber plates.

Sample No.	Materials Type	Absorbed Energy (J)
M1	Mixed Carbon–Glass (2 Carbon plies in the middle)	10.82
M2	Mixed Carbon–Glass (2 Carbon plies in the bottom)	11.51
M3	Mixed Carbon–Glass (2 Carbon plies in the top)	11.28

3.5. Effect of Resin Type

A satisfying result was obtained for absorbed energy time given in Table 6 for glass and carbon-fiber plates using different types of resin. It was found that the plate with phenolic resin gives the highest absorbed energy when compared to the epoxy and polyester resins for both glass and carbon composite plates. Moreover, when the force-time history is investigated, the results indicate that composite plates with phenolic resins have the highest resisting force (low peak load), while the composite samples with epoxy and polyester resins have the lowest resisting forces.

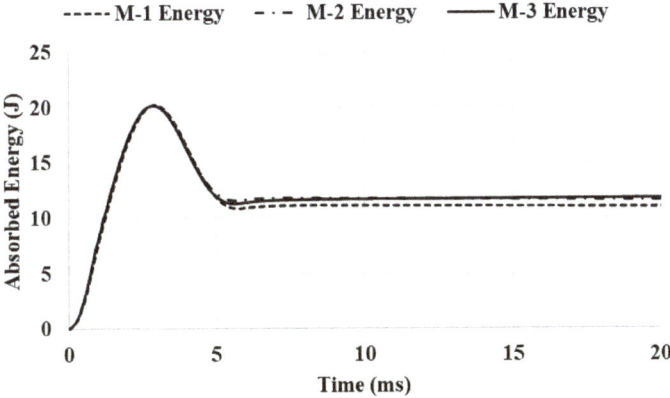

Figure 9. Absorbed energy for different mixed (glass/carbon) fiber epoxy plates.

Table 6. Low-velocity impact test conditions and results for glass- and carbon-fiber plates with different resins.

Sample No.	Fiber Type	Glue Type (Resin)	Peak Load (kN)	Absorbed Energy (J)
G1	Glass	Epoxy	7.59	9.00
G1 PH	Glass	PH	5.53	13.20
G1 PL	Glass	PL	7.57	9.49
C4	Carbon	Epoxy	7.66	12.75
C4 PH	Carbon	PH	5.62	13.78
C4 PL	Carbon	PL	8.34	11.18

Figure 10 shows the surface of the glass-fiber composite plates with different resins (epoxy, polyester, and phenolic) that was tested at the same energy levels. The figure reveals that the plate with phenolic resin shows minimal damage in both the top and bottom sides of the plates. The damage of the glass fiber with epoxy and PL resin samples is observed on both sides of the plate. It is also observed that the damage distribution is much more abundant in the glass-fiber plate with polyester resin than in the glass-fiber epoxy plate.

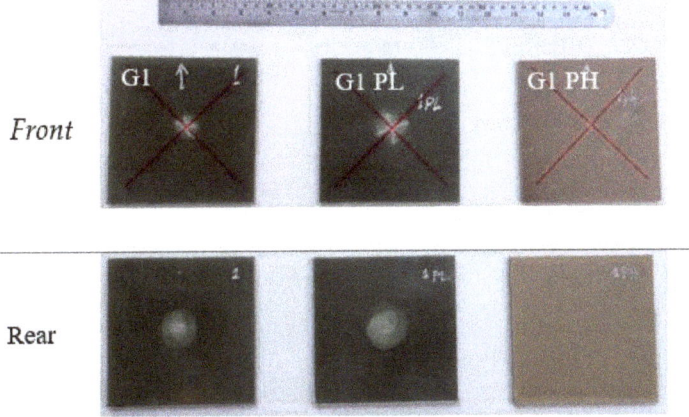

Figure 10. Glass-fiber plates with different resins.

4. Finite Element Modeling

The composite plates were modeled using 3-D shell elements as an area without thickness. The thicknesses and orientations were given as the composite layup data using ANSYS ACP (Canonsburg, PA, USA). The striker was modeled as a 3-D rigid body. Frictionless contact between striker and plate was considered. The amount of damage was calculated as the loss in the kinetic energy of the striker. The FE model is shown in Figure 11.

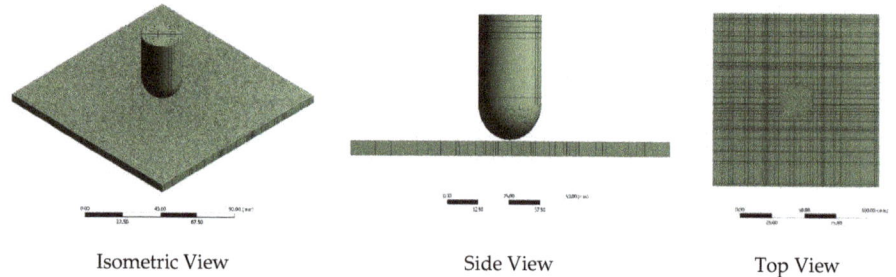

Isometric View   Side View   Top View

Figure 11. Impact testing modeling using ANSYS LS-DYNA.

In this work, the fixed composite plate was subjected to impact loading represented by the dropped object at a velocity of 2.06 m/s. The impactor (12.7 mm in diameter) was modeled as a rigid hemispherical body. The mechanical properties of the glass and carbon fiber-reinforced composites are listed in Table 7. The initial velocity and mass of the striker were set depending on the energy level considered. The impactor was constrained to movement within 5 degrees of freedom (x and y translations and 3 rotations) and was allowed to move only in the -y-direction. A sufficient density of 7860 kg/m^3 was assigned to the rigid impactor.

Table 7. Material properties of epoxy e-glass woven and epoxy carbon woven [37,38].

Property	Epoxy E-Glass Woven	Epoxy Carbon Woven
Density (kg/m^3)	2000	1540
Elastic Properties		
E_1 (MPa)	50,000	400,000
E_2 (MPa)	8000	30,000
E_3 (MPa)	8000	30,000
G_{12} (MPa)	5000	6000
G_{23} (MPa)	1000	4000
G_{13} (MPa)	5000	6000
ν_{12}	0.3	0.3
ν_{23}	0.4	0.4
ν_{13}	0.3	0.3
Ply Strengths		
X_t (MPa)	750	5000
Y_t (MPa)	270	2500
Z_t (MPa)	270	5000
X_c (MPa)	750	5000
Y_c (MPa)	270	2500
Z_c (MPa)	270	5000
S_{12} (MPa)	70	400
S_{23} (MPa)	30	200
S_{13} (MPa)	70	400

The Probabilistic Design System (PDS) module of the commercial finite element software ANSYS was used for the Monte Carlo simulation. A total of 1000 analysis loops are run to obtain the output parameters as a function of the set of random input variables. The 19 plates were manufactured and then tested based on previous work using the Monte Carlo method for random variables to evaluate the effect of variability in the governing parameters for the outcome of the experiment. We selected the von Mises equivalent stress as the outcome of the numerical experiment for the new proposed plates design. To determine the mesh size of elements in finite element modeling, a convergence test was conducted on several cases of the models where Table 8 shows the convergence case for 20 J impact testing. To determine the size of elements in finite element modeling, a convergence test was conducted on several cases of the models where Table 8 shows the case for 20 J impact testing. The analysis showed that, the optimum mesh size was when we selected the axial edge sizing to be 100 divisions with a bias factor of 2.

Table 8. Mesh convergence check for 20 J case.

No of Division for the Axial Edge Sizing	Bias Factor	No. of Elements	Finite Element Modelling (FEM) Absorbed Energy (J)
10	3	100	12.89
25	3	625	11.47
50	3	2500	11.37
75	3	5625	11.13
100	2	10,000	10.68
125	3	12,500	10.68
150	3	15,000	10.68

The simulation was accomplished through a concept known as birth and death of elements in ANSYS. To achieve the "element death" effect, the ANSYS program does not actually remove "killed" elements. Instead, it deactivates them by multiplying their stiffness by a severe reduction factor (ESTIF). This factor is set to 1.0×10^{-6} by default. An element's strain is also set to zero as soon as that element is killed. In like manner, when elements are "born", they are not actually added to the model; they are simply reactivated. When an element is reactivated, its stiffness, mass, element loads, etc. return to their full original values.

Contact between the impactor and the whole laminate composite was simulated using the automatic-surface-to-surface penalty-based contact algorithm to accommodate impact initiation and progress. A contact criterion based on 0.01 mm of the normal distance between the contact surfaces was adopted for the simulation. The loads and boundary conditions are shown in Figure 12. The finite element model is generated in ANSYS ACP for angle-ply laminate having different stacking sequences.

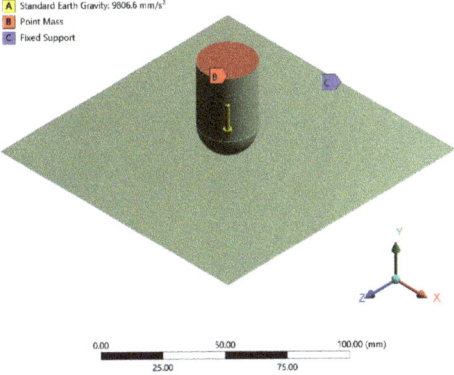

Figure 12. Loading and boundary conditions.

Table 9 lists a comparison of the absorbed energy response, while Figure 13 shows the absorbed energy-time history of the fiberglass plate G4 and Carbon-fiber plate C1 for the case of 20 J obtained by the experiment and by the FE model. The FE results are very close to the experimental values for the final absorbed energy, with an error of less than 9%. This means that the test results validate the FE model. Hereafter, the FE model can be used to perform a parametric analysis; this indicates that the developed FE model can reasonably predict the actual behavior of any composite plate under low-velocity impact loading. The difference in the starting energy time is due to the placement of the impactor close to the plate during modeling to optimize the computational time, whereas in the experiments the impactor falls from a height based on the required energy, which is 20 J in this case. The effect of gravity was considered on the calibration cases for C1 and G1 plates. This is done by including the gravity as an initial condition for the system then by suppressing this feature during analysis. It was found that it is acceptable to ignore gravity as the impact process was performed in a fraction of a second. Including the gravity affects the absorbed energy steady-state line as the bouncing of the impactor decreases with time due to the gravity, and hence the slight inclination in the steady-state line of the absorbed energy.

Table 9. Comparison between the experimental and FE measurement for the absorbed energies.

Sample No.	Experimental Absorbed Energy (J)	FEM Absorbed Energy (J)	ΔE (J)	Percentage %
C1	15.68	16.90	1.28	8.20
G1	10.16	10.68	0.52	5.07

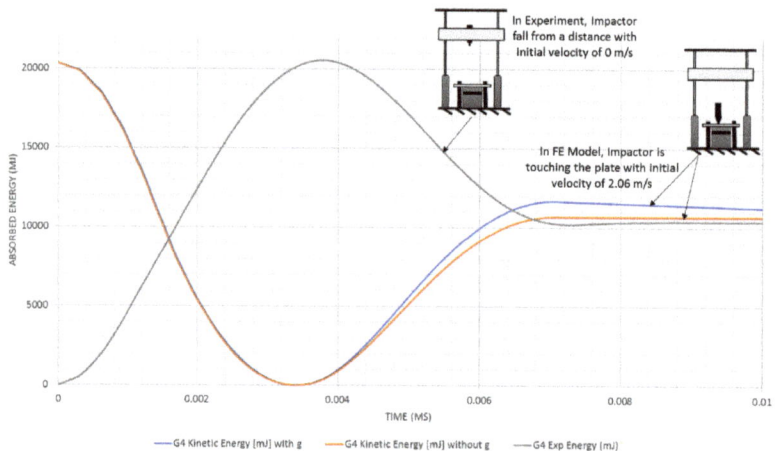

Figure 13. Energy-time curves comparison of the experimental test data and finite element analysis (FEA) data of the G4 fiberglass plate impacted with 20 J.

5. Proposed Design for Composite Plates

It can be concluded from the 19 plate designs and the low-velocity impact tests that the best plate designs with high impact resistance (high absorbed energy), as shown in Table 10, have fewer layers (smaller thickness) with a stacking sequence of [90/0/45/−45]s or [60/45/−45/−60]s. The use of phenolic as a resin gave high impact resistance among all the impacted plates. However, the use of phenolic resin was avoided for the proposed cases as it is costly and not easy to use in manufacturing. Therefore, all cases were modeled with glass-fiber epoxy, as presented in Table 11. The following combinations were adopted for the new designs:

- 16 layers and 24 layers with 3.8 mm and 5 mm plate thickness, respectively.

- Stacking sequence of [90/0/45/−45]s and [60/45/−45/−60]s.

Table 10. Summary of the experimental results for high-performing cases with high absorbed energies.

Sample No.	Peak Load (kN)	Deflection at Peak Load (mm)	Absorbed Energy (J)
C2	4.10	4.69	16.27
C1	4.48	4.97	15.68
C8	11.44	2.73	14.16
G1 PH	5.53	7.47	13.20
M2	8.74	3.53	11.51
G6	9.66	3.19	11.08
G4	7.45	4.76	10.22

Table 11. Proposed new plate designs.

Case No.	No. of Layers	Thickness (mm)	Staking Sequence	Resin/Fiber
Plate-1	16	3.8	90/0/45/−45	Glass Epoxy Woven
Plate-2	16	3.8	60/45/−45/−60	Glass Epoxy Woven
Plate-3	16	5.144	90/0/45/−45	Glass Epoxy Woven
Plate-4	16	5.144	60/45/−45/−60	Glass Epoxy Woven
Plate-5	24	3.8	90/0/45/−45	Glass Epoxy Woven
Plate-6	24	3.8	60/45/−45/−60	Glass Epoxy Woven
Plate-7	24	5.144	90/0/45/−45	Glass Epoxy Woven
Plate-8	24	5.144	60/45/−45/−60	Glass Epoxy Woven

Modeling of the proposed new plate designs reveals that, as shown in Table 12, Plate-7 (with 24 fiberglass plies and a stacking sequence of 90/0/45/−45) has the best design from the impact point of view with predicted absorbed energy of 11.753 J. Figure 14 shows the maximum von Mises (VM) stress distribution for all proposed plate designs, where plates with [60/45/−45/−60]s have lower von Mises stresses when they compared with the plates of [90/0/45/−45]s stacking sequence. In general, [90/0/45/−45]s stacking sequence performed better in impact resistance than [60/45/−45/−60]s for all the simulated cases. Moreover, this design is better than the manufactured fiberglass plates.

Table 12. Summary of the predicted absorbed energy for the new proposed designs.

Case No.	FE Absorbed Energy (J)
Plate-1	9.99
Plate-2	7.47
Plate-3	11.29
Plate-4	9.54
Plate-5	8.87
Plate-6	8.59
Plate-7	11.75
Plate-8	10.44

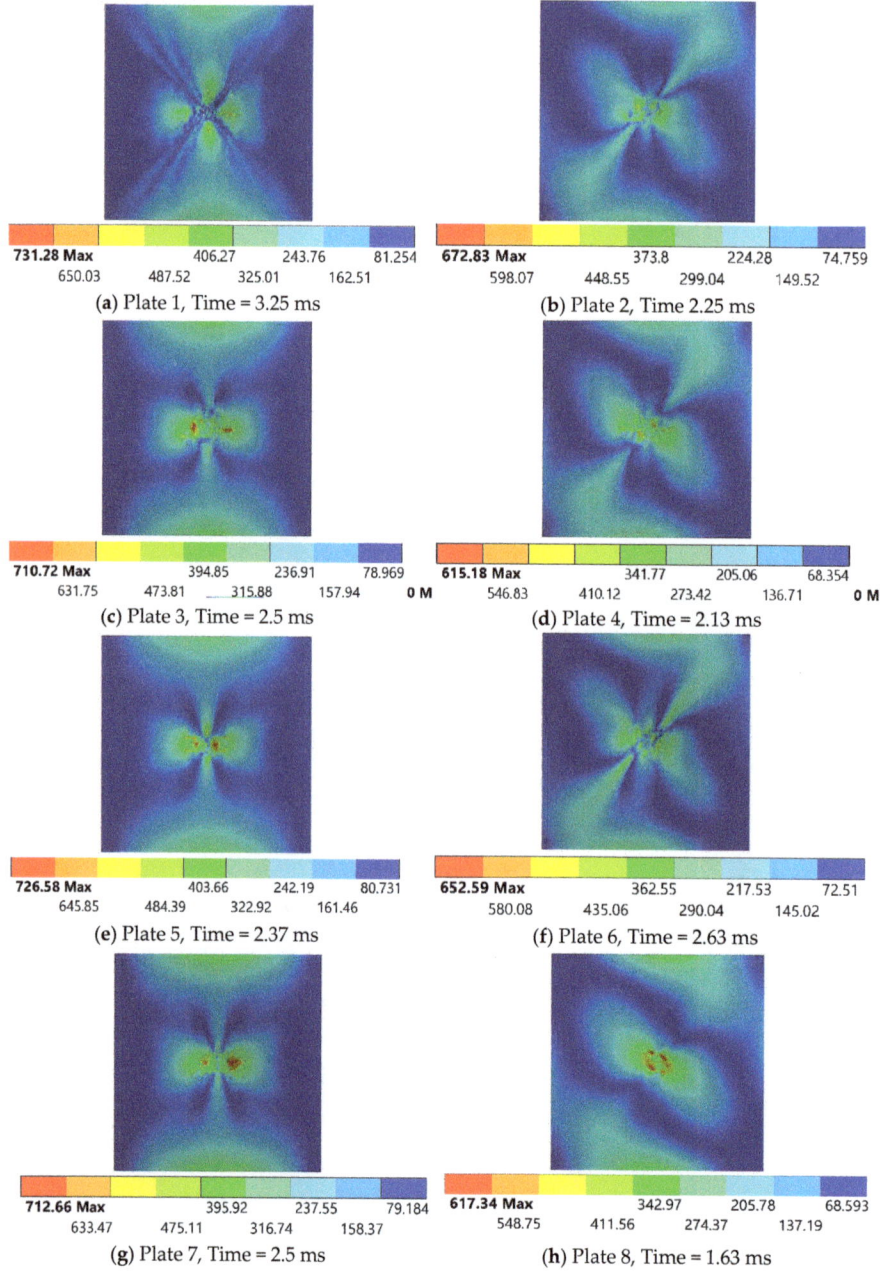

Figure 14. Von Mises stress distribution for all plate designs.

6. Conclusions

In this study, the low-velocity impact behavior of laminated carbon-, glass-, and mixed-fiber plates with different types of resin was investigated experimentally and numerically. The results presented in the current study give an insight into the effects of the considered parameters on the impact-resistance performance. Several types of stacking sequence at constant impact energy are considered in order to investigate the behavior of composite structures. The main conclusions of the study are:

The amount of energy absorbed (impact performance) varies significantly for the variations in the thickness of a single layer, number of layers, and stacking sequence.

The experimental test data showed an increased energy absorption for the composite plates made with phenolic resin.

The carbon-fiber/epoxy composite plate has better impact resistance compared to the glass-fiber/epoxy composite plate due to the higher measured absorbed energies of the carbon-fiber/epoxy.

Visual inspections showed a large extent of damage was observed on the polyester and epoxy resin plates when they are compared with the same plates made with phenolic resin.

The effect of the carbon-fiber plies location for the mixed plates was not exceptionally pronounced.

The stacking sequence with [90/0/45/−45]s was better than [60/45/−45/−60]s in terms of impact resistance, as concluded from the simulated cases.

The results that were captured from the current work provide a motivation to further examine the samples in the future in order to determine the damage morphology under scanning electron microscopy. Moreover, the results from this study help researchers in designing composite laminated plates with better impact resistance. The results allow for a more methodical approach in selecting the parameters to vary in order to achieve better impact performance of composite laminates against low-velocity impact loadings.

Author Contributions: Conceptualization, A.F.M.A. and A.S.A.; methodology, A.S.A.; software, A.S.A.; validation A.S.A., K.S.A.-A., and A.F.M.A.; formal analysis; writing—original draft preparation, A.S.A.; writing—review and editing, K.S.A.-A.; supervision, A.F.M.A.; project administration, F.A.A.-S.; All authors have read and agreed to the published version of the manuscript.

Funding: This research received no external funding.

Acknowledgments: The authors gratefully acknowledge the support of F.A.A.-S for providing the composite sample plates.

Conflicts of Interest: The authors declare no conflict of interest.

References

1. ASTM D7136/D7136M, Standard Test. *Method for Measuring the Damage Resistance of a Fiber-Reinforced Polymer Matrix Composite to a Drop-Weight Impact Event*; ASTM International: West Kanshohoken, PA, USA, 2005.
2. Khan, S.H.; Sharma, A.; Parameswaran, V. An Impact Induced Damage in Composite Laminates with Intra-layer and Inter-laminate Damage. *Procedia Eng.* **2017**, *173*, 409–416. [CrossRef]
3. Khan, S.U.; Kim, J.-K. Impact and Delamination Failure of Multiscale Carbon Nanotube-Fiber Reinforced Polymer Composites: A Review. *Int. J. Aeronaut. Space Sci.* **2011**, *12*, 115–133. [CrossRef]
4. Rawat, P.; Singh, K.K. Damage Tolerance of Carbon Fiber Woven Composite Doped with MWCNTs under Low-velocity Impact. *Procedia Eng.* **2017**, *173*. [CrossRef]
5. Selver, E.; Potluri, P.; Hogg, P.; Soutis, C.; Soutis, C. Impact damage tolerance of thermoset composites reinforced with hybrid commingled yarns. *Compos. Part B Eng.* **2016**, *91*, 522–538. [CrossRef]
6. Hosur, M.; Adbullah, M.; Jeelani, S. Studies on the low-velocity impact response of woven hybrid composites. *Compos. Struct.* **2005**, *67*, 253–262. [CrossRef]
7. Yang, F.; Cantwell, W. Impact damage initiation in composite materials. *Compos. Sci. Technol.* **2010**, *70*, 336–342. [CrossRef]
8. KERŠYS, A.; Keršien, N.; Žiliukas, A. Experimental Research of the Impact Response of E-Glass/Epoxy and Carbon/Epoxy Composite Systems. *Mater. Sci.* **2010**, *16*, 324–329.
9. Rilo, N.F.; Ferreira, L.M.S. Experimental study of low-velocity impacts on glass-epoxy laminated composite plates. *Int. J. Mech. Mater. Des.* **2008**, *4*, 291–300. [CrossRef]
10. Asaee, Z.; Shadlou, S.; Taheri, F. Low-velocity impact response of fiberglass/magnesium FMLs with a new 3D fiberglass fabric. *Compos. Struct.* **2015**, *122*, 155–165. [CrossRef]

11. Asaee, Z.; Taheri, F. Experimental and numerical investigation into the influence of stacking sequence on the low-velocity impact response of new 3D FMLs. *Compos. Struct.* **2016**, *140*, 136–146. [CrossRef]
12. Boria, S.; Santulli, C.; Sarasini, F.; Tirillò, J.; Caruso, P.; Infantino, M. Potential of wool felts in combination with glass fibers: Mechanical and low-velocity impact assessment. *Compos. Part B Eng.* **2017**, *118*, 158–168. [CrossRef]
13. Caprino, G.; Carrino, L.; Durante, M.; Langella, A.; Lopresto, V. Low impact behavior of hemp fiber reinforced epoxy composites. *Compos. Struct.* **2015**, *133*, 892–901. [CrossRef]
14. Rafiq, A.; Merah, N.; Boukhili, R.; Al-Qadhi, M. Impact resistance of hybrid glass fiber reinforced epoxy/nanoclay composite. *Polym. Test.* **2017**, *57*, 1–11. [CrossRef]
15. Haro, E.; Odeshi, A.; Szpunar, J.A. The energy absorption behavior of hybrid composite laminates containing nano-fillers under ballistic impact. *Int. J. Impact Eng.* **2016**, *96*, 11–22. [CrossRef]
16. Fragassa, C.; Pavlovic, A.; Santulli, C. Mechanical and impact characterization of flax and basalt fiber vinyl ester composites and their hybrids. *Compos. Part B Eng.* **2018**, *137*, 247–259. [CrossRef]
17. Lee, J.; Bhattacharyya, D.; Zhang, M.Q.; Yuan, Y.C. Mechanical properties of a self-healing fiber reinforced epoxy composites. *Compos. Part B Eng.* **2015**, *78*, 515–519. [CrossRef]
18. Dhand, V.; Mittal, G.; Rhee, K.Y.; Park, S.-J.; Hui, D. A short review on basalt fiber reinforced polymer composites. *Compos. Part B Eng.* **2015**, *73*, 166–180. [CrossRef]
19. Kubiak, T.; Kaczmarek, L. Estimation of load-carrying capacity for thin-walled composite beams. *Compos. Struct.* **2015**, *119*, 749–756. [CrossRef]
20. Malik, M.H.; Arif, A.F.M.; Al-Sulaiman, F.A.a.; Khan, Z. Impact resistance of composite laminate flat plates–A parametric sensitivity analysis approach. *Compos. Struct.* **2013**, *102*, 138–147. [CrossRef]
21. Malik, M.; Arif, A. ANN prediction model for composite plates against low velocity impact loads using finite element analysis. *Compos. Struct.* **2013**, *101*, 290–300. [CrossRef]
22. Artero-Guerrero, J.A.; Pernas-Sánchez, J.; Martín-Montal, J.; Varas, D.; Lopez, J. The influence of laminate stacking sequence on ballistic limit using a combined Experimental/FEM/Artificial Neural Networks (ANN) methodology. *Compos. Struct.* **2018**, *183*, 299–308. [CrossRef]
23. Riccio, A.; Di Felice, G.; Saputo, S.; Scaramuzzino, F. Stacking Sequence Effects on Damage Onset in Composite Laminate Subjected to Low Velocity Impact. *Procedia Eng.* **2014**, *88*, 222–229. [CrossRef]
24. Kessler, M.R. *Advanced Topics in Characterization of Composites*; Trafford Publishing (UK) Ltd.: Victoria, BC, Canada, 2004.
25. Cucinotta, F.; Guglielmino, E.; Risitano, G.; Sfravara, F. Assessment of Damage Evolution in Sandwich Composite Material Subjected to Repeated Impacts by Means Optical Measurements. *Procedia Struct. Integr.* **2016**, *2*, 3660–3667. [CrossRef]
26. Gunasegaran, V.; Prashanth, R.; Narayanan, M. Experimental Investigation and Finite Element Analysis of Filament Wound GRP Pipes for Underground Applications. *Procedia Eng.* **2013**, *64*, 1293–1301. [CrossRef]
27. Yang, R.; He, Y. Optically and non-optically excited thermography for composites: A review. *Infrared Phys. Technol.* **2016**, *75*, 26–50. [CrossRef]
28. Rios, A.D.S.; De Amorim Júnior, W.F.; De Moura, E.P.; De Deus, E.P.; Feitosa, J.P.D.A. Effects of accelerated aging on mechanical, thermal and morphological behavior of polyurethane/epoxy/fiberglass composites. *Polym. Test.* **2016**, *50*, 152–163. [CrossRef]
29. Castellano, A.; Fraddosio, A.; Piccioni, M.D. Ultrasonic goniometric immersion tests for the characterization of fatigue post-LVI damage induced anisotropy superimposed to the constitutive anisotropy of polymer composites. *Compos. Part B Eng.* **2017**, *116*, 122–136. [CrossRef]
30. De Fenza, A.; Petrone, G.; Pecora, R.; Barile, M. Post-impact damage detection on a winglet structure realized in composite material. *Compos. Struct.* **2017**, *169*, 129–137. [CrossRef]
31. Farooq, U.; Myler, P. Prediction of load threshold of fiber-reinforced laminated composite panels subjected to low-velocity drop-weight impact using efficient data filtering techniques. *Results Phys.* **2015**, *5*, 206–221. [CrossRef]
32. Fang, H.; Gutowski, M.; DiSogra, M.; Wang, Q. A numerical and experimental study of woven fabric material under ballistic impacts. *Adv. Eng. Softw.* **2016**, *96*, 14–28. [CrossRef]
33. Zhikharev, M.V.; Sapozhnikov, S.B. Two-scale modeling of high-velocity fragment GFRP penetration for assessment of ballistic limit. *Int. J. Impact Eng.* **2017**, *101*, 42–48. [CrossRef]

34. Usta, F.; Mullaoglu, F.; Türkmen, H.S.; Balkan, D.; Mecitoglu, Z.; Kurtaran, H.; Akay, E. Effects of Thickness and Curvature on Impact Behaviour of Composite Panels. *Procedia Eng.* **2016**, *167*, 216–222. [CrossRef]
35. Jang, B.; Kowbel, W.; Jang, B. Impact behavior and impact-fatigue testing of polymer composites. *Compos. Sci. Technol.* **1992**, *44*, 107–118. [CrossRef]
36. Abrate, S. *Impact engineering of composite structures*; Springer: Berlin, Germany, 2011.
37. ANSYS. *ANSYS LS-DYNA User's Guide, Release 12.1*; ANSYS Inc.: Canonsburg, PA, USA, 2009.
38. Card, M.F.; Wall, L.D., Jr. *NASA-TN-D-6140, L-7058: Torsional Shear Strength of Filament-Wound Glass-Epoxy Tubes*; National Aeronautics and Space Administration: Washington, DC, USA, 1971.

© 2020 by the authors. Licensee MDPI, Basel, Switzerland. This article is an open access article distributed under the terms and conditions of the Creative Commons Attribution (CC BY) license (http://creativecommons.org/licenses/by/4.0/).

Article

Adhesion of Multifunctional Substrates for Integrated Cure Monitoring Film Sensors to Carbon Fiber Reinforced Polymers

Alexander Kyriazis [1,*], Kais Asali [1], Michael Sinapius [1], Korbinian Rager [2] and Andreas Dietzel [2]

[1] Institut für Adaptronik und Funktionsintegration, Technische Universität Braunschweig, 38106 Braunschweig, Germany; k.asali@tu-braunschweig.de (K.A.); m.sinapius@tu-braunschweig.de (M.S.)
[2] Institut für Mikrotechnik, Technische Universität Braunschweig, 38124 Braunschweig, Germany; k.rager@tu-braunschweig.de (K.R.); a.dietzel@tu-braunschweig.de (A.D.)
* Correspondence: a.kyriazis@tu-braunschweig.de

Received: 10 August 2020; Accepted: 15 September 2020; Published: 17 September 2020

Abstract: During fiber composite production, the quality of the manufactured parts can be assured by measuring the progress of the curing reaction. Dielectric film sensors are particularly suitable for this measurement task, as they can quantify the degree of curing very specifically and locally. These sensors are usually manufactured on PI films, which can lead to delaminations after integration. Other authors report that this negative influence can be reduced by miniaturization and a suitable shaping of the sensors. This article pursues as an alternative, a novel approach to achieve a material closure instead of a geometrically generated form closure by choosing suitable thermoplastic materials. Thermoplastic films made of PEI, PES and PA6 are proposed as carrier substrates for thin film sensors. They are investigated with regard to their mechanical effects in FRP. The experiments show that the integration of PES and PEI in FRP has the best shear strength, but PA6 leads to a higher critical energy release rate during crack propagation in mode I. For PI, a locally strongly scattering critical energy release rate was observed. Neither in tensile nor in Compression After Impact (CAI) tests a significant influence of the films on these characteristic values could be proven.

Keywords: fiber reinforced polymer; sensor integration; interdigital sensor; curing process monitoring; epoxy; curing; film sensor; flexible sensor; thermoplast

1. Introduction

Fiber reinforced polymers offer a number of advantages over the classic construction materials, such as their high specific stiffness or the potential to increase the stiffness of the component specifically in the direction of the loads that are encountered during operation. These materials consist of a fiber component, which is essentially responsible for the load-bearing properties and the stiffness and a plastic matrix, which stabilizes the fibers and is responsible for the load introduction into the fibers and the load transfer between the fibers. Thermosetting plastics are most commonly used as matrix materials, especially epoxy resin. During production, thermosetting plastics cure to their final shape and transform from a liquid monomer mixture into a solid polymer network.

The mechanical properties of fiber reinforced polymer structures depend on the degree of completion of the curing reaction. Incomplete curing may result in the structure having a lower cross-linking density, which impairs elastic modulus and strength. Monitoring of the production process is necessary for quality assurance, since errors during curing can lead to failure of the part during operation. The data obtained allows process parameters to be optimized so that fewer faulty parts are produced. In the future, it would also be conceivable to control production by selectively post-curing areas of the component where the curing reaction takes place more slowly.

The curing reaction can be monitored via various physical effects, since a wide variety of properties change during polymerization [1,2]. Structure-borne sound methods are based on the change of mechanical properties [3], refractive methods use the change of the refractive index [4], strain-based methods such as Fiber Bragg Gratings measure the chemical shrinkage during the curing reaction [5], thermodynamic methods such as Differential Scanning Calorimetry measure the heat released during the exothermic reaction and dielectric methods are based on the change of dielectric properties of the epoxy resin [6]. Dielectric monitoring of the curing reaction using film sensors has several advantages over the other methods mentioned. For example, dielectric measurement of the curing reaction allows a very fine spatial resolution of the curing reaction and is therefore not dependent on the global fiber volume content like structure-borne sound methods. Unlike strain-based methods, a dielectric method records the curing reaction even before the gel point, when the liquid epoxy resin is not yet able to transfer forces to the fibers. Compared to thermodynamic methods, dielectric measurement has the advantage that it works outside a laboratory environment. In addition, there are standard processes, like e.g., flexible circuit board technology for the production of film sensors which allow a reduction of the manufacturing costs as already used by Yang et al. for Polyimide (PI) substrates [7].

The fact that the dielectric properties change during the curing of epoxy resins has been known and used to observe the curing reaction since the middle of the last century [8]. The dielectric measurement of the curing reaction by means of film sensors became possible by using interdigital structures. Dielectric curing monitoring using 100 µm thin PI-copper sensors was already carried out in 1995 by Maistros and Partridge [6]. Based on so-called flexible printed circuit boards made of a PI substrate with a copper coating, miniaturized interdigital sensors can be designed. The sensitivity of these sensors strongly depends on the geometry of the interdigital electrodes [7]. Sensitivity and immunity to interference can be further enhanced by electrically shielding the sensor with a copper coating on the bottom of the PI film used [9].

In order to use the full potential of film sensors, they should be integrated into the fiber composite polymer. The embedding of dielectric film sensors raises several questions, such as the influence on the resin flow, the signal transmission out of the fiber composite or the bonding of the sensors to the rest of the composite, see Figure 1. The article at hand addresses the influence of the integrated thermoplastic films on the mechanical properties of the composite. The planar structure of the sensors poses a problem, since the sensors can promote delamination of the fiber composite structure and subsequent crack propagation if the substrate is not suitable. Dumstorff et al. speak about a wound in the matrix material which can be reduced by a high adhesion strength between sensor and structure and an adapted stiffness between sensor and structure [10]. Miniaturization of the sensors helps to reduce the wound effect. In addition to the miniaturization of the sensors, the insertion of holes in the substrate film also has improved the interlaminar shear strength [11]. Here, the increase in adhesion strength is achieved by the microgeometry of the interface by creating a form closure. By replacing PI with plasticized RTM6, a doubling of the adhesive strength can be achieved [12]. Because the already cured film does not cross-link with the epoxy matrix, the adhesion strength does not come close to the strength of the reference samples without a film.

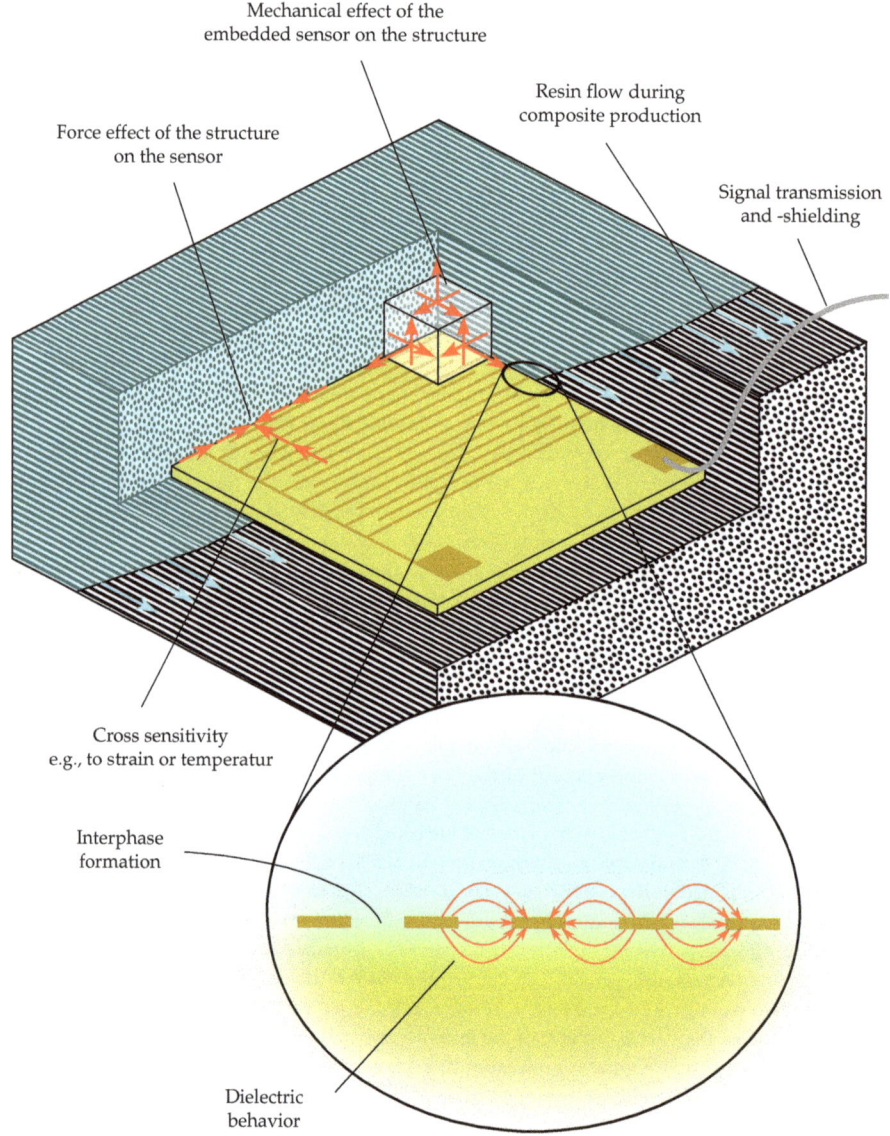

Figure 1. Questions arising from the integration of film sensors.

As an alternative to form closure, the function of the bonding can also be achieved by material closure. For this purpose, sensor substrates have to be used which can cross-link with the epoxy resin or form a strong bond with the epoxy resin. If the selected substrates also have a similar stiffness as the surrounding bond, the wound effect can be avoided. One way to increase binding is a surface treatment of PI film, in which PI is superficially converted to polyamic acid by a chemical treatment, resulting in a broader interphase due to chain entanglement and chemical reactions [13]. An increase in the critical energy release rate from 25 J m^{-2} to 100 J m^{-2} is reported [14]. An alternative is to use substrates that show a good adhesion to epoxy without surface treatment. The thermoplastics polyetherimide (PEI) and polyethersulfone (PES) are known as partially soluble toughening modifiers and it has been proven that they form an interphase with the epoxy resin [15]. Polyamide 6 (PA6) is used as a yarn filament in fiber layups and in this respect also has a practical significance in combination with fiber composites. In addition, chemical reactions, hydrogen bonding and mechanical entanglement of the polymer chains can result in high adhesion strength between epoxy resin and PA6 [16].

The suitability of the above materials as a sensor substrate with minimal wound effect is tested using various mechanical testing methods. Tensile tests perpendicular to the fiber orientation are used to demonstrate that 25 µm thin thermoplastic films do not have a negative effect on the tensile strength of the matrix resin even if the thermoplastic could dissolve and diffuse into the epoxy matrix. The effects of the sensor edges are also investigated using tensile specimens by embedding strips of the film materials used. Interlaminar shear (ILS) tests serve both to determine the shear strength of the epoxy-thermoplastic compound and to evaluate the gold-epoxy compound and the effect of sensor edges. Double Cantilever Beam (DCB) samples are used to determine the effect of the films on the delamination tendency for the particularly critical Mode I crack propagation due to peel stresses. Compression After Impact (CAI) tests are used to determine the sensitivity of laminates with the above-mentioned film materials to impact damage. In addition to the area of damage caused by an impact, CAI tests also provide information about the residual strength of damaged fiber composite plastic structures.

2. Materials and Methods

Commercial HexPly 8552 prepreg material from Hexcel with unidirectional AS4 carbon fibers with a filament count of 12 k was used for all specimens presented in this article. According to the manufacturer, it is an amine-curing, toughened epoxy resin system. The nominal thickness of a cured layer is 130 µm and the nominal fiber volume content is 57.42%. The cure cycle shown in Figure 2 was used to cure all specimens. This differs from the curing cycle from the data sheet only in that the vacuum is not reduced to 0.2 bar after application of the autoclave pressure but is kept at 1 bar for technical reasons.

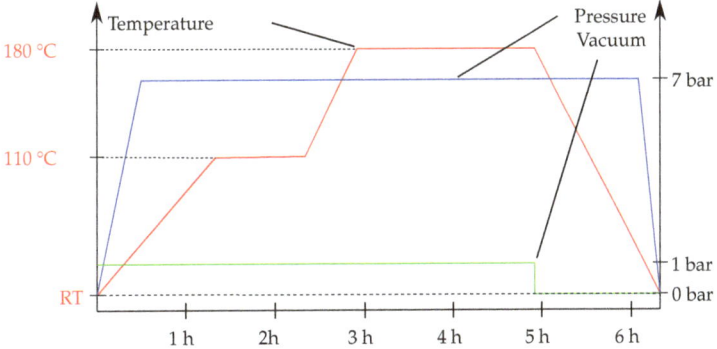

Figure 2. Autoclave curing cycle for all laminates in this articles.

The thermoplastic films in this study were a PA6 film, a PEI film, a PES film and a PI film. The PA6 film used is a 25 µm thin cast polyamide film which was kindly provided by mf-folien GmbH. A 25 µm thin Ultem 1000 film from Goodfellow was selected as PEI. The PES film utilized is the 25 µm thin Lite S film from Lipp Terler. The PI film is a 50 µm thin Kapton film HN type from Dupont.

The ILS specimens are made of 16 layers of prepreg and manufactured according to DIN EN 2563 geometry with unidirectional fiber orientation in the length direction, see Figure 3. Except for the reference sample, a 25 µm thick thermoplastic film of PEI, PES or PA6 is inserted between the eighth and ninth layer in the middle plane. The calculated thickness of the laminates is therefore $16 \cdot 130$ µm $+ 25$ µm $= 2.105$ mm. To coat the thermoplastic films with gold, they are first sputtered with chromium for 50 s at a power of 50 W and then with gold for 300 s. This creates a chromium adhesion layer on the thermoplastic film that is about 20 nm thin and a gold layer about 150 nm thin. The film stripes inserted in some of the test specimens have a length of 5 mm and a width of 10 mm. For each of the configurations in Figure 3b, seven specimens were tested. According to DIN EN 2563, the test fixture has a bearing distance of 10 mm. The compression test is carried out on a Instron universal testing machine with a force range up to 30 kN and a load cell with a measuring range up to 10 kN.

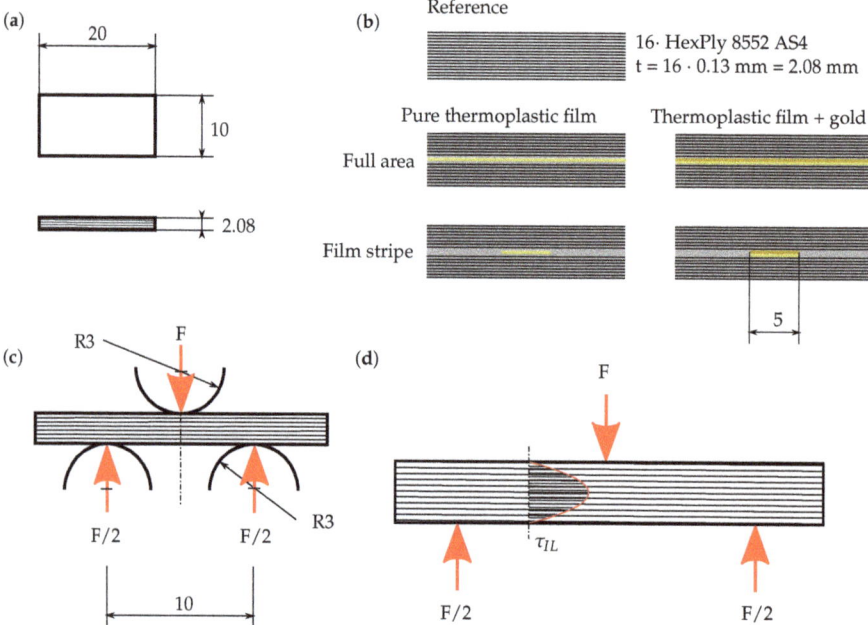

Figure 3. ILS testing device and specimen layup: (**a**) Specimen geometry. (**b**) Prepreg layers and integrated thermoplastic films with or without gold coating. (**c**) Geometry of the testing device. (**d**) Shear stress due to specimen loading.

From the measured force F the acting interlaminar shear stress τ_{ILS} is calculated according to DIN EN 2563 from the specimen width b and the specimen thickness h. The calculation formula proposed by the standard is given in the following and is basically only valid for the case in which no plastic deformations occur. Because the epoxy matrix is very brittle and the volume content of the ductile thermoplastics is very low, the formula is assumed to be valid:

$$\tau_{ILS} = \frac{3 \cdot F}{4 \cdot b \cdot h} \quad (1)$$

The differential compliance δ is calculated by deriving the traverse path x by the acting compressive force F. It includes the compliance of the specimen as well as that of the testing machine, but this is small compared to the compliance of the specimen and is also the same for all specimens.

$$\delta = \frac{\partial x}{\partial F} \quad (2)$$

For the tensile tests, coupons are made from 8 layers of prepreg with a fiber orientation of 90° to the tensile direction respectively the longest dimension of the specimen, resulting in a laminate thickness of about 1 mm, see Figure 4. Each plate is cut into seven specimens, which are 15 mm wide and correspond to the type A specimen described in DIN EN ISO 527. Between the fourth and fifth layer, 25 µm thin films of PEI, PES or PA6 are inserted over the entire surface or film strips of these thermoplastics with a width of 20 mm. To apply the loads in the tensile testing machine, GFRP cap strips about 1 mm thick with a 45°/−45° layer structure are glued onto the samples with a Loctite EA 9466 epoxy adhesive. For this purpose the specimens are manufactured with peel ply on the surface. In order to glue the cap strips to the laminates, the peel ply was removed from the laminate and the cap strip, then the adhesive was applied and the adhesive cured at room temperature for one day. No further sanding process was performed.

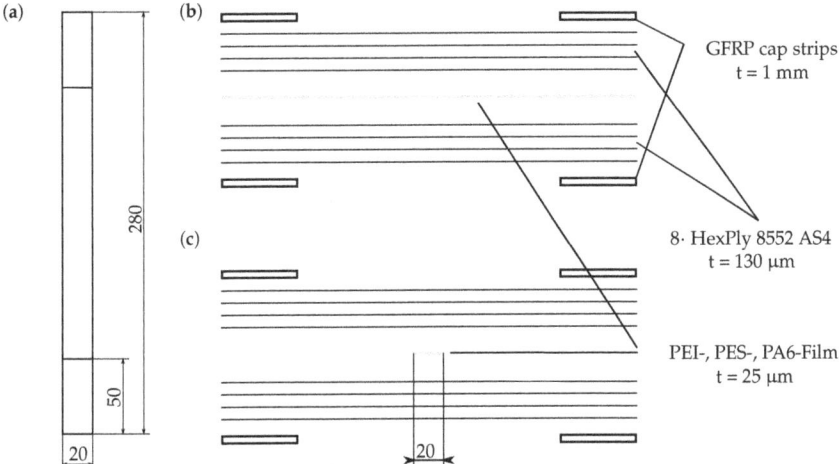

Figure 4. Geometry and layup of tensile test specimens. (**a**) Specimen geometry. (**b**) Specimen with integrated thermoplastic film. (**c**) Specimen with integrated thermoplastic film strip.

The specimens are tested using a universal testing machine with a force range up to 30 kN with a load cell with a measuring range up to 10 kN. The tensile stress is calculated in accordance with DIN EN ISO 527 as the force measured by the tensile testing machine in relation to the specimen cross-section measured before the test. The maximum tensile stress occurring in the test is specified as the tensile strength.

The manufacturing of the DCB specimens is carried out according to the dimensions and recommendations mentioned in DIN EN 6033, see Figure 5. The specimens are 250 mm long, 25 mm wide and about 3 mm thick corresponding to 24 prepreg layers. The initial crack has a length of 25 mm, which is created by the insertion of a separating film. First of all, 12 layers of prepreg are laminated and then the thermoplastic film and the release film are introduced. After that, the remaining 12 layers are laminated and the whole specimen plate is cured in the autoclave. Seven specimens for each configuration are cut out of the manufactured plate and piano hinges (25 mm × 25 mm × 1 mm) are glued onto the samples as a load introduction using Loctite EA 9466 epoxy adhesive. On the laminate surfaces a peel ply was used to achieve a rough surface and the piano hinges were sanded with 180 grit sand paper. To ensure that the peel strength of the adhesive is not exceeded, the hinges are glued on in reverse in comparison to DIN EN 6033, see Figure 5.

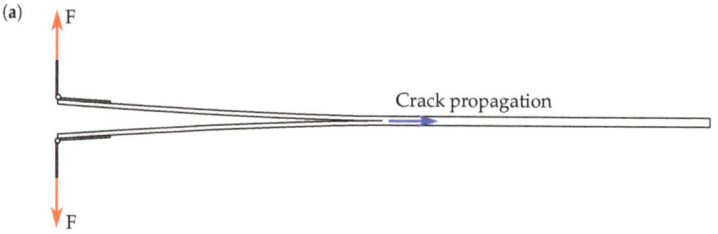

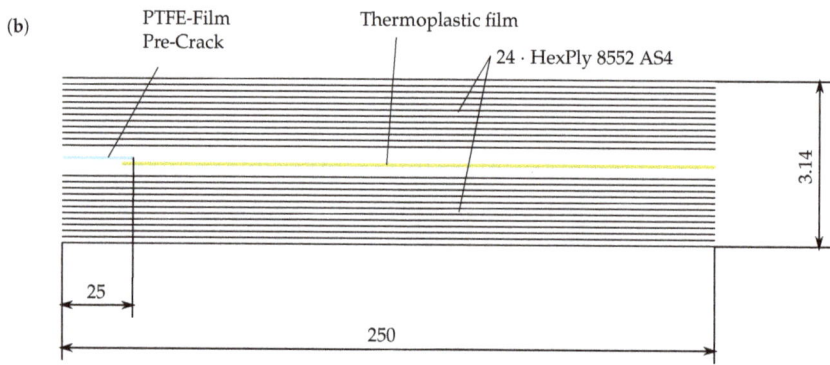

Figure 5. Testing and layup of double cantilever beam test specimens: (**a**) Shematic of a Double Cantilever Beam (DCB) specimen with glued piano hinges in the test setup. (**b**) Layup and geometry DCB specimen.

For the purpose of testing, a universal testing machine is used. The procedure for the testing process also complies with the recommendations of DIN EN 6033. Initially the testing machine is operated with a traverse speed of 2 mm min^{-1} for one minute to induce an initial crack of 10 to 15 mm length to prevent the effect of the precrack, then the length of the initial crack is marked. The testing machine is then driven at a traverse speed of 10 mm min^{-1} until a predetermined displacement value is reached. For the reference samples the displacement value was set to 22 mm, due to the higher crack resistance the displacement value was adjusted to 28 mm for the samples with PEI and PES film and 29 mm for the samples with integrated PA6 film. In the case of the samples with integrated PI film, the displacement value was adjusted several times so the maximum displacement values lie between 20 and 28 mm. After that the final crack length is marked. During the testing of the DCB specimens, force and traverse path are measured and a video of crack propagation is recorded.

The obtained data is evaluated according to both DIN EN 6033 and ASTM D5528. The purpose of using two different standards is to determine whether there may be a difference in the resulting G_{IC} values. The ASTM D5528 considers large displacement effects, which influence the results and must be considered. The G_{IC} value according to DIN EN 6033 is calculated using the following equation:

$$G_{IC} = \frac{W}{b \cdot \Delta a} \tag{3}$$

where Δa is the difference of the initial crack and the crack at the end of the experiment and b is the width of the specimen. W represents the energy under the force traverse path curve $F(x)$ during loading, crack propagation and unloading. It can be calculated from the following circulation integral:

$$W = \oint F \cdot dx \tag{4}$$

ASTM D5528 calculates the G_{IC} value via Modified Beam Theory (MBT), Compliance Calibration (CC) method or Modified Compliance Calibration (MCC) method. It should be noted here that the G_{IC} value is a function of the crack length a for all three methods of ASTM D5528, so it is necessary to take a video recording of the crack propagation through the specimens during the test procedure.

$$\text{MBT: } G_{IC} = \frac{3 \cdot F \cdot x}{2 \cdot b \cdot (a + |\Delta|)} \cdot K \tag{5}$$

$$\text{CC: } G_{IC} = \frac{n \cdot F \cdot x}{2 \cdot b \cdot a} \cdot K \tag{6}$$

$$\text{MCC: } G_{IC} = \frac{3 \cdot F^2 \cdot C^{\frac{2}{3}}}{2 \cdot A_1 \cdot b \cdot h} \cdot K \tag{7}$$

where C is the compliance of the DCB specimens, calculated by dividing the traverse path x by the force F: $C = x/F$. The correction length Δ, the stiffness exponent n and MCC coefficient A_1 can be calculated from a linear regression of the calculated compliance over crack length curves.

To account for large displacement effects ASTM D5528 uses a correction factor K, which was proposed by Williams [17] and has to be multiplied with the calculated G_{IC} value:

$$K = 1 - 0.3 \cdot \left(\frac{x}{a}\right)^2 - 1.5 \cdot \frac{x \cdot t}{a^2} \tag{8}$$

where $t = \frac{h}{4} + h_{ph}$ is defined as a quarter of the specimen thickness h plus the thickness of the piano hinges $h_{ph} = 1$ mm.

The CAI test specimens are sawn from a coupon made of 32 layers of prepreg. During production, 25 µm thick films of PEI, PES and PA6 are laminated over the entire surface between the fourth and fifth layer and between the 28th and 29th layer, so that the samples are symmetrical overall. To evaluate the influence, a reference sample without integrated films is also produced. The layer sequence is $(45°, 0°, -45°, 90°)x4|sym$, see Figure 6. Calculated, this results in a coupon thickness of 4.16 mm for the reference sample or 4.21 mm for samples with integrated films. A thickness measurement shows that all coupons have a thickness of about 4.08 mm with a fluctuation range of 40 µm. From each of the four coupons, six specimens with an outer dimension of 102 mm × 152 mm are sawn out and then milled with a diamond cutter to the final dimension of 100 mm × 150 mm. According to DIN ISO 18352, close dimensional and parallelism tolerances between the edges must be fulfilled.

Three specimens for each configuration were subjected to an impact energy of 20 J and the remaining three specimens for each configuration were subjected to an impact energy of 15 J. The impacts were performed on an Amsler HIT600F drop tower from Zwick/Roell with a drop height and drop mass of 0.574 m and 3.553 kg (20 J) and 0.604 m and 2.533 kg (15 J) respectively. The resulting impact damage is first assessed optically and then measured with an ultrasonic examination. From the ultrasonic images the damage areas are determined. A Zwick 1484 universal testing machine with a load cell up to 250 kN is used for testing the residual compressive strength.

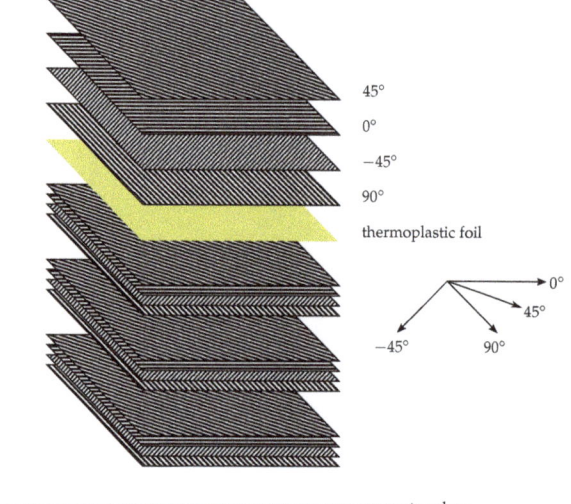

Figure 6. Layup of Compression After Impact (CAI) coupon with integrated thermoplastic film.

3. Results

The tests performed allow conclusions about the effects of the integrated thermoplastic films on the matrix tensile strength, adhesion strength under shear load, peel strength and damage tolerance of fiber composite laminates. The error bars in all diagrams shown throughout the following indicate the respective confidence interval for an assumed Gosset distribution.

The tensile tests show no significant differences between the tensile strengths with and without the embedded thermoplastic films. Evenly integrated film strips and the associated edges have no significant influence on the matrix tensile strength. The nonsignificance can be observed in Figure 7 on the basis of the overlapping of the confidence intervals. Taking into account an alpha error correction from the significance level 5% to 0.57% because of multiple comparisons, the nonsignificance of the differences can also be shown in a pairwise comparison with the reference sample in two sample t-tests.

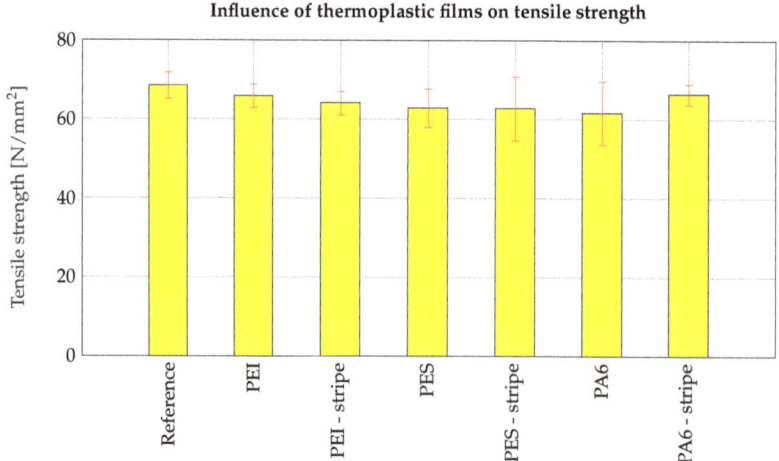

Figure 7. Influence of thermoplastic films and sensor edges on the tensile strength perpendicular to the fibers.

Figure 8 shows the influences of the pure thermoplastic film, the gold coating and integrated film stripes on the interlaminar shear strength (ILSS). When considering the influence of the film material, it is shown that PEI and PES do not result in a significant reduction of ILSS. In tendency the PEI and PES films seem to slightly improve the ILSS, but a statistical significance cannot be proven. For films made of PA6, a small but significant reduction of the interlaminar shear strength is observed.

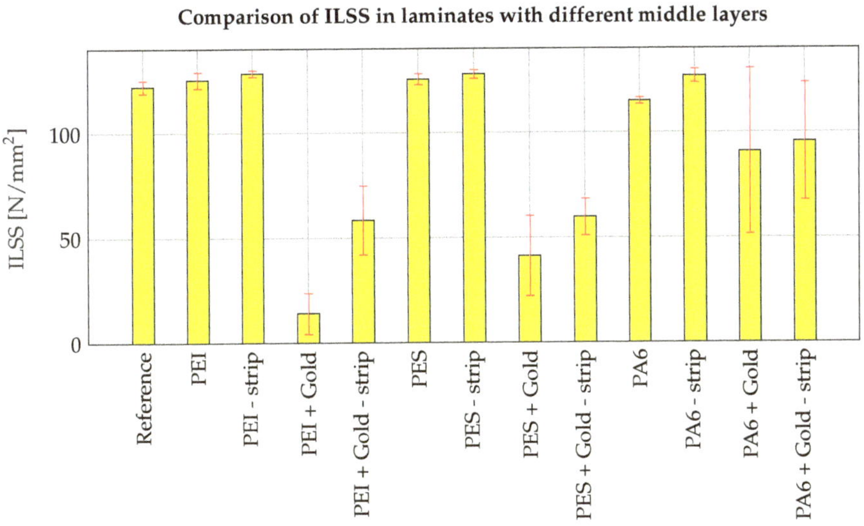

Comparison of ILSS in laminates with different middle layers

[Bar chart showing ILSS [N/mm²] for: Reference, PEI, PEI-strip, PEI+Gold, PEI+Gold-strip, PES, PES-strip, PES+Gold, PES+Gold-strip, PA6, PA6-strip, PA6+Gold, PA6+Gold-strip]

Figure 8. Influence of thermoplastic films, sensor edges and gold coating on the interlaminar shear strength.

The samples with full-surface film with gold coating show a considerable reduction and strong scattering in the measured ILSS. To calculate the ILSS, the force value at the first failure, i.e., before the first drop in the traverse path-force characteristic curve, is used in accordance with DIN EN 2563. For films without gold coating, no significant influence of the sensor edges can be observed. In the case of gold-coated film strips, there is an improvement in ILSS compared to the full-surface film, which is significant at least for PEI films.

Further insights can be gained from the ILS data by calculating the differential compliance curves from the force-traverse movement curves. To obtain smooth curves, the x and F curves are each smoothed over 5 measurements and the course of δ is also smoothed. This results in a compromise between low falsification of the measured data and smooth curves. In laminates with integrated PA6 films, a sigmoidal increase in differential compliance at low forces from 1.5 kN can be observed, in the case of PES from 1.7 kN, see Figure 9. Laminates with PEI film do not show such a sigmoidal increase but achieve a slightly higher differential compliance at a force of 3 kN compared to the reference laminates.

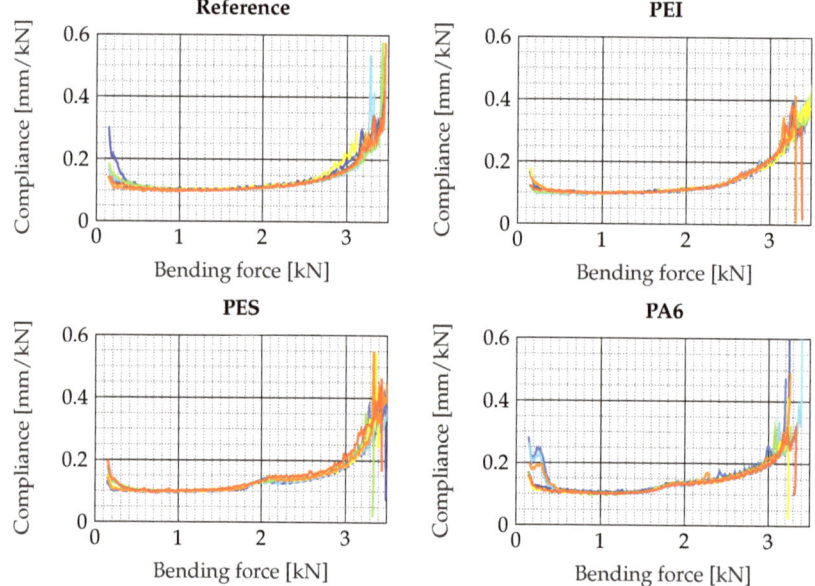

Figure 9. Influence of thermoplastic films on differential compliance in the interlaminar shear test. Every line shows the course of the compliance function of one single specimen.

The critical energy release rate for mode I crack propagation is obtained from DCB tests. The results show a significant improvement in the energy release rate for specimens with PEI, PES and PA6 film compared to the reference specimens, whereas the specimens with PI film show no significant improvement in the energy release rate compared to the reference specimens, see Figure 10.

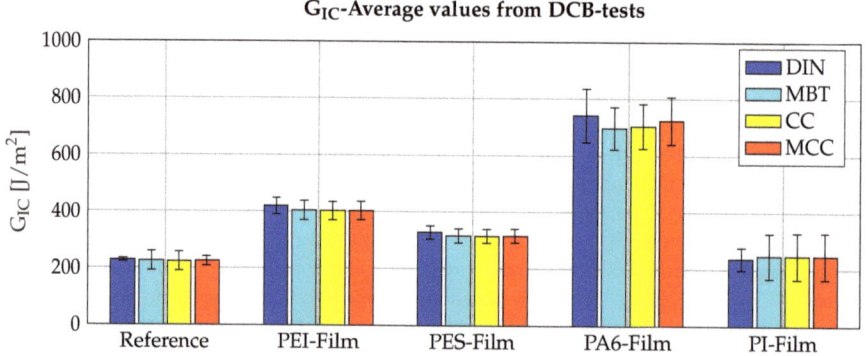

Figure 10. Influence of thermoplastic films on the mode I critical energy release rate.

Video recordings of the crack propagation during the testing of the specimens show different scenarios of crack propagation. For example, the crack propagation in the reference specimens and the specimens with PEI film was quasi-stable and continuous, whereby in the PES specimens, the crack propagation had a stick-slip form. The crack growth in the PA6 specimens propagates via crack jumping. In the case of the specimens with PI film, the crack propagation had a different shape. It was found that the force increased for some specimens after the first drop in force (after the crack has been created). The optical investigations show zones in the fracture surface of PI specimens where very high adhesion occurs. This can be seen by a steep increase of the normalized G_{IC} in the

R-curves. For example, the green curve for the PI specimens in Figure 11 indicates that the crack stops at a crack length of about 40 mm. In the force traverse path curve in Figure 12, a stop in crack propagation can be recognized by the transition from a decreasing force curve to an increasing force curve. This progression is not only apparent for PI but also for the samples with integrated PA6 film, which show a pronounced crack jumping. The normalization in Figure 11 is done by dividing the G_{IC} values calculated according to MBT by the G_{IC} values calculated according to DIN EN 6033, which are not a function of crack length a. The division gives unitless values that are a function of the crack length a.

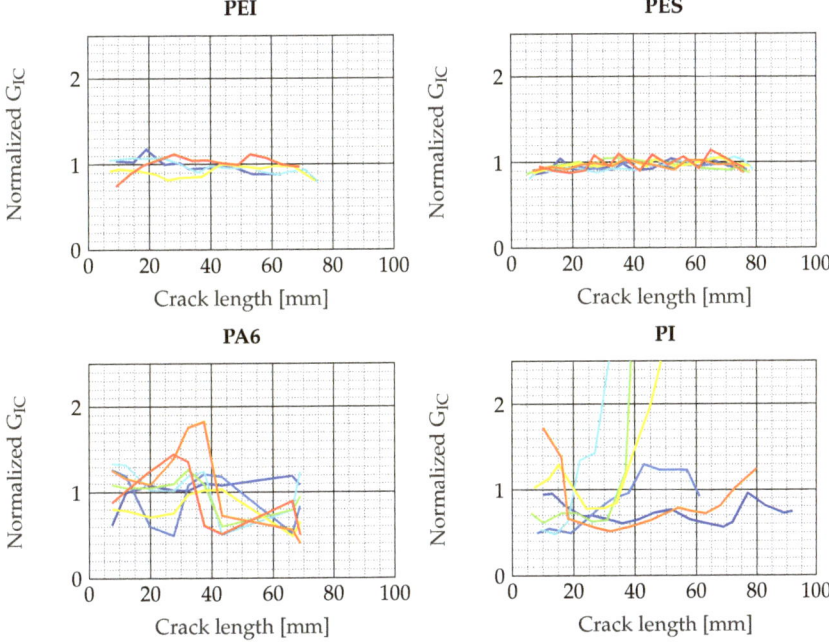

Figure 11. R-curves of different thermoplastic films in the DCB test. Every line shows the R-curve of one single specimen.

In CAI tests on laminates with integrated PEI, PES and PA6 films, no significant influence of the integrated film compared to the reference can be demonstrated, neither when considering the damaged area nor when considering the residual compressive strength, see Figure 13. Especially the damaged area does not seem to be strongly influenced by the impact energy or by the inserted films. On the basis of the available data, however, it seems unlikely that the films have led to a degradation of the CAI characteristics. An analysis of the data with ANOVA shows that no significant influence of the films can be proven, but the influence of the impact energy is as expected: higher impact energies lead to larger damage areas and lower residual compressive strengths. The PES sample series impacted at 20 J is noticeable by a high scatter of residual compressive strengths. In contrast, the samples with integrated PEI films show the narrowest confidence interval.

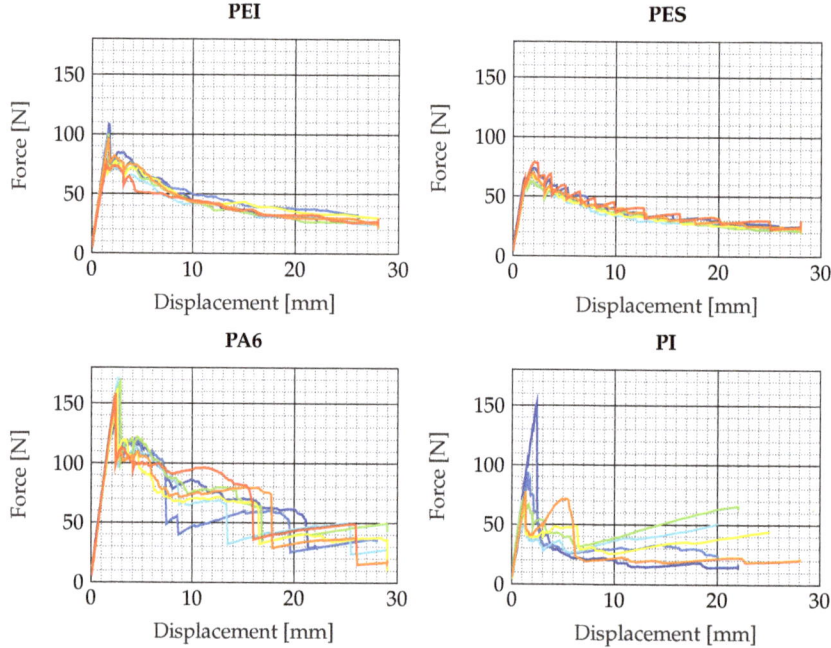

Figure 12. Force-displacement curves of laminates with different thermoplastic films in the DCB test. Every line shows the force-displacement curve of one single specimen.

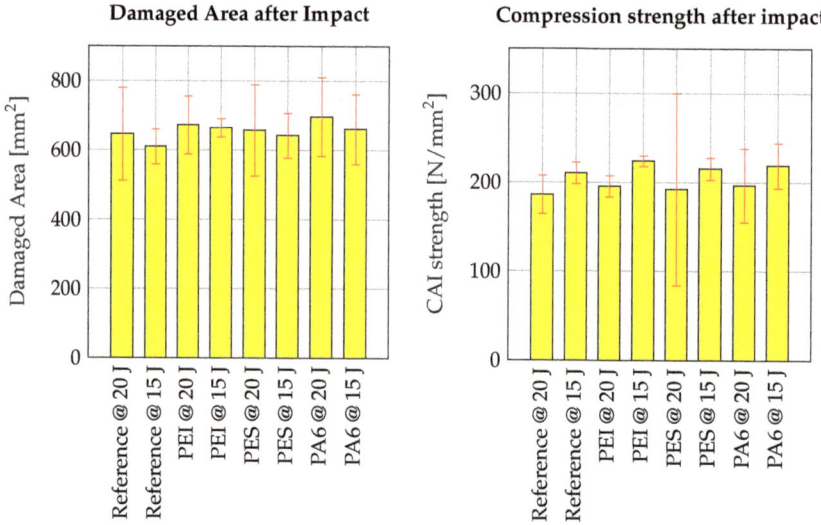

Figure 13. Influence of thermoplastic films on the damaged area and compression strength after impact. The left diagram shows the damaged area after impact, the right diagram indicates the compression strengths after impact.

4. Discussion

No influence of the integrated films on the matrix tensile strength can be deduced from the tensile test data obtained, see Figure 7. The tensile specimens are found to break almost ideally elastically without warning and to fragment into several small pieces, whereby no fibers are torn but only the matrix between the fibers, see Figure 14a. This can be attributed to the fiber orientation of 90° with respect to the loading direction. It is believed that due to the peel ply, the surface roughness acts like an array of small notches, which lead to excessive stress in the surface. In combination with the extreme brittleness of the epoxy resin, it is plausible that both effects together lead to scattering of the breaking forces and thus of the strength values calculated from them. The tensile tests show that the film material used has no significant influence on the matrix tensile strength, even if it is partially soluble in the epoxy resin, and that even possible local stiffness jumps at sensor edges do not play a role in the tensile test. Since the stiffness jumps are small due to the low thickness of the films of 25 µm, the small influence of the sensor edges is not surprising.

ILS tests show that both PES and PEI do not compromise interlaminar shear strength, but PA6 causes a slight reduction in interlaminar shear strength. The adhesion of the gold layer to the epoxy resin is weaker than that to the thermoplastic substrate because a chromium adhesion layer is used as an additional bonding layer between the gold layer and the thermoplastic. While for the samples with gold coating a complete separation of the specimens in the middle plane can be observed, the samples without gold coating show a pronounced crack near the middle plane but continue to hold together in the area outside the force introduction, see Figure 14b,c. Figure 14b shows that the gold stayed almost completely on the thermoplastic film. This also indicates a very weak bond between the gold layer and the epoxy resin. Figure 14c shows that in the specimens with fully integrated PEI film, the cracks avoided the interface between the thermoplastic and the epoxy matrix and spread through the surrounding epoxy matrix. The same behavior was observed with PES and PA6 films. This indicates that the ILSS obtained from the experiments only defines the lower limit for the true shear strength between thermoplastic and epoxy resin. Figure 14d shows a crack that propagated between the stripe of the gold-coated PES film and the epoxy matrix, leading to delamination in the mid-plane. Since the crack propagated from the center into only one direction, the specimens continued to hold together. A mixed behavior was observed in the specimens with film stripes without gold coating. In some specimens the crack avoided the thermoplastic film as shown in Figure 14c, but in other specimens the crack spread from the end of the thermoplastic film stripe similar to Figure 14d. However, as no degradation of ILSS is observed, it seems unlikely that the edges of thermoplastic films will affect the shear strength.

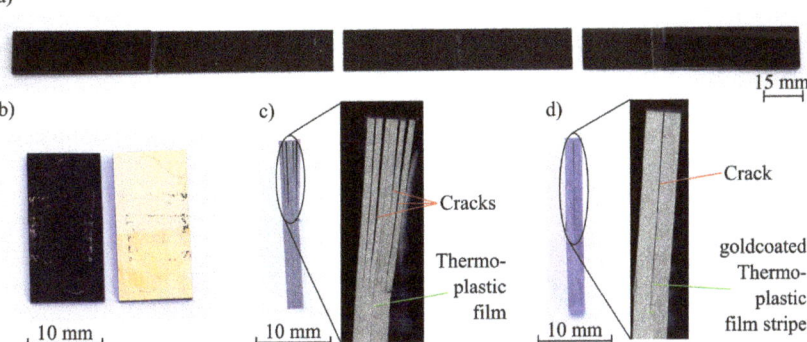

Figure 14. Typical fracture behavior of ILS and tensile specimens: (**a**) tensile specimen fragmented into several small pieces, (**b**) ILS specimen broken in the middle plane between gold and epoxy resin, (**c**) cracks in an ILS specimen with integrated PEI film and (**d**) crack propagated from a gold coated PES film stripe embedded in the center of an ILS specimen.

The strong scattering of the ILSS in specimens with gold-coated film could be explained by the fact that stress concentration occurs during the experiment at statistically distributed microscopic bonding defects in the contact plane, which can hardly be compensated by plastic yielding. Since the weak interphase between gold and epoxy is not able to dissipate as high amounts of energy as the thermoplastic/epoxy interphase, it fails completely starting from stress concentrations. The proposed explanation also shows reasons why the measured ILSS of the gold coating is significantly higher for ductile PA6 films than for comparatively brittle PEI films, see Figure 7. In the ductile PA6 film, stress concentrations are better reduced by plastic yielding than in PEI film, so the test specimens failed at higher external loads resulting in higher strengths. However, the results show that the interfaces between gold and epoxy are an issue for adhesion and form a weak point. These weak interfaces have to be compensated by the surrounding thermoplastic in an integrated sensor by plastic yielding and strong adhesion between thermoplastic and epoxy.

From the investigations on the thermoplastic film stripes without gold coating, it can be concluded that no impairment of the ILSS is to be expected from the edges of the sensors either. The investigations on gold-coated film stripes show that the surrounding compound even has a stabilizing effect on the sensors. The integration of the films leads to a reduction of the mechanical stiffness above a certain force level, see Figure 9. This sigmoidal increase in compliance indicates a plastic flow of PES and PA6 that occurs well before the actual failure. In contrast, the strong scattering of the compliance curves in the lower force range is mainly due to the penetration of the test die into the irregular surface of the specimens. The investigations on full-area gold coating and pure film only cover unrealistic limiting cases of integrated sensor technology. To examine the influences of gold patterns that are typical for sensors, further investigations will be carried out. In these investigations the influence of PI on ILSS will also be compared with the results from this work.

For the DCB tests there are several hypotheses that can give an explanation to the results. The PA6-Film possesses the largest elongation at break among the considered thermoplastics. It is noteworthy that this mechanical property describes the ability of the material to dissipate mechanical energy into plastic deformation.

Another explanation is based on the polarity theory, which suggests that the polarity plays an important role in the quality of an adhesive because of the force between the dipoles of the atoms/molecules [18]. According to Deng et al. [16] covalent bonds, hydrogen bonds via the amide group and entanglements of the molecular chain contribute to strong adhesion between PA6 and epoxy. PA6 can form hydrogen bonds via the amide group and with the hydrogen atoms of the carbon chain. The combination of good adhesion as is shown in the ILS tests with high ductility yields the very high cricital energy release rate for PA6. Another mechanism that promotes high energy release rates is the uneven surface caused by curling of the PA6 film seen in Figure 15a. Due to the larger surface area between PA6 and epoxy, the crack has to expose a larger area, which leads to higher crack energy consumption. The reason for crack jumping in PA6 specimens can be attributed to the fact that PA6 consists of amorphous and crystalline zones. The interface between these two areas is considered as a weak point in the structure, where cracks can easily spread.

Crack propagation in the form of stick-slip in specimens with PES film can be attributed to the phenomenon of fiber bridging. During the optical examination of the video recording, it was found that the fibers intertwine and form interface bridges, resulting in the curves seen in Figures 10 and 12. Figure 15b shows that the crack propagation in a stick-slip pattern has also left traces on the fracture surfaces. In the figure, alternating dark and light areas can be observed on both sides of the foil in the crack propagation direction. It is worth noting that the alternating pattern ends with the crack propagation in the experiment (red line in Figure 15b). During manual peeling after the experiment, no such structures are observed in the surface.

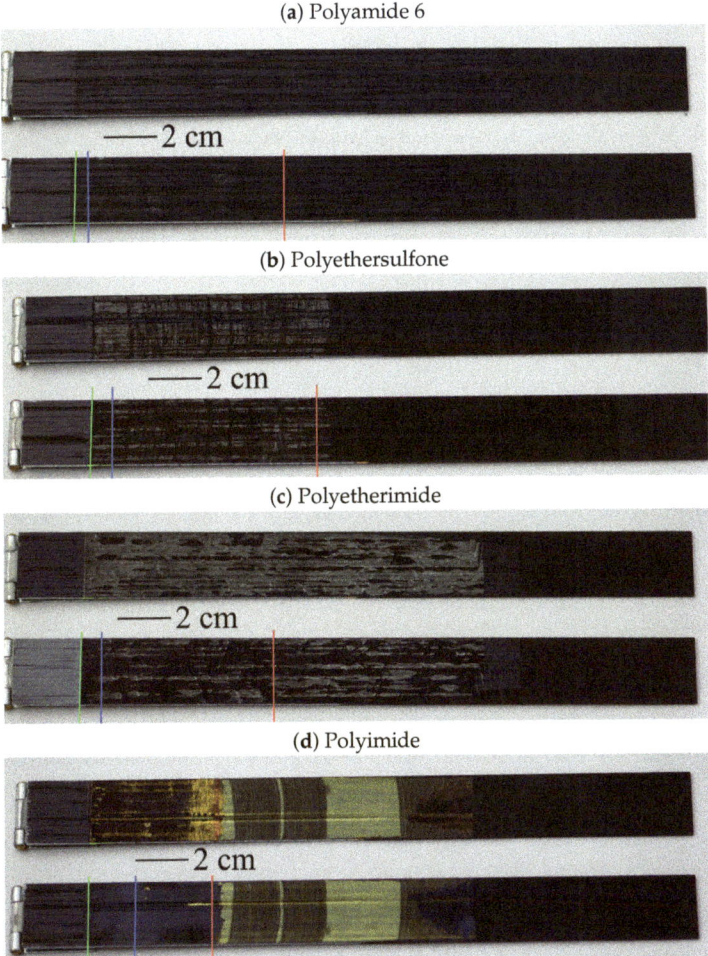

Figure 15. Fracture surfaces of DCB specimens with different thermoplastic films. The green line indicates the end of the inserted separating film, the blue line indicates the end of the precrack and the red line indicates the end of the crack after crack propagation in the experiment.

PES and PEI which are able to dissolve in epoxy show a significant improvement of the G_{IC}-value but compared to the PA6 film the improvement due to the PES and PEI films is far smaller. This can be attributed to the smaller plastic strain of PES and PEI. The better adhesion proved by the ILS tests is not decisive for the influence on the energy release rate. In Figure 15c, residues (bright structures) of the film can be observed on both sides of the crack shore. Thus, the crack must have crossed the foil several times.

PI shows weak adhesion in some zones of the specimens and strong adhesion in other zones. The reasons for the strong variation are not completely understood. The observation of the fracture surfaces in Figure 15d shows that three different zones can be distinguished. In the first zone directly behind the crack (right of the green line), the crack spreads between the PI film and the matrix. In the second zone, which starts about one centimetre before the end of the crack propagation in the experiment (1 cm left of the red line), the PI film still detaches itself from the matrix on one side but shows fine line-shaped structures, which make the area on the photo appear lighter. It is assumed that these line-shaped structures indicate a plastic deformation in the film. In the third zone to the right of the red line in Figure 15d, film residues can be seen on both sides of the specimen, which suggests that the crack has spread through the PI film during manual peeling after the experiments. In this zone, similar alternating structures as observed in the samples with PES film in Figure 15b are also visible in places. In the R-curves in Figure 11, the difference between the zones can be seen from the fact that the crack propagation resistance increases greatly at the transition from the first zone to the second zone and the crack even stops completely at the transition to zone 3. What causes crack propagation into the thermoplastic film is an open question. In further investigations, thinner PI films from a spin coating process will be investigated. Spin-coating allows the production of thinner layers, which are more similar to those in integrated sensors. Whether crack propagation into the laminate can be observed for such thin films is an open question for follow-up investigations. However, the large scattering in the G_{IC} value and the weak adhesion in the PI-epoxy interface indicate that integrating PI into carbon fiber reinforced polymers causes an increased crack propagation risk.

The examination of the CAI specimens shows that a deterioration of the damage tolerance due to the integrated films is improbable. However, for a statistically significant result, more than six specimens per film material would have to be examined. As the adhesion of the thermoplastic films to the epoxy resin is good, as already shown by the ILS and DCB tests, a negative effect of the pure films seems to be unlikely. For possible positive effects of the inserted thermoplastics, the volume fraction of the thermoplastic component is decisive. Since as little thermoplastic material as possible should be introduced in order to avoid a wound effect, no significant positive effect of the introduced thermoplastic films is to be expected. In order to choose a thermoplastic material as a sensor substrate, CAI tests are not relevant.

5. Conclusions

The ILS tests show that PEI and PES have excellent shear strength in combination with the prepreg used. In comparison, PA6 exhibits slightly but significantly lower binding but dissipates more energy during crack propagation in the DCB test. However, PEI and PES are also suitable for improving the crack propagation resistance of epoxy resin. The classically used PI shows a locally strongly differing adhesion in the DCB test. Overall, no significant influence of the PI film can be detected in the DCB tests due to the large scattering. Studies on the shear strength of the PI-epoxy interface are still to be carried out.

Neither in the tensile tests nor in the CAI tests a significant influence of any of the inserted film materials can be proven. In the case of the tensile tests, the nonsignificant influences show that the integration of thin thermoplastic films does not noticeably influence the tensile strength of the epoxy resin matrix. Scattering caused by notch effects due to the irregular surface due to the peel ply distinctly overweigh possible influences of the thermoplastic films. This result is not surprising, but it was important to exclude such an influence by means of the tests. In the CAI test, the nonsignificance can be attributed to the small amount of thermoplastic material introduced. If, for example, films were integrated between all prepreg layers, as is the case with interleafing, a significant influence could possibly be observed, but this case does not reflect the situation that occurs with sensor integration.

In further work it is planned to replace the integrated films by film sensors and thus measure the curing reaction. For this reason, the influence of sensor-typical gold patterns on the energy release rate and ILSS will have to be determined in future work. The mechanical influence of the interconnections will also be investigated in further work. Since some of the thermoplastics used in this article are soluble, the interphase formed between epoxy resin and the thermoplastic substrate during curing will also be the subject of further investigations. In addition, the integration of sensors raises manufacturing and electrical questions, which will also have to be addressed.

Author Contributions: Conceptualization, A.K. and M.S.; Data curation, A.K. and K.A.; Formal analysis, A.K. and K.A.; Funding acquisition, M.S. and A.D.; Investigation, A.K. and K.A.; Methodology, A.K.; Project administration, M.S. and A.D.; Resources, K.R.; Supervision, M.S.; Visualization, A.K.; Writing—Original draft, A.K., K.A. and

M.S.; Writing—Review and editing, A.K. All authors have read and agreed to the published version of the manuscript.

Funding: This research was funded by Deutsche Forschungsgemeinschaft (DFG) Grant number 397053684, "Eingebettete multifunktionale Sensoren zur Steuerung des Aushärteprozesses von Faserverbunden".

Conflicts of Interest: The authors declare no conflict of interest. The funders had no role in the design of the study; in the collection, analyses, or interpretation of data; in the writing of the manuscript, or in the decision to publish the results.

Abbreviations

The following abbreviations are used in this manuscript:

CAI	Compression after impact
CC	Complianca calibration
DCB	Double cantilever beam
FRP	Fiber reinforced polymer
ILS	Interlaminar Shear
ILSS	Interlaminar Shear Strength
MBT	Modified beam theory
MCC	Modified compliance calibration
PA6	Polyamide 6
PEI	Polyetherimide
PES	Polyethersulfon
PI	Polyimide

References

1. Heider, D. Cure Monitoring and Control. *ASM Handb.* **2001**, *21*, 692–698.
2. Konstantopoulos, S. Monitoring the production of FRP composites: A review of in-line sensing methods. *Express Polym. Lett.* **2014**, *11*, 823–840.
3. Lionetto, F. Monitoring the Cure State of Thermosetting Resins by Ultrasound. *Materials* **2013**, *6*, 3783–3804. [CrossRef]
4. Giordano, M. A Fiber Optic Thermoset Cure Monitoring Sensor. *Polym. Compos.* **2000**, *21*, 523–530. [CrossRef]
5. Prussak, R. Evaluation of residual stress development in FRP-metal hybrids using fiber Bragg grating sensors. *Prod. Eng.* **2018**, *12*, 259–267. [CrossRef]
6. Maistros, G. Dielectric monitoring of cure in a commercial carbon-fibre composite. *Compos. Sci. Technol.* **1995**, *53*, 355–359. [CrossRef]
7. Yang, Y. Design and fabrication of a flexible dielectric sensor system for in situ and real-time production monitoring of glass fibre reinforced composites. *Sens. Actuators A* **2016**, *243*, 103–110.
8. Delmonte, J. Electrical Properties of Epoxy Resins During Polymerization. *J. Appl. Polym. Sci.* **1959**, *4*, 108–113. [CrossRef]
9. Yang, Y. Design and Fabrication of a Shielded Interdigital Sensor for Noninvasive In Situ Real-Time Production Monitoring of Polymers. *J. Appl. Polym. Sci.* **2016**, *54*, 2028–2037.
10. Dumstorff, G. Integration Without Disruption: The Basic Challenge of Sensor Integration. *IEEE Sens. J.* **2014**, *14*, 2102–2111. [CrossRef]
11. Kahali Moghaddam, M. Embedding Rigid and Flexible Inlays in Carbon Fiber Reinforced Plastics. In Proceedings of the IEEE/ASME International Conference on Advanced Intelligent Mechatronics (AIM), Besacon, France, 8–11 July 2014.
12. Kahali Moghaddam, M. Sensors on a plasticized thermoset substrate for cure monitoring of CFRP production. *Sens. Actuators A Phys.* **2017**, *267*, 560–566. [CrossRef]
13. Kim, S.H. Improvement in the adhesion of polyimide/epoxy joints using various curing agents. *J. Appl. Polym.* **2002**, *86*, 812–820. [CrossRef]
14. Gurumurthy, C. Controlling Interfacial Interpenetration and Fracture Properties of Polyimide/Epoxy. *J. Adhes.* **2006**, *82*, 239–266. [CrossRef]

15. Hodgkin, J.H. Thermoplastic Toughening of Epoxy Resins: A Critical Review. *Polym. Adv. Technol.* **1998**, *9*, 3–10. [CrossRef]
16. Deng, S. Thermoplastic-epoxy interactions and their potential applications in joining composite structures—A review. *Compos. Part A* **2014**, *68*, 121–132. [CrossRef]
17. Williams, J.G. The fracture mechanics of delamination tests. *J. Strain Anal.* **1989**, *24*, 207–214. [CrossRef]
18. Habenicht, G. *Kleben Grundlagen, Technologien, Anwendungen*; Springer: Heidelberg, Germany, 2009.

© 2020 by the authors. Licensee MDPI, Basel, Switzerland. This article is an open access article distributed under the terms and conditions of the Creative Commons Attribution (CC BY) license (http://creativecommons.org/licenses/by/4.0/).

Article

Impact Performance and Bending Behavior of Carbon-Fiber Foam-Core Sandwich Composite Structures in Cold Arctic Temperature

M.H. Khan [1], Bing Li [2] and K.T. Tan [1,*]

1. Department of Mechanical Engineering, The University of Akron, Akron, OH 44325-3903, USA; mhk11@zips.uakron.edu
2. School of Aeronautics, Northwestern Polytechnical University, Xi'an 710072, China; bingli@nwpu.edu.cn
* Correspondence: ktan@uakron.edu; Tel.: +1-330-972-7184; Fax: +1-330-972-6027

Received: 15 August 2020; Accepted: 8 September 2020; Published: 10 September 2020

Abstract: This study investigates the impact performance, post-impact bending behavior and damage mechanisms of Divinycell H-100 foam core with woven carbon fiber reinforced polymer (CFRP) face sheets sandwich panel in cold temperature Arctic conditions. Low-velocity impact tests were performed at 23, −30 and −70 °C. Results indicate that exposure to low temperature reduces impact damage tolerance significantly. X-ray microcomputed tomography is utilized to reveal damage modes such as matrix cracking, delamination and fiber breakage on the CFRP face sheet, as well as core crushing, core shearing and debonding in the Polyvinyl Chloride (PVC) foam core. Post-impact bending tests reveal that residual flexural properties are more sensitive to the in-plane compressive property of the CFRP face sheet than the tensile property. Specifically, the degradation of flexural strength strongly depends on pre-existing impact damage and temperature conditions. Statistical analyses based on this study are employed to show that flexural performance is dominantly governed by face sheet thickness and pre-bending impact energy.

Keywords: composite sandwich structure; impact performance; bending behavior; arctic temperature

1. Introduction

Reduction in Arctic sea ice in the region over the last three decades has opened more efficient sailing routes [1]. New seaways through the Northern route have resulted in the increased deployment of marine and naval vessels in extreme low-temperature Arctic conditions. This requires advanced materials to combat the fundamental challenges associated with operating in such a cold and harsh environment. Therefore, a better understanding of how materials and structures behave and perform at extremely low temperatures is of utmost importance.

Advanced composite materials are widely used in automotive, aerospace, wind energy and marine industries. Sandwich composites are commonly employed in aircraft structures, ship hulls, wind turbine blades and bridge decks due to their enhanced bending stiffness, low weight, excellent thermal insulation, and acoustic damping capabilities. Sandwich structures typically consist of two thin and stiff skins (face sheets), which are separated by a thick, light, and shear-resistant core. Foam core sandwich structures are preferred to honeycomb structure for marine applications due to their low water absorption properties. However, one of the major concerns in the use of sandwich composites is their susceptibility to impact damage, which may occur during service and maintenance conditions. Impact loadings like bird strike, tool drop, hailstones, and debris impact by hurricane or tsunami can significantly reduce residual strength of the sandwich composites [2]. Low-velocity impact (less than 10 m/s) is commonly used to analyze the impact-induced damage phenomenon, because damage incurred is barely visible, yet extremely detrimental to the post-impact health of the composite structure [3]. Raju et al. [4] and Xue et al. [5] studied how the thickness of a honeycomb core affected the impact tolerance of sandwich panels. Atas and Potoglu [6] and James et al. [7] examined how the carbon fiber reinforced polymer (CFRP) thickness improved low-velocity impact resistance composite structures, and confirmed that improvement to impact damage can be made by using a thicker core and face sheet. Zhang and Tan [8], Huo et al. [9], Xin et al. [10] investigated how the shape of the impactor head affected the impact performance of sandwich composites. Papa et al. [11] studied the impact resistance and flexural behavior dependency on the stacking sequence of fiber hybrid composites. A constitutive model was proposed to understand the failure features and strain rate dependency of composite structures by Long et al. [12].

Jia et al. [13] found out that CFRP composites have enhanced flexural strength, maximum deflection, and energy absorption at a lower temperature. However, this study was limited only to composite laminates. Khan et al. [14] investigated impact performance and damage modes of Polyvinyl Chloride (PVC) foam core sandwich structures subjected to low-velocity impact damage. However, the study of post-impact structural integrity and damage mechanisms associated with after-impact loading was lacking. Schubel et al. [15] studied low-velocity impact and the post-impact compressive strength of composite sandwich panels at room temperature. However, the work lacked any study of how post-impact loading, such as bending, influenced the damage tolerance of sandwich structures in cold temperature Arctic conditions.

Yang et al. [16] explored how temperature influences impact behavior in foam-core sandwich composites under low-velocity range and concluded that low temperature results in reduced damage area and indentation depth. However, the understanding of important features with respect to impact resistance such as peak force and energy absorption was not analyzed. Erickson et al. [17] studied the effect of temperature on low-velocity impact and the post-impact bending behavior of composite sandwich panels. Although the low-velocity impact tests were executed at different temperatures, the post-impact bending tests were performed constantly at room temperature. This approach has failed to provide a good understanding of the relationship between temperature and post-impact bending performance.

In this study, low-velocity impact response, impact induced damage mechanisms and the post-impact flexural behavior of carbon-fiber foam-core sandwich composites are investigated at room and low temperature conditions. Section 1 provides an introduction, covering related published papers and their research gaps that have motivated the current work. Section 2 describes the materials and methods used in the current investigation, including the impact test setup, X-ray tomography technique, three-point bending test setup, and the statistical design of experiment approach. Section 3 presents the results and discussion, in terms of impact performance, impact damage mechanisms, residual flexural strength after impact, post-bending micro-computed tomography, and statistical analysis for understanding the influence of the factors. Section 4 ends with a conclusion for the current work. This work provides a better understanding of the impact dynamic response of composite sandwich structures at extremely low Arctic temperatures.

2. Materials and Methods

2.1. Materials

The sandwich composite is made of Divinycell Polyvinyl Chloride (PVC) H-100 foam core (6.35 mm thick) between 0/90° woven carbon fiber reinforced polymer (CFRP) face sheets. The lay-up schedule of the sandwich composite is 3 layers of plain weave carbon fiber, Divinycell H-100 Foam and another 3 layers of plain weave carbon fiber. A wet layup process was used and co-cured all together, peel ply was put down on tool, each fabric ply is laid down with resin applied, then core, then top face sheet plies with resin. Breather ply, caul plate and part was vacuum bagged. After vacuum was pulled, the specimen was cured in the oven, debagged and trimmed to final size. Epoxy resin accounted for approximately 50% of the composition. As the specimens are made by DragonPlate company, the exact epoxy resin and fiber types remain proprietary.

Two types of specimens were tested: thinner face sheet (only one woven CFRP lamina) and thicker face sheet (three woven laminae) on each skin. Each lamina was 0.25 mm in thickness. The sandwich panel was then cut into 150 mm × 100 mm sample size. For low-temperature testing, specimens were first cooled down to −23 °C in a freezer over a day period. −23 °C was the lowest temperature that the freezer could achieve. Subsequently, before impact testing, they were further cooled to −30 and −70 °C in the environmental chamber of an Instron CEAST 9350 impact machine using liquid nitrogen gas. The preconditioning in the freezer allowed the specimens to cool gradually from room temperature to low temperature, so that the specimens did not experience rapid cooling that might have caused damage to the specimens due to sudden thermal shrinkage.

2.2. Impact Test

The Instron CEAST 9350 drop tower, having a 16 mm diameter hemispherical impactor as shown in Figure 1, was used for the impact. The specimen was placed on a cylindrical support frame with a window of 76 mm diameter. Specimen was clamped with 100 N clamping force to avoid any tilt or distortion during impact. A 3.482 kg striker was raised to a height of 0.234 m and 0.117 m and subsequently dropped with impact velocities of 2.14 m/s and 1.52 m/s, respectively, to generate 8 J and 4 J impact energy. These energy levels were chosen such that the lower energy of 4 J damaged only the front face sheet, and 8 J impact penetrated the front face sheet and incurred damage to the core. Strain gauges were present inside the tup that was connected to the steel impactor. The strain measurements over the duration of the impact event provided the displacement of the impactor (deflection of the specimen), which were subsequently integrated to calculate the velocity and acceleration profiles, and thereby the force measurements. All these data were captured and measured by the DAS64K data acquisition system of Instron CEAST 9350 impact machine. Specimens were tested at 23, −30 and −70 °C. The temperature of 23 °C was selected as the benchmark case for room temperature, while −30 and −70 °C are the average and lowest temperature in the Arctic region, respectively [18]. A thermostatic chamber cooled by liquid nitrogen was used for low-temperature testing. At least four specimens were tested for each case.

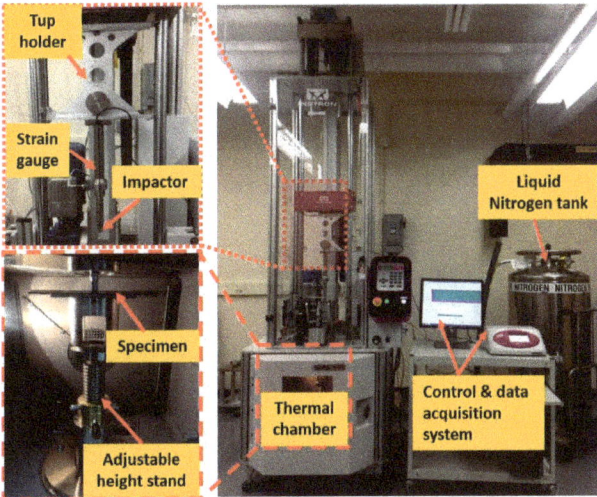

Figure 1. Experimental setup for impact test with Instron CEAST 9350 drop tower that is equipped with a thermal chamber for low-temperature testing and a control and data acquisition system to capture force, time, displacement and energy impact data.

2.3. X-ray Tomography for Damage Inspection

In this study, 3D renderings and cross-sectional images of the samples were acquired for both post-impact, and post-bending test specimens using a Nikon Metrology XTH 320 LC X-ray micro-computed tomography system. Numerous projection images were captured, as the sample rotated a complete 360° revolution. A 3D reconstruction was done using software CT Pro provided by Nikon Metrology. A 225 kV microfocus X-ray source penetrated the specimen for tomography with X-ray emission scanning at 90 keV and 50 µA. Each rotational image was averaged twice, acquiring 1800 scans. The collected data was 41.7 µm in voxel size. VG Studio Max software was then used for 3D image reconstruction.

2.4. Post-Impact Three-Point Bending Test

Post-impact bending tests were conducted to investigate the residual flexural strength of sandwich structures as shown in Figure 2, using the Instron 5582 machine following ASTM C 393 standard with a crosshead speed of 0.5 mm/min. Samples were tested at a room temperature of 23 °C and cold temperatures of −30 and −70 °C using an environmental chamber cooled by liquid nitrogen. The specimen was tested such that the impacted face sheet experienced either compression (inward) or tension (outward). These orientations were chosen to resemble actual service conditions, whereby depending on the supports, the composite structure could either bend concave or convex after being subjected to impact. In all cases, at least two specimens were tested under each condition.

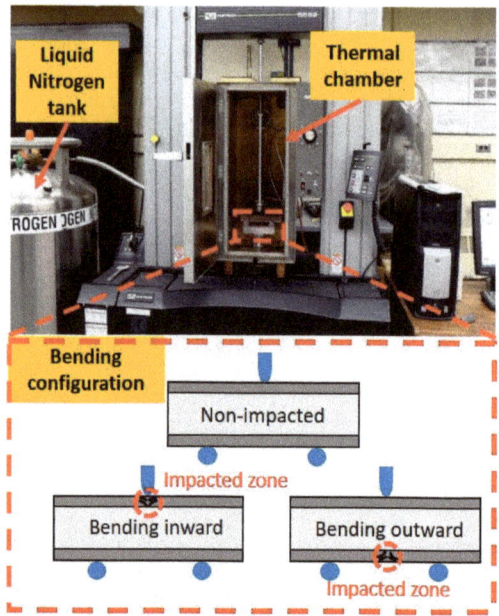

Figure 2. Experimental setup for the bending test with an Instron 5582 universal testing machine that is equipped with thermal chamber for low-temperature static testing. Schematic illustrates three bending configurations during bending test: nonimpacted case; bending inward case where impact damage is at top face sheet; bending outward case where impact damage is at bottom face sheet.

2.5. Statistical Design of Experiment

Design of Experiments (DoE) is a statistical tool that defines how input factors influence an outcome [19–21]. Standard factorial design dictates all variables having the same number of levels. In this study, two face sheet thickness (0.25 mm and 0.75 mm); three different temperatures (23, −30 and

−70 °C); two impact energies (0 J and 8 J); two bending configurations (bending inward and outward), were considered as the main factors. Table 1 presents the two-level DoE used for the four factors. The aim is to understand how these factors affect after-impact flexural performances. The 0 J impact energy was considered as the nonimpacted bending configuration and 4 J and 8 J were considered the impacted bending configurations. For each configuration of nonimpacted, inward and outward cases, testing was repeated at least twice and the peak force values were tabulated.

Table 1. Four factorial two levels Design of Experiment (DoE).

Factors	Factor Levels	
	Low Level (−1)	High Level (+1)
Face sheet thickness	0.25 mm	0.75 mm
Temperature	−70 °C	23 °C
Impact energy	0 J	8 J
Bending configuration	Inward	Outward

3. Results and Discussion

3.1. Impact Performance

Representative force-displacement plots for both thick and thin specimens impacted with 4 J and 8 J at different temperatures 23, −30, and −70 °C are presented in Figure 3. The sampling rate used during the impact test was 100 kHz for data acquisition, which gave sufficient precision and accuracy in the measurement of low-velocity impact tests. The markers in Figure 3 are intentionally spaced out at fixed intervals to make the curves distinguishable for different temperatures. The force-displacement curves showed linear behavior up to a certain point (defined as critical force for front face sheet damage) whereby sudden load drop indicated the failure of the front carbon fiber/epoxy composite face sheet (Figure 3a–c), except in a thick specimen impacted with low and insufficient impact energy of 4 J (Figure 3d), in which significant recovery happened along the displacement axis due to rebound of the impactor. However, for thin specimens, 4 J energy was enough to fail the front face sheet (Figure 3b). The critical forces for front face sheet penetration are shown in Figure 4. As the temperature decreased the force required for front face sheet penetration decreased for both thick and thin specimens due to the increased brittleness of the carbon fiber reinforced polymer (CFRP) face sheet. Low temperature severely degraded the composite face sheet's impact tolerance, the epoxy matrix of face sheet became exceptionally brittle, thus requiring less force for matrix cracks and brittle fracture [14,15]. A thicker face sheet offered higher resistance than a thin face sheet [5], thus thin specimen front sheets were penetrated at lower force than thick specimens. After the sudden load drop, as seen in Figure 3a–c, impact force continued to rise over time due to foam densification and finally declined until the end of the impact event with significant permanent deformation (Figure 3a–c) and rebound (3d). Moreover, with a decrease in temperature, the permanent displacement induced by impact for thin (4 J and 8 J) and 8 J for thick specimens of force-displacement plots generally increased, indicating more damage at lower cold temperatures. The 4 J thick specimen did not show this trend because the impact did not induce sufficient damage to the front face sheet.

The area under the curve in Figure 3 represents the energy absorbed by the damaged specimen and is shown in Figure 5. With the increase of impact energy, Figure 5a, and a decrease in temperature, Figure 5b, the percentage of energy absorbed increased, thereby indicating greater damage induced by impact at low temperature. The absorbed energy was also more sensitive to temperature change at low impact energy level (4 J) compared to higher impact energy (8 J) (Figure 5b) where absorbed energy had reached a high consistent percentage of more than 90%.

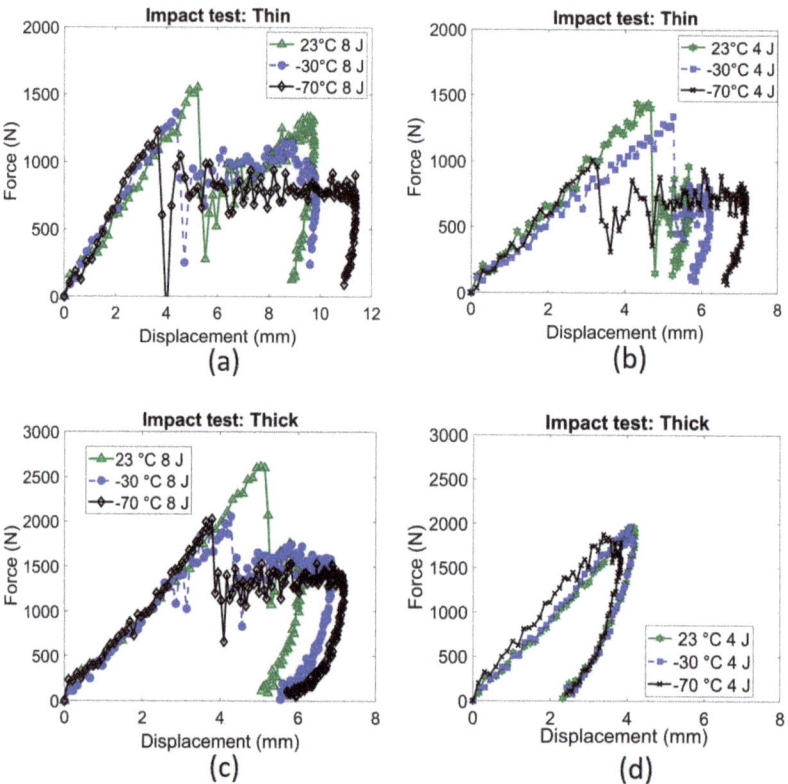

Figure 3. Impact force-displacement plots for (**a**) thin 8 J; (**b**) thin 4 J; (**c**) thick 8 J; (**d**) thick 4 J.

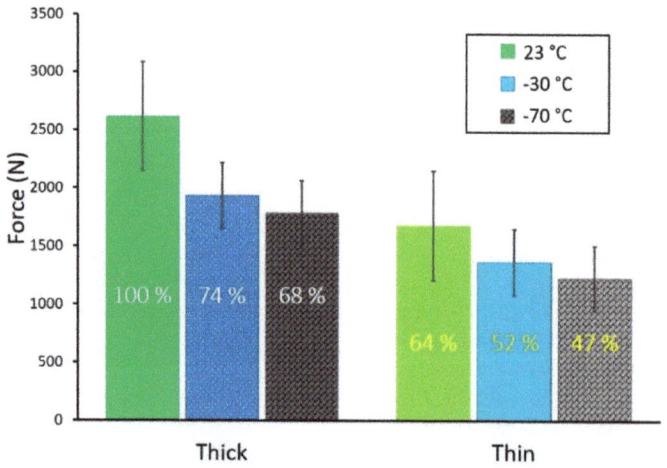

Figure 4. Critical front face sheet damage force for 8 J impact energy.

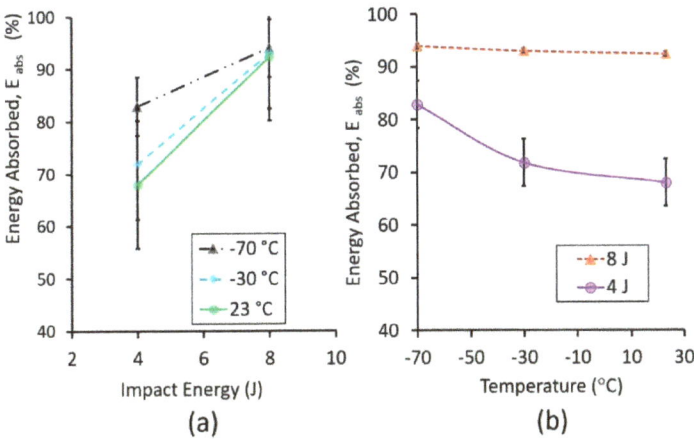

Figure 5. Thick specimens impacted with specified impact energy: (**a**) energy absorbed against impact energy; (**b**) energy absorbed against test temperature.

3.2. Impact Damage Mechanisms

Visible impact damage for thin specimens impacted with 8 J impact energy are shown in Figure 6. At 23 °C (Figure 6a), the damage was limited to localized matrix cracking only; while at −30 °C, matrix cracks were propagated deeper and longer both vertically and horizontally along the 0/90° woven fiber directions of the face sheet (Figure 6b). At the most extreme condition of −70 °C (Figure 6c), severe damage of matrix cracks and fiber breakage occurred, indicating face sheet penetration. X-ray µCT images are exhibited in Figure 7. At 23 °C (Figure 7a) the front face sheet suffered delamination and matrix failure, though partial penetration was observable. This ultimately led to fiber breakage at the front face sheet. Beyond front face sheet penetration, foam core damage was characterized by two mechanisms: core densification and core shearing. Core densification occurs due to high compressive normal stress loading under the impact region. Core densification is induced either by front face sheet compression to the core or by impactor tip compressing the core.

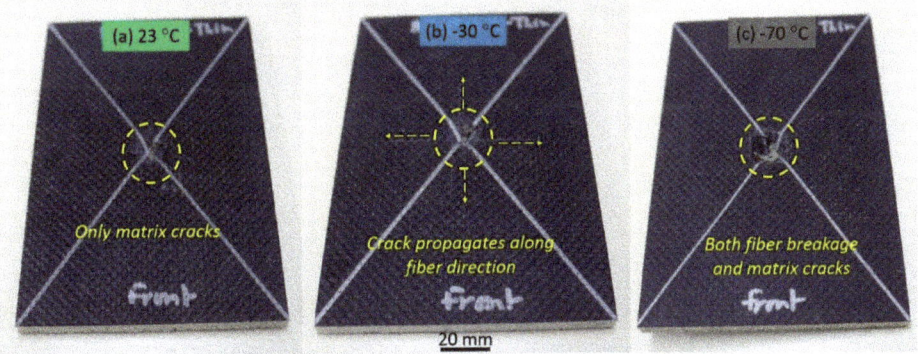

Figure 6. Impact damage (8 J) for thin face sheet sandwich composites at: (**a**) 23 °C; (**b**) −30 °C; (**c**) −70 °C.

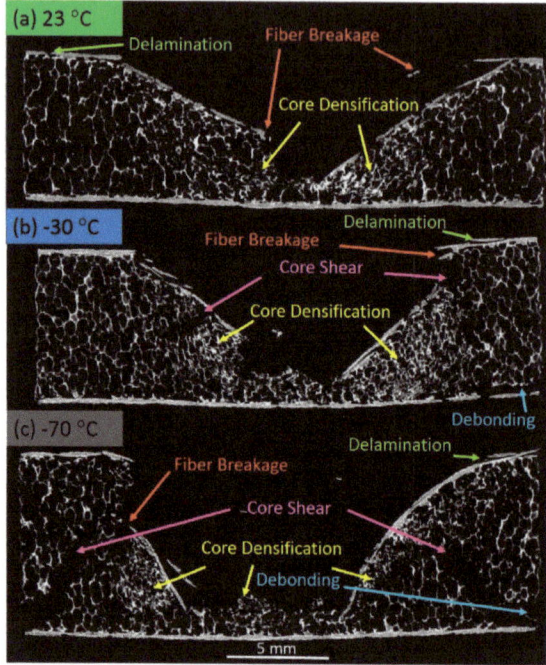

Figure 7. Post-impact µCT images of impact center for thin face sheet sandwich composites impacted with 8 J impact energy at (**a**) 23 °C; (**b**) −30 °C; (**c**) −70 °C.

As depicted in Figure 7a,b, core densification was dominated by the impacted front face sheet compression at 23 and −30 °C, whereas densification occurred near the bottom face sheet due to impactor tip compression (no severe back face sheet penetration) at −70 °C (Figure 7c). At −30 °C, core densification became prominent and led to the onset of core shearing (Figure 7b). Core shear is defined by the propagation of crack forming shear bands with a 45° angle across the core thickness. Thus, the shear bands propagate conically from the impact face sheet side into the foam core layer, as portrayed in Figure 7b,c. Core shear eventually results in debonding of the core from the back face sheet (Figure 7b,c). As the temperature decreases, foam core material becomes extremely brittle whereby cracks can propagate easily [22], and core shear stress increases [23], which subsequently leads to severe core shear failure at extreme low-temperature −70 °C (Figure 7c). The damage modes of core crushing and densification become more evident with the decrease of temperature. The debonding area also increases with the decrease of temperature in Arctic conditions as described by Elamin et al. [24]. It is worth mentioning that this study differs from [16] with the aim to relate core shear intensity to residual flexural strength after impact.

3.3. Residual Flexural Strength after Impact

The bending test results for post-impacted specimens compared to nonimpacted specimens are shown in Figure 8. The force-displacement curves were characterized by an initial linear elastic regime followed by a load drop. Nonimpacted specimens showed a trend of increasing peak force or flexural strength for decreasing temperature. The thick specimens broke catastrophically for bending outward cases at −70 °C (Figure 8c), while at other temperatures, the specimens typically exhibited a gradual decline in flexural force after the peak force. This is due to the embrittlement of the PVC core at low temperature −70 °C that facilitates core shear. Figure 9 plots the flexural peak load against pre-bending impact energy. Generally, when impact energy increases, flexural peak load decreases.

Bending outward cases perform better due to an undamaged top face sheet matrix, which can resist compressive load during bending. Although the bottom face sheet suffers impact damage, the damage is typically in the form of matrix cracking, thus the bottom face sheet can still withstand tensile load by the intact carbon fibers that are not severely damaged. However, for bending inward cases, the top face sheet has reduced compressive properties due to prior impact that damages the matrix. Therefore, the top face sheet was unable to withstand compressive stresses during three-point bending test and exhibited reduced flexural strength.

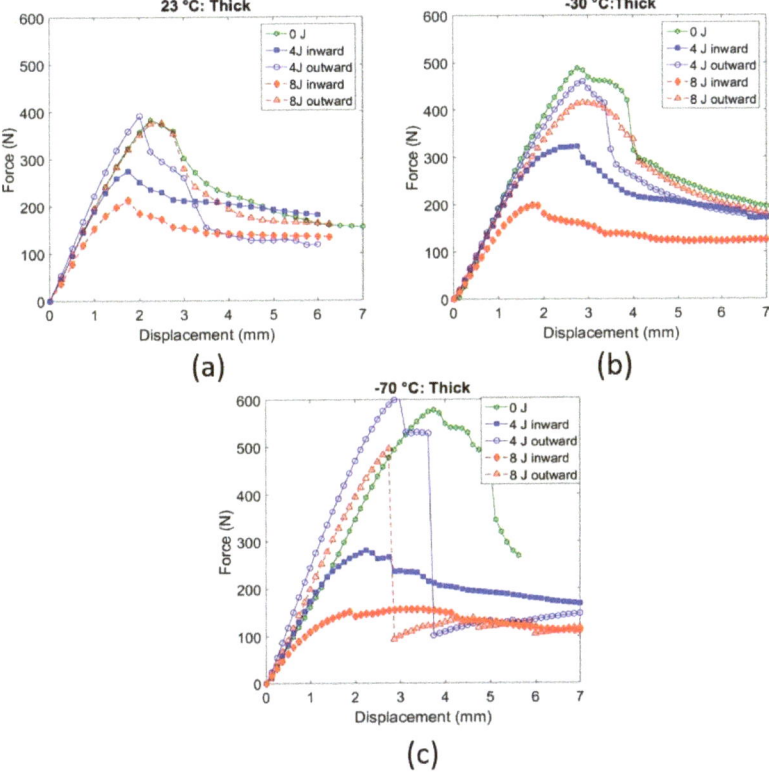

Figure 8. Force against displacement plots for thick specimens under three-point bending at (a) 23 °C; (b) −30 °C; (c) −70 °C.

Figure 10 shows the effect of temperature on flexural peak force of the nonimpacted beams. It is interesting to note that the bending behavior seemed to exhibit more "resistance" with decreasing temperatures. This is attributed to the fact that at low temperature, the compressive strength of the front CFRP face sheet increases, which makes the nonimpacted materials perform better under flexural load. Flexural strength is also found to be increased at Arctic subzero temperature for foam structures [25].

3.4. Post-Bending Micro-Computed Tomography

Figure 11 depicts the plan view of µCT images taken at 3.5 mm from the top face sheet, showing contrast in core shear intensity at impact and bending after impact cases. From the plan view, core shear appears as a ring of empty space with variable radius along the core thickness. This substantiates the previous discussion that at low temperature (−70 °C), prominent and severe core shearing occurs with an approximate diameter of 18.2 mm circulating the impact zone and gradually approaches to the back face sheet forming a 3-D conical shape (Figure 11a). However, after post-impact bending the core shear front diameter has increased from 18.2 mm to 20.9 mm (Figure 11b), by a significant +15 %. This increase in core shearing is not associated with any new damage mechanisms. Rather, core shear front expansion occurs only due to flexural bending of the specimen.

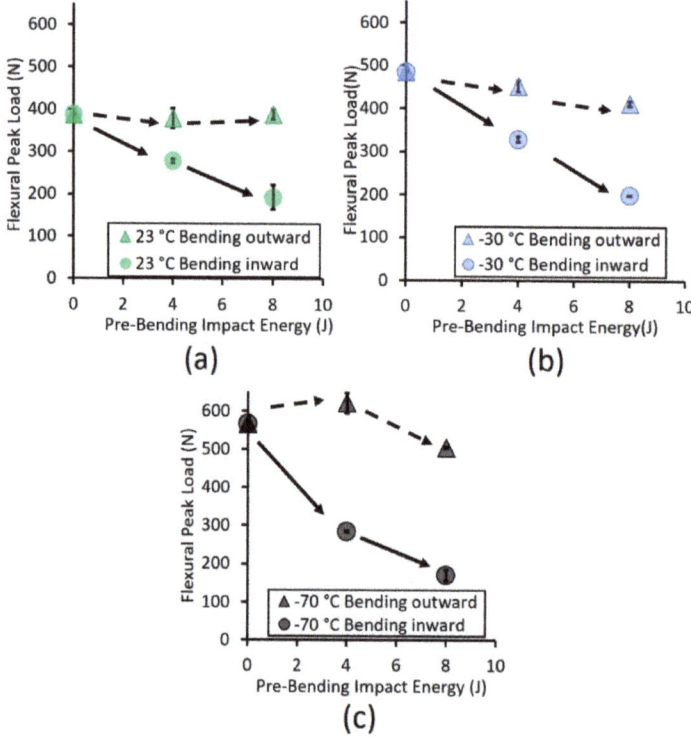

Figure 9. Effect of temperature on flexural peak load for thick impacted specimens at (a) 23 °C; (b) −30 °C; (c) −70 °C.

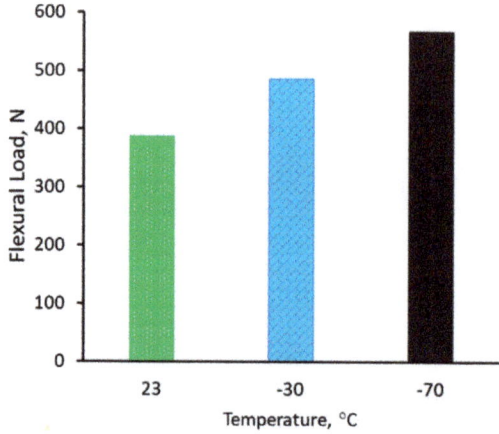

Figure 10. Flexural peak force against temperature for nonimpacted thick specimens at 23, −30 and −70 °C.

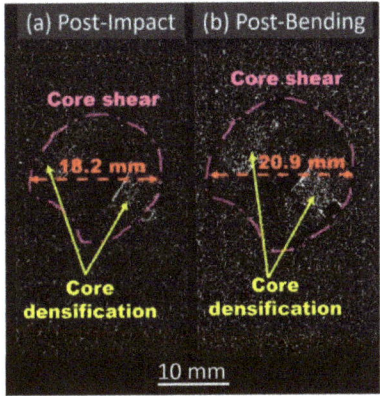

Figure 11. μCT images of foam core damage in thin face sheet composites (8 J, −70 °C): plan view at 3.5 mm from front face sheet for (**a**) post-impact; (**b**) post-bending condition.

3.5. Statistical Analysis for Understanding the Influence of Factors

The Pareto chart obtained from Minitab shows the absolute significance of the standardized effects from the biggest to the smallest. The reference line shows the effects which are statistically significant. Statistical software Minitab 18 yields the Pareto charts as shown in Figure 12a which indicates that face sheet thickness is the most significant factor that governs the flexural peak force, followed by pre-bend impact energy and temperature. Earlier observation in Figures 8 and 9 showed similar observations that these three factors were the prominent factors controlling flexural performance for the composite sandwich panel. Bending configuration had a much smaller effect than the other three.

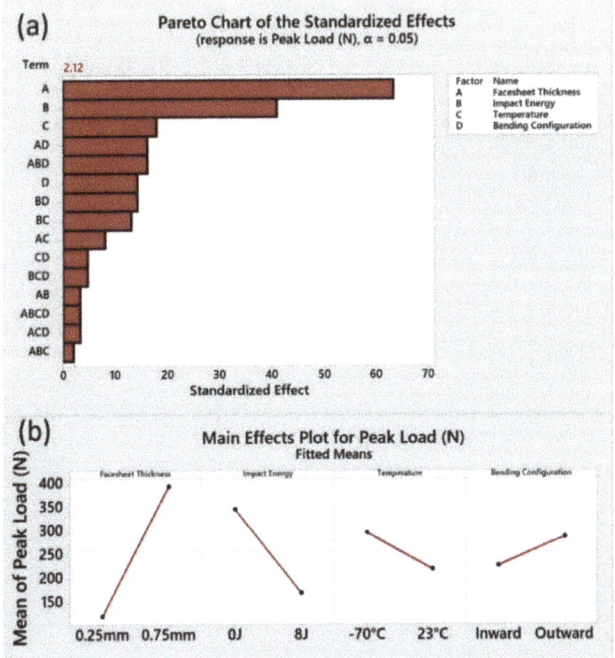

Figure 12. (**a**) Pareto chart; (**b**) main effects plot for post-impact flexural peak load.

The main effects plots, given in Figure 12b, which identified the influence of individual factors further validated the findings from the Pareto charts. The main effects plot provides the average data at the low and high factors. The gradient of the straight line corresponds to the significance of the factor directly. It is evident that the flexural peak force increases with increase in face sheet thickness (Th) whereas it decreases with increase in impact energy (Energ) and temperature. Inward bending shows inferior flexural performance compared to outward bending configurations. The flexural peak force is mainly influenced by face sheet thickness and impact energy whereas temperature and configuration (Conf) both have mild influences. Figure 13 portrays the interaction plot, which interprets the two-way interaction of the factors. If the slope of the two lines is not the same, there exists some interaction. It is concluded that apart from face sheet thickness and impact energy, which both are strong independent factors, all the other factors have interactions with each other.

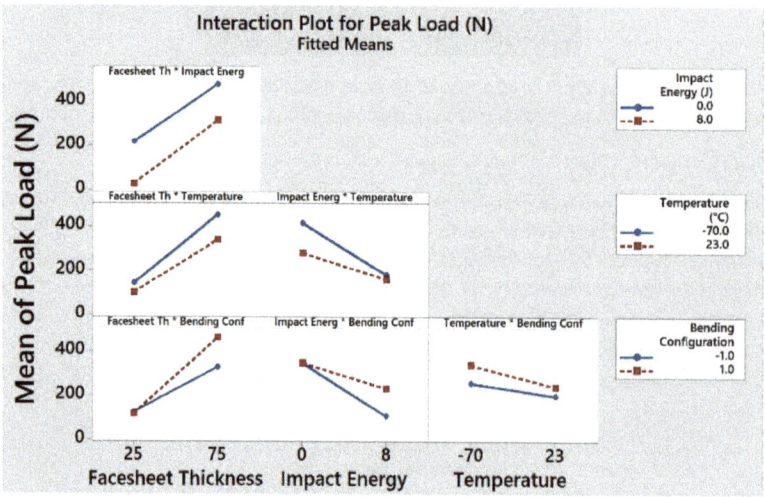

Figure 13. Interaction plot for flexural peak load, showing the interaction of face sheet thickness, impact energy and temperature on flexural force.

Contour plots are obtained by the constant response line projections from the response surface into the two-dimensional factor plane. Thus, the contour plot gives the prediction of impact energy and temperature on the x- and y-axes, respectively, along with contour lines for the peak load. At a specific impact energy, the contour plot can be used to graphically predict the flexural force for a particular temperature. Figure 14a shows that increasing impact energy will significantly reduce peak force for a given temperature for the bending inward case, whereas for the outward case, impact energy has a gentler influence (Figure 14b). This substantiates the experimental observation in Figure 9, as discussed earlier. Analysis of variance (ANOVA) is a statistical tool that calculates data based on the difference between two or more means. The variance within the groups and variances between the groups are compared. Subsequently, a regression equation can be obtained to describe the relationship between the response and the variables or factors. The algebraic representation of the regression equation for a linear model can be represented as:

$$Y = c + m_1 x_1 \tag{1}$$

where Y is the response variable, c is the constant or intercept of y-axis, m_1 is the slope of the line and x_1 is the value of the factor. The regression equation with more than one factor can be represented as:

$$Y = c + m_1 x_1 + m_2 x_2 + \ldots\ldots\ldots + m_n x_n \tag{2}$$

where $m_1, m_2, \ldots, m_n$ are the coefficients of the $x_1, x_2, \ldots, x_n$ factors. From ANOVA, considering the factors having p-values less than 0.05 only, the regression equation for the bending peak force, P is as follows:

$$\begin{aligned}P(N) = 75.91+ & \ 4.743t - 23.29IE - 0.501 \times T - 0.000 \times C + 0.0951 t \times IE \\ & -0.01906\ t \times T + 0.000 \times t \times C + 0.1060T \times IE - 0.000 \times T \times C \\ & -9.32\ C \times IE + 0.000965\ t \times T \times IE + 0.3160\ t \times C \times IE \\ & +0.0000T \times t \times C + 0.0225\ C \times T \times IE \cdot - 0.001553\ t \times T \times C \times IE \end{aligned} \quad (3)$$

here, t is the face sheet thickness, IE is the impact energy, T is the temperature and C is the bending configuration. Graphical prediction based on contour plot and statistical prediction based on regression Equation (3) are further compared with experimental results for flexural peak force for 4 J pre-bend impact energy at 23, −30 and −70 °C, as shown in Figure 15. Excellent agreement between experimental data and statistical prediction was observed. The greatest differences were for the −70 °C cases, where the statistical predictions deviate from experimental data. In reality, the impacted front face sheet matrix which was in compression during inward bending was severely embrittled at −70 °C [22,23] that led to flexural failure of the specimen at a much lower force. However, for the outward case, linear assumptions made by the statistical analysis underpredicted the experimental peak load, whereby the undamaged front face sheet compressive properties were enhanced at low temperature (−70 °C) and bottom face sheet tensile properties were uncompromised due to little or no damage during impact.

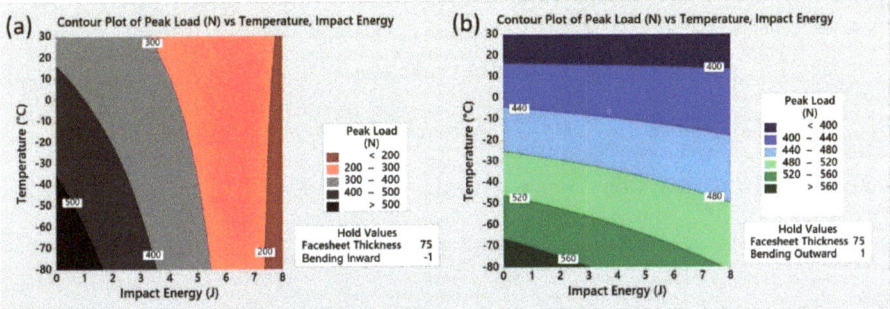

Figure 14. Contour plot prediction of flexural peak load for 0.75 mm face sheet sandwich composites under (**a**) bending inward; (**b**) bending outward configuration.

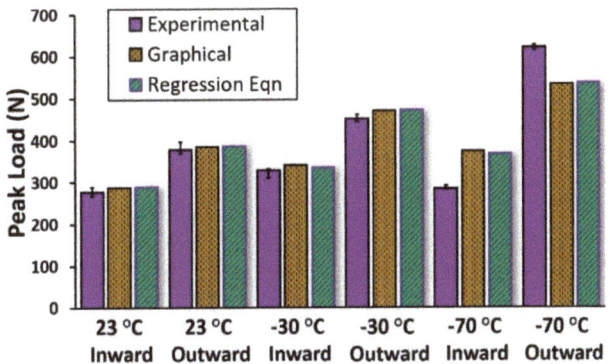

Figure 15. Statistical analysis and prediction of post-impact flexural peak load for 0.75 mm face sheet sandwich composites under 4 J impact energy.

4. Conclusions

The impact performance and bending behavior of carbon-fiber foam-core composite sandwich structures in low-temperature Arctic conditions were investigated. Impact performance is affected by impact test temperature and drastically lowered at extremely low-temperature Arctic conditions. At 23 °C, matrix cracking and delamination are dominant damage mechanisms; while at −70 °C, core shear, debonding, fiber breakage and severe face sheet penetration are the dominating failure modes. The intensity of core shear and debonding increases with decrease in temperature. Post-impact flexural strength generally decreases with increase in pre-bend impact energy. Flexural performance is superior in the case of bending outward rather than in the case of bending inward, due to reduced compressive properties at the impacted face. Flexural damage mechanisms associated with post-impact and post-bending conditions at −30 and −70 °C are dominated by core shear and debonding. Statistical techniques were employed to investigate flexural peak force by understanding the effect of the factors. It was revealed that flexural performance was mainly driven by temperature, face sheet thickness and pre-bend impact induced damage or impact energy. Graphical contour plots and regression equation were derived for predicting the bending collapse or peak force and further compared with experimental data. A good agreement was achieved between experimental and statistical results. The findings from this work provided a better understanding on the impact behavior and post-impact flexural performance of carbon-fiber foam-core sandwich composites when employed in extremely cold Arctic temperatures.

Author Contributions: Conceptualization, M.H.K. and K.T.T.; methodology, M.H.K.; validation, B.L. and K.T.T.; investigation, M.H.K.; resources, K.T.T.; data curation, M.H.K., B.L. and K.T.T.; writing—original draft preparation, M.H.K.; writing—review and editing, B.L. and K.T.T.; visualization, M.H.K.; supervision, K.T.T.; project administration, K.T.T.; funding acquisition, K.T.T. All authors have read and agreed to the published version of the manuscript.

Funding: This research was funded by the U.S. Office of Naval Research (ONR Solid Mechanics Program Manager: Yapa Rajapakse), grant number N00014-18-1-2546.

Acknowledgments: The authors would like to thank Lingyan Li for her help with the X-ray micro-CT scanning of the specimens.

Conflicts of Interest: The authors declare no conflict of interest.

References

1. Vihma, T. Effects of arctic sea ice decline on weather and climate: A review. *Surv. Geophys.* **2014**, *35*, 1175–1214. [CrossRef]
2. Sabah, S.A.; Kueh, A.B.H.; Al-Fasih, M. Comparative low-velocity impact behavior of bio-inspired and conventional sandwich composite beams. *Compos. Sci. Technol.* **2017**, *149*, 64–74. [CrossRef]
3. Feng, D.; Aymerich, F. Finite element modelling of damage induced by low-velocity impact on composite laminates. *Compos. Struct.* **2014**, *108*, 161–171. [CrossRef]
4. Raju, K.; Smith, B.; Tomblin, J.; Liew, K.; Guarddon, J. Impact damage resistance and tolerance of honeycomb core sandwich panels. *J. Compos. Mater.* **2008**, *42*, 385–412. [CrossRef]
5. Xue, X.; Zhang, C.; Chen, W.; Wu, M.; Zhao, J. Study on the impact resistance of honeycomb sandwich structures under low-velocity/heavy mass. *Compos. Struct.* **2019**, *226*. [CrossRef]
6. Atas, C.; Potoğlu, U. The effect of face-sheet thickness on low-velocity impact response of sandwich composites with foam cores. *J. Sandw. Struct. Mater.* **2015**, *18*, 215–228. [CrossRef]
7. James, R.; Giurgiutiu, V.; Flores, M.; Mei, H.; Haider, M.F. Challenges of generating controlled one-inch impact damage in thick CFRP composites. In *AIAA Scitech 2020 Forum*; American Institute of Aeronautics and Astronautics (AIAA): Reston, VA, USA, 2020; p. 0723.
8. Zhang, C.; Tan, K. Low-velocity impact response and compression after impact behavior of tubular composite sandwich structures. *Compos. Part B Eng.* **2020**, *193*, 108026. [CrossRef]
9. Huo, X.; Liu, H.; Luo, Q.; Sun, G.; Li, Q. On low-velocity impact response of foam-core sandwich panels. *Int. J. Mech. Sci.* **2020**, *181*, 105681. [CrossRef]

10. Xin, Y.; Yan, H.; Cheng, S.; Li, H. Drop weight impact tests on composite sandwich panel of aluminum foam and epoxy resin. *Mech. Adv. Mater. Struct.* **2019**, 1–14. [CrossRef]
11. Papa, I.; Boccarusso, L.; Langella, A.; LoPresto, V. Carbon/glass hybrid composite laminates in vinylester resin: Bending and low velocity impact tests. *Compos. Struct.* **2020**, *232*, 111571. [CrossRef]
12. Long, S.; Yao, X.; Wang, H.; Zhang, X. A dynamic constitutive model for fiber-reinforced composite under impact loading. *Int. J. Mech. Sci.* **2020**, *166*, 105226. [CrossRef]
13. Jia, Z.; Li, T.; Chiang, F.-P.; Wang, L. An experimental investigation of the temperature effect on the mechanics of carbon fiber reinforced polymer composites. *Compos. Sci. Technol.* **2018**, *154*, 53–63. [CrossRef]
14. Khan, M.; Elamin, M.; Li, B.; Tan, K. X-ray micro-computed tomography analysis of impact damage morphology in composite sandwich structures due to cold temperature arctic condition. *J. Compos. Mater.* **2018**, *52*, 3509–3522. [CrossRef]
15. Schubel, P.M.; Luo, J.-J.; Daniel, I.M. Low velocity impact behavior of composite sandwich panels. *Compos. Part A: Appl. Sci. Manuf.* **2005**, *36*, 1389–1396. [CrossRef]
16. Yang, P.; Shams, S.S.; Slay, A.; Brokate, B.; Elhajjar, R. Evaluation of temperature effects on low velocity impact damage in composite sandwich panels with polymeric foam cores. *Compos. Struct.* **2015**, *129*, 213–223. [CrossRef]
17. Erickson, M.D.; Kallmeyer, A.R.; Kellogg, K.G. Effect of temperature on the low-velocity impact behavior of composite sandwich panels. *J. Sandw. Struct. Mater.* **2005**, *7*, 245–264. [CrossRef]
18. National Geographic Arctic. Available online: https://www.nationalgeographic.org/encyclopedia/arctic/ (accessed on 28 March 2018).
19. Sutherland, L. The effects of test parameters on the impact response of glass reinforced plastic using an experimental design approach. *Compos. Sci. Technol.* **2003**, *63*, 1–18. [CrossRef]
20. Alonso, M.V.; Auad, M.L.; Nutt, S.R. Modeling the compressive properties of glass fiber reinforced epoxy foam using the analysis of variance approach. *Compos. Sci. Technol.* **2006**, *66*, 2126–2134. [CrossRef]
21. De Vivo, B.; Lamberti, P.; Spinelli, G.; Tucci, V.; Guadagno, L.; Raimondo, M.; Vertuccio, L.; Vittoria, V. Improvement of the electrical conductivity in multiphase epoxy-based MWCNT nanocomposites by means of an optimized clay content. *Compos. Sci. Technol.* **2013**, *89*, 69–76. [CrossRef]
22. Ibekwe, S.I.; Mensah, P.F.; Li, G.; Pang, S.-S.; Stubblefield, M.A. Impact and post impact response of laminated beams at low temperatures. *Compos. Struct.* **2007**, *79*, 12–17. [CrossRef]
23. Salehi-Khojin, A.; Mahinfalah, M.; Bashirzadeh, R.; Freeman, B. Temperature effects on Kevlar/hybrid and carbon fiber composite sandwiches under impact loading. *Compos. Struct.* **2007**, *78*, 197–206. [CrossRef]
24. Elamin, M.; Li, B.; Tan, K. Impact damage of composite sandwich structures in arctic condition. *Compos. Struct.* **2018**, *192*, 422–433. [CrossRef]
25. Garcia, C.D.; Shahapurkar, K.; Doddamani, M.; Kumar, G.C.M.; Prabhakar, P. Effect of arctic environment on flexural behavior of fly ash cenosphere reinforced epoxy syntactic foams. *Compos. Part B Eng.* **2018**, *151*, 265–273. [CrossRef]

© 2020 by the authors. Licensee MDPI, Basel, Switzerland. This article is an open access article distributed under the terms and conditions of the Creative Commons Attribution (CC BY) license (http://creativecommons.org/licenses/by/4.0/).

Article

Effect of Power Ultrasonic on the Expansion of Fiber Strands

Frederik Wilhelm *, Sebastian Strauß, Raffael Weigant and Klaus Drechsler

Fraunhofer Institute for Casting, Composite and Processing Technology IGCV, 86159 Augsburg, Germany; sebastian.strauss@igcv.fraunhofer.de (S.S.); raffael.weigant@igcv.fraunhofer.de (R.W.); klaus.drechsler@igcv.fraunhofer.de (K.D.)
* Correspondence: frederik.wilhelm@igcv.fraunhofer.de

Received: 29 March 2020; Accepted: 30 April 2020; Published: 8 May 2020

Abstract: The present study investigates the effect of power ultrasonic on the expansion of fiber strands. A potential application of such expansion is in the production process known as closed injection pultrusion. The fiber strand in the pultrusion injection chamber is in compacted form, and so, any expansion of the fiber strand resulting from power ultrasonic should lead to improved fiber wetting. To investigate this, a wetted fiber strand was clamped on two sides and sonicated in the middle from below. The potential expansion of the fiber strand was visually determined through an observation window. The study concluded that power ultrasonic has a minimal to virtually negligible effect on the expansion of both glass and carbon fiber. The degree of expansion remains within a range of 3% maximum, with a standard deviation in the respective midpoint tests of up to 60% for glass fiber and over 100% for carbon fiber. This shows that the fibers are limited in their freedom of movement, and so no expansion can be achieved using power ultrasonic. A further increase in amplitude does not lead to any further expansion but to the destruction of the fibers.

Keywords: power ultrasonic; expansion; carbon fiber; glass fiber; pultrusion; closed-injection pultrusion

1. Introduction

The use of pultrusion in component production has attracted great attention in recent years, due to such excellent properties as high strength-to-weight ratio and stiffness, and also due to the high potential for automation. In most cases, the fibers are impregnated in an open impregnation bath. Here, the choice is limited to matrix systems with a long processing time or pot life at room temperature [1].

However, in order to further increase productivity, it is necessary to use highly reactive matrix systems, such as polyurethanes [2,3]. Fiber impregnation then involves the use of a closed injection and impregnation chamber (ii-chamber) rather than an open impregnation bath. A number of different types of ii-chambers are already in use in industry [4–7]. However, the complete and uniform impregnation of a fiber strand with a resin system, particularly in the case of carbon fibers, is still a major challenge with this system [8]. While in an open impregnation bath, the fiber strands can be spread out, thus giving the resin system a short flow path, in an ii-chamber, the fibers have to be impregnated as a compressed strand. The effect of this is to reduce and significantly scatter the mechanical properties of the finished component [9].

One possible way of solving this issue is to integrate power ultrasonic (US) into the ii-chamber. US has many applications, such as mixing, homogenizing and dispersing, and it is already in use in numerous sectors (including the biological and chemical industries) [10–12]. The intended effect of US is to expand the fiber strands through their transient cavitation. Transient cavitation occurs at sound pressures with a power density of 10 W/cm^2. The tensile strength of the liquid is thereby exceeded [13], so that cavities/bubbles that are either empty (true cavitation) or filled with gas or water vapor are formed in the liquid [10,12]. Newton first coined the term 'cavitation' in 1704. It had previously been identified as material spalling in the early years of ship propellers [14–16]. The gas bubbles expand over several oscillation periods until their radius more than doubles within half a period before collapsing within a few milliseconds [17]. The conditions prevailing inside the gas bubbles include speeds of 50 to 150 m/s, temperatures of up to 5000 K, and pressures in the range of 10^9 to 10^{10} Pa [10,13,18,19]. Figure 1 contains a schematic representation of the process.

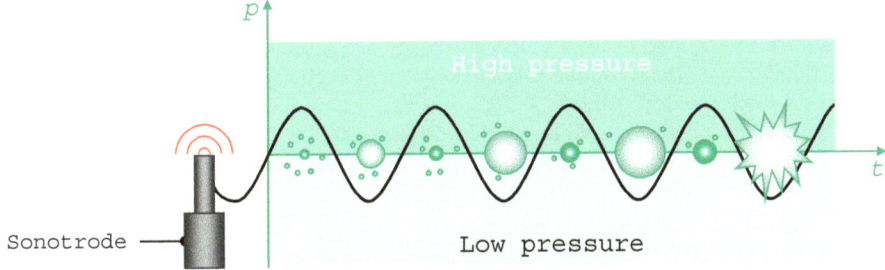

Figure 1. Formation of a cavitation bubble during sonication.

The effect of US on composites manufacturing has been previously investigated. These have demonstrated such benefits as enhanced mechanical properties due to better impregnation, improved fiber-matrix adhesion, and increased glass transition temperature [20–23]. Improved component properties were also demonstrated by the use of US in pultrusion [24]. However, few publications have considered the influence of US on fiber distribution. For excitations in the low frequency range, a fiber strand can be compacted further than would be possible using static compaction. Kruckenberg et al. investigated the compacting curve for glass fiber and carbon fiber fabrics at an excitation frequency of 1 to 10 Hz [25]. They found that an increase in fiber volume content of up to 16% is possible for glass fiber fabrics, with up to 6% for carbon fiber fabrics. Gutiérrez et al. conducted similar investigations with comparable results [26]. Glass fiber fabrics were stimulated during compaction and the compaction curve recorded for a frequency range of 10 to 300 Hz and amplitude of 50 to 100 µm. A higher compaction was achieved than with static compaction, especially at low frequencies, accompanied by an increase in fiber volume content of about 10%. The greater compaction was assumed to be due to the more homogeneous distribution of the fibers [27]. The vibration stimulation led to better spreading of the fibers and in turn, to a more uniform fiber distribution [28].

Yamahira et al. showed that US can be employed to attain specific fiber alignments [29]. A beaker was filled with an aqueous sugar solution containing polystyrene fibers, which were then irradiated at frequencies of 25 and 46 kHz. This formed a standing wave in the aqueous sugar solution, and the fibers aligned themselves with this wave.

The aforementioned publications presented the positive effect of the vibration excitation or sonication on the fiber distribution. Investigation of the effect was limited to the compaction and orientation of the fibers. However, they did not consider fiber movement during acoustic irradiation to quantify the possible expansion, which would result in the local increase in permeability referred to at the beginning.

This study therefore focuses on the direct, visual expansion of the fiber strand during sonication, correlated to total area. This enables us to understand the effect of US on fiber strand expansion and, in turn, permeability.

2. Materials and Methods

2.1. Materials

Preliminary investigations have shown that the Biresin® CR141 epoxy resin system made by Sika, Bad Urach, Germany, which is used as standard in the pultrusion process, becomes cloudy after only a short period of sonication [30]. Any expansion of the fiber strand that might have occurred is then no longer visible.

In order to avoid clouding, a silicone oil (KORASILON® M100 made by Obermeier in Bad Berleburg, Germany) is used instead. It has a viscosity of 100 mPa·s and is thus, comparable to the epoxy resin system. The substitute medium has already been used successfully, as reported in publications in similar fields [1,31].

The type of glass fiber to be used here is SE 3030 with 4800 tex, made by 3B in Herve, Belgium, while the carbon fiber is CT50-4.0/240-E100 with 50k, made by SGL Carbon, Wiesbaden, Germany.

2.2. Methods

A test chamber was developed to visually determine the effect of US on fiber distribution and quantify the extent to which US is able to expand a fiber strand. Figure 2 shows the schematic (left) and the real-life assembly (right) of the test chamber. In addition, the schematic representation shows the manipulated variables mass m, amplitude u and number of rovings n. The expansion B is shown as the test variable. It indicates the extent to which the area of the fiber strand is expanded by US, expressed as a percentage.

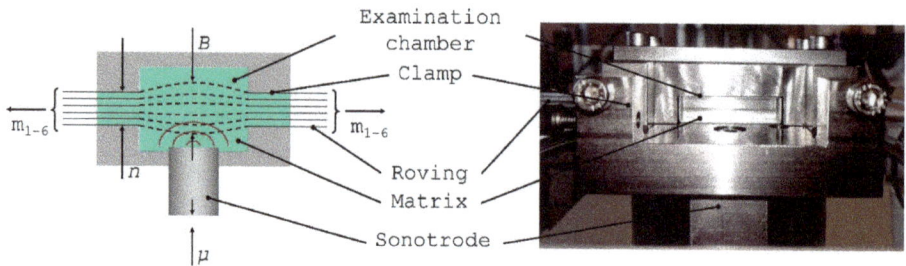

Figure 2. Test chamber: schematic representation (**left**) and real-life assembly (**right**, [32]).

Below the test chamber, US is coupled into the examination chamber with a frequency of 20 kHz. The diameter of the sonotrode is 20 mm. The fiber strand is clamped on the left and right and has a fiber volume content of 65% in this area. The fiber volume content is determined by the pultrusion process.

The manipulated variables for the test chamber result from the following parameters:

- Mass per roving m_n: The rovings are pulled continuously during pultrusion and are therefore under tension. In the test chamber, the preload on the rovings is simulated by employing an additional weight per roving. This corresponds to the actual stress occurring in the chamber geometries (conical and drop-shaped) [4,6]. The stress is based on the values from the preliminary investigation and varies within the range 250 to 500 g per roving [30].
- Number of rovings n: The layer thickness is adjusted by varying the number of rovings. A number of rovings of between 2 and 6 corresponds to a layer thickness of between 2 and 6 mm. The number of rovings determines the average profile thickness in pultrusion.
- Amplitude u: The amplitude range can be varied within the range from 12 to 48 µm. Any further increase in amplitude will cause direct damage to the fiber [33].

Silicone oil is injected into the test chamber at a pressure of 1 bar by means of a pneumatic pressure pot. The fiber strand in the test chamber can be viewed through the observation window

(Figure 3). A camera (EOS 500D, Canon, Tokyo, Japan, 15.1 megapixels) records the condition before and during sonication. Afterwards, the percentage changes in the areas with and without sonication are determined and evaluated. One millimeter corresponds to 76 pixels on the digital image.

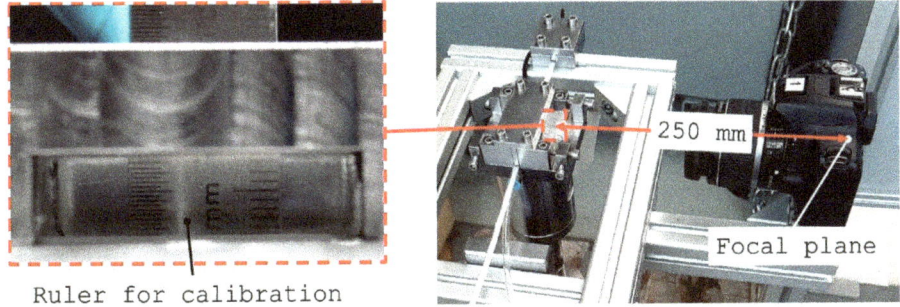

Figure 3. Evaluation of expansion B using a camera system (**right**), showing the viewfinder image of the observation chamber (**left**), where a calibration ruler can be seen.

The manipulated variables—weight per roving, amplitude, and number of rovings—determine the experiment setup. For the manipulated variables and the glass fiber and carbon fiber materials, the experiment test plan is as shown in Table 1.

Table 1. Test plan for the test rig with corresponding manipulated variables.

Manipulated Variables	Variable	Unit	Level 1	Level 2	Midpoint
Weight per roving	m	g	250	500	375
Amplitude	u	µm	12	48	30
Number of rovings	n	-	2	6	3

All tests are performed once, except for the midpoint test, which is repeated three times. This enables a statement to be made about scatter in the experiment setup. All tests are carried out in the sequence given in Table 2.

Table 2. Sequence of test steps.

Section	Activity
Setup	Link rovings to weights Insert rovings into test chamber Close test rig Calibrate US
Start	Open ball valve of pressure pot Flush test chamber with silicone oil for 30 s at a pressure of 1 bar Start self-timer (5 images after 10 s)
$u = 12$ µm	5 s no US 5 s with US 5 images at an amplitude of 12 µm; deactivate sonication
$u = 48$ µm	5 s no US 5 s with US 5 images at an amplitude of 48 µm; deactivate sonication
End	Close ball valve of pressure pot
Dismantling	Open test chamber and clean

3. Results

The aim of the test chamber is to characterize the effect of US on the expansion of a fiber strand. When it expands, the permeability of the fiber strand increases, thus, facilitating impregnation. The test rig enables explicit investigation of the effect of expansion, by allowing fiber movements to be viewed through an observation window (Figure 3).

3.1. Expansion of Glass and Carbon Fiber Strand

Figure 4 shows the test results for glass fiber (left) and carbon fiber (right). The measuring points are as follows: fiber type[G ≙ Glass, C ≙ Carbon]_mass[g]_number of rovings[-]_amplitude[μm].

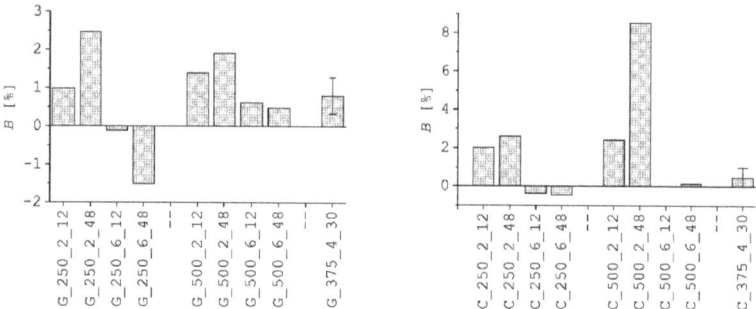

Figure 4. Measurements resulting from the investigation of glass fiber expansion (**left**) and carbon fiber expansion (**right**).

The expansion range for glass fiber extends from −1.5% to 2.5% and for carbon fiber from −0.5% to 8.5%. With the exception of an outlier in the carbon fiber test series, the amount of expansion is in the low single-digit range. The midpoint test shows that the measurements can be expected to display a high degree of scatter. A standard deviation of up to 60% for glass fiber and over 100% for carbon fiber were determined in the respective midpoint tests.

3.2. Analysing the Effect of Manipulated Variables

In order to determine the effect of the individual manipulated variables, all results were analyzed in effect diagrams. Figure 5 is an example of how effect diagrams (2 and 3) of individual manipulated variables can be formed from a test setup (1) with two manipulated variables (*a* and *b*) and the test variable *y*. The multi-dimensional test setup is reduced to the dimension of the manipulated variable and the mean value per stage determined from the individual measuring points. The effect of a particular manipulated variable is determined by linear interpolation. The effect is deemed positive if the test variable increases as the manipulated variable increases, as shown in the effect diagram (2). Otherwise, the effect is negative (see (3) in Figure 5).

Figure 6 shows the effect of US on the glass fiber for the test variable (expansion B). The effect remains within a range of 3%. Transferred to the reference area without sonication, the area undergoes only minimal change as a result of US. The manipulated variable weight per roving has the greatest positive effect on expansion, followed by amplitude. The effect decreases with increasing layer thickness. The midpoint test shows an average build-up of 0.8%, with a minimum value of 0.5% and a maximum value of 1.4%. In comparison, the effect on expansion for all manipulated variables is largely hidden in the scatter of the test setup.

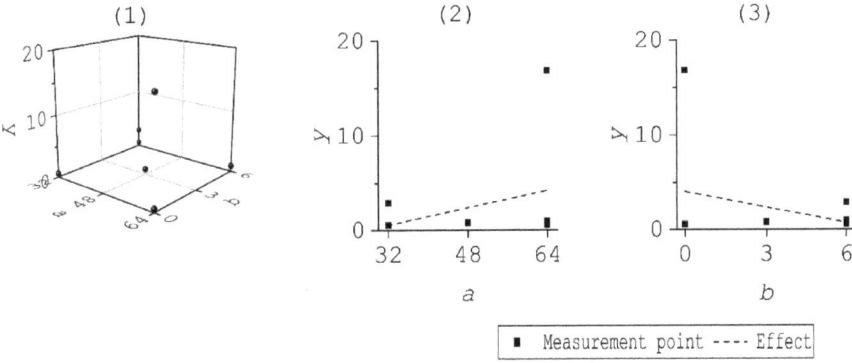

Figure 5. Transfer of test space (**1**) to effect diagrams (**2**) and (**3**).

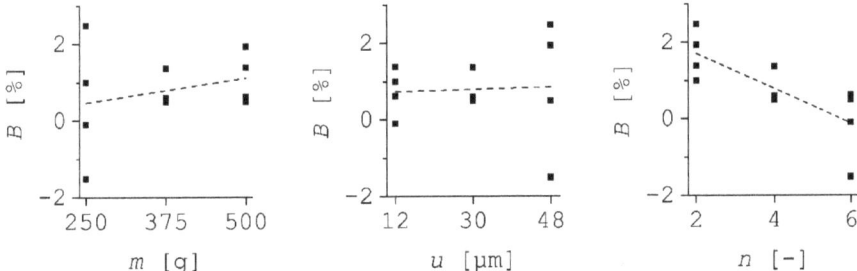

Figure 6. Effect diagram for the expansion of glass fiber [30]. Manipulated variables from left to right: mass, amplitude and number of rovings.

A similar situation can be observed in the effect diagram for carbon fiber in Figure 7. The manipulated variable weight per roving has, like the amplitude, the greatest positive effect on expansion. As in the glass fiber tests, the effect on expansion decreases with increasing layer thickness.

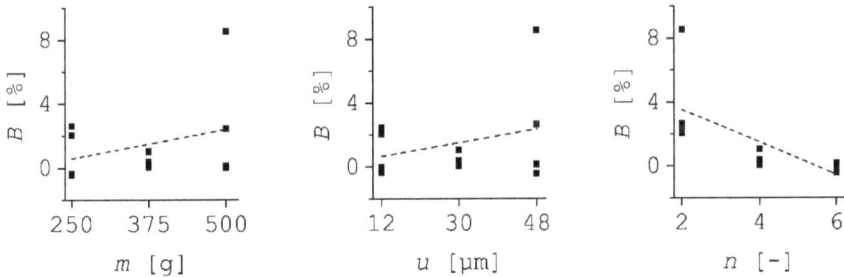

Figure 7. Effect diagram for the expansion of carbon fiber [30]. Manipulated variables from left to right: mass, amplitude and number of rovings.

4. Discussion

This study investigated the effect of US on the expansion of a fiber strand. The focus of the study was on quantifying the effect during sonication.

The investigation demonstrates that both ultrasonic amplitude and fiber tension have a similarly positive effect on the expansion of a fiber strand. The fact that the expansion increases with increasing fiber tension cannot be explained directly. This is because as the fiber tension increases, any potential expansion would require correspondingly higher external forces. However, since a relative increase in the fiber strand was indeed determined, the fiber strand is correspondingly more compact at a higher fiber tension. However, US has a small to virtually negligible effect on the expansion of both glass and carbon fiber. The expansion lies within a range of maximum 3%, with a standard deviation of up to 60% for glass fiber and over 100% for carbon fiber, as determined in the midpoint tests. This means that the fibers are limited in their freedom of movement, with the result that no expansion can be achieved by US. Any further increase in amplitude will not lead to further expansion but to destruction of the fibers. The fiber volume content is in direct correlation to the expansion. A minimal reduction in the fiber volume content in the compaction area investigated is equivalent to a minimal increase in permeability. Applied to pultrusion, the mechanical expansion of the fiber strand by US causes only a slight improvement in impregnation. Any improvement to the mechanical properties of the pultruded components as a result of US, as shown in Paper [24], can thus not be attributed to expansion of the fiber strand but to an impact on the resin system.

Author Contributions: Conceptualization, F.W.; methodology, F.W., S.S.; validation, F.W., S.S., and R.W.; investigation, S.S., R.W.; data curation, F.W., R.W.; writing—review and editing, F.W., S.S., R.W., K.D.; visualization, F.W., R.W.; project administration, F.W.; funding acquisition, F.W. All authors have read and agreed to the published version of the manuscript.

Funding: This research and development project is partly funded by the German Federal Ministry for Economic Affairs and Energy (BMWi) within the Framework Concept "Central Innovation Program for SMEs" and managed by the Project Management Agency AiF Projekt GmbH. The facilities and key technology equipment in Augsburg are funded by the Region of Bavaria; City of Augsburg; BMBF and European Union (in the context of the program "Investing In Your Future"—European Regional Development Fund).

Conflicts of Interest: The authors declare no conflict of interest.

References

1. Bezerra, R. Modelling and Simulation of the Closed Injection Pultrusion Process. Ph.D. Thesis, Karlsruher Institut für Technologie (KIT), Karlsruhe, Germany, 2017.
2. Connolly, M.; King, J.; Shidaker, T.; Duncan, A. Pultruding Polyurethane Composite Profiles: Practical Guidelines for Injection Box Design, Component Metering Equipment and Processing. In *COMPOSITES*; United States of America: Columbus, OH, USA, 2005.
3. Strauß, S.; Senz, A.; Ellinger, J. Comparison of the Processing of Epoxy Resins in Pultrusion with Open Bath Impregnation and Closed-Injection Pultrusion. *J. Compos. Sci.* **2019**, *3*, 87. [CrossRef]
4. Brown, R.J.; Kharchenko, S.; Coffee, H.D.; Huang, L. System for Producing Pultruded Components. U.S. Patent 8597016 B2, 3 December 2013.
5. Goldsworthy, W.B. Pultrusion Machine and Method. U.S. Patent 3556888, 19 January 1971.
6. Koppernaes, C.; Nolet, S.G.; Fanucci, J.P. Method and Apparatus for Wetting Fiber Reinforcements with Matrix Materials in the Pultrusion Process Using Continuous in-Line Degassing. U.S. Patent 53150890, 17 December 1991.
7. Thorning, H. *Fiberline Design Manual*; Fiberline Composites: Kolding, Denmark, 2003.
8. Wilhelm, F. Closed injection pultrusion. In Proceedings of the Travelling Conference ReHCarbo, Shanghai, China; Jeonju, Korea; Bangkok, Thailand, 23–26 October 2017.
9. Wilhelm, F.; Wiethaler, J.; Karl, R. Power ultrasonic in closed injection pultrusion. In Proceedings of the ECCM18—18th European Conference on Composite Materials, Athen, Greece, 24–28 June 2018.
10. Peuker, U.A.; Hoffmann, U.; Wietelmann, U.; Bandelin, S.; Jung, R. Sonochemistry. In *Book Ullmann's Encyclopedia of Industrial Chemistry*; Wiley-VCH Verlag GmbH & Co. KGaA: Weinheim, Germany, 2012.
11. Bogoeva-Gaceva, G.; Heraković, N.; Dimeski, D.; Stefov, V. Ultrasound assisted process for enhanced interlaminar shear strength of carbon fiber/epoxy resin composites. *Maced. J. Chem. Chem. Eng.* **2010**, *29*, 149–155. [CrossRef]
12. Mason, T.J.; Lorimer, J.P. Applied Sonochemistry. In *The Uses of Power Ultrasound in Chemistry and Processing*; WILEY-VCH Verlag GmbH & Co. KGaA: Weinheim, Germany, 2002.
13. Hayek-Boelingen, V.M. Wege zum Kontaminationstoleranten Kleben. Ph.D. Thesis, Universität Bundeswehr, München, Germany, 2004.

14. Chen, D.; Sharma, S.K.; Mudhoo, A. Handbook on Applications of Ultrasound. In *Book Sonochemistry for Sustainability*; CRC Press Taylor & Francis Group: Boca Raton, FL, USA, 2012.
15. Noltingk, B.E.; Neppiras, E.A. Cavitation produced by Ultrasonics. *Proc. Phys. Soc. B* **1950**, *63*, 674–685. [CrossRef]
16. Newton, I. *Opticks or a Treatise of the Reflections, Refractions, Inflections & Colours of Light*; Dover Publ: New York, NY, USA, 1979.
17. Ohl, S.W.; Klaseboer, E.; Khoo, B.C. Bubbles with shock waves and ultrasound: A review. *Interface Focus* **2015**, *5*, 1–15. [CrossRef] [PubMed]
18. Suslick, K.S.; Hammerton, D.A.; Cline, R.E. The sonochemical hot spot. *J. Am. Chem. Soc.* **1986**, *108*, 5641–5642. [CrossRef]
19. Santos, H.M.; Lodeiro, C.; Capelo-Martinez, J.-L. Ultrasound in Chemistry. In *Analytical Applications. The Power of Ultrasound*; WILEY-VCH Verlag GmbH & Co. KGaA: Weinheim, Germany, 2009.
20. Huang, Y.D.; Liu, L.; Qiu, J.H.; Shao, L. Influence of ultrasonic treatment on the characteristics of epoxy resin and the interfacial property of its carbon fiber composites. *Compos. Sci. Technol.* **2002**, *62*, 2153–2159. [CrossRef]
21. Liu, L.; Shao, L.; Huang, Y.; Jiang, B.; Zhang, Z. Effect of ultrasound on epoxy resin system and interface property. In Proceedings of the 13th International Conference on Composite Materials (ICCM13), Beijing, China, 25–29 June 2001.
22. Qiao, J.; Li, Y.; Li, L. Ultrasound-assisted 3D printing of continuous fiber-reinforced thermoplastic (FRTP) composites. *Addit. Manuf.* **2019**, *30*, 100926. [CrossRef]
23. Bogoeva-Gaceva, G.; Dimeski, D.; Heraković, N. Effect of sonication applied during production of carbon fiber/epoxy resin composites evaluated by differential scanning calorimetry and thermo-gravimetric analysis. *Maced. J. Chem. Chem. Eng.* **2011**, *30*, 189–195. [CrossRef]
24. Tessier, N.J.; Kiernan, D.; Madenjian, A.; Moulder, G. Epoxy matrix pultrusions enhanced by ultrasonics. *Mod.Plast.* **1986**, *63*, 86–90.
25. Kruckenberg, T.; Ye, L.; Paton, R. Static and vibration compaction and microstructure analysis on plain-woven textile fabrics. *Composites Part A* **2008**, *39*, 488–502. [CrossRef]
26. Gutiérrez, J.; Ruiz, E.; Trochu, F. High-frequency vibrations on the compaction of dry fibrous reinforcements. *Adv. Comp. Mat.* **2013**, *22*, 13–27. [CrossRef]
27. Gutiérrez, J.; Ruiz, E.; Trochu, F. Exploring the behavior of glass fiber reinforcements under vibration-assisted compaction. *J. Tex. Inst.* **2013**, *104*, 980–993. [CrossRef]
28. Meier, R. Über das Fließverhalten von Epoxidharzsystemen und Vibrationsunterstützte Harzinfiltrationsprozesse. Ph.D. Thesis, Technische Universität München, München, Germany, 2017.
29. Yamahira, S.; Hatanaka, S.-I.; Kuwabara, M.; Asai, S. Orientation of Fibers in Liquid by Ultrasonic Standing Waves. *Jap. Jou. Ap. Phy.* **2000**, *39*, 3683–3687. [CrossRef]
30. Weigant, R. Analyse der Auswirkungen von Leistungsultraschall auf die Aufbauschung von Glas- und Kohlenstofffasern. Bachelor's Thesis, Hochschule München, München, Germany, 2018.
31. Bezerra, R.; Wilhelm, F.; Henning, F. Compressibility and permeability of fiber reinforcements for pultrusion. In Proceedings of the ECCM16–16th European Conference on Composite Materials, Seville, Spain, 22–26 June 2016.
32. Karl, R. Untersuchung der Auswirkung von Ultraschall in der Pultrusion zur Verbesserung der Durchtränkung von Faserpaketen. Master's Thesis, Hochschule München, München, Germany, 2017.
33. Christensen, S.; Stober, E.J. Vibration Assisted Processing of Viscous Thermoplastics. U.S. Patent 6592799, 15 July 2003.

© 2020 by the authors. Licensee MDPI, Basel, Switzerland. This article is an open access article distributed under the terms and conditions of the Creative Commons Attribution (CC BY) license (http://creativecommons.org/licenses/by/4.0/).

MDPI
St. Alban-Anlage 66
4052 Basel
Switzerland
www.mdpi.com

Journal of Composites Science Editorial Office
E-mail: jcs@mdpi.com
www.mdpi.com/journal/jcs

Disclaimer/Publisher's Note: The statements, opinions and data contained in all publications are solely those of the individual author(s) and contributor(s) and not of MDPI and/or the editor(s). MDPI and/or the editor(s) disclaim responsibility for any injury to people or property resulting from any ideas, methods, instructions or products referred to in the content.

www.ingramcontent.com/pod-product-compliance
Lightning Source LLC
LaVergne TN
LVHW070233100526
838202LV00015B/2123